MATRIX-ANALYTIC METHODS
THEORY AND APPLICATIONS

PROCEEDINGS OF THE FOURTH INTERNATIONAL CONFERENCE

MATRIX-ANALYTIC METHODS
THEORY AND APPLICATIONS

Adelaide, Australia 14 – 16 July 2002

edited by

Guy Latouche
Université Libre de Bruxelles, Belgium

Peter Taylor
The University of Melbourne, Australia

World Scientific
New Jersey • London • Singapore • Hong Kong

Published by

World Scientific Publishing Co. Pte. Ltd.

P O Box 128, Farrer Road, Singapore 912805

USA office: Suite 1B, 1060 Main Street, River Edge, NJ 07661

UK office: 57 Shelton Street, Covent Garden, London WC2H 9HE

British Library Cataloguing-in-Publication Data
A catalogue record for this book is available from the British Library.

MATRIX-ANALYTIC METHODS: THEORY AND APPLICATIONS

Printed in Singapore by World Scientific Printers (S) Pte Ltd

Preface

Matrix-analytic methods are fundamental to the analysis of a family of Markov processes rich in structure and of wide applicability. They are extensively used in the modelling and performance analysis of computer systems, telecommunication networks, network protocols and many other stochastic systems of current commercial and engineering interest.

Following the success of three previous conferences held in Flint (Michigan), Winnipeg (Manitoba) and Leuven (Belgium), the Fourth International Conference on Matrix-Analytic Methods in Stochastic Models was held in Adelaide (Australia) in July 2002. The conference brought together the top researchers in the field who presented papers dealing with new theoretical developments and applications. This volume contains a selection of papers presented at the conference. The papers were subject to a rigorous refereeing process comparable to that which would be used by an international journal in the field.

The papers fall into a number of different categories. Approximately a third deals with various aspects of the theory of block-structured Markov chains. They demonstrate how the specific structure of transition matrices can be exploited. Another third of the papers deals with the analysis of complex queueing models. The final third deals with parameter estimation and specific applications to such areas as cellular mobile systems, FS-ALOHA, the Internet and production systems.

Three leading researchers in the field were invited to present a lecture to the conference: Masakiyo Miyazawa presented a paper on Markov additive processes in the context of matrix-analytic methods, V. Ramaswami discussed a number of applications and G.W. Stewart shared his vast experience on numerical methods for block Hessenberg matrices. A paper based on Masakiyo Miyazawa's talk is included in this volume.

In order to encourage young researchers to attend the conference, the organisers implemented a streamlined procedure, accepting submissions from students long after the general deadline. As a consequence, some late submissions by students are not included in these proceedings. They were, nevertheless, part of the official conference programme.

We would like to thank Kathryn Kennedy, David Green, Angela Hoffmann and Michael Green for their help in preparing the manuscripts for final publication. We acknowledge with gratitude financial assistance from several sources, specifically, the Australian Mathematical Society, the Teletraffic Research Centre, the Department of Applied Mathematics at the University of

Adelaide and the University of Adelaide itself.

Finally, it is a pleasure to acknowledge that the workshop could not have been held had it not been for the active involvment of reviewers and the authors who were all very good at respecting deadlines.

Guy Latouche
Peter Taylor

Contents

Author Index

Organisers

Conference chair
David Green, University of Adelaide, Australia
Programme co-chairs
Guy Latouche, Université Libre de Bruxelles, Belgium
Peter Taylor, University of Adelaide, Australia
Organising committee
Nigel Bean, University of Adelaide, Australia
Mark Fackrell, University of Adelaide, Australia
Barbara Gare, University of Adelaide, Australia
Angela Hoffmann, University of Adelaide, Australia
Kathryn Kennedy, University of Adelaide, Australia
Scientific advisory committee
Attahiru Alfa, University of Windsor, Canada
Dieter Baum, University of Trier, Germany
Nigel Bean, University of Adelaide, Australia
Dario Bini, University of Pisa, Italy
Lothar Breuer, University of Trier, Germany
Srinivas Chakravarthy, Kettering University, United States of America
David Green, University of Adelaide, Australia
Qi-Ming He, Dalhousie University, Canada
Dirk Kroese, University of Queensland, Australia
Herlinde Leemans, Catholic University of Leuven, Belgium
Yuanlie Lin, Tsinghua University, China
Naoki Makimoto, The University of Tsukuba, Japan
Beatrice Meini, University of Pisa, Italy
Marcel F. Neuts, The University of Arizona, United States of America
Bo Friis Nielsen, Technical University of Denmark, Denmark
Shoichi Nishimura, Science University of Tokyo, Japan
Phil Pollett, University of Queensland, Australia
V. Ramaswami, AT&T Labs, United States of America
Marie-Ange Remiche, Université Libre de Bruxelles, Belgium
Werner Scheinhardt, University of Twente, The Netherlands
Mark Squillante, IBM T.J. Watson Research Centre, United States of America
Yukio Takahashi, Tokyo Institute of Technology, Japan
Miklós Telek, Technical University of Budapest, Hungary
Erik van Doorn, University of Twente, The Netherlands
Qiang Ye, University of Kentucky, United States of America

Reviewers

Attihuru Alfa
Dieter Baum
Nigel Bean
Dario Bini
Lothar Breuer
Srinivas Chakravrathy
Mark Fackrell
David Green
Boudewijn Haverkort
Qi-Ming He
Dirk Kroese
Guy Latouche
Yuanlie Lin
Naoki Makimoto
Beatrice Meini
Marcel Neuts
Bo Friis Nielsen
Shoichi Nishimura
Phil Pollett
V Ramaswami
Marie-Ange Remiche
Mark Squillante
Yukio Takahashi
Peter Taylor
Miklos Telek
Erik van Doorn
Qiang Ye

Sponsors

Australian Mathematical Society
Teletraffic Research Centre
Department of Applied Mathematics, University of Adelaide
University of Adelaide

MATRIX-ANALYTIC METHODS
THEORY AND APPLICATIONS

A NEW ALGORITHM FOR COMPUTING THE RATE MATRIX OF GI/M/1 TYPE MARKOV CHAINS

ATTAHIRU SULE ALFA

Department of Industrial and Manufacturing Systems Engineering, University of Windsor, Windsor, Ontario, Canada, N9B 3P4

E-mail: alfa@uwindsor.ca

BHASKAR SENGUPTA

C&C Research Labs., NEC USA Inc., 4 Independence Way, Princeton NJ 08540, U.S.A.

Email: bhaskar@nec-lab.com

TETSUYA TAKINE

Department of Applied Mathematics and Physics, Graduate School of Informatics, Kyoto University, Kyoto 606-8501, Japan

Email: takine@amp.i.kyoto-u.ac.jp

JUNGONG XUE

Department of Industrial and Manufacturing Systems Engineering, University of Windsor, Windsor, Ontario, Canada, N9B 3P4

E-mail: jxue@server.uwindsor.ca

In this paper, we present a new method for finding the R matrix which plays a crucial role in determining the steady-state distribution of Markov chains of the GI/M/1 type. We formulate the problem as a non-linear programming problem. We first solve this problem by a steepest-descent-like algorithm and point out the limitations of this algorithm. Next, we carry out a perturbation analysis and develop a new algorithm which circumvents the limitations of the earlier algorithm. We perform numerical experiments and show that our algorithm performs better than what we call the "standard method" of solution.

1 Introduction

Consider a Markov chain $\{(X_\nu, N_\nu);\ \nu = 0, 1, \ldots\}$ in which X_ν takes a countable number of values $0, 1, 2, \ldots$ and N_ν takes a finite number of values $1, \ldots, m$. The transition probability matrix in block partition form is given by

$$\begin{bmatrix} B_0 & A_0 & & & & & \\ B_1 & A_1 & A_0 & & & & \\ B_2 & A_2 & A_1 & A_0 & & & \\ B_3 & A_3 & A_2 & A_1 & A_0 & & \\ \cdot & \cdot & \cdot & \cdot & \cdot & \cdot & \cdot \\ \cdot & \cdot & \cdot & \cdot & \cdot & \cdot & \cdot \cdot \end{bmatrix} \tag{1}$$

where A_i and B_i for $i = 0, 1, \ldots$ are all $m \times m$ matrices. This is the type of chain referred to as a Markov chain of the GI/M/1 type (see Neuts [15]). If it is stable, the steady-state distribution of this Markov chain is known to have the matrix-geometric form. Let π_k be a $1 \times m$ vector whose elements π_{kj} represent the steady state probability that $X_\nu = k$ and $N_\nu = j$ for $k = 0, 1, \ldots$ and $j = 1, \ldots, m$. Then the solution is given by $\pi_k = \pi_0 R^k$, where R is the minimal nonnegative solution to the non-linear matrix equation

$$R = \sum_{k=0}^{\infty} R^k A_k \tag{2}$$

and π_0 is the left invariant eigenvector (corresponding to the eigenvalue of 1) of $\sum_{k=0}^{\infty} R^k B_k$ when normalized by the equation $\pi_0 (I - R)^{-1} \mathbf{e} = 1$. Throughout the paper, $\mathbf{e}$ is an $m \times 1$ vector of ones.

The computation of R plays a crucial role in queuing analysis and has attracted considerable attention from many researchers (see Neuts [15], Grassmann and Heyman [7], Gün [8], Kao [9], Latouche [10,11], Lucantoni and Ramaswami [14], Sengupta [19], Akar and Sohraby [1]). Numerous algorithms have been designed to compute the R matrix. In [15], Neuts suggests these two iteration schemes

$$X_0 = 0, \qquad X_{k+1} = \sum_{v=0}^{\infty} X_k^v A_k, \quad k \geq 0, \tag{3}$$

and

$$X_0 = 0, \qquad X_{k+1} = \left(\sum_{v=0, v \neq 1}^{\infty} X_k^v A_k \right)(I - A_1)^{-1}, \quad k \geq 0, \tag{4}$$

which are shown to be such that $0 \leq X_k \uparrow R$ as $k \uparrow \infty$. It is pointed out that the iteration Eq. (4) converges faster than Eq. (3). However, these schemes all suffer from slow convergence when η, the Perron eigenvalue of R, is close to 1. To speed up convergence in this case, one can use the Newton method, which can be described as

$$X_{k+1} = X_k + Y_k, \tag{5}$$

where Y_k is the unique solution to the linear system

$$Y_k = (\sum_{v=0}^{\infty} X_k^v A_v - X_k) + \sum_{v=1}^{\infty}\sum_{j=1}^{v-1} X_k^j Y_k X_k^{v-1-j} A_k, \tag{6}$$

see [18]. Although Newton method converges in far less number of steps, it could actually be more time-consuming than even the direct method Eq. (3), because of the need to solve the large linear system Eq. (6) at each iteration. To this end, some modifications of Newton method are suggested, where Y_k is approximated. Different approximation strategies lead to different iterative methods. Usually, more accurate approximations take more time to compute, but result in fewer iteration steps. It is not easy to resolve the trade-off between them. We refer to [18] for a comprehensive survey.

Several breakthroughs have been achieved in recent years for some special cases of GI/M/1 type Markov chains, among them are the logarithmic reduction algorithm by Latouche and Ramaswami [12] for QBD and invariant subspace method by Akar and Sohraby [1] for those with rational generation function. Even though some efficient quadratically convergent algorithms, see [5,6,13], have been designed for computing the G matrix of general M/G/1 type Markov chains, the same is not true for the computation of the R matrix for general GI/M/1 type chains."

In an earlier paper Alfa, Sengupta and Takine [2] developed a non-linear programming method for finding the R and G matrices in the GI/M/1 and M/G/1 type Markov chains, respectively. In that paper the Karush-Kuhn-Tucker (KKT) conditions were obtained for these two non-linear programming problems. While the non-linear matrix equations resulting from the KKT conditions may be solved using Newton iterates, the resulting algorithm is not efficient. The paper later focuses on the M/G/1 type chains and develops an efficient algorithm for the G matrix using a simpler formulation. In the current paper, we focus on the GI/M/1 type Markov chain and develop a simple and efficient algorithm for the R matrix. We formulate the problem of finding the R matrix as a non-linear programming problem, then we design a steepest-descent-like method to solve it. At each iteration, a line search problem is required to be solved. Instead of a time-consuming process to find an optimal solution for this line search problem, we compute a nearly optimal solution with very little effort.

Throughout the paper we assume that the Markov chain is stable. We also assume that the following two conditions hold, which is true in most applications of interest:

1. Every row of the matrix A_0 has at least one positive element.

2. $A = \sum_{v=0}^{\infty} A_v$ is stochastic and $\sum_{v=1}^{\infty} A_v$ is irreducible.

These two conditions guarantee that the rate matrix R is irreducible, and thus the eigenvector of R corresponding to the Perron eigenvalue has entries with the same sign. We will explore this fact to prove that R is the unique solution to the non-linear programming problem.

Throughout this paper, we denote by $\|*\|_1$ and $\|*\|_\infty$ the 1-norm and ∞-norm, respectively. We let B^T denote the transpose of matrix B.

This paper is organized as follows. In Section 2, we formulate the problem of finding the R matrix as a non-linear programming problem and present a steepest-descent-like method to solve it. In Section 3, we carry out a perturbation analysis to overcome the limitations of the steepest-descent-like algorithm and develop a new algorithm. In Section 4, we report the numerical results.

2 The Non-linear Programming Problem

In this section, we formulate a non-linear programming problem, which leads to the solution of the R matrix for the GI/M/1 paradigm.

Let $A(z) = \sum_{k=0}^{\infty} A_k z^k, \quad |z| \leq 1$, and let $\chi(z)$ be the eigenvalue with maximal real part associated with $A(z)$. It is well-known from Neuts [15] that η, the Perron eigenvalue of the matrix R, is the smallest positive solution to the equation $z = \chi(z)$ and that $\mathbf{u}$, the left eigenvector (of dimension $1 \times m$) of R associated with η, is also the eigenvector of $A(\eta)$ associated with η. Since R is irreducible, $\mathbf{u}$ can be chosen to be positive. There exist simple methods for computing η and $\mathbf{u}$ (Neuts [15]). In what follows, we assume that $\mathbf{ue} = 1$. For broad classes of GI/M/1 Markov chains, this kind of computation takes very little time, see [17].

Let X be any $m \times m$ matrix and let $f(X) = \sum_{k=0}^{\infty} X^k A_k$. For two matrices Y and Z, let $Y \circ Z$ denote their elementwise product. We define the function $H(X)$ as

$$H(X) = \sum_{i,j=1}^{m} ([f(X)]_{ij} - X_{ij})^2 = \mathbf{e}^T((f(X) - X) \circ (f(X) - X))\mathbf{e}.$$

Theorem 1 *If the transition matrix of the GI/M/1 system is positive recurrent, then the R matrix is the unique optimal solution to the following non-linear programming problem:*

$$\text{minimize } H(X) \tag{7}$$

$$\text{subject to } \mathbf{u}X = \eta\mathbf{u} \tag{8}$$

$$X \geq 0, \tag{9}$$

Proof: First, we observe that R satisfies the constraints and has an objective function value of zero. Therefore, it is the optimal solution to Eq. (7-9). Now we prove it is the unique solution. Suppose there exists another optimal solution Z. From $H(Z) = 0$, we have $f(Z) = Z$. Since R is the minimal nonnegative solution to the equation $f(X) = X$, we have $Z \geq R$. Thus $\mathbf{u}Z \geq \mathbf{u}R = \eta\mathbf{u}$. Because of the fact that $\mathbf{u}$ is positive and $Z \neq R$, $\mathbf{u}Z \neq \eta\mathbf{u}$, which contradicts constraint Eq. (8). ♯

Now let us discuss how to solve this non-linear programming problem. Suppose X is a nonnegative approximation for R satisfying $\mathbf{u}X = \eta\mathbf{u}$. We can come up with a "better" approximation (*i.e.*, one with a lower value of the objective function) by adding to X a correction in the direction $d = f(X) - X$. This leads to the following line search problem:

$$\begin{aligned} \text{Minimize} \quad & H(X + \theta d) \\ \text{Subject to} \quad & \mathbf{u}(X + \theta d) = \eta\mathbf{u} \\ & X + \theta d \geq 0. \end{aligned} \tag{10}$$

Since $\mathbf{u}f(X) = \mathbf{u}A(\eta) = \eta\mathbf{u}$, we have $\mathbf{u}d = 0$ and thus $\mathbf{u}(X + \theta d) = \eta\mathbf{u}$ for any θ. We denote the (i,j)th elements of the matrices X and d by X_{ij} and d_{ij} respectively. To make $X + \theta d$ nonnegative, θ is required to be in the interval $[\theta_{min}, \theta_{max}]$ where

$$\theta_{max} = min_{ij}\left\{\left|\frac{X_{ij}}{d_{ij}}\right| : d_{ij} < 0\right\} \quad and \quad \theta_{min} = -min_{ij}\left\{\left|\frac{X_{ij}}{d_{ij}}\right| : d_{ij} > 0\right\}.$$

Then problem Eq. (10) is equivalent to the following problem:

$$\begin{aligned} \text{minimize} \quad & H(X + \theta d) \\ \text{subject to} \quad & \theta_{min} \leq \theta \leq \theta_{max} \end{aligned}$$

This method is the modification of steepest-descent method with the gradient search direction replaced by the search direction given by the residual d. See page 300 of Bazaraa, Sherali and Shetty [3] for details on the steepest descent method for solving nonlinear programming problems. The advantage of modifying the traditional steepest descent method is that the search direction is easy to compute and the search is limited to nonnegative matrices with η as the Perron eigenvalue and $\mathbf{u}$ as the associated left eigenvector. This method is called a steepest-descent-like method and provides an initial method for the computation of R. The complete details of this algorithm are:

Steep-Descent-Like Method

1. Calculate η and $\mathbf{u}$.

2. Choose $X_0 = A(\eta)$ and set $k = 0$ and stopping threshold ϵ.

3. Stop if $\|f(X_k) - X_k\|_1 \leq \epsilon$, otherwise let $d_k = f(X_k) - X_k$. Let θ_k be the optimal solution to the following line search problem:

$$\text{minimize} \;\; H(X_k + \theta d_k) \tag{11}$$

$$\text{subject to} \;\; \theta_{min} \leq \theta \leq \theta_{max}, \tag{12}$$

where

$$\theta_{max} = min_{ij}\left\{\left|\frac{X_{ij}^{(k)}}{d_{ij}^{(k)}}\right| : \; d_{ij}^{(k)} < 0\right\} \quad and \quad \theta_{min} = -min_{ij}\left\{\left|\frac{X_{ij}^{(k)}}{d_{ij}^{(k)}}\right| : \; d_{ij}^{(k)} > 0\right\}.$$

Here we denote by $X_{ij}^{(k)}$ and $d_{ij}^{(k)}$ the entries of X_k and d_k.

4. Let $X_{k+1} = X_k + \theta_k d_k$ and set k to $k+1$. Go to Step 3.

The steepest-descent-like method is of practical use only when the line search problem Eq. (11-12) can be solved efficiently. However, some difficulties arise in doing so. First, the objective function $H(X_k + \theta d_k)$ is not *unimodal* in the interval $[\theta_{min}, \; \theta_{max}]$, and therefore, some popular methods, such as golden search method and Fibonacci search method [3] can not be used. Second, as X_k converges, the entries of $d_k = f(X_k) - X_k$ tend to zero, and it can be expected that the search interval $[\theta_{min}, \; \theta_{max}]$ becomes very large, which means that the computational burden for the search problem becomes large. In the next section, we present a new algorithm which circumvents these difficulties of the steepest-descent-like method.

3 The New Algorithm

The main idea of the new algorithm is to find a nearly optimal solution to the line search problem Eq. (11-12) efficiently instead of solving it exactly. Besides, this approximation should not affect the convergence of the objective function to zero. Our idea is described as follows. When X_k converges to R, $d_k = f(X_k) - X_k$ tends to zero and $E_k = X_k - R$, the error of X_k, also tends to zero. Thus if $\|d_k\|_1$ is sufficiently small, the optimal solution to the line search problem Eq. (11-12) can be obtained by moving away from X_k a very short distance in the direction d_k, which means θd_k is very small and $f(X_k + \theta d_k)$ is well approximated by the first order expansion

$$f(X_k + \theta d_k) \approx f(X_k) + \theta \sum_{v=1}^{\infty} \sum_{j=0}^{v-1} X_k^j d_k X_k^{v-1-j} A_v \tag{13}$$

$$= f(X_k) + \theta S_k, \tag{14}$$

where

$$S_k = \sum_{v=1}^{\infty} \sum_{j=0}^{v-1} X_k^j d_k X_k^{v-1-j} A_v.$$

Accordingly the objective function $H(X + \theta d_k)$ can be approximated as

$$\widetilde{H}(X + \theta d_k) = \mathbf{e}^T((d_k + \theta(S_k - d_k)) \circ (d_k + \theta(S_k - d_k)))\mathbf{e}.$$

Then the optimal solution to the modified search problem

$$\text{minimize} \quad \widetilde{H}(X_k + \theta d_k) \tag{15}$$

$$\text{subject to} \quad \theta_{min} \leq \theta \leq \theta_{max}, \tag{16}$$

can be viewed as a nearly optimal solution to the line search problem Eq. (11-12). Fortunately, this modified search problem can be solved explicitly. We have

$$\widetilde{H}(X_k + \theta d_k) = \theta^2 \mathbf{e}^T((S_k - d_k) \circ (S_k - d_k))\mathbf{e}$$
$$+ 2\theta \mathbf{e}^T((S_k - d_k) \circ d_k)\mathbf{e} + \mathbf{e}^T(d_k \circ d_k)\mathbf{e}.$$

The function $\widetilde{H}(X_k + \theta d_k)$ attains its minimum if θ is chosen as

$$\theta_0 = \frac{\mathbf{e}^T(d_k \circ (d_k - S_k))\mathbf{e}}{\mathbf{e}^T((d_k - S_k) \circ (d_k - S_k))\mathbf{e}}. \tag{17}$$

Therefore, if $X_k + \theta_0 d_k$ is nonnegative, θ_0 is the optimal solution to Eq. (15-16).

Now the idea of the new algorithm can be clarified. It consists of two stages. In the first stage, we iteratively compute X_k such that $\mathbf{u}X_k = \eta\mathbf{u}$ and $\|d_k\|_1 = \|f(X_k) - X_k\|_1 > \delta$ for all k, where δ is some prespecified threshold. When $\|d_k\|$ falls to a sufficiently low value, we invoke the second stage of the algorithm. In this stage, using the last computed value of X_k as the starting point, we iteratively carry out the steepest-descent-like method, where we solve the line search problem Eq. (15-16). Note that in each iteration, we attempt to solve the line search problem approximately, to reduce the computational burden. At this point, two questions arise: 1. How do we perform the iterations of the first stage? 2. How small should $\|d_k\|_1$ be so that we invoke the steepest-descent-like method?

Although the iterations Eq. (3) and Eq. (4) converge very slowly as η tends to 1, it is known that a properly chosen X_0 can greatly speed up the convergence (see [18]). Neuts [16] recommends that X_0 be chosen such that $\mathbf{u}X_0 = \eta\mathbf{u}$, which has the advantage of limiting the search to nonnegative matrices with η as the Perron eigenvalue and $\mathbf{u}$ as the associated left eigenvector. Extensive numerical experience shows that this strategy works very well. Using this knowledge, we adopt the following iterative scheme for the first stage

$$X_0 = A(\eta) \qquad \text{and} \quad X_{k+1} = (\sum_{v=0, v\neq 1}^{\infty} X_k^v A_v)(I - A_1)^{-1} = g(X_k) \tag{18}$$

for $k = 0, 1, \cdots$. Obviously, $\mathbf{u}X_k = \eta \mathbf{u}$ for all k.

We now discuss how small $\|f(X_k) - X_k\|_1$ should be so that the iteration Eq. (4) switches to steepest-descent-like method. We know that the optimal solution to Eq. (15-16) can be viewed as a nearly optimal solution to Eq. (11-12) only when the approximation in Eq. (13) is sufficiently accurate. This requires $\|E_k\|_1 = \|X_k - R\|_1$ to be small. Even though E_k itself is unknown, we can bound $\|E_k\|_1$ by carrying out a perturbation analysis which bounds $\|E_k\|_1$ in terms of $\|d_k\|_1$. We first investigate the property of a matrix which plays a crucial role in the perturbation analysis. Note that a matrix Y is an M-matrix if Y can be written as $\rho I - Z$, where Z is a nonnegative matrix and ρ is greater than or equal to the Perron eigenvalue of Z.

Lemma 1 *Let $U_R = \sum_{v=1}^{\infty} \sum_{j=0}^{v-1} \eta^j R^{v-1-j} A_v$ then $I - U_R$ is a nonsingular M-matrix. Suppose $\mathbf{w} = \sum_{v=2}^{\infty} A_v \mathbf{e}$ is a positive vector and $\alpha = \min_i \mathbf{w}_i$, then*

$$\|(I - U_R)^{-1}\|_\infty \leq \frac{1}{\alpha(1-\eta)}. \tag{19}$$

Proof. Let

$$T_R = \sum_{v=1}^{\infty} \sum_{j=0}^{v-1} R^{v-1-j} A_v.$$

Then T_R is a stochastic matrix since

$$\begin{aligned} T_R \mathbf{e} &= \sum_{v=1}^{\infty} \sum_{j=0}^{v-1} R^{v-1-j} A_v \mathbf{e} \\ &= (I - R)^{-1} \sum_{v=1}^{\infty} (I - R^v) A_v \mathbf{e} \\ &= (I - R)^{-1} (A - R) \mathbf{e} \\ &= (I - R)^{-1} (I - R) \mathbf{e} \\ &= \mathbf{e}. \end{aligned}$$

Noting $T_R \geq \sum_{v=1}^{\infty} A_v$ and the assumption that $\sum_{v=1}^{\infty} A_v$ is irreducible, we conclude that T_R is irreducible. Because $T_R \geq U_R$ and $T_R \neq U_R$, the Perron eigenvalue of U_R is less than 1, which implies $I - U_R$ is a nonsingular matrix and $(I - U_R)^{-1} \geq 0$, see [4]. If $\alpha > 0$, then

$$\begin{aligned}(I - U_R)\mathbf{e} &= \sum_{v=1}^{\infty}\sum_{j=0}^{v-1}(1-\eta^j)R^{v-1-j}A_v\mathbf{e} \\ &\geq \sum_{v=2}^{\infty}(1-\eta^{v-1})A_v\mathbf{e} \\ &\geq (1-\eta)\mathbf{w} \\ &\geq \alpha(1-\eta)\mathbf{e},\end{aligned}$$

which leads to

$$\|(I - U_R)^{-1}\|_\infty \leq \frac{1}{\alpha(1-\eta)}.$$

♯

When $d_k = f(X_k) - X_k$ is sufficiently small, $E_k = X_k - R$ is tiny so that

$$T(E_k) = f(R + E_k) - f(R) - \sum_{v=1}^{\infty}\sum_{j=0}^{v-1}(R^j E_k R^{v-1-j})A_v,$$

the truncation error of the first order expansion of $f(R+E_k)$ at R, is of order $O(\|E_k\|_1^2)$ and is small compared to E_k. With an assumption on the bound of $T(E_k)$, we present the perturbation result.

Theorem 2 *Let* $E_k = X_k - R$ *and* $\mathbf{u}R = \eta\mathbf{u}$. *Let* $\beta = \max_i \mathbf{u}_i$ *and* $\gamma = \min_i \mathbf{u}_i$. *If* $\|d_k\|_1$ *is sufficiently small so that*

$$\|\mathbf{u}|T(E_k)|(I - U_R)^{-1}\|_1 \leq 0.5\|\mathbf{u}|E_k|\|_1,$$

then

$$\|E_k\|_1 \leq \frac{2\beta}{\gamma}\|(I - U_R)^{-1}\|_1\|d_k\|_1. \tag{20}$$

Proof. We have

$$E_k = f(X_k) - d_k - R = f(R + E_k) - d_k - R.$$

Performing a first order expansion of $f(R + E_k)$ around R, we get

$$E_k = f(R) + \sum_{v=1}^{\infty}\sum_{j=0}^{v-1}(R^j E_k R^{v-1-j})A_v - d_k - R + T(E_k)$$

$$= \sum_{v=1}^{\infty} \sum_{j=0}^{v-1} (R^j E_k R^{v-1-j}) A_v - d_k + T(E_k). \tag{21}$$

In what follows, for any matrix Y, we denote by $|Y|$ is the matrix whose entries are the absolute values of those of Y. This leads to

$$|E_k| \leq \sum_{v=1}^{\infty} \sum_{j=0}^{v-1} (R^j |E_k| R^{v-1-j}) A_v + |d_k| + |T(E_k)|.$$

Pre-multiplying both sides by $\mathbf{u}$ yields

$$\mathbf{u}|E_k| \leq \mathbf{u}|E_k| U_R + \mathbf{u}|d_k| + \mathbf{u}|T(E_k)|.$$

Since $I - U_R$ is an M-matrix, $(I - U_R)^{-1}$ is nonnegative, we have

$$\mathbf{u}|E_k| \leq \mathbf{u}|d_k|(I - U_R)^{-1} + \mathbf{u}|T(E_k)|(I - U_R)^{-1}.$$

Taking 1-norm on both sides we have

$$||\mathbf{u}|E_k|||_1 \leq ||\mathbf{u}|d_k|||_1 ||(I - U_R)^{-1}||_1 + ||\mathbf{u}|T(E_k)|(I - U_R)^{-1}||_1.$$

With the assumption

$$||\mathbf{u}|T(E_k)|(I - U_R)^{-1}||_1 \leq 0.5||\mathbf{u}|E_k|||_1,$$

we arrive at

$$0.5||\mathbf{u}|E_k|||_1 \leq ||\mathbf{u}|d_k|||_1 ||(I - U_R)^{-1}||_1.$$

Noting that

$$\gamma ||E_k||_1 \leq ||\mathbf{u}|E_k|||_1 \quad \text{and} \quad ||\mathbf{u}|d_k|||_1 \leq \beta ||d_k||_1,$$

we complete the proof. ♯

Since the 1-norm and ∞-norm are equivalent, or more precisely, since

$$\frac{1}{m}||(I - U_R)^{-1}||_1 \leq ||(I - U_R)^{-1}||_\infty \leq m||(I - U_R)^{-1}||_1,$$

we have

$$||E_k||_1 \leq \frac{2m\beta}{\alpha\gamma} \frac{||d_k||_1}{1 - \eta}. \tag{22}$$

This is the error bound for E_k in terms of d_k.

Now we discuss how to choose the switching point. We note that when the smallest entry of $\mathbf{w}$ in Lemma 1 is very small compared to the largest entry, the error bound Eq. (19) is pessimistic. Similarly, if the smallest entry of $\mathbf{u}$ is very small compared to the largest one, the term β/γ in Eq. (20) becomes very large and is an an excessive over-estimation since $||\mathbf{u}|E_k|||_1 \geq \gamma ||E_k||_1$

is too pessimistic when γ is small. Consequently, the coefficient $2m\beta/\alpha\gamma$ in error bound Eq. (22) is usually an overestimate. We thus care little about this coefficient and take $1-\eta$ as the main factor determining the sensitivity of E_k to d_k. According to our perturbation analysis, when $(1-\eta)^{-1}\|d_k\|_1$ is small enough, the switching is invoked. In practice, we observe that spending too much time in the first stage to make $(1-\eta)^{-1}d_k$ small enough for switching is not productive. This is because even if $(1-\eta)^{-1}d_k$ is not small, $X_k+\theta_0 d_k$ (where θ_0 is given by Eq. (17)) could be a better approximation of R than $g(X_k)$. It suggests that the switching be invoked earlier than our perturbation analysis predicts. To make the switching safe, we impose a switching condition on each iteration after the switching. If $H(X_k+\theta_k d_k) < H(f(X_k))$ and $X_k+\theta_k d_k$ is nonnegative, then let $X_{k+1} = X_k+\theta_k d_k$, otherwise, let $X_{k+1} = f(X_k)$. From our numerical experiments on broad classes of GI/M/1 type Markov chains , we recommend choosing $\delta = \min\{10(1-\eta),\ 0.01\}$. When $\|d_k\|_1$ is less than δ, we switch from the iterations Eq. (18) to the steepest-descent-like method. The new algorithm can be described as follows.

New Algorithm

1. Compute η and $\mathbf{u}$.

2. Set $X_0 = A(\eta)$, $\delta = \min\{0.01, 10(1-\eta)\}$ and $k=0$. Choose a stopping threshold ϵ.

3. While $\|d_k\|_1 > \delta$,

$$X_{k+1} = g(X_k)$$

$$d_{k+1} = f(X_{k+1}) - X_{k+1}$$

$$k \leftarrow k+1$$

end

4. While $\|d_k\|_1 > \epsilon$,

$$\theta_k = \frac{\mathbf{e}^T(d_k^T \circ (d_k - S_k))\mathbf{e}}{\mathbf{e}^T((d_k-S_k)\circ(d_k-S_k))\mathbf{e}}.$$

$$X_{k+1} = \begin{cases} X_k+\theta_k d_k & X_k+\theta_k d_k \ge 0 \text{ and } H(X_k+\theta_k d_k) \le H(f(X_k)) \\ g(X_k) & \text{otherwise} \end{cases}$$

$$k \leftarrow k+1$$

end

Remark: The computational burden of this new algorithm is dominated by the the computation of S_k, $f(X_k)$ and $g(X_k)$, which can be done simultaneously by the following method. Suppose both S_k, $f(X_k)$ and $g(X_k)$ are truncated at A_N, where N is the minimum integer satisfying

$$\| \sum_{j=N+1}^{\infty} A_v \|_1 \leq \epsilon_1,$$

where ϵ_1 is a small threshold and can be assumed to be negligible. Using this assumption, we have

$$S_k \approx \sum_{v=1}^{N} \sum_{j=0}^{v-1} X_k^j d_k X_k^{v-1-j} A_v.$$

and

$$f(X_k) \approx \sum_{v=0}^{N} X_k^v A_v \quad \text{and} \quad g(X_k) \approx \left(\sum_{v=0, v\neq 1}^{N} X_k^v A_v \right)(I - A_1)^{-1}.$$

Noting that

$$S_k \approx \sum_{j=0}^{N-1} X_k^j d_k \left(\sum_{v=j+1}^{N} X_k^{v-1-j} A_v \right),$$

we can perform the iteration:

Set $f^{(0)} = A_N$ and $S^{(0)} = d_k f^{(0)}$
For $j = 1, \cdots, N-1$
  $f^{(j)} = X_k f^{(j-1)} + A_{N-j}$
  $S^{(j)} = X_k S^{(j-1)} + d_k f^{(j)}$
End.

It is easy to show that $f^{(j)} = \sum_{v=N-j}^{N} X_k^{v-N+j} A_v$, and therefore,

$$f(X_k) \approx X_k f^{(N-1)} + A_0 \quad \text{and} \quad S_k \approx S^{(N-1)}.$$

and

$$g(X_k) \approx (A_0 + X_k f^{(N-2)})(I - A_1)^{-1}$$

Implemented in this way, the computational time of each iterative step in the second stage of the new algorithm is roughly three or four times, if the switching condition is violated, of that in the first stage.

4 Numerical Examples

In this section, we report on numercial experiments on broad classes of GI/M/1 type Markov chains to compare our algorithm against what we call the standard algorithm, which has the form

$$X_{k+1} = \sum_{v \neq 1}^{\infty} X_k^v A_k (I - A_1)^{-1}.$$

with starting guess $X_0 = A(\eta)$. In all the examples, we truncate the sequences of $\{A_l\}_{l=1}^{\infty}$ at N, where $\| \sum_{l=N+1}^{\infty} A_1 \|_1 \leq 1.0 \times 10^{-10}$. The stopping criteria in our algorithm is

$$\|X_k - \sum_{l=0}^{N} X_k^{l-1} A_l\| \leq 10^{-10}.$$

Thus we can expect the norm of the exact residual $X_k - \sum_{l=0}^{\infty} X_k^{l-1} A_l$ is of order 10^{-10}. For each example, we compare our new algorithm against the standard algorithm for many different traffic intensities. Both algorithms are coded in *MATLAB* and run on a IBM PC with speed 800 MHz. All the cpu times reported are obtained by using the command "cputime".

In the two tables displaying our results, we denote by ρ the traffic intensity, by T_{new} and T_{std} the cpu time for the new algorithm and the standard algorithm, respectively, and by K_{new} and K_{std} the number of iterations.

Example 1

We first consider a D/PH/1 queue, where interarrival times are constant and service times are i.i.d. according to a PH distribution with representation $(\mathbf{a}, T)$. Let h denote the length of the interarrival time. Then the rate matrix R associated with the imbedded process immediately before arrivals satisfies

$$R = \sum_{k=0}^{\infty} R^k A_k,$$

where the A_k satisfies

$$\sum_{k=0}^{\infty} A_k z^k = \exp[\{T + z(-T\mathbf{e}\mathbf{a})h\}].$$

We test both algorithms on the Markov chains obtained by randomly generating $(\mathbf{a}, T)$ with different traffic intensities. We observe for this Markov chain, the spectral radius of R increases slowly with the growth of traffic intensity and the standard algorithm converges fast even when the intensity is near to

Table 1.

ρ	0.70	0.80	0.90	0.95	0.97
η	0.4676	0.6292	0.8073	0.9019	0.9407
T_{std}	0.77	0.88	0.93	0.94	0.99
T_{new}	0.77	0.71	0.77	0.71	0.71
K_{std}	11	14	16	18	18
K_{new}	5	5	6	6	6

1. In this case, our algorithm slightly outperforms the standard one. We denote by m the order of T. Table 1 reports the numerical result for a randomly generated $(\mathbf{a}, T)$, for which $m = 40$.

Example 2

Now we consider a single server queue in which interarrival times are deterministic (and equal 1). The service process is governed by an $m-$state Markov chain with states $1, 2, \cdots, m$. We assume that customers are served according to the exponential distribution with rate $i\mu$ when the Markov chain is in state i. The state transitions occur upon arrivals. The transitions between states of this m-state Markov chain are governed by the matrix $P = [p_{i,j}]$, where $p_{i,j}$ denotes the transition probability from state i to state j. Let

$$p_{i,i} = 0.7, \quad i = 1, 2, \cdots, m,$$

$$p_{1,2} = p_{m,m-1} = 0.3,$$

$$p_{i,i+1} = p_{i,i-1} = 0.15, \quad i = 2, \cdots, m-1,$$

and

$$p_{i,j} = 0, \quad \text{otherwise.}$$

This is a GI/M/1 type Markov chain with

$$A_k = D_k P \qquad k = 0,\ 1, \cdots$$

where D_k is a diagonal matrix whose ith diagonal element is given by $(i\mu)^k e^{-i\mu}/k!$. For this example, m was chosen to be 70. This Markov chain has the interesting property that η tends to 1 very fast as the traffic intensity increases. The standard algorithm converges slowly even for a mild traffic intensity. In this case, our algorithm converges much faster than the standard one. Note that the effectiveness of our algorithm is more pronounced when spectral radius is close to 1.

Table 2.

ρ	0.75	0.80	0.85	0.90	0.95
η	0.99978	0.99982	0.99989	0.99991	0.99995
T_{std}	53.50	61.95	74.04	76.18	85.24
T_{new}	18.61	27.96	27.03	31.26	32.41
K_{std}	1282	1472	1873	1973	2240
K_{new}	211	415	378	428	551

References

1. N. Akar and K. Sohraby, An invariant subspace approach in M/G/1 and G/M/1 type Markov chains, *Stochastic Models*, Vol 13, No. 3, 1997.
2. A. S. Alfa, B. Sengupta and T. Takine, The use of non-linear programming in matrix analytic methods, *Stochastic Models*, vol. 14, Nos. 1 & 2, 351-367 (1998).
3. M. S. Bazaraa, H. D. Sherali and C. M. Shetty, *Nonlinear programming: Theory and algorithms (second edition)*, Wiley, New York (1993).
4. A. Berman and R. J. Plemmons, *Nonnegative Matrices in the Mathematical Science,* Academic Press, 1979.
5. D. Bini and B. Meini, On the solution of a nonlinear matrix equation arising in queuing problems, *SIAM J. Matrix Anal. Appl.*, Vol. 17, 906-926 (1996).
6. D. Bini and B. Meini, Improved cyclic reduction for solving queueing problems, *Numerical Algorithms,* Vol. 15, 57-74 (1997).
7. W. Grassmann and D. Heyman, Equilibrium distribution of block structured Markov chains with repeating rows, *JAP*, Vol. 27, 557-576, 1990.
8. L. Gün, Experimental results on matrix-analytical solution techniques - Extensions and comparisons, *Stochastic Models*, Vol. 5, 669-682 (1989).
9. E. P. C. Kao, Using state reduction for computing steady state probabilities of queues of GI/PH/1 types, *ORSA J. on Computing*, Vol. 3, 231-240 (1991).
10. G. Latouche, A note on two matrices occuring in the solution of quasi-birth and death processes, *Stochastic Models*, Vol. 3, 251-257 (1987).
11. G. Latouche, Algorithms for infinite Markov chains with repeating columns, *IMA Workshop on Linear Algebra, Markov Chains and Queueing Models* (1992).
12. G. Latouche and V. Ramaswami, A logarithmic reduction algorithm for quasi birth and death processes, *JAP*, Vol. 30, 650-674 (1993).

13. G. Latouche and G. W. Stewart, Numerical methods for M/G/1 type queues, *in Proc. of the Second International Workshop on Numerical Solutions of Markov Chains, Raleigh, NC,* 571-581 (1995).
14. D. M. Lucantoni and V. Ramaswami, Efficient algorithms for solving the non-linear matrix equations arising in phase type queues, *Stochastic Models*, Vol. 1, 29-52 (1985).
15. M. F. Neuts, *Matrix-geometric solutions in stochastic models: An algorithmic approach*, Johns Hopkins, Baltimore (1981).
16. M. F. Neuts, Matrix-analytic methods in queuing theory, *European Journal of Operational Research*, vol. 15, 2-12 (1984).
17. M. F. Neuts, The Caudal Characteristic curve of Queue, Adv. Appl. Prob., vol. 18, 221-254(1986).
18. V. Ramaswami, Nonlinear matrix equations in applied probability - Solution techniques and open problems, *SIAM Rev.*, vol. 30, 256-263 (1988).
19. B. Sengupta, Markov processes whose steady state distribution is matrix-exponential with an application to the GI/PH/1 queue, *AAP*, Vol. 21, 159-180 (1989).

DECAY RATES OF DISCRETE PHASE-TYPE DISTRIBUTIONS WITH INFINITELY-MANY PHASES

N.G. BEAN

Department of Applied Mathematics, University of Adelaide, Australia 5005.
nbean@maths.adelaide.edu.au

B.F. NIELSEN

Informatics and Mathematical Modelling, Technical University of Denmark, DK-2800 Kgs. Lyngby, Denmark. bfn@imm.dtu.dk

In this paper we investigate the factors that determine the decay rate of discrete phase-type distributions when there are a countably-infinite number of phases.

A discrete phase-type distribution is any distribution that can be described as the time to absorption of a discrete-time Markov chain on a finite state space with a substochastic transition matrix T and honest initial probability distribution $\boldsymbol{\alpha}$. In this situation it has been known for a long time that the decay rate is always given by the maximal eigenvalue of T, regardless of the choice of initial distribution $\boldsymbol{\alpha}$.

In this paper we consider the same setting, but allow for the state space to consist of a countably-infinite number of phases. We find that the behaviour of the decay rate is now significantly more interesting.

We specifically consider phase type distributions where the transition matrix T is such that absorption can occur through only a finite number of phases and where T can be permuted to a form with a block upper-triangular structure. We explicitly investigate the situation where T has the structure of a level-dependent Quasi-Birth-and-Death process (QBD) and then extend this to the block upper-triangular structure.

Under these assumptions, we show that the decay-rate is always determined by either the convergence radius of the transition matrix, T, or the convergence radius of a series constructed from the initial distribution $\boldsymbol{\alpha}$ and certain properties of T.

Key words: Decay rate, phase-type distribution, quasi-birth-and-death process, convergence radius.

1 Introduction

The idea of phase-type distributions has been around since 1917 when A.K. Erlang[1] introduced what has now become known as the Erlang distribution. However, it is much more recently that they have become the focus of intense research.

Any discrete distribution, $\{p_k\}$, that can be represented by the time to absorption of a discrete-time Markov chain on $m+1$ states (phases) is a discrete phase-type distribution. Throughout this paper we shall assume that all distributions are discrete, however, we note that the extension to the continuous-time setting is reasonably natural.

We denote the absorbing state by state 0 and the non-absorbing states by the states $\{1, 2, \ldots, m\}$. The process starts according to an honest initial probability distribution $(\alpha_0, \alpha_1, \ldots, \alpha_m)$, that is, it starts in state k with probability α_k. We shall generally assume that $\alpha_0 = 0$, however, this is not necessary. We also let $\boldsymbol{\alpha} = (\alpha_1, \alpha_2, \ldots, \alpha_m)$.

The Markov chain is governed by a transition matrix T which is a substochastic $m \times m$ matrix. We also identify the $m \times 1$ column vector $\boldsymbol{T_0} = \boldsymbol{e} - T\boldsymbol{e}$, where $\boldsymbol{e}$ is an $m \times 1$ column vector consisting entirely of ones. This vector $\boldsymbol{T_0}$ records the one-step probability of the Markov chain being absorbed, for every state.

Neuts[2] states on page 46 that the probability density $\{p_k\}$ is given by,

$$p_0 = \alpha_0, \tag{1.1}$$

$$p_k = \boldsymbol{\alpha} T^{k-1} \boldsymbol{T_0}, \quad k \geq 1, \tag{1.2}$$

and the probability generating function ($P(z) = \sum_{k=0}^{\infty} p_k z^k$) by

$$P(z) = \alpha_0 + \sum_{n=1}^{\infty} \boldsymbol{\alpha} T^{n-1} \boldsymbol{T_0} z^n. \tag{1.3}$$

One of the more interesting recent innovations has been to be extend the class of phase-type distributions to allow a countably-infinite number of states. That is, to allow $m = \infty$. This does not affect the fundamental understanding of the phase-type distributions, nor the equations given above, but adds significantly to the flexibility of the class. For example, it opens up the possibility of distributions with a heavy tail[3]. For some results on the relationships of this and related distributional classes see, for example, the papers by Shi and co-workers[4,5].

Throughout this paper we assume that the mean of the phase-type distribution is finite and so the mean time to absorption of the Markov process is finite. We also assume that the representation of the phase-type distribution, $(\boldsymbol{\alpha}, T)$ is irreducible, as otherwise the states that cannot be reached can be eliminated from the representation. Note that this does not imply that T is irreducible, instead the required communication can be provided by the initial vector $\boldsymbol{\alpha}$. For simplicity of exposition, we further assume that T is irreducible, however, the results can be extended to apply more generally to the situation where every state communicates with at least one state in level 1.

In this paper we consider the decay-rate of discrete phase-type distributions where the number of phases is countably infinite. Throughout, we denote the decay-rate by η and define it to be the reciprocal of the abscissa of convergence of the probability generating function (1.3) of the distribution. This follows the ideas in Neuts[6]. Of course, the concept of a convergence radius applies to any series and not just to a probability generating function. It is convenient to further generalize the concept of a convergence radius to any irreducible matrix. Let $\Theta(A)$ represent the convergence radius of the irreducible matrix A, defined by

$$\Theta(A) = \sup\left\{\delta \geq 0 : \sum_{n=0}^{\infty} A_{ij}^{(n)} \delta^n < \infty\right\}, \tag{1.4}$$

where $A_{ij}^{(n)} = [A^n]_{ij}$ is the (i, j)th element of the nth power of A. Note that the convergence radius is independent of the indices i and j and that if the matrix A is finite-dimensional then

$$\Theta(A) = \frac{1}{\chi(A)},$$

where $\chi(A)$ is the dominant eigenvalue of A.

In the next section we briefly investigate the decay rate under the restriction that the phase space is finite in order to set the scene for the later sections where the phase space is countably infinite. In Section 3 we briefly discuss the general theory when the phase space is countably infinite. In Section 4 we explicitly consider the situation where the governing transition matrix T represents a level-dependent quasi-birth-and-death process. In Section 5 we generalize these results to the situation where T represents a process of M/G/1 type, in other words where T has a block upper-triangular structure. In Sections 6 and 7 we specialize the results of Section 4 to the situations where T represents a level-independent quasi-birth-and-death process and a level-independent birth-and-death process, respectively. We present these special cases to simplify the consideration of examples in Section 8. In Section 9 we summarize the contribution of this paper.

2 Phase-Type Distributions on a Finite Phase Space

Neuts[6] showed that the decay rate of a finite-space discrete phase-type distribution is given by the dominant eigenvalue of T, $\chi(T)$. Thus, the decay rate of every finite-space phase-type distribution with a common transition matrix T must have the same decay-rate. Specifically, if we let $\boldsymbol{\alpha}$ be the left-eigenvector of T associated with the dominant eigenvalue $\chi(T)$, normalised so that the elements sum to one, then the phase-type distribution will be the geometric distribution with parameter $\chi(T)$. However, with different $\boldsymbol{\alpha}$ there may be many different distributions.

Consider, for example, the very simple transition matrix

$$T = \begin{pmatrix} 1/4 & 1/4 \\ 1/2 & 1/4 \end{pmatrix}$$

which has dominant eigenvalue $\chi(T) = \dfrac{\sqrt{2}+1}{4}$ and associated left eigenvector $\boldsymbol{u}$, normalised so that $\boldsymbol{ue} = 1$, given by $\boldsymbol{u} = \left(\dfrac{\sqrt{2}}{1+\sqrt{2}}, \dfrac{1}{1+\sqrt{2}}\right)$.

If we let $\boldsymbol{\alpha} = \boldsymbol{u}$ then the resultant distribution is the geometric distribution with parameter $\chi(T)$. However, if we let $\boldsymbol{\alpha}$ be any other initial distribution, say $\boldsymbol{\alpha} = (1, 0)$, then we get a different phase-type distribution, but with the same decay-rate $\chi(T)$. In the table below, we give the first ten elements of the probability mass functions for these two distributions.

Time	0	1	2	3	4	5	6	7	8	9
Geometric	0	0.396	0.239	0.144	0.087	0.053	0.032	0.019	0.012	0.007
Other	0	0.500	0.188	0.125	0.074	0.045	0.027	0.016	0.010	0.006

Table 1. Probability Mass Functions for the two examples.

Of course, the situation is not always that clear. If T is chosen so that $\boldsymbol{T_0}$ has all entries the same, then regardless of the choice of $\boldsymbol{\alpha}$ the phase-type distribution will always be the geometric distribution with parameter $\chi(T)$. For an indication of the interesting behaviour possible in these circumstances see Bean and Green[7].

3 Phase-Type Distributions on a Countably-Infinite Phase Space

As mentioned above, Neuts[6] proved that the decay rate of a finite-space discrete phase-type distribution is given by the dominant eigenvalue of T, $\chi(T)$. We would like a similar statement when the phase-space consists of a countably-infinite number of states.

On closer inspection of the proof, we see that the result is more generally stated as

$$\eta = \lim_{z \to 0^+} \chi(T + z\boldsymbol{T_0}\boldsymbol{\alpha}). \tag{3.1}$$

The fact that $\lim_{z \to 0^+} \chi(T + z\boldsymbol{T_0}\boldsymbol{\alpha}) = \chi(T)$ is a property of finite-dimensional matrices.

The ideas behind the proof can be generalized to the situation where the matrices are no longer of finite dimension with the aid of some notational changes. Namely, the dominant eigenvalue-eigenvector pair is replaced by the reciprocal of the convergence radius and the associated r-**sub**invariant measure (see for example Seneta[8]). The reason we need to use the associated r-**sub**invariant measure is that there is no guarantee (in general) that an r-invariant measure exists, however it is known that an r-**sub**invariant measure exists for all r less than or equal to the convergence radius. If we then pursue the generalization of the argument in Neuts, we find that

$$\eta \leq \frac{1}{\lim_{z \to 0^+} \Theta(T + z\boldsymbol{T_0}\boldsymbol{\alpha})},$$

or more conveniently, that

$$\frac{1}{\eta} \geq \lim_{z \to 0^+} \Theta(T + z\boldsymbol{T_0}\boldsymbol{\alpha}) \equiv \tau. \tag{3.2}$$

We have tried to prove that $1/\eta = \tau$ for the general case where the phase-space may be countably-infinite, but we have been unsuccessful, except to prove this bound.

Nonetheless, in the specific circumstances of this paper, by directly calculating the probability generating function of the phase-type distribution, we are able to identify the decay rate of the phase-type distribution.

It is worth noting for the special case considered in this paper, see Appendix A, that $1/\eta = \tau$ and so the bound given above is in fact exact. Of course, r-invariant measures have been shown to exist for all QBDs[9], and the matrix $T + z\boldsymbol{T_0}\boldsymbol{\alpha}$ is a minor variant of a QBD. So it would seem likely that r-invariant measures would exist for this structure, which is sufficient to explain the exactness of the bound.

However, for the general case of phase-type distributions on a countably-infinite phase space we know of no way to guarantee the existence of r-invariant measures. Therefore, whether $1/\eta = \tau$ in general for phase-type distributions on a countably-infinite phase space is an open question.

4 Transition Matrices that represent Level Dependent QBDs

4.1 Notation and Assumptions

In this section we consider the situation where the transition matrix T represents a level-dependent quasi-birth-and-death process and has the block-partitioned form

$$T = \begin{pmatrix} A_1^{(1)} & A_0^{(1)} & 0 & 0 & 0 & \cdots \\ A_2^{(2)} & A_1^{(2)} & A_0^{(2)} & 0 & 0 & \ddots \\ 0 & A_2^{(3)} & A_1^{(3)} & A_0^{(3)} & 0 & \ddots \\ 0 & 0 & A_2^{(4)} & A_1^{(4)} & A_0^{(4)} & \ddots \\ \vdots & \ddots & \ddots & \ddots & \ddots & \ddots \end{pmatrix}, \tag{4.1}$$

where $A_0^{(k)}$, $A_1^{(k)}$ and $A_2^{(k)}$, for all $k \geq 1$ are such that T is a substochastic matrix. The block-partitioning arises from the fact that we think of the state-space as being two-dimensional, where the first index is allowed to range over the nonnegative integers and is known as the *level*, while the second index is allowed to range over a finite set that depends on the level k, $\{1, 2, \ldots, M_k\}$, and is known as the *phase*. The blocks then represent the phases within the levels. Thus, for example, $A_2^{(k)}$ is an $M_k \times M_{k-1}$ matrix that governs the transitions from the phases in level k down to the phases in level $k-1$.

The level-independent quasi-birth-and-death process is then a special case where all levels consist of the phases $\{1, 2, \ldots, M\}$ and the block-matrices are square matrices and do not depend on the level k, so they are denoted A_0, A_1 and A_2. The birth-and-death process is a further special case where the phase set is a singleton and so the block-matrices are simply scalars. This applies to the level-dependent and level-independent cases.

As mentioned in Section 1, for simplicity of exposition, we assume that T is irreducible. However, the results can be extended to apply to the situation where every state communicates with at least one state in level 1.

We shall be making use of the results of Bean, Pollett and Taylor[9] where they specifically assume that absorption is only possible from level 1 and so we can only consider the class where the number of states from which absorption is possible is finite. Therefore, we assume that $\boldsymbol{T_0} = \boldsymbol{e} - \boldsymbol{Te}$ is given by

$$\boldsymbol{T_0} = \begin{pmatrix} A_2^{(1)}\boldsymbol{e} \\ 0 \\ 0 \\ \vdots \end{pmatrix}.$$

We denote the initial probability distribution by

$$\boldsymbol{\alpha} = (\boldsymbol{\alpha_1}, \boldsymbol{\alpha_2}, \dots),$$

where $\boldsymbol{\alpha_k}$ represents the probabilities of starting in the particular phases in level k.

As we observed in Section 3 we must work directly with the probability generating function of the phase-type distribution (1.3), as there is no other convenient expression to work with. That is, we directly need to determine the convergence radius of

$$P(z) = \sum_{n=1}^{\infty} \boldsymbol{\alpha} T^{n-1} \boldsymbol{T_0} z^n + \alpha_0.$$

We choose to rewrite this as

$$P(z) = z\boldsymbol{\alpha} \left(\sum_{n=1}^{\infty} T^{n-1} z^{n-1} \right) \boldsymbol{T_0} + \alpha_0; \tag{4.2}$$

such a rearrangement is justified since all the terms are nonnegative.

In order to gain a deeper understanding of the behaviour of $P(z)$ we wish to study the expression

$$N(z) = \sum_{n=0}^{\infty} T^n z^n. \tag{4.3}$$

It is well known, for any probability transition matrix T, that the $(i,j)^{\text{th}}$ element of T^m represents the probability that the process is in state j exactly m time units after it started in state i. Therefore the $(i,j)^{\text{th}}$ element of $N(1) = \sum_{n=0}^{\infty} T^n$ represents the expected number of visits to state j given that the process started in state i. If T is substochastic then this must be finite. Now, introduce $z \geq 1$ as a reward factor at each time step. That is, any event that occurs at time point k is paid a reward of z^k. Then we find that the $(i,j)^{\text{th}}$ element of $N(z)$ represents the expected total reward paid on visits to state j given that the process started in state i. It is this expression, which is finite at $z = 1$ and increasing in z, that plays a major role in the decay rate of the phase-type distribution. Of course, the convergence radius of $N(z)$ is exactly the convergence radius of the probability generating function, when the phase space is finite, but this is not necessarily the case when the phase-space is countably-infinite.

It is natural to decompose the matrix $N(z)$ into blocks, and so we write it as

$$N(z) = \begin{pmatrix} N_{11}(z) & N_{12}(z) & N_{13}(z) & \cdots \\ N_{21}(z) & N_{22}(z) & N_{23}(z) & \ddots \\ N_{31}(z) & N_{32}(z) & N_{33}(z) & \ddots \\ \vdots & \ddots & \ddots & \ddots \end{pmatrix}. \tag{4.4}$$

In order to further investigate these finite-dimensional matrix blocks, we introduce some matrix families that are related to the standard matrices in the matrix-analytic methods literature, namely, G and U.

4.2 The families of matrices $G_k(\delta)$ and $U_k(\delta)$

The matrix family $\{G_k(\delta),\ k \geq 1\}$ is the minimal (elementwise) nonnegative solution to the set of matrix-recurrence equations

$$G_k(\delta) = \delta\left[A_2^{(k)} + A_1^{(k)} G_k(\delta) + A_0^{(k)} G_{k+1}(\delta) G_k(\delta)\right], \quad k \geq 1. \tag{4.5}$$

It is not generally possible to analytically determine the $G_k(\delta)$, but efficient numerical schemes exist[9].

We can also define the family of matrices

$$U_k(\delta) = \delta\left[A_1^{(k)} + A_0^{(k)} G_{k+1}(\delta)\right], \quad k \geq 1. \tag{4.6}$$

When $\delta = 1$, these are the families of matrices $\{G_k,\ k \geq 1\}$ and $\{U_k,\ k \geq 1\}$ seen in the literature on level-dependent quasi-birth-and-death processes[10]. These are themselves a generalization of the well-known G and U matrices used in the literature on matrix-analytic methods[2,11,12]. When $\delta > 1$, we interpret the factor δ as a multiplicative reward, earned every step of the Markov process. Probabilistic arguments can now be recovered, as described in the text following equation (4.3). The expected reward matrices, $G_k(\delta)$ and $U_k(\delta)$, have the following probabilistic interpretation. Let the $(i,j)^{\text{th}}$ element of the matrix $G_k^{(n)}$ be the probability that the process *first* visits level $k-1$ at time point n, and does so in phase j, given that it starts in phase i of level k at time point 0. Then, $G_k(\delta)$ is $\sum_{n=1}^{\infty} G_k^{(n)} \delta^n$. Similarly, let the $(i,j)^{\text{th}}$ element of the matrix $U_k^{(n)}$ be the probability that the process *first* returns to level k at time point n, and does so in phase j, given that it starts in phase i of level k at time point 0. Then, $U_k(\delta)$ is $\sum_{n=1}^{\infty} U_k^{(n)} \delta^n$. For more details on these ideas see the papers by Bean, Pollett, Taylor and co-authors[13,14,9] and the paper by Ramaswami[15].

It then follows from the physical interpretations, that

$$N_{kk}(\delta) = \sum_{n=0}^{\infty} [U_k(\delta)]^n, \quad k \geq 1, \tag{4.7}$$

$$N_{mk}(\delta) = \left(\prod_{\ell=k+1}^{m} G_\ell(\delta) \right) N_{kk}(\delta), \quad m > k \geq 1, \tag{4.8}$$

where throughout we assume that an empty product of matrices is the identity matrix of appropriate order and that $\prod_{j=k}^{m} G_j(\delta)$ is interpreted as $G_m(\delta)G_{m-1}(\delta) \cdots G_{k+1}(\delta)G_k(\delta)$.

It then follows[9] that the convergence radius of T is

$$\beta = \sup \{\delta \geq 1 : \chi(U_1(\delta)) \leq 1\}, \tag{4.9}$$

and this is also the supremum for which $N_{11}(\delta)$ is finite. Using Bean, Pollett and Taylor[9], it is easy to see that the supremum of the set of values δ for which a nonnegative solution for the family of matrices $\{G_k(\delta),\ k \geq 1\}$ exists, is also β. Note that the supremum of the set of values δ for which a nonnegative solution for the family of matrices $\{G_k(\delta),\ k \geq 2\}$, or equivalently the family $\{U_k(\delta),\ k \geq 1\}$, exists is at least as large as β and in some circumstances is strictly greater than β.

4.3 Determination of η

Let $\widehat{\boldsymbol{\alpha} \circ G}(\delta)$ denote the series

$$\widehat{\boldsymbol{\alpha} \circ G}(\delta) = \sum_{k=1}^{\infty} \left(\boldsymbol{\alpha}_k \prod_{j=2}^{k} G_j(\delta) \right). \tag{4.10}$$

By recalling that $\boldsymbol{T}_0$ is a column vector with the first block given by $A_2\boldsymbol{e}$, and all remaining blocks filled with zeros, we can see from equations (4.2), (4.4), (4.7) and (4.8), that

$$P(\delta) = \delta \widehat{\boldsymbol{\alpha} \circ G}(\delta) N_{11}(\delta) A_2^{(1)} \boldsymbol{e} + \alpha_0. \tag{4.11}$$

Theorem 1 *The decay rate of the PH-distribution* $(T, \boldsymbol{\alpha})$ *is given by*

$$1/\eta = \min(\beta, \gamma), \tag{4.12}$$

where β *is the convergence radius of* T *and* γ *is the convergence radius of* $\widehat{\boldsymbol{\alpha} \circ G}(\delta)$.

Proof: By equation (4.11) we see that $1/\eta$ can be no larger than β since β is the convergence radius of $N_{11}(\delta)$. Therefore, if $\beta \leq \gamma$ then $1/\eta = \beta$. However, it could be that $\widehat{\boldsymbol{\alpha} \circ G}(\delta)$ is infinite at $\delta = \beta$. If $\gamma < \beta$, then $P(\delta)$ must be infinite for all $\delta > \gamma$ and finite for all $\delta < \gamma$, and so $1/\eta = \gamma$. ■

Therefore, the decay rate of the phase-type distribution $(T, \boldsymbol{\alpha})$ is given by the maximum of

1. the reciprocal of β, the convergence radius of the transition matrix T, and

2. the reciprocal of γ, where γ is the convergence radius of the matrix-series $\widehat{\boldsymbol{\alpha} \circ G}(\delta)$. Here, both the convergence properties of the initial probability vector $\boldsymbol{\alpha}$ and the properties of the transition matrix T *combine* to determine the decay-rate of the phase-type distribution $(T, \boldsymbol{\alpha})$.

4.4 Single Unifying Condition

These two conditions to describe the decay rate can be summarised neatly in a single condition. Bean, Pollett and Taylor[9] state that $G_1(\delta) = \delta \sum_{n=0}^{\infty} [U_1(\delta)]^n A_2^{(1)} = \delta N_{11}(\delta) A_2^{(1)}$, and so we can rewrite equation (4.11) as

$$P(\delta) = \sum_{k=1}^{\infty} \boldsymbol{\alpha_k} \left(\prod_{j=1}^{k} G_j(\delta) \right) \boldsymbol{e} + \alpha_0. \tag{4.13}$$

The decay-rate, η, of the phase-type distribution $(T, \boldsymbol{\alpha})$ is therefore given by the reciprocal of the convergence radius of this function.

The probabilistic interpretation of this expression is, of course, exactly that of $P(\delta)$: it represents the expected reward for visiting the absorbing state 0, given that the process starts according to the initial probability distribution $\boldsymbol{\alpha}$ at time point 0, where a visit to the absorbing state at time n earns reward δ^n. We then require the convergence radius for this expected reward, and then η is its reciprocal.

The reason that we don't work with this expression directly in the above derivation is that $N_{11}(\beta)$ can be finite or infinite. Thus, detecting (numerically or analytically) the convergence radius is a hard task. Instead it is much easier to explicitly identify β and then work with a series where all the terms themselves are guaranteed to be finite, in other words to work with $\widehat{\alpha \circ G(\delta)}$. To identify β it is easier to use an expression that is strictly less than one if and only if $N_{11}(\delta)$ is finite. This is exactly what we have done above by defining β in terms of $\chi(U_1(\delta))$ as in equation (4.9).

5 Processes with Block Upper-Triangular Matrices

Having developed the results in the previous section, we can also apply them to the situation where the transition matrix T has block upper-triangular form. In other words, the transition matrix represents a process of M/G/1 type. The only difference in the mathematical statements is that the matrix family $\{G_k(\delta),\ k \geq 1\}$ is now defined as the minimal nonnegative solutions to the family of equations

$$G_k(\delta) = \delta \sum_{n=0}^{\infty} A_n^{(k)} \left(\prod_{\ell=k}^{n+k-1} G_\ell(\delta) \right), \quad k \geq 1,$$

and the matrix family $\{U_k(\delta),\ k \geq 1\}$ is now given by

$$U_k(\delta) = \delta \sum_{n=1}^{\infty} A_n^{(k)} \left(\prod_{\ell=k+1}^{n+k-1} G_\ell(\delta) \right), \quad k \geq 1.$$

6 Transition Matrices that represent Level Independent QBDs

In this section we consider the special case of Section 4 where the behaviour of the process is independent of the level in which the process currently resides. Thus $A_0^{(k)} = A_0$, $A_1^{(k)} = A_1$ and $A_2^{(k)} = A_2$, for all $k \geq 1$.

In this situation the analysis simplifies quite considerably as we simply require the one matrix $G(\delta)$ that is the minimal nonnegative solution to the matrix-quadratic equation

$$G = \delta \left[A_2 + A_1 G + A_0 G^2 \right], \tag{6.1}$$

and the one matrix

$$U(\delta) = \delta \left[A_1 + A_0 G \right]. \tag{6.2}$$

Some consequences of this include:

- The composite function $\widehat{\boldsymbol{\alpha} \circ G}(\delta)$ and the generating function $P(\delta)$ now denote the simpler matrix-series

$$\widehat{\boldsymbol{\alpha} \circ G}(\delta) = \sum_{k=1}^{\infty} \boldsymbol{\alpha_k} G(\delta)^{k-1}, \tag{6.3}$$

and

$$P(\delta) = \sum_{k=1}^{\infty} \boldsymbol{\alpha_k} G(\delta)^k + \alpha_0. \tag{6.4}$$

The remaining arguments remain as in Section 4.

7 Transition Matrices that represent Level Independent Birth-and-Death Processes

In this section we consider the special case of Section 6 where the phase space at each level is a singleton and so the matrices A_0, A_1, A_2 are replaced by the scalars a_0, a_1 and a_2. In this special case, other methods of progress are possible[16,17], but we find it more convenient to continue by specialising the results in the previous sections.

In this situation the analysis simplifies quite considerably as we no longer need to deal with matrices. Consequently, the scalar $g(\delta)$ that is the minimal nonnegative solution to the quadratic equation

$$g = \delta\left[a_2 + a_1 g + a_0 g^2\right], \tag{7.1}$$

can be deduced analytically to be

$$g(\delta) = \frac{(1 - \delta a_1) - \sqrt{(1 - \delta a_1)^2 - 4\delta^2 a_0 a_2}}{2\delta a_0} \tag{7.2}$$

for $\delta \leq \beta$, where β is given by

$$\beta = \frac{1}{a_1 + 2\sqrt{a_0 a_2}}. \tag{7.3}$$

In fact, $g(\beta) = \sqrt{\dfrac{a_2}{a_0}}$.

Some consequences of this include:

- The function $\chi(\cdot)$ is no longer required.
- $u(\delta) = \delta a_1 + \dfrac{(1 - \delta a_1) - \sqrt{(1 - \delta a_1)^2 - 4\delta^2 a_0 a_2}}{2}$.

- The composite function $\widehat{\alpha \circ g}(\delta)$ and the generating function $P(\delta)$ now denote the simple series (in fact generating functions of $\boldsymbol{\alpha}$) given by

$$\widehat{\alpha \circ g}(\delta) = \sum_{k=1}^{\infty} \alpha_k g(\delta)^{k-1}, \tag{7.4}$$

and

$$P(\delta) = \sum_{k=1}^{\infty} \alpha_k g(\delta)^k + \alpha_0. \tag{7.5}$$

The remaining arguments remain as in Section 4.

8 Examples

The suite of examples that we consider are the level-independent birth-and-death processes as we can proceed analytically and treat classes of problems at one time. It is a reasonably straightforward matter to numerically treat any arbitrary QBD process, however, we do not proceed along those lines as it would involve treating one numerical problem at a time.

The following theorem is a well-known result and, in fact, one that is proved in greater generality on page 138 of van Doorn and Schrijner[17]. However, we still present the theorem here, as the proof is an example of the application of the main results of this paper.

Theorem 2 *If T is of the form of a level-independent birth-and-death process and if we choose the vector $\boldsymbol{\alpha}$ so that, for some $\sigma \in (1, \Theta(T)]$*

$$\boldsymbol{\alpha} T = \sigma \boldsymbol{\alpha}, \tag{8.1}$$

then the resultant phase-type distribution is geometric with parameter $\eta = \dfrac{1}{\sigma}$. In fact, such a distribution exists for all $\sigma \in (1, \Theta(T)]$.

Proof: Let $\beta = \Theta(T)$. By Theorems 6 and 9 of Bean, Pollett and Taylor[14], we are able to identify suitable vectors $\boldsymbol{\alpha}$ for all $\sigma \in (1, \beta]$.

If $\sigma < \beta$, then

$$\alpha_j = c\left(r_2(\sigma)^j - r_1(\sigma)^j\right), \tag{8.2}$$

where c is a normalising constant, $r_1(\sigma)$ and $r_2(\sigma)$ are the two solutions to the quadratic equation

$$r = \sigma\left[a_0 + r a_1 + r^2 a_2\right],$$

and are given by

$$\frac{(1-\sigma a_1) \pm \sqrt{(1-\sigma a_1)^2 - 4\sigma^2 a_0 a_2}}{2\sigma a_2}.$$

We denote the larger solution by $r_2(\sigma)$ and the smaller solution by $r_1(\sigma)$.

A fact that will be very useful is that for any value $\delta \leq \beta$,

$$g(\delta) = r_1(\delta)\frac{a_2}{a_0}\ . \tag{8.3}$$

If $\sigma = \beta$, then

$$\alpha_j = c\left(j r_1(\sigma)^{j-1}\right), \tag{8.4}$$

where c is a normalising constant and $r_1(\sigma)$ is as above.

In this case, we know that $r_1(\sigma) = r_1(\beta) = \sqrt{\frac{a_0}{a_2}}$, by equation (8.3), and so $\alpha_j = cj\left(\frac{a_0}{a_2}\right)^{(j-1)/2}$. Consequently,

$$\begin{aligned}
\widehat{\alpha \circ g}(\beta) &= \sum_{k=1}^{\infty} \alpha_k g(\beta)^{k-1}, \\
&= c\sum_{k=1}^{\infty} k\left(\frac{a_0}{a_2}\right)^{(k-1)/2}\left(\frac{a_2}{a_0}\right)^{(k-1)/2}, \\
&= c\sum_{k=1}^{\infty} k,
\end{aligned}$$

which clearly does not converge. Therefore, we must identify γ the convergence radius of this series. It turns out that $\gamma = \beta$ as $g(\delta)$ is increasing and so for any $\delta < \beta$ we have that $g(\delta) < g(\beta)$ and so

$$\widehat{\alpha \circ g}(\delta) = c\sum_{k=1}^{\infty} k w^{k-1}$$

where $w < 1$ and hence the series converges. We can therefore conclude that $\eta = \frac{1}{\gamma} = \frac{1}{\beta} = \frac{1}{\sigma}$ and so the decay rate of the distribution is given by $\frac{1}{\sigma}$. That the distribution is geometric follows from equation (8.1).

Now let us consider the case where $\sigma < \beta$. We first show that $\widehat{\alpha \circ g}(\sigma)$ diverges and hence, of course, so must $\widehat{\alpha \circ g}(\beta)$. By use of equations (8.2) and (8.3) we have that

$$\begin{aligned}
\widehat{\boldsymbol{\alpha} \circ g}(\sigma) &= \sum_{k=1}^{\infty} \alpha_k g(\sigma)^{k-1}, \\
&= c\sum_{k=1}^{\infty} \left[r_2(\sigma)^k g(\sigma)^{k-1} - r_1(\sigma)^k g(\sigma)^{k-1}\right], \\
&= c\sum_{k=1}^{\infty} \left[r_2(\sigma)^k r_1(\sigma)^{k-1}\left(\frac{a_2}{a_0}\right)^{k-1} - r_1(\sigma)^k r_1(\sigma)^{k-1}\left(\frac{a_2}{a_0}\right)^{k-1}\right], \\
&= c r_1(\sigma)^{-1}\frac{a_0}{a_2}\sum_{k=1}^{\infty} \left[(r_2(\sigma) r_1(\sigma))^k \left(\frac{a_2}{a_0}\right)^{k} - r_1(\sigma)^{2k}\left(\frac{a_2}{a_0}\right)^{k}\right],
\end{aligned}$$

Now, since $\sigma < \beta$ and $r_1(\cdot)$ is an increasing function, we know that $r_1(\delta) < \sqrt{\frac{a_0}{a_2}}$ for all $\sigma \leq \delta < \beta$. Also, it is easy to show that $r_2(\delta) r_1(\delta) = \frac{a_0}{a_2}$ for all $1 \leq \delta < \beta$. Therefore, we can conclude that

$$\widehat{\boldsymbol{\alpha} \circ g}(\sigma) = c r_1(\sigma)^{-1}\frac{a_0}{a_2}\sum_{k=1}^{\infty}\left[1 - w^k\right],$$

where $w < 1$, and so $\widehat{\boldsymbol{\alpha} \circ g}(\sigma)$ diverges. Consequently, we must identify γ the convergence radius of this series. It turns out that $\gamma = \sigma$ as $g(\delta)$ is increasing and so for any $\delta < \sigma$ we have that $g(\delta) < g(\sigma)$ and so

$$\widehat{\boldsymbol{\alpha} \circ g}(\delta) = c g(\delta)^{-1}\sum_{k=1}^{\infty}\left[w_2^k - w_1^k\right]$$

where $w_1 < w_2 < 1$ and hence the series converges. We can therefore conclude that $\eta = \frac{1}{\gamma} = \frac{1}{\sigma}$ and so the decay rate of the distribution is given by $\frac{1}{\sigma}$. That the distribution is geometric again follows from equation (8.1). ■

If we choose an initial vector $\boldsymbol{\alpha}$ that does not obey equation (8.1) then the resultant phase-type distribution will generally not be a geometric distribution. Nonetheless, the decay rate of the distribution will be as described in this paper. For example, consider the following theorem.

Theorem 3 *Let* $\alpha_j = (1-p)p^{j-1}$, $j \geq 1$, *for some* $p \in (0,1)$. *If* p *is such that* $p\sqrt{\frac{a_2}{a_0}} < 1$ *then* $\eta = \frac{1}{\beta} = a_1 + 2\sqrt{a_0 a_2}$. *On the other hand, if* $p\sqrt{\frac{a_2}{a_0}} \geq 1$

then $\eta = \dfrac{1}{\gamma}$ *where* γ *obeys the equation*

$$\gamma\left[a_0 + a_1 p + a_2 p^2\right] = p.$$

Proof: If $\alpha_j = (1-p)p^{j-1}$, $j \geq 1$, for some $p \in (0,1)$, then

$$\widehat{\alpha \circ g}(\beta) = (1-p)\sum_{k=1}^{\infty}\left[p\sqrt{\frac{a_2}{a_0}}\right]^{k-1},$$

and so converges if and only if $p\sqrt{\dfrac{a_2}{a_0}} < 1$. In this case we know that $\eta = \dfrac{1}{\beta} = a_1 + 2\sqrt{a_0 a_2}$. On the other hand, if $p\sqrt{\dfrac{a_2}{a_0}} \geq 1$ then $\widehat{\alpha \circ g}(\beta)$ diverges and we need to identify the value γ that is the radius of convergence of $\widehat{\alpha \circ g}(\delta)$. Now $\widehat{\alpha \circ g}(\delta) = (1-p)\sum_{k=1}^{\infty}\left[pg(\delta)\right]^{k-1}$, and so γ must be chosen so that $pg(\gamma) = 1$. After some arithmetic, it turns out that this is equivalent to requiring γ to obey the equation

$$\gamma\left[a_0 + a_1 p + a_2 p^2\right] = p.$$

■

We have left the equation in this form, as it reveals that we can consider this question from another viewpoint. If you want to use an initial vector $\boldsymbol{\alpha}$ of the form given above, and require a decay rate of $\eta = \dfrac{1}{\gamma}$, then you need to choose p to obey this same equation, in other words you need $p = r_1(\gamma)$ or $p = r_2(\gamma)$, and this exists for all $\gamma \in (1, \beta]$. Since $r_i(\beta) = \sqrt{\dfrac{a_0}{a_2}}$ for $i = 1, 2$ and $r_1(\cdot)$ is increasing and $r_2(\cdot)$ is decreasing, it turns out that $r_1(\gamma)\sqrt{\dfrac{a_2}{a_0}} < 1$ and $r_2(\gamma)\sqrt{\dfrac{a_2}{a_0}} > 1$ for all $\gamma < \beta$. Consequently, you need $p = r_2(\gamma)$ to achieve a decay rate of $\eta = \dfrac{1}{\gamma}$. When $p = r_1(\gamma)$ the decay rate is $\eta = \dfrac{1}{\beta}$.

In the table below we give the first 10 elements of the probability mass functions for such an example. We choose $a_0 = 1/8$, $a_1 = 3/8$, $a_2 = 1/2$ and require a decay rate of $\eta = \dfrac{1}{\gamma} = 0.9$. For this example, we find that $\beta = 8/7$ and that $r_1(\gamma) = \dfrac{21 - \sqrt{21^2 - 20^2}}{40} \approx 0.36492$ and $r_2(\gamma) = \dfrac{21 + \sqrt{21^2 - 20^2}}{40} \approx 0.68508$. The decay rates can be verified numerically, however, very high precision arithmetic and very large matrices are required when $p = r_1(\gamma)$.

Time	0	1	2	3	4	5	6	7	8	9
$p = r_1(\gamma)$	0	0.318	0.177	0.119	0.085	0.063	0.048	0.037	0.029	0.023
$p = r_2(\gamma)$	0	0.158	0.113	0.091	0.076	0.065	0.056	0.049	0.043	0.038

Table 2. Probability Mass Functions for the two examples.

9 Summary

In this paper we have considered the decay-rate of phase-type distributions with an underlying state space that is countably-infinite. We have explicitly considered the class of distributions where the governing transition matrix has a block upper-triangular structure. This has allowed us to proceed using the framework of matrix-analytic methods to efficiently identify the crucial elements.

When a phase-type distribution has a countably infinite underlying phase-space with a given transition matrix T, there is a fundamental minimum decay-rate that can be achieved, $1/\beta$. By choosing initial vectors $\boldsymbol{\alpha}$ appropriately, it is usually possible to create many different distributions. However, any decay-rate between this extreme value and 1 can usually be achieved by choosing other initial vectors. Again, by choosing the initial vectors appropriately, it is usually possible to create many different distributions with the same decay-rate. It is almost impossible to make general statements about the nature of distributions that can be achieved for a given matrix T as, even in the finite phase-space situation, this is an incredibly difficult problem; see, for example, Bean and Green[7]. However, if we restrict attention to level-independent birth-and-death processes, then we have shown in Theorem 2 that geometric distributions exist with parameters $\eta = 1/\delta$ for all $\delta \in (1, \beta]$. A similar result can also be proven for level-independent quasi-birth-and-death processes by using Theorems 6 and 9 of Bean, Pollett and Taylor[14] to provide the required δ-invariant measures.

ACKNOWLEDGEMENTS

Nigel Bean would like to thank the University of Adelaide for financial support during his Special Studies Program. This support allowed him to visit Denmark, where the work in this paper was initiated. He would also like to acknowledge the support of the Australian Research Council through Discovery Grant DP0209921.

Appendix

A The Decay Rate Bound is Exact

As we observed in Section 3, the decay rate, η, of the phase-type distribution defined by T and $\boldsymbol{\alpha}$ is bounded by

$$\frac{1}{\eta} \geq \lim_{z \to 0^+} \Theta\left(T + z\boldsymbol{T_0}\boldsymbol{\alpha}\right).$$

As mentioned in Sections 1 and 4, for simplicity of exposition, we assume throughout that T is irreducible. However, the results in the main body of the paper can be extended to apply to the situation where every state communicates with at least one state in level 1. In this appendix, we also assume that the matrix T with the first block of rows and columns removed, known as ${}_2T$, is irreducible; but we note that the results can be extended to apply to the situation where every state of the process represented by ${}_2T$ communicates with at least one state in level 2.

In this section, we show for the special circumstances considered in this paper, that $\tau \equiv \lim_{z\to 0^+} \Theta\left(T + z\boldsymbol{T_0}\boldsymbol{\alpha}\right) = \min(\beta, \gamma) = 1/\eta$, and hence that the bound is exact.

Consider the matrix $V(z) = T + z\boldsymbol{T_0}\boldsymbol{\alpha}$, which is specifically given in block-partitioned form by

$$V(z) = \begin{pmatrix} W_1 & W_2 & W_3 & W_4 & W_5 & \cdots \\ A_2^{(2)} & A_1^{(2)} & A_0^{(2)} & 0 & 0 & \ddots \\ 0 & A_2^{(3)} & A_1^{(3)} & A_0^{(3)} & 0 & \ddots \\ 0 & 0 & A_2^{(4)} & A_1^{(4)} & A_0^{(4)} & \ddots \\ \vdots & \ddots & \ddots & \ddots & \ddots & \ddots \end{pmatrix}, \qquad \text{(A.1)}$$

where $W_1 = A_1^{(1)} + zA_2^{(1)}\boldsymbol{e}\boldsymbol{\alpha_1}$, $W_2 = A_0^{(1)} + zA_2^{(1)}\boldsymbol{e}\boldsymbol{\alpha_2}$ and $W_j = zA_2^{(1)}\boldsymbol{e}\boldsymbol{\alpha_j}$ for all $j \geq 3$. When $z = 1$, note that $V(z)$ is a stochastic matrix, and for all $0 \leq z < 1$, $V(z)$ is a strictly substochastic matrix.

Following the results in Bean, Pollett and Taylor[9], we shall consider the substochastic matrix that results from deleting the row and column corresponding to level 1. This is the transition matrix of the process restricted to levels $\{2, 3, \dots\}$ and is given by

$$ {}_2V = {}_2T = \begin{pmatrix} A_1^{(2)} & A_0^{(2)} & 0 & 0 & \ddots \\ A_2^{(3)} & A_1^{(3)} & A_0^{(3)} & 0 & \ddots \\ 0 & A_2^{(4)} & A_1^{(4)} & A_0^{(4)} & \ddots \\ \vdots & \ddots & \ddots & \ddots & \ddots & \ddots \end{pmatrix}. \tag{A.2} $$

Note that ${}_2T$ is independent of z and is just the transition matrix of a simple absorbing level-dependent quasi-birth-and-death process, where absorption can only occur through the bottom level. Accordingly, we can apply the results given in Section 4.1 to show that $\beta_2 \equiv \Theta({}_2T)$ is given by

$$ \beta_2 = \sup\{\delta \geq 1 : \chi(U_2(\delta)) \leq 1\}. \tag{A.3} $$

Note that $\beta_2 \geq \beta$ and it is fairly easy to contruct examples[9] where $\beta_2 = \beta$ and where $\beta_2 > \beta$.

Now, consider the expected reward on first returning to level 1 given that the process starts in level 1. This is essentially the $U_1(\delta)$ matrix, but for the process $V(z)$. Specifically, let $U(\delta, z)_{ij}$ be the expected reward on first returning to level 1 and that return occurs in phase j, given that the process starts in phase i of level 1. Then the matrix $U(\delta, z)$ is given by

$$ U(\delta, z) = \delta\,[W_1 + W_2 G_2(\delta) + W_3 G_3(\delta) G_2(\delta) + \cdots]. \tag{A.4} $$

Following the ideas behind Bean, Pollett and Taylor[9], we can see that the convergence radius of $V(z)$, denoted $\beta(z)$, is given by

$$ \beta(z) = \sup\{\delta \geq 1 : \chi(U(\delta, z)) \leq 1\}. \tag{A.5} $$

Now, substitute for the particular forms of W_j, to find that

$$ U(\delta, z) = U_1(\delta) + \delta z A_2^{(1)} \widehat{\boldsymbol{e}\boldsymbol{\alpha} \circ G}(\delta). \tag{A.6} $$

In order to find $\beta(z)$ we need to have a good understanding of $\chi(U(\delta, z))$ as a function of δ. It is clear that the value of $\chi(U(\beta, z))$ will be highly significant, as $U(\delta, z)$ does not exist for $\delta > \beta$.

Consider, $U(\delta, z)$ as the sum of two nonnegative matrices, $U_1(\delta)$ and $\delta\left[z A_2^{(1)} \widehat{\boldsymbol{e}\boldsymbol{\alpha} \circ G}(\delta)\right]$. We know that

$$ \chi(U_1(\delta)) \text{ is increasing for all } 1 \leq \delta \leq \beta_2, \tag{A.7} $$

$$ \chi(U_1(\beta)) \leq 1, \tag{A.8} $$

and

$$U_1(\delta) \text{ is infinite for all } \delta > \beta_2. \tag{A.9}$$

Here, and throughout, when we say that a matrix is infinite, we mean that at least one element has a non-finite value.

Consequently, we need to consider the two cases: **Case 1:** $\widehat{\boldsymbol{\alpha} \circ G}(\beta) < \infty$ and **Case 2:** $\widehat{\boldsymbol{\alpha} \circ G}(\beta) = \infty$.

- **Case 1:** Let us first consider the situation where $\chi(U_1(\beta_2)) < 1$ and so $\beta = \beta_2$. Since $\widehat{\boldsymbol{\alpha} \circ G}(\beta) < \infty$, for sufficiently small z we have that $\chi(U(\beta, z)) < 1$ and so $\beta(z) \geq \beta$. However, there can be no finite nonnegative solution of equation (4.5) with $\delta > \beta_2 = \beta$ and so, for sufficiently small z, $\beta(z) = \beta$. Hence $\tau = \beta$.

  If $\chi(U_1(\beta)) = 1$ and so $\beta \leq \beta_2$ then the argument becomes slightly more delicate. Consider $\delta < \beta$. Since $\widehat{\boldsymbol{\alpha} \circ G}(\delta) < \infty$, there exists a $z_\delta > 0$ such that if $z < z_\delta$ then equation (A.7) implies that $\chi(U(\delta, z)) < 1$ and so $\beta(z) \geq \delta$. This applies for all $\delta < \beta$ and since $\beta(z) \leq \beta$ for all $z > 0$, it is clear that $\lim_{z \to 0+} \beta(z) = \beta$. Again $\tau = \beta$.

- **Case 2:** Since $\widehat{\boldsymbol{\alpha} \circ G}(\beta) = \infty$, it is clear for all $z > 0$ that $\chi(U(\beta, z)) = \infty > 1$. This will hold whenever $\widehat{\boldsymbol{\alpha} \circ G}(\delta) = \infty$. Therefore, let γ be the supremum of the interval over which the matrix-series $\widehat{\boldsymbol{\alpha} \circ G}(\delta)$ is finite. Note that $\widehat{\boldsymbol{\alpha} \circ G}(\gamma)$ could be finite or infinite and that $\chi(U_1(\gamma)) < 1$.

  If $\widehat{\boldsymbol{\alpha} \circ G}(\gamma) < \infty$ then for sufficiently small z we have that $\chi(U(\gamma, z)) < 1$ and so $\beta(z) \geq \gamma$. However, for any $\delta > \gamma$ and $z > 0$ we know that $\chi(U(\delta, z)) = \infty > 1$ and so for all $z > 0$ we have that $\beta(z) \leq \gamma$. Hence, for sufficiently small z, $\beta(z) = \gamma$ and hence $\tau = \gamma$.

  Now, if $\widehat{\boldsymbol{\alpha} \circ G}(\gamma) = \infty$ then it is again clear that $\beta(z)$ can be no greater than γ for all $z > 0$. In fact, for all $z > 0$, $\chi(U(\gamma, z)) = \infty > 1$. Consider $\delta < \gamma$. Since $\widehat{\boldsymbol{\alpha} \circ G}(\delta) < \infty$, there exists a $z_\delta > 0$ such that if $z < z_\delta$ then equation (A.7) implies that $\chi(U(\delta, z)) < 1$ and so $\beta(z) \geq \delta$. This applies for all $\delta < \gamma$ and since $\beta(z) \leq \gamma$ for all $z > 0$, it is clear that $\lim_{z \to 0+} \beta(z) = \gamma$. Hence $\tau = \gamma$.

Therefore, we have shown that $\tau = \min(\beta, \gamma) = 1/\eta$.

References

1. A.K. Erlang. Solution of some problems in the theory of probabilities of significance in automatic telephone exchanges. *The Post Office Electrical Engineer's Journal*, 10:189–197, 1917-18.
2. M.F. Neuts. *Matrix Geometric Solutions in Stochastic Models.* The John Hopkins University Press, Baltimore, 1981.
3. B.F. Nielsen and P.G. Taylor. Matrix-Geometric Modelling of Heavy-Tailed Queues. In preparation.
4. D.H. Shi, J. Guo, and L. Liu. SPH-Distributions and the Rectangle-Iterative Algorithm. In Attahiru S. Alfa and Srinivas R. Chakravarthy, editors, *Matrix-Analytic Methods in Stochastic Models*, volume 183 of *Lecture Notes in Pure and Applied Mathematics*, pages 207–224. Marcel Dekker, New York, 1997.
5. D. Shi and D. Liu. Markovian Models for Non-Negative Random Variables. In Attahiru S. Alfa and Srinivas R. Chakravarthy, editors, *Advances in Matrix-Analytic Methods for Stochastic Models*, pages 403–428. Notable Publications Inc., NJ, USA, 1998.
6. M.F. Neuts. The Abscissa of Convergence of the Laplace-Stieltjes Transform of a PH-Distribution. *Communications in Statistics: Simulation and Computation*, 13:367–373, 1984.
7. N.G. Bean and D.A. Green. When is a MAP Poisson? *Mathematical and Computer Modelling*, 31:31–46, 2000.
8. E. Seneta. *Non-negative Matrices and Markov Chains.* Springer-Verlag, New York, 1981.
9. N.G. Bean, P.K. Pollett, and P.G. Taylor. Quasistationary Distributions for Level-Dependent Quasi-Birth-and-Death Processes. *Stochastic Models*, 16:511–541, 2000.
10. L. Bright and P.G. Taylor. Calculating the Equilibrium Distribution in Level Dependent Quasi-Birth-and-Death Processes. *Stochastic Models*, 11:497–526, 1995.
11. M.F. Neuts. *Structured Stochastic Matrices of M/G/1 type and their Applications.* Marcel Dekker, New York, 1989.
12. G. Latouche and V. Ramaswami. *Introduction to Matrix Analytic Methods in Stochastic Modelling.* ASA-SIAM Series on Statistics and Applied Probability. Society for Industrial and Applied Mathematics, Philadelphia, USA, 1998.
13. N.G. Bean, L. Bright, G. Latouche, C.E.M. Pearce, P.K. Pollett, and P.G. Taylor. The Quasistationary Behaviour of Quasi-Birth-and-Death Processes. *Annals of Applied Probability*, 7:134–155, 1997.

14. N.G. Bean, P.K. Pollett, and P.G. Taylor. The Quasistationary Distributions of level-independent Quasi-Birth-and-Death Processes. *Stochastic Models (Special Issue in Honour of Marcel Neuts)*, 14:389–406, 1998.
15. V. Ramaswami. Matrix Analytic Methods: a Tutorial Overview with Some Extensions and New Results. In Attahiru S. Alfa and Srinivas R. Chakravarthy, editors, *Matrix-Analytic Methods in Stochastic Models*, volume 183 of *Lecture Notes in Pure and Applied Mathematics*, pages 261–296. Marcel Dekker, New York, 1997.
16. P. Schrijner. *Quasi-Stationarity of Discrete-Time Markov Chains.* PhD thesis, Faculty of Applied Mathematics, University of Twente, The Netherlands, 1995.
17. E.A. van Doorn and P. Schrijner. Geometric Ergodicity and Quasi-Stationarity in Discrete-Time Birth-Death Processes. *Journal of the Australian Mathematical Society, Series B*, 37:121–144, 1995.

DISTRIBUTIONS OF REWARD FUNCTIONS ON CONTINUOUS-TIME MARKOV CHAINS

MOGENS BLADT
Department of Statistics, IIMAS, Universidad Nacional Autonoma de Mexico Apartado Postal 20-726, 01000 Mexico, D.F. Mexico
E-mail: bladt@sigma.iimas.unam.mx

BEATRICE MEINI
Dipartimento di Matematica, Università di Pisa, via Buonarroti 2, 56127 Pisa, Italy
E-mail: meini@dm.unipi.it

MARCEL F. NEUTS
Department of Systems and Industrial Engineering, The University of Arizona, Tucson, AZ 85721, USA
E-mail: marcel@sie.arizona.edu

BRUNO SERICOLA
IRISA - INRIA, Campus universitaire de Beaulieu, 35042 Rennes Cedex, France
E-mail: sericola@irisa.fr

We develop algorithms for the computation of the distribution of the total reward accrued during $[0, t)$ in a finite continuous-parameter Markov chain. During sojourns, the reward grows linearly at a rate depending on the state visited. At transitions, there can be instantaneous rewards whose values depend on the states involved in the transition. For moderate values of t, the reward distribution is obtained by implementing a series representation, due to Sericola, that is based on the uniformization method. As an alternative, that distribution can also be computed by the numerical inversion of its Laplace-Stieltjes transform. For larger values of t, we implement a matrix convolution product to compute a related semi-Markov matrix efficiently and accurately.

1 Introduction

We consider an irreducible, continuous-time m-state Markov chain $\{J(t)\}$ with generator Q. Our objective is to develop the theory of and numerical procedures for various probability distributions associated with a reward function defined on the Markov chain.

There is a continuous reward in that, for every unit of time spent in state j, a reward a_j accrues. In addition, there are instantaneous rewards associated with the various transitions. At each transition $h \to r$, an instantaneous, finite reward c_{hr} is received. We do not impose restrictions on the signs of the quantities $\{c_{hr}\}$.

We start by defining several random variables of interest and by introducing notation. The random variables $N_{hk}(t)$ are the numbers of transitions $h \to k$ during $[0,t)$. We make the convention that $N_{hh}(t) = 0$, for $1 \leq h \leq m$. For use in transforms, we let Z be a matrix with elements z_{hk} where $z_{hh} = 1$, for $1 \leq h \leq m$. We recall that the *Schur product*, $A \bullet B$, of $m \times m$ matrices A and B is the matrix with elements $A_{hk}B_{hk}$.

The piecewise constant random function $a_{J(t)}$ takes the value a_j when $J(t) = j$. The total continuous reward $R_j(t)$ earned during sojourns in the state j over an interval $[0,t)$ is given by

$$R_j(t) = \int_0^t a_{J(u)} 1_{\{J(u)=j\}} du, \quad \text{for} \quad 1 \leq j \leq m,$$

where $1_{\{c\}}$ equals 1 if condition c holds and 0 otherwise, and the total continuous reward $R(t)$ over an interval $[0,t)$ is given by

$$R(t) = \int_0^t a_{J(u)} du.$$

In the context of dependability analysis of fault-tolerant computer systems, the random variable $R(t)$ is referred to as a performability measure, see e.g. [3] and [2] and the references therein.

We shall derive a concise expression for the joint Laplace-Stieltjes transform and generating function of the random variables $\{R_j(t)\}$ and $\{N_{hk}(t)\}$ taking the initial and final states $J(0)$ and $J(t)$ of the Markov chain into account. By $\mathbf{s}$ we denote the vector with components $s_1, \ldots, s_m$. By $\Delta(\mathbf{s})$, we denote an $m \times m$ diagonal matrix with the quantities $s_1, \ldots, s_m$ as its diagonal elements. We are interested in the transform

$$V^*_{ij}(\mathbf{s}, Z; t) = E\left\{ \exp[-\sum_{h=1}^{m} s_h R_h(t)] \prod_{h,k} z_{hk}^{N_{hk}(t)} 1_{\{J(t)=j\}} \middle| J(0) = i \right\},$$

for $1 \leq i, j \leq m$. By $V^*(\mathbf{s}, Z; t)$, we denote the $m \times m$ matrix with elements $V^*_{ij}(\mathbf{s}, Z; t)$. For all $t \geq 0$, the matrix $V^*(\mathbf{s}, Z; t)$ is well-defined and analytic for all complex values of z_{hk} and s_ν, for $1 \leq h, k \leq m$, $1 \leq \nu \leq m$.

The remainder of the paper is organized as follows. In the next section, we present the main theorem which gives the expression of the transform $V^*(\mathbf{s}, Z; t)$. That theorem is used in Section 3 to derive formulas for the first two moments of various measures combining linear and instantaneous rewards. Section 4 is devoted to the total continuous reward distribution over $[0,t)$.

We first develop an algorithm based on explicit formulas leading to a stable method whose precision can be specified in advance. Secondly, we compute that distribution by the numerical inversion of Laplace-Stieltjes transform and we compare these two methods through numerical examples. Finally, we develop a new method based on a matrix convolution product. This method uses the explicit solution for moderate values of t and implements on that basis the matrix convolution product for larger values of t.

2 The Main Theorem

Theorem 2.1 *For $t \geq 0$, the matrix $V^*(\mathbf{s}, Z; t)$ is given by*

$$V^*(\mathbf{s}, Z; t) = \exp\{[Q \bullet Z - \Delta(\mathbf{a})\Delta(\mathbf{s})]t\}. \tag{1}$$

Proof. The conditional probability

$$P\{R_\nu(t) \leq x_\nu, 1 \leq \nu \leq m; N_{hk}(t) = K_{hk}, 1 \leq h, k \leq m; J(t) = j | J(0) = i\}$$

depends on t, on the m nonnegative variables $\{x_\nu\}$, and on the $m(m-1)$ nonnegative integer-valued variables $\{K_{hk}\}$. We concisely denote that probability mass-function by $V_{ij}(\mathbf{x}, K; t)$. Moreover, let $\mathbf{e}_i$ be the unit vector with i-th component equal to one and denote by $J(i,r)$ an $m \times m$ matrix with a single non-zero element equal to one at the indices i, r. The notation $U(\mathbf{x}-\mathbf{b})$ signifies the m-variate degenerate distribution at $\mathbf{b}$.

Now distinguishing the cases where the state of the Markov chain does not change in $[0, t)$ and where there is a first state change at some time u, $0 \leq u \leq t$, and applying a standard first passage argument, we obtain the equation

$$\begin{aligned} V_{ij}(\mathbf{x}, K; t) &= \delta_{ij} \exp(Q_{ii}t) U(\mathbf{x} - a_i t \mathbf{e}_i) \\ &\quad + \sum_{r \neq i} \int_0^t \exp(Q_{ii}u) Q_{ir} V_{rj}(\mathbf{x} - a_i u \mathbf{e}_i, K - J(i,r); t-u) du. \end{aligned} \tag{2}$$

By a simple change of variable, the second term may be rewritten as

$$\sum_{r \neq i} \int_0^t \exp[Q_{ii}(t-u)] Q_{ir} V_{rj}(\mathbf{x} - a_i(t-u)\mathbf{e}_i, K - J(i,r); u) du.$$

To facilitate the derivation of equation (1) we introduce and evaluate the transforms

$$V_{ij}^0(\mathbf{s}, Z; t) = \int_0^\infty \cdots \int_0^\infty \sum_K \prod_{h,k} z_{hk}^{K_{hk}} \exp\left(-\sum_{\nu=1}^m s_\nu x_\nu\right) V_{ij}(\mathbf{x}, K; t) dx_1 \cdots dx_m.$$

With respect to the variables $x_1, \ldots, x_m$, these are Laplace, not Laplace-Stieltjes, transforms. Equation (2) leads to

$$V_{ij}^0(\mathbf{s}, Z; t) = \delta_{ij} \exp[(Q_{ii} - a_i s_i)t](s_1 \cdots s_m)^{-1}$$
$$+\sum_{r \neq i} \sum_{K} \prod_{h,k} z_{hk}^{K_{hk}} \int_0^t \exp(Q_{ii}u) Q_{ir} du \int_0^\infty \cdots \int_{a_i u}^\infty \cdots \int_0^\infty \exp\left(-\sum_{\nu=1}^m s_\nu x_\nu\right)$$
$$\times V_{rj}(\mathbf{x} - a_i u \mathbf{e}_i, K - J(i,r); t-u) dx_1 \cdots dx_m.$$

By routine changes of variables that reduces to

$$V_{ij}^0(\mathbf{s}, Z; t) = \delta_{ij} \exp[(Q_{ii} - a_i s_i)t](s_1 \cdots s_m)^{-1}$$
$$+\sum_{r \neq i} \int_0^t \exp[(Q_{ii} - a_i s_i)u] Q_{ir} z_{ir} V_{ij}^0(\mathbf{s}, Z; t-u) du. \qquad (3)$$

In equation (3) we multiply both sides by $\exp[(a_i s_i - Q_{ii})t]$ and we differentiate the resulting equation with respect to t. Routine simplifications lead to the differential equations

$$\frac{d}{dt} V_{ij}^0(\mathbf{s}, Z; t) = -a_i s_i V_{ij}^0(\mathbf{s}, Z; t) + [(Q \bullet Z) V^0(\mathbf{s}, Z; t)]_{ij}, \qquad (4)$$

for $1 \leq i, j \leq m$ with initial conditions $V_{ij}^0(\mathbf{s}, Z; 0) = \delta_{ij}(s_1 \cdots s_m)^{-1}$.

The Laplace-Stieltjes transforms $V_{ij}^*(\mathbf{s}, Z; t)$ are related to the Laplace transforms $V_{ij}^0(\mathbf{s}, Z; t)$ by $V_{ij}^*(\mathbf{s}, Z; t) = s_1 \cdots s_m V_{ij}^0(\mathbf{s}, Z; t)$. They satisfy the same differential equations with constant coefficients as in (4) but with initial conditions $V_{ij}^*(\mathbf{s}, Z; 0) = \delta_{ij}$. Integrating these equations we obtain (1). ∎

Corollary 2.2 *The joint Laplace-Stieltjes transform of the total continuous rewards $R_\nu(t)$ and the total instantaneous rewards $c_{hk} N_{hk}(t)$ is given by the matrix $V^*(\mathbf{s}, \Xi; t)$, where the matrix Ξ is obtained by setting $z_{hk} = \exp(-c_{hk}\xi_{hk})$, for $1 \leq h, k \leq m$. The ξ_{hk} are the transform variables corresponding to the total instantaneous rewards $c_{hk} N_{hk}(t)$.*

Proof. Obvious from the definition of the Laplace-Stieltjes transform and the fact that

$$\exp\{-c_{hk} N_{hk}(t) \xi_{hk}\} = [\exp(-c_{hk}\xi_{hk})]^{N_{hk}(t)},$$

for $1 \leq h, k \leq m$. ∎

3 Moment Formulas

We derive formulas for the mean and variance of the total reward accrued during an interval $[0, t)$ in the stationary version of the process. Using special choices of the quantities a_i and c_{hk}, we can immediately obtain the first two moments of various interesting quantities associated with finite Markov chains.

The matrix $\Xi^o(s)$ has elements $\exp(-c_{hk}s)$. The vector $\boldsymbol{\theta}$ is the stationary probability vector of the matrix Q and $\mathbf{e}$ is the column vector with all components equal to 1. The Laplace-Stieltjes transform of the total reward in the interval $[0, t)$ is then given by $\psi(s) = \boldsymbol{\theta}V^*(s, \Xi^o(s); t)\mathbf{e}$, where

$$V^*(s, \Xi^o(s); t) = \exp\{[Q \bullet \Xi^o(s) - s\Delta(\mathbf{a})]t\}.$$

The computation of the mean and variance amounts to evaluating the first two derivatives of $\psi(s)$ at zero. However, because of the matrix functions involved, that computation requires manipulations that we need to present in some detail. These are similar to those in Narayana and Neuts [4]. First some preliminaries: the quantity ω^* is defined by

$$\omega^* = \boldsymbol{\theta}[Q \bullet C + \Delta(\mathbf{a})]\mathbf{e} = \boldsymbol{\theta}(Q \bullet C)\mathbf{e} + \boldsymbol{\theta}\mathbf{a},$$

in which $C = \{c_{hk}\}$, where by convention, $c_{hh} = 0$ for $1 \leq h \leq m$. ω^* is the steady-state instantaneous reward rate. The first term is the contribution of the instantaneous rewards; the second term corresponds to the piecewise linear rewards. It is well-known that the matrix $\mathbf{e}\boldsymbol{\theta} - Q$ is invertible and that

$$\int_0^t \exp(Qu)du = \mathbf{e}\boldsymbol{\theta}t + [I - \exp(Qt)](\mathbf{e}\boldsymbol{\theta} - Q)^{-1}. \tag{5}$$

Theorem 3.1 *The mean total reward in $[0, t)$ is given by $\mu_1'(t) = \omega^* t$, and the corresponding variance $\sigma^2(t)$ by*

$$\begin{aligned}\sigma^2(t) &= t\boldsymbol{\theta}(Q \bullet C \bullet C)\mathbf{e} \\ &\quad + 2t\left\{\boldsymbol{\theta}[Q \bullet C + \Delta(\mathbf{a})](\mathbf{e}\boldsymbol{\theta} - Q)^{-1}[Q \bullet C + \Delta(\mathbf{a})]\mathbf{e} - \omega^{*2}\right\} \\ &\quad - 2\boldsymbol{\theta}[Q \bullet C + \Delta(\mathbf{a})](\mathbf{e}\boldsymbol{\theta} - Q)^{-1}[I - \exp(Qt)](\mathbf{e}\boldsymbol{\theta} - Q)^{-1}[Q \bullet C + \Delta(\mathbf{a})]\mathbf{e}.\end{aligned}$$

Proof. We introduce the matrices $M_1(t)$ and $M_2(t)$, defined by

$$M_1(t) = -[\frac{\partial}{\partial s}V^*(s, \Xi^o(s); t)]_{s=0}, \qquad M_2(t) = [\frac{\partial^2}{\partial s^2}V^*(s, \Xi^o(s); t)]_{s=0}.$$

We twice differentiate with respect to s in the differential equation

$$\frac{\partial}{\partial t}V^*(s, \Xi^o(s); t) = V^*(s, \Xi^o(s); t)[Q \bullet \Xi^o(s) - s\Delta(\mathbf{a})],$$

we set $s = 0$, and we notice that

$$[\frac{\partial}{\partial s}\Xi^o(s)]_{s=0} = -C, \quad [\frac{\partial^2}{\partial s^2}\Xi^o(s)]_{s=0} = C \bullet C,$$

to obtain the differential equations

$$\frac{d}{dt}M_1(t) = M_1(t)Q + \exp(Qt)[Q \bullet C + \Delta(\mathbf{a})],$$

and

$$\frac{d}{dt}M_2(t) = M_2(t)Q + 2M_1(t)[Q \bullet C + \Delta(\mathbf{a})] + \exp(Qt)(Q \bullet C \bullet C).$$

We postmultiply by $\exp(-Qt)$ in both equations and integrate. That leads to

$$M_1(t) = \int_0^t \exp(Qu)[Q \bullet C + \Delta(\mathbf{a})]\exp[Q(t-u)]du, \tag{6}$$

and

$$\begin{aligned} M_2(t) - M_1(t) = & \int_0^t [2M_1(u) - \exp(Qu)][Q \bullet C + \Delta(\mathbf{a})]\exp[Q(t-u)]du \\ & + \int_0^t \exp(Qu)(Q \bullet C \bullet C)\exp[Q(t-u)]du. \end{aligned} \tag{7}$$

We premultiply by $\boldsymbol{\theta}$ in (6) and invoke the integration formula (5) to obtain that

$$\boldsymbol{\theta}M_1(t) = \omega^*\boldsymbol{\theta}t + \boldsymbol{\theta}[Q \bullet C + \Delta(\mathbf{a})][I - \exp(Qt)](\mathbf{e}\boldsymbol{\theta} - Q)^{-1}. \tag{8}$$

The equality (8) readily implies that $\boldsymbol{\theta}M_1(t)\mathbf{e} = \omega^* t$. Premultiplying by $\boldsymbol{\theta}$ in (7) leads to

$$\boldsymbol{\theta}M_2(t)\mathbf{e} = \boldsymbol{\theta}(Q \bullet C \bullet C)\mathbf{e}t + 2\boldsymbol{\theta}\int_0^t M_1(u)du[Q \bullet C + \Delta(\mathbf{a})]\mathbf{e}. \tag{9}$$

The integral is evaluated by using formulas (8) and (5) and performing routine simplifications. We obtain that

$$\begin{aligned} \boldsymbol{\theta}\int_0^t M_1(u)du = & \frac{1}{2}\omega^* t^2\boldsymbol{\theta} + \boldsymbol{\theta}[Q \bullet C + \Delta(\mathbf{a})](\mathbf{e}\boldsymbol{\theta} - Q)^{-1}t - \omega^* t\boldsymbol{\theta} \\ & - \boldsymbol{\theta}[Q \bullet C + \Delta(\mathbf{a})][I - \exp(Qt)](\mathbf{e}\boldsymbol{\theta} - Q)^{-2}. \end{aligned}$$

Upon substitution into the formula (9) for the second moment, the stated formula for the variance is obtained after simplifications. ∎

For selected choices of the parameter a_i and c_{hk}, we obtain moment formulas of special interest. For example, setting all $c_{hk} = 0$, and $a_i = 1$, for i belonging to a set B of indices and 0 otherwise, we obtain the moments of the total sojourn time of the Markov chain in the set of states B. Setting all $a_i = 0$, and $c_{hk} = 1$ if k belongs to B, and 0 otherwise, we obtain moments of the total number of visits to the set B during $[0, t)$.

4 The Total Continuous Reward Distribution

We recall that henceforth all the instantaneous rewards c_{hk} are zero. In this section we consider the semi-Markov matrix $W(x,t) = \Big(W_{ij}(x,t)\Big)$ where

$$W_{ij}(x,t) = P\{R(t) \leq x, J(t) = j | J(0) = i\}.$$

We partition the state space $S = \{1, \dots, m\}$ of the Markov chain $\{J(t)\}$ into disjoint subsets containing the states with the same reward rates. The number of distinct rewards is $\phi + 1$ and their different values are

$$r_0 < r_1 < \cdots < r_{\phi-1} < r_\phi.$$

States in the subsets B_l, $l = 0, \dots, \phi$, have the same reward rate r_l. That is, for $l = 0, \dots, \phi$,

$$B_l = \{i \in S | a_i = r_l\}.$$

We then have $R(t) \in [r_0 t, r_\phi t]$ with probability one. Without loss of generality, we may set $r_0 = 0$. That can be done by considering the random variable $R(t) - r_0 t$ instead of $R(t)$ and the reward rates $r_l - r_0$ instead of r_l.

We denote by P the transition probability matrix of the uniformized discrete time Markov chain associated to the Markov chain $\{J(t)\}$, with the same initial distribution. The matrix P is related to the generator Q by $P = I + Q/\lambda$, where I is the identity matrix and λ satisfies $\lambda \geq \max\{-Q_{ii},\ i \in S\}$.

Using the partition B_0, ..., B_ϕ, the matrices Q, P, and $W(x,t)$ can be written, for $u, v = 0, \dots, \phi$, as

$$Q = \{Q_{B_u B_v}\}, \ P = \{P_{B_u B_v}\} \text{ and } W(x,t) = \{W_{B_u B_v}(x,t)\}.$$

The distribution of $R(t)$ has at most $\phi+1$ jumps at the points $r_0 t = 0, r_1 t, \dots, r_\phi t$. For $t > 0$, the jump at $x = r_l t$ is the probability that the Markov chain $\{J(t)\}$, starting in subset B_l, stays in that set during all of $[0, t)$. Therefore, for $t > 0$, and $0 \leq l \leq \phi$,

$$P\{R(t) = r_l t, J(t) = j | J(0) = i\} = \begin{cases} (e^{Q_{B_l B_l} t})_{ij} & \text{if } i, j \in B_l, \\ 0 & \text{otherwise,} \end{cases} \tag{10}$$

which can also be written as

$$P\{R(t) = r_l t, J(t) = j | J(0) = i\} = \sum_{n=0}^{\infty} e^{-\lambda t} \frac{(\lambda t)^n}{n!} (P_{B_l B_l})^n_{ij} 1_{\{i,j \in B_l\}}.$$

4.1 Explicit Formulas

An explicit formula for the matrix $W(x,t)$, is given by the following theorem. It is derived in [5].

Theorem 4.1 *For every $t > 0$, and $x \in [r_{h-1}t, r_h t)$, for $1 \le h \le \phi$,*

$$W(x,t) = \sum_{n=0}^{\infty} e^{-\lambda t} \frac{(\lambda t)^n}{n!} \sum_{k=0}^{n} \binom{n}{k} x_h^k (1 - x_h)^{n-k} C^{(h)}(n,k), \tag{11}$$

where $x_h = \dfrac{x - r_{h-1}t}{(r_h - r_{h-1})t}$ *and* $C^{(h)}(n,k) = \left(C^{(h)}_{B_u B_v}(n,k)\right)_{0 \le u,v \le \phi}$ *are matrices given by the recurrence relations:*
For $h \le u \le \phi$, and $0 \le v \le \phi$:

for $n \ge 0$: $C^{(1)}_{B_u B_v}(n,0) = 0_{B_u B_v}$, $C^{(h)}_{B_u B_v}(n,0) = C^{(h-1)}_{B_u B_v}(n,n)$, *for* $h > 1$;

for $1 \le k \le n$:

$$C^{(h)}_{B_u B_v}(n,k) = \frac{r_u - r_h}{r_u - r_{h-1}} C^{(h)}_{B_u B_v}(n,k-1) + \frac{r_h - r_{h-1}}{r_u - r_{h-1}} \sum_{w=0}^{\phi} P_{B_u B_w} C^{(h)}_{B_w B_v}(n-1,k-1). \tag{12}$$

For $0 \le u \le h-1$, and $0 \le v \le \phi$:

for $n \ge 0$: $C^{(\phi)}_{B_u B_v}(n,n) = (P^n)_{B_u B_v}$, $C^{(h)}_{B_u B_v}(n,n) = C^{(h+1)}_{B_u B_v}(n,0)$, *for* $h < \phi$;

for $0 \le k \le n-1$:

$$C^{(h)}_{B_u B_v}(n,k) = \frac{r_{h-1} - r_u}{r_h - r_u} C^{(h)}_{B_u B_v}(n,k+1) + \frac{r_h - r_{h-1}}{r_h - r_u} \sum_{w=0}^{\phi} P_{B_u B_w} C^{(h)}_{B_w B_v}(n-1,k). \tag{13}$$

Proof. See [5] ■

In what follows, we denote by $W'(x,t)$ the partial derivative of $W(x,t)$ with respect to x. That matrix, defined only for $t > 0$ and $x \ne r_l t$, $l = 0, \ldots, \phi$, is given in the following corollary.

Corollary 4.2 *For $t > 0$, and $x \in (r_{h-1}t, r_h t)$, for $1 \le h \le \phi$, we have*

$$W'(x,t) = \frac{\lambda}{r_h - r_{h-1}} \sum_{n=0}^{\infty} e^{-\lambda t} \frac{(\lambda t)^n}{n!} \sum_{k=0}^{n} \binom{n}{k} x_h^k (1-x_h)^{n-k} \times \left[C^{(h)}(n+1,k+1) - C^{(h)}(n+1,k)\right]. \qquad (14)$$

Proof. Obvious from relation (11). ■

Note that in (12), that is for $h \le u$, we have

$$0 \le \frac{r_u - r_h}{r_u - r_{h-1}} = 1 - \frac{r_h - r_{h-1}}{r_u - r_{h-1}} \le 1,$$

and in (13), that is for $u \le h-1$, we have

$$0 \le \frac{r_{h-1} - r_u}{r_h - r_u} = 1 - \frac{r_h - r_{h-1}}{r_h - r_u} \le 1.$$

The following corollary gives some properties of the matrices $C^{(h)}(n,k)$. If M and K are square matrices of the same order, $M \le K$ signifies element-wise inequality.

Corollary 4.3 *For every $n \ge 0$, $0 \le k \le n$, and $1 \le h \le \phi$,*

$$0 \le C^{(h)}(n,k) \le P^n, \quad 0 \le k \le n,$$

$$C^{(h)}(n,k) \le C^{(h)}(n,k+1), \quad 0 \le k \le n-1.$$

Proof. For the first inequality, see [5]. The same recurrence mechanism is used to prove the second one. ■

These considerations yield a computational method that avoids numerical problems since, except for the ratio $\lambda/(r_h - r_{h-1})$ in (14), all the computed quantities are between 0 and 1 and require only additions and multiplications of nonnegative quantities. This leads to a stable algorithm whose precision can be specified in advance.

Let ε be the desired precision for the computation of $W(x,t)$. We define the integer N by

$$N = \min\left\{ n \ge 0 \;\middle|\; \sum_{i=0}^{n} e^{-\lambda t} \frac{(\lambda t)^i}{i!} \ge 1 - \varepsilon \right\}. \qquad (15)$$

We thus have

$$W(x,t) = \sum_{n=0}^{N} e^{-\lambda t} \frac{(\lambda t)^n}{n!} \sum_{k=0}^{n} \binom{n}{k} x_h^k (1-x_h)^{n-k} C^{(h)}(n,k) + e(N).$$

From the first inequality of Corollary 4.3, we obtain that the remainder of the series $e(N)$ satisfies $e_{ij}(N) \le \varepsilon$, for every $i, j \in S$.

With regards to $W'(x,t)$, again from Corollary 4.3, we have that

$$0 \le C^{(h)}(n+1,k+1) - C^{(h)}(n+1,k) \le P^{n+1}, \quad 0 \le k \le n,$$

and so we obtain that

$$W'(x,t) = \frac{\lambda}{r_h - r_{h-1}} \sum_{n=0}^{\infty} e^{-\lambda t} \frac{(\lambda t)^n}{n!} \sum_{k=0}^{n} \binom{n}{k} x_h^k (1-x_h)^{n-k} \times \left[C^{(h)}(n+1,k+1) - C^{(h)}(n+1,k)\right] + e^1(N),$$

where the remainder of the series $e^1(N)$ is such that, for every $i, j \in S$,

$$e_{ij}^1(N) \le \frac{\lambda\varepsilon}{r_h - r_{h-1}} \le \frac{\lambda\varepsilon}{r},$$

and $r = \min\{r_h - r_{h-1},\ h = 1, \dots, \phi\}$.

4.1.1 Algorithmic aspects

In this section we consider the computation of matrix $W(x,t)$. The main effort goes into the computation of the matrices $C^{(h)}(n,k)$. With regards to storage requirements, since the values of the $C^{(h)}(n,k)$ at step n depend only on their values at step $n-1$, we need to store only two arrays of $(N+1)\phi$ matrices. At step n, we need to compute $n+1$ matrices for each $h = 1, \dots, \phi$. That can easily be seen from the algorithmic description in Table 2. The procedure **Accumulate**(n) is used to compute the approximate matrix $W^\varepsilon(x,t)$ defined, for $h = 1, \dots, \phi$ and $x \in [r_{h-1}t, r_h t)$, by

$$W^\varepsilon(x,t) = \sum_{n=0}^{N} e^{-\lambda t} \frac{(\lambda t)^n}{n!} \sum_{k=0}^{n} \binom{n}{k} x_h^k (1-x_h)^{n-k} C^{(h)}(n,k).$$

By the definition of N in (15) and from Corollary 4.3, we have, for a given value of the precision ε, that

$$\sup_{i \in S} \sum_{j \in S} (W_{ij}(x,t) - W_{ij}^\varepsilon(x,t)) \le \varepsilon.$$

The procedure **Accumulate**(n), described in Table 1, involves a fixed value $t > 0$ and M distinct values of x, denoted by $x(i)$, $1 \le i \le M$. We initialize $W^\varepsilon(x(i),t) = 0$, we denote by h_i the index such that $x(i) \in [r_{h_i-1}t, r_{h_i}t)$, and we define

$$x_{h_i} = \frac{x(i) - r_{h_i-1}t}{(r_{h_i} - r_{h_i-1})t}.$$

Table 1. The procedure **Accumulate**(n)

for $i = 1$ **to** M **do**

$$W^\varepsilon(x(i), t, n) = e^{-\lambda t}\frac{(\lambda t)^n}{n!}\sum_{k=0}^{n}\binom{n}{k}x_{h_i}^k(1-x_{h_i})^{n-k}C^{(h_i)}(n,k)$$

$$W^\varepsilon(x(i), t) = W^\varepsilon(x(i), t) + W^\varepsilon(x(i), t, n)$$

endfor

Table 2. Computation of the matrices $C^{(h)}(n,k)$ and $W(x,t)$

for $h = 1$ **to** ϕ **do** $\forall\, u, v = 0, \ldots, \phi$, $C^{(h)}_{B_uB_v}(0,0) = 0_{B_uB_v}$ **endfor**
for $h = 1$ **to** ϕ **do** $\forall\, u = 0, \ldots, h-1$, $C^{(h)}_{B_uB_u}(0,0) = I_{B_uB_v}$ **endfor**
Accumulate(0)
for $n = 1$ **to** N **do**
  $\forall\, u = 1, \ldots, \phi$, $\forall\, v = 0, \ldots, \phi$, $C^{(1)}_{B_uB_v}(n,0) = 0_{B_uB_v}$
  for $h = 1$ **to** ϕ **do**
    for $k = 1$ **to** n **do**
      $\forall\, u = h, \ldots, \phi$, $\forall\, v = 0, \ldots, \phi$, compute relation (12)
    endfor
    $\forall\, u = h+1, \ldots, \phi$, $\forall\, v = 0, \ldots, \phi$, $C^{(h+1)}_{B_uB_v}(n,0) = C^{(h)}_{B_uB_v}(n,n)$
  endfor
  $\forall\, u = 0, \ldots, \phi-1$, $\forall\, v = 0, \ldots, \phi$, $C^{(\phi)}_{B_uB_v}(n,n) = (P^n)_{B_uB_v}$
  for $h = \phi$ **downto** 1 **do**
    for $k = n-1$ **downto** 0 **do**
      $\forall\, u = 0, \ldots, h-1$, $\forall\, v = 0, \ldots, \phi$, compute relation (13)
    endfor
    $\forall\, u = 0, \ldots, h-2$, $\forall\, v = 0, \ldots, \phi$, $C^{(h-1)}_{B_uB_v}(n,n) = C^{(h)}_{B_uB_v}(n,0)$
  endfor
  Accumulate(n)
endfor

Note that the integer N, defined in (15), is an increasing function of t, say $N(t)$. So, if the matrix $W(x,t)$ is to be computed at L different $t-$values, say $t_1 < \ldots < t_L$, we need only evaluate the matrices $C^{(h)}(n,k)$ for $n = 0, 1, \ldots, N(t_L)$, as these matrices do not depend on the values of $t_1, \ldots, t_L$.

The main effort required for the computation of matrices $W(x,t)$ or $W'(x,t)$ is in the computation of matrices $C^{(h)}(n,k)$. We use for matrix P

a compact storage. If d denotes the connectivity degree of matrix P, that is the maximum number of nonzero entries in each row, then the computational cost of one matrix $C^{(h)}(n,k)$ is $O(dm^2)$. The number of such matrices that have to be computed (see Table 2) is equal to $\phi(N+1)(N+2)/2$, The total computational effort required is thus $O(\phi dm^2 N^2/2)$. Concerning the storage requirements, it is easy to see, from Table 2, that we need to store two arrays of $\phi(N+1)$ matrices for the recursive computation of matrices $C^{(h)}(n,k)$. Thus the storage complexity is $O(\phi m^2 N)$.

Note also that if one only wants to compute the distribution $P\{R(t) \leq x\}$, there is no need to evaluate the matrices $C^{(h)}(n,k)$. It then suffices to evaluate the vectors $b^{(h)}(n,k) = C^{(h)}(n,k)\mathbf{e}$. The algorithm thereby becomes more efficient, as the matrix-matrix products are replaced by matrix-vector products. In that case, the end product of the algorithm is the vector $G(x,t) = W(x,t)\mathbf{e}$ and the complexity is reduced by a factor m.

4.1.2 Numerical examples

Consider the Markov chain with $S = \{1,2,3\}$, the generator Q and the reward vector a, given by

$$Q = \begin{pmatrix} -1 & 1 & 0 \\ 0.5 & -1 & 0.5 \\ 0 & 1 & -1 \end{pmatrix} \quad \text{and} \quad a = \begin{pmatrix} 2 & 1 & 0 \end{pmatrix}.$$

We thus have $\lambda = 1$, $\phi = 2$ and $0 \leq R(t) \leq 2t$ with probability 1. For the error tolerance $\varepsilon = 10^{-10}$, we obtain the following results.

For $t = 1$

$$W(0,1) = \begin{pmatrix} 0.0000000000 & 0.0000000000 & 0.0000000000 \\ 0.0000000000 & 0.0000000000 & 0.0000000000 \\ 0.0000000000 & 0.0000000000 & 0.3678794412 \end{pmatrix}.$$

Note the high precision of the algorithm: the element $W_{33}(0,1)$ is the jump corresponding to the MC staying in the state 3 up to time $t = 1$. That is also equal to $\exp(-1) \approx 0.36787944117$.

$$W(1-10^{-12},1) = \begin{pmatrix} 0.0010069669 & 0.0163117922 & 0.0499470501 \\ 0.0081558961 & 0.0998941002 & 0.2080102831 \\ 0.0499470501 & 0.4160205662 & 0.4667665745 \end{pmatrix},$$

$$W(1,1) = \begin{pmatrix} 0.0010069669 & 0.0163117922 & 0.0499470501 \\ 0.0081558961 & 0.4677735414 & 0.2080102831 \\ 0.0499470501 & 0.4160205662 & 0.4667665745 \end{pmatrix}.$$

Again, note the high precision of the algorithm: all the elements of $W(1-10^{-12},1)$ and $W(1,1)$ are equal except for the element of indices $(2,2)$. The difference between these two values is 0.3678794412; it corresponds to the jump at $x=1$, the probability that the MC stays in state 2 beyond time $t=1$, or $\exp(-1)\approx 0.36787944117$.

$$W(2^-,1)=\begin{pmatrix} 0.0998941002 & 0.4323323583 & 0.0998941002 \\ 0.2161661792 & 0.5676676416 & 0.2161661792 \\ 0.0998941002 & 0.4323323583 & 0.4677735414 \end{pmatrix}.$$

The elements of the matrix $W(2^-,1)$ are

$$W_{ij}(2^-,1)=P\{R(t)<2, J(1)=j \mid J(0)=i\}.$$

Since $W(2,1)=e^Q$, we easily obtain the jump

$$\begin{aligned} P\{R(t)=2, J(1)=1 \mid J(0)=1\} &= \sum_{j=1}^{3} W_{1j}(2,1)-\sum_{j=1}^{3} W_{1j}(2^-,1) \\ &= 1-\sum_{j=1}^{3} W_{1j}(2^-,1)=0.3678794412, \end{aligned}$$

whose value is $\exp(-1)\approx 0.36787944117$.

For $t=100$, we obtain $W_{ij}(0,100)=0$ for every $1\le i,j\le 3$, and

$$W(100,100)=\begin{pmatrix} 0.1050278103 & 0.2299261526 & 0.1250000000 \\ 0.1149630763 & 0.2500000000 & 0.1350369237 \\ 0.1250000000 & 0.2700738474 & 0.1449721896 \end{pmatrix},$$

$$W(200^-,100)=\begin{pmatrix} 0.2500000000 & 0.5000000000 & 0.2500000000 \\ 0.2500000000 & 0.5000000000 & 0.2500000000 \\ 0.2500000000 & 0.5000000000 & 0.2500000000 \end{pmatrix}.$$

Note that in this case the jumps are invisible since $\exp(-100)\approx 0.372\times 10^{-43}$.

4.2 Numerical transform inversion

The joint Laplace transform and generating function $V^*(s,Z;t)$ can be numerically inverted, at least in some special cases. We only consider the case where we want to find the distribution (density) of the total continuous reward earned in some subset of the state–space, A say. In this case the joint transform of course reduces to a Laplace–transform for this reward. Since the density of interest is concentrated on the positive real axis we can use the

Bromwich inversion integral as follows. For simplicity let $V^*(t)$ denote the Laplace transform for the total continuous reward. The ij'th element is hence the Laplace transform corresponding to the case where the Markov jump process initiates in state i and is in state j at time t. Let the corresponding (defective) density of total reward earned in the set A be $\boldsymbol{f}(t)$, whose i, j's element corresponds to the conditional density given initiation of the Markov jump process in state i, and subject to being in state j at time t. Then the Bromwich inversion integral is

$$\boldsymbol{f}(t) = \frac{2}{\pi} \int_0^\infty \mathrm{Re}(V^*(iu)) \cos(ut) du.$$

This integral is hence solved by numerical integration (trapezoidal rule) choosing discretization such that the cosine term becomes $(-1)^k$ and we approximate the integral by an alternating series, which in turn is calculated by Euler summation to approximate the infinite series. See Abate and Whitt (1992) [1] for details.

The transform $V^*(iu)$ in the integral above is essentially a matrix–exponential of a complex matrix. Such a matrix–exponential is obviously the solution to a system of linear differential equations, and we solve for the matrix–exponential by solving the system of differential equation using a fourth order Runge–Kutta method. In order to speed up the procedure there is also a scaling consideration involved where we use the property of the matrix–exponential $\exp(\boldsymbol{\Gamma}t) = \exp(\boldsymbol{\Gamma}t/n)^n$. If we choose n to be a power of 2, $n = 2^k$ say, then the power of the exponential is particularly fast to calculate by repeated squaring of the exponential k times with itself.

The numerical inversion of the transform requires the evaluation of the exponential of a matrix, here carried out by the fourth order Runge–Kutta method, the complexity of which is $O(m^2 n_1)$, where n_1 is the number of discretization steps for solving the differential equations. The storage requirement for evaluating the transform is $O(m^2)$ as we need to store the intensity matrix $(m \times m)$ and reward vector (dimension m). The numerical integration depends linearly on the number of steps involved when we consider the total reward earned in some states. If we let n_2 denote the number of integration steps and n_3 the number of density points to be produced, the complexity of the total algorithm is $O(m^2 n_1 n_2 n_3)$, while the storage requirements remain $O(m^2)$.

4.3 Comparison of the Two Methods

Next, we compare the performance of the explicit method and the numerical transform inversion by means of two examples. We refer to the former as the *exact* method and to the latter as the *inversion* method.

4.3.1 A 3-state model

We again consider the example of section 4.1.2. We compute the conditional density $W'_{13}(x, 10)$ for various values of x. For the *exact* method, the precision ε was set to 10^{-10}, while the error for the *inversion* method is estimated in the course of the computation. We obtained $W'_{13}(x, 10)$ for many x; only a few numerical results are shown in Table 3. The third column lists the estimated error for the *inversion* method. Among all the computed values, the maximum absolute difference between the results of both methods is 1.3×10^{-7}. In particular, this suggests that the error estimates in the *inversion* method are not accurate. All the values in the third column of Table 3 are much smaller than 1.3×10^{-7}. Given the high accuracy of the *exact* method, it appears that the error in the *inversion* method is larger than reported.

Table 3. Numerical results for the 3-state model

x	*exact*	*inversion*	Error Estim. on Inv.
1	6.90841D-05	6.9084099823223D-05	6.7762635780344D-21
5	8.0198771D-03	8.0198758928317D-03	1.0842021724855D-18
10	3.39710030D-02	3.3970937106725D-02	1.9185555921730D-10
15	8.0198771D-03	8.0198749959861D-03	1.6532381453410D-10
16	3.9259414D-03	3.9259417172244D-03	1.8769383399939D-10
17	1.5424709D-03	1.5424716317178D-03	1.6531584738286D-10
18	4.395360D-04	4.3953734438919D-04	1.7517860938427D-10
19	6.90841D-05	6.9086110168773D-05	9.4898452709599D-11
20	0.0000000000	1.3247682929818D-07	1.9573838184247D-10

4.3.2 A stiff model

Consider a system with N processors that, independently of each other, are subject to failure and repair. The times to failure and the repair times of each processor are exponential, respectively, with parameters β and μ. There is a single repairman. We denote by $J(t)$ the number of operational processors at time t. The transition rates of the Markov chain $J(t)$ are shown in Figure 1. We assign a reward equal to 1 to states N and $N - 1$ and equal to 0 to all other states.

Such a model is called stiff when the ratio between the largest and the

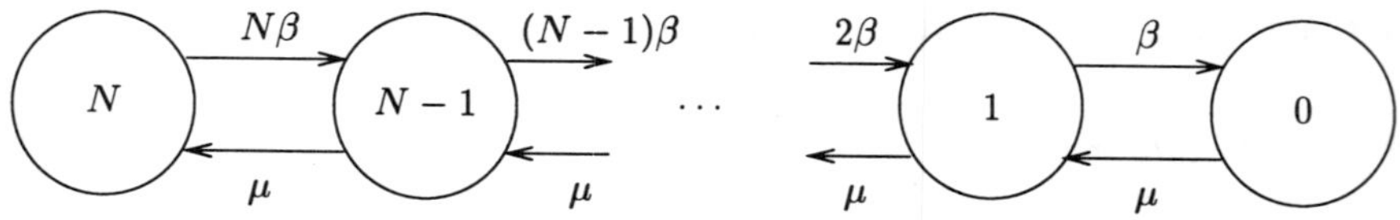

Figure 1. A stiff model

smallest transition rates is very large. By choosing $\beta = 10^{-9}$ and $\mu = 1$, we obtain an example of a *stiff* model.

We compute the conditional density of the total continuous reward earned up to time $t = 100$, that the state is then N, given that 1 is the initial state. For this example, $R(t)$ is called *the interval availability* over $[0, t)$. We computed $W'_{1N}(x, 100)$ for various values of the accumulated reward x and for $N = 10$. As before, for the *exact* method, the precision was specified as $\varepsilon = 10^{-10}$, while the error for the *inversion* method is estimated as we go.

We obtained $W'_{1N}(x, 100)$ for many values of x and, as for the preceding example, Table 4 lists only some representative values.

Among all computed values, the maximum absolute difference between the results of both methods is 5.5×10^{-6}. As for the previous example, this suggests that the error estimates in the *inversion* method are not accurate. All the values in the third column of Table 4 are much smaller than 5.5×10^{-6}. Given the high accuracy of the *exact* method, it appears again that the error in the *inversion* method is larger than reported.

Table 4. Numerical results for the stiff model

x	*exact*	*inversion*	Error Estim. on Inv.
70	4.061D-07	3.6675885125778D-07	1.0268233628745D-10
75	1.68185D-05	1.6742507306377D-05	1.4984542538130D-10
80	5.234676D-04	5.2333975210878D-04	1.7149709717655D-10
85	1.03702940D-02	1.0369952661451D-02	1.5675087616795D-10
90	9.00792255D-02	9.0078490790866D-02	1.0241073961081D-10
95	1.044448605D-01	1.0443933140437D-01	4.8018394815941D-11
96	0.0595403611D-02	5.9534948246358D-02	9.9827379002049D-11
97	0.0216040309D-02	2.1604262271277D-02	1.0371982179375D-10
98	0.0034370865D-03	3.4426169921810D-03	7.2716614413909D-11
99	0.0000729920D-05	7.7315641585370D-05	7.5063514621719D-11
100	0.0000000000	-2.2215910573491D-07	1.5172630167441D-10

These two examples show, as expected, that the *exact* method has a high precision that can be given in advance. So we can evaluate beforehand the time needed execute the corresponding algorithm. This execution time can be very important for large values of the mission time t and also for a large number of distinct rewards. Concerning the *inversion* method, it is not so accurate and the error estimated is not reliable, so we are not sure that it gives the correct result. The main advantage of that method is that the execution time is independent of the mission time t and of the number of distinct rewards ϕ.

4.4 Convolution Method

In this section we spell out the convolution properties of the matrix $W(x,t)$ and we use these properties to develop a new algorithm to deal with large values of t. The *exact* algorithm developed from the explicit formulas serves as a starting point for the convolution method. We thus initiate the convolutions with data having very high precision.

To simplify notation, we denote by Q_l the matrix $Q_{B_l B_l}$, for $0 \leq l \leq \phi$. For any real numbers a and b, we define $a \wedge b = \min(a,b)$. Recall that for $x \geq r_\phi(t+s)$, we have $W(x,t+s) = e^{Q(t+s)}$, and for $x = 0$, we have, $W(0,t+s) = e^{Q(t+s)}_{ij} 1_{\{i,j \in B_0\}}$.

Theorem 4.4 *For* $0 < x < r_\phi(t+s)$, *we have that*

$$W_{ij}(x,t+s) = \sum_{k \in S} \int_0^{x \wedge r_\phi t} W'_{ik}(v,t) W_{kj}(x-v,s)dv$$
$$+ \sum_{l=0}^{\phi} \sum_{k \in B_l} e^{Q_l t}_{ik} W_{kj}(x - r_l t, s) 1_{\{i \in B_l\}} 1_{\{x \geq r_l t\}}. \qquad (16)$$

Proof. By $R(t,t+s)$, we denote the total continuous reward over the interval $(t,t+x]$. We thus have $R(t) = R(0,t)$ and $R(t+s) = R(t) + R(t,t+s)$. Using this relation, we have

$$W_{ij}(x,t+s) = P\{R(t+s) \leq x, J(t+s) = j \mid J(0) = i\}$$
$$= \sum_{k \in S} P\{R(t) + R(t,t+s) \leq x, J(t+s) = j, J(t) = k \mid J(0) = i\}$$
$$= \sum_{k \in S} \int_{v \geq 0} P\{R(t,t+s) \leq x - v, J(t+s) = j | J(t) = k, R(t) = v, J(0) = i\}$$
$$\times dP\{R(t) \leq v, J(t) = k \mid J(0) = i\}$$

$$
\begin{aligned}
&= \sum_{k\in S}\int_{v\geq 0} P\{R(t,t+s)\leq x-v, J(t+s)=j \mid J(t)=k\} \\
&\qquad\qquad\qquad\qquad \times dP\{R(t)\leq v, J(t)=k \mid J(0)=i\} \\
&= \sum_{k\in S}\int_{v\geq 0} P\{R(s)\leq x-v, J(s)=j \mid J(0)=k\} \\
&\qquad\qquad\qquad\qquad \times dP\{R(t)\leq v, J(t)=k \mid J(0)=i\} \\
&= \sum_{k\in S}\int_{v\geq 0} dW_{ik}(v,t)W_{kj}(x-v,s).
\end{aligned}
$$

The fourth equality is due to the Markov property and the fifth comes from the homogeneity of J. The jumps arising in $W_{ik}(v,t)$ are described in relation (10). Using that relation, we have

$$
\begin{aligned}
W_{ij}(x,t+s) &= \sum_{k\in S}\int_{v\geq 0} W'_{ik}(v,t)W_{kj}(x-v,s)dv \\
&\quad + \sum_{k\in S}\sum_{l=0}^{\phi} \Pr\{R(t)=r_l t, J(t)=k \mid J(0)=i\}W_{kj}(x-r_l t,s) \\
&= \sum_{k\in S}\int_{v\geq 0} W'_{ik}(v,t)W_{kj}(x-v,s)dv \\
&\quad + \sum_{k\in S}\sum_{l=0}^{\phi} e^{Q_l t}_{ik} 1_{\{i,k\in B_l\}} W_{kj}(x-r_l t,s) \\
&= \sum_{k\in S}\int_{v\geq 0} W'_{ik}(v,t)W_{kj}(x-v,s)dv \\
&\quad + \sum_{l=0}^{\phi}\sum_{k\in B_l} e^{Q_l t}_{ik} W_{kj}(x-r_l t,s)1_{\{i\in B_l\}},
\end{aligned}
$$

and the result follows since $W_{kj}(x-v,s)=0$, for $v>x$, $W'_{ik}(v,t)=0$, for $v>r_\phi t$ and, $W_{kj}(x-r_l t,s)=0$, for $x<r_l t$. ■

The following corollary is a simplified version of relation (16). As usual, we define $x^+=\max(0,x)$, for any real number x.

Corollary 4.5 *For* $0<x<r_\phi(t+s)$, *we have*

$$
W_{ij}(x,t+s) = \sum_{k\in S}\int_{(x-r_\phi s)^+}^{x\wedge r_\phi t} W'_{ik}(v,t)W_{kj}(x-v,s)dv
$$

$$+\sum_{k\in S} W_{ik}((x-r_\phi s)^+,t)e^{Qs}_{kj}$$

$$+\sum_{l=0}^{\phi}\sum_{k\in B_l} e^{Q_l t}_{ik} W_{kj}(x-r_l t,s)1_{\{i\in B_l\}}1_{\{r_l t\leq x<r_l t+r_\phi s\}}, \quad (17)$$

Proof. Consider relation (16) and denote by β the integral part and by α the other part corresponding to the jumps. We thus have

$$\alpha=\sum_{l=0}^{\phi}\sum_{k\in B_l} e^{Q_l t}_{ik} W_{kj}(x-r_l t,s)1_{\{i\in B_l\}}1_{\{x\geq r_l t\}},$$

and

$$\beta=\sum_{k\in S}\int_0^{x\wedge r_\phi t} W'_{ik}(v,t)W_{kj}(x-v,s)dv.$$

Since $W_{kj}(x-r_l t,s)=e^{Qs}_{kj}$, if $x\geq r_l t+r_\phi s$ and as $x<r_\phi(t+s)$, we get

$$\begin{aligned}\alpha&=\sum_{l=0}^{\phi}\sum_{k\in B_l} e^{Q_l t}_{ik} W_{kj}(x-r_l t,s)1_{\{i\in B_l\}}1_{\{r_l t\leq x<r_l t+r_\phi s\}}\\&+\sum_{l=0}^{\phi}\sum_{k\in B_l} e^{Q_l t}_{ik} e^{Qs}_{kj}1_{\{i\in B_l\}}1_{\{x\geq r_l t+r_\phi s\}}\\&=\sum_{l=0}^{\phi}\sum_{k\in B_l} e^{Q_l t}_{ik} W_{kj}(x-r_l t,s)1_{\{i\in B_l\}}1_{\{r_l t\leq x<r_l t+r_\phi s\}}\\&+\sum_{l=0}^{\phi-1}\sum_{k\in B_l} e^{Q_l t}_{ik} e^{Qs}_{kj}1_{\{i\in B_l\}}1_{\{x\geq r_l t+r_\phi s\}}.\end{aligned} \quad (18)$$

In the same way, since $W_{kj}(x-v,s)=e^{Qs}_{kj}$, if $v\leq x-r_\phi s$, we get

$$\begin{aligned}\beta&=\sum_{k\in S}\left(\int_0^{(x-r_\phi s)^+} W'_{ik}(v,t)dv\right)e^{Qs}_{kj}\\&+\sum_{k\in S}\int_{(x-r_\phi s)^+}^{x\wedge r_\phi t} W'_{ik}(v,t)W_{kj}(x-v,s)dv.\end{aligned}$$

Let us denote by θ the integral arising in the first sum. We have

$$\theta=\int_0^{(x-r_\phi s)^+} W'_{ik}(v,t)dv$$

$$= W_{ik}((x - r_\phi s)^+, t) - \sum_{h=1}^{\phi} \sum_{l=0}^{h-1} e_{ik}^{Q_l t} 1_{\{i,k \in B_l\}} 1_{\{(x - r_\phi s)^+ \in [r_{h-1}t, r_h t)\}}$$

$$= W_{ik}((x - r_\phi s)^+, t) - \sum_{l=0}^{\phi-1} \sum_{h=l+1}^{\phi} e_{ik}^{Q_l t} 1_{\{i,k \in B_l\}} 1_{\{(x - r_\phi s)^+ \in [r_{h-1}t, r_h t)\}}$$

$$= W_{ik}((x - r_\phi s)^+, t) - \sum_{l=0}^{\phi-1} e_{ik}^{Q_l t} 1_{\{i,k \in B_l\}} 1_{\{x \geq r_l t + r_\phi s\}}.$$

Finally, we obtain that

$$\beta = \sum_{k \in S} W_{ik}((x - r_\phi s)^+, t) e_{kj}^{Q_s} - \sum_{l=0}^{\phi-1} \sum_{k \in B_l} e_{ik}^{Q_l t} e_{kj}^{Q_s} 1_{\{i \in B_l\}} 1_{\{x \geq r_l t + r_\phi s\}}$$

$$+ \sum_{k \in S} \int_{(x - r_\phi s)^+}^{x \wedge r_\phi t} W'_{ik}(v, t) W_{kj}(x - v, s) dv. \tag{19}$$

By adding the expressions (18) and (19), we obtain the desired result. ■

4.4.1 Algorithmic aspects

We wish to compute $W_{i,j}(x, 2t)$ for some values of x, where $0 < x < 2r_\phi t$, assuming that we know $W'_{ij}(x, t)$, $W_{ij}(x, t)$ at the same points x. For this purpose, we use the relations (16) and (17) for $s = t$. Let us consider formula (16). The main difficulty consists in the evaluation of matrices $U(x, t)$ defined by

$$U_{i,j}(x, t) = \sum_{k \in S} \int_0^{x \wedge r_\phi t} W'_{ik}(v, t) W_{kj}(x - v, t) dv.$$

In order to simplify notations, let us write $W'_{ij}(x)$ instead of $W'_{ij}(x, t)$, $W_{ij}(x)$ instead of $W_{ij}(x, t)$ and $U_{ij}(x)$ instead of $U_{ij}(x, t)$. Moreover, we can write that

$$U(x) = \int_0^{x \wedge r_\phi t} W'(v) W(x - v) dv$$

where the integral of a matrix function is the matrix whose entries are the integrals of the entries of the matrix function.

Let us partition the interval $[0, 2r_\phi t]$ into subintervals $[x_i, x_{i+1}]$, $i = 0, \ldots, N-1$, where $x_i = 2ir_\phi t/N$, $i = 0, \ldots, N$. In this way $x_0 = 0$, $x_{N/2} = r_\phi t$, $x_N = 2r_\phi t$.

We wish to approximate $U(x_i)$ by using the trapezoidal rule with knots x_k, $k = 0, \ldots, \min(i, N/2)$ (the trapezoidal rule can be naturally extended to the case of matrix functions).

Let us first consider the case $i \geq N/2$, i.e., $x_i \geq r_\phi t$. Then

$$U(x_i) = \int_0^{r_\phi t} W'(v)W(x_i - v)dv$$

can be approximated by

$$\widetilde{U}(x_i) = \frac{2r_\phi t}{N}\left[W'(x_0)W(x_i) + W'(x_{N/2})W(x_i - x_{N/2})\right] + \frac{4r_\phi t}{N}\sum_{k=1}^{N/2-1} W'(x_k)W(x_i - x_k). \quad (20)$$

Now, $x_i - x_j = (2t(i-j)r_\phi)/N = x_{i-j}$. Thus, if we set $w_i = W(x_i)$, $w'_i = W'(x_i)$, and $\widetilde{u}_i = \widetilde{U}(x_i)$, $i = 0, \ldots, N$, we can write (20) as

$$\widetilde{u}_i = \frac{2r_\phi t}{N}\left(w'_0 w_i + w'_{N/2} w_{i-N/2} + 2\sum_{k=1}^{N/2-1} w'_k w_{i-k}\right), \quad i = N/2, \ldots, N-1. \quad (21)$$

If $i = 1, \ldots, N/2 - 1$, then

$$U(x_i) = \int_0^{x_i} W'(v)W(x_i - v, s)dv$$

can be approximated by

$$\widetilde{U}(x_i) = \frac{2r_\phi t}{N}\left(W'(x_0)W(x_i) + W'(x_i)W(x_0) + 2\sum_{k=1}^{i-1} W'(x_k)W(x_i - x_k)\right),$$

which yields

$$\widetilde{u}_i = \frac{2r_\phi t}{N}\left(w'_0 w_i + w'_i w_0 + 2\sum_{k=1}^{i-1} w'_k w_{i-k}\right), \quad i = 1, \ldots, N/2 - 1. \quad (22)$$

If we write (21,22) in matrix form we obtain that

$$\begin{bmatrix} \widetilde{u}_1 \\ \widetilde{u}_2 \\ \vdots \\ \vdots \\ \widetilde{u}_{N-2} \\ \widetilde{u}_{N-1} \end{bmatrix} = \frac{2r_\phi t}{N}\left(\begin{bmatrix} w'_1 \\ \vdots \\ w'_{N/2} \\ 0 \\ \vdots \\ 0 \end{bmatrix} w_0 + T_N \begin{bmatrix} w_1 \\ w_2 \\ \vdots \\ \vdots \\ w_{N-2} \\ w_{N-1} \end{bmatrix}\right),$$

where

$$T_N = \begin{bmatrix} w'_0 & & & & & & \bigcirc \\ 2w'_1 & w'_0 & & & & & \\ \vdots & 2w'_1 & \ddots & & & & \\ 2w'_{N/2-1} & \vdots & \ddots & \ddots & & & \\ w'_{N/2} & 2w'_{N/2-1} & & \ddots & \ddots & & \\ 0 & \ddots & \ddots & & \ddots & \ddots & \\ \vdots & \ddots & \ddots & \ddots & & \ddots & w'_0 \\ 0 & \dots & 0 & w'_{N/2} & 2w'_{N/2-1} & \dots\ 2w'_1 & w'_0 \end{bmatrix}$$

is a block $(N-1) \times (N-1)$ banded lower triangular block Toeplitz matrix.

The product between T_N and the block vector $w_N = (w_i)_{i=1,N-1}$ can be computed by means of FFT's of length N with a computational cost of $O(m^2 N \log N + m^3 N)$ arithmetic operations, where the size of the blocks w_i is equal to m, the size of the state space of the Markov chain $\{J(t)\}$. More specifically, we may define the $N \times N$ block triangular Toeplitz matrix T_N^*, obtained by adding a block row and a block column to T_N, and consider the problem of computing the product $T_N^* w_N^*$, where $w_N^* = (w_i^*)_{i=1,N}$ is the block vector obtained by appending to w_N a null block component: in fact, the vector $T_N w_N$ is given by the first $N-1$ block components of $T_N^* w_N^*$. Assuming that N is a power of 2, we may partition the matrix T_N^* into a 2×2 block matrix,

$$T_N^* = \begin{bmatrix} T_{N,1} & 0 \\ T_{N,2} & T_{N,1} \end{bmatrix},$$

where $T_{N,1}$ and $T_{N,2}$ are respectively block lower and upper triangular block Toeplitz matrices with block size $N/2$, and write the vector $T_N^* w_N^*$ as

$$T_N^* w_N^* = \begin{bmatrix} T_{N,1} w_{N,1} \\ T_{N,2} w_{N,1} + T_{N,1} w_{N,2} \end{bmatrix},$$

where $w_{N,1} = (w_i^*)_{i=1,N/2}$, $w_{N,2} = (w_i^*)_{i=N/2+1,N}$. The block vectors $T_{N,1} w_{N,1}$, $T_{N,2} w_{N,1}$ and $T_{N,1} w_{N,2}$ can be obtained by means of the relation

$$C_N \left[\begin{array}{c|c} w_{N,1} & 0 \\ 0 & w_{N,2} \end{array}\right] = \left[\begin{array}{c|c} T_{N,1} w_{N,1} & T_{N,2} w_{N,2} \\ T_{N,2} w_{N,1} & T_{N,1} w_{N,2} \end{array}\right], \tag{23}$$

where

$$C_N = \begin{bmatrix} T_{N,1} & T_{N,2} \\ T_{N,2} & T_{N,1} \end{bmatrix}$$

is the block circulant matrix defined by the first block column of T_N^*. Since C_N is block circulant, the right hand side in (23) can be computed according to the following scheme:

1. *DFT associated with C_N:* Evaluate the matrix polynomial $w'_N(z) = w'_0 + 2\sum_{i=1}^{N/2-1} w'_i z^i + w'_{N/2} z^{N/2}$ at the N-th roots of 1, by means of m^2 FFT's of length N;

2. *DFT's associated with $w_{N,1}$ and $w_{N,2}$*: evaluate the matrix polynomials $w_{N,1}(z) = \sum_{i=0}^{N/2-1} w_{i+1} z^i$ and $w_{N,2}(z) = \sum_{i=N/2}^{N-2} w_{i+1} z^i$ at the N-th roots of 1, by means of $2m^2$ FFT's of length N;

3. *Convolution:* Compute the values of the matrix polynomials $p_1(z) = w'_N(z)w_{N,1}(z)$ and $p_2(z) = w'_N(z)w_{N,2}(z)$ at the N-th roots of 1 by means of $2N$ matrix products;

4. *IDFT's*: Interpolate the values obtained at the previous step by means of $2m^2$ FFT's of length N, thus obtaining the block coefficients of $p_1(z)$ and $p_2(z)$; the block coefficients of $p_1(z)$ and $p_2(z)$ coincide with the block entries of the vectors $\begin{bmatrix} T_{N,1}w_{N,1} \\ T_{N,2}w_{N,1} \end{bmatrix}$ and $\begin{bmatrix} T_{N,2}w_{N,2} \\ T_{N,1}w_{N,2} \end{bmatrix}$ in (23), respectively.

Thus the overall computational cost amounts to $O(m^2 N \log N + m^3 N)$ arithmetic operations.

Concerning the number N of knots which are sufficient to have a good approximation of the integral, we can use the estimates of the approximation error of the trapezoidal rule. In particular

$$\max_{i=0,\ldots,N-1} |u_i - \tilde{u}_i| \leq \frac{r_\phi t}{6N^2}\gamma,$$

where γ is an upper bound to the maximum norm of the second derivative of the argument of the integral.

References

1. J. Abate and W. Whitt, The Fourier-series method for inverting transforms of probability distributions, Queueing Systems, 10, 5-88, 1992.
2. E. de Souza e Silva and H. R. Gail, An algorithm to calculate transient distributions of cumulative rate and impulse based reward, Stochastic Models, 14(3), 1998.
3. H. Nabli and B. Sericola, Performability analysis: A new algorithm, IEEE Transactions on Computers, 45(4), April 1996.

4. S. Narayana and M. F. Neuts, The first two moments matrices of the counts for the Markovian arrival process, Stochastic Models, 8, 459-477, 1992.
5. B. Sericola, Occupation times in Markov processes, Stochastic Models, 16, 479-510, 2000.

Acknowledgments

This research of M. F. Neuts was supported in part by NSF Grant Nr. DMI-9988749.

A BATCH MARKOVIAN QUEUE WITH A VARIABLE NUMBER OF SERVERS AND GROUP SERVICES

SRINIVAS R. CHAKRAVARTHY
Department of Industrial and Manufacturing Engineering and Business, Kettering University, Flint, MI 48439, USA
E-mail: schakrav@kettering.edu

ALEXANDER N. DUDIN
Department of Applied Mathematics and Computer Science, Belarussian State University, Minsk, Belarus
E-mail: dudin@bsu.by

In this paper, we consider a multi-server queuing model with a finite buffer in which customers arrive according to a batch Markovian arrival process ($BMAP$). These customers are served in groups of varying sizes ranging from a predetermined value L through a maximum size, K. The service times are exponentially distributed. The number of servers in the system at any given time varies between a lower limit and an upper limit. The steady state analysis of the model is performed by exploiting the structure of the coefficient matrices. Some interesting numerical examples are discussed.

1 INTRODUCTION

Often times in queuing systems, the server utilization is inversely proportional to the performance requirements of the customers. For example, in a finite capacity queuing model to guarantee that an admitted customer will have to wait no longer than a predetermined value with a certain probability, one to has to increase the service rate or increase the number of servers. In either case the server utilization will be much smaller compared to a system without the restriction. One of the ways to balance this conflict between the server utilization and the performance requirements is to adjust the number of servers present in the system dynamically [1, 16, 17].

In the context of an $M/M/2$ queueing model, Bell [2] investigated the optimal policy by allowing the number of servers to be varied. Using different cost structure such as switching cost, holding cost and service cost, in Zhang, et al [27], the number of servers is dynamically varied so as to minimize the expected cost over an infinite horizon. In the case of a single server queueing model with Poisson arrivals and general services, optimal policies under random vacation using diffusion approximations are presented in Okamura, et al [22]. Dudin and Khalaf [15] considered an $M/M/N$ queue in which all servers are removed once the system becomes empty. The economic behavior of a removable and non-reliable server in the context of a finite and infinite capacity Markovian queueing system was studied in [23-26]. Recently, Li and Yang [18] studied an $M/M/s$ queueing system in which the number of servers at any time varied between a lower limit and an upper limit. Under the assumption that the server release times and search times are exponential, they show that the steady state solution is of matrix-geometric type.

In the context of a queueing system (with a fixed number of servers)with a finite buffer of size K, a different type of service scheme was introduced in Chakravarthy [4]. The pre-assigned number $L \geq 1$, called the *threshold* operates as follows. An idle server finding fewer than L customers in the queue remains idle until the queue size builds up to L or more. However, when i, $L \leq i \leq K$, customers present, the idle server initiates a service for the entire group. Service scheme of this type in the context of finite capacity $GI/PH/1$ and $MAP/G/1$ with single arrivals, and $BMAP/M/c$ models were investigated in the papers [4-13]. Some potential applications of this type of service mechanism in computer communications and manufacturing processes were outlined in those papers. For example, in computer and communications engineering, requests(messages or terminals or satellites) that involve information of a general nature such as access to a common data base or a common input-output device such as a laser printer or a color plotter, can be handled in groups. Another example is in load balancing using probing in distributed processing. When jobs arrive into the dispatcher, it probes the distributed system for the type of load (heavy, moderate or light) and accordingly the jobs are distributed to balance the load among various processors. In all of the above applications, we see that the customers can be processed in groups of varying sizes, which motivates the need for the type of service mechanism considered here.

In this paper we introduce the removable server concept in the context of a finite capacity multiserver queueing system with the above mentioned service scheme. The paper is organized as follows. In Section 2 the mathematical model and the service control mechanism are described. The steady state analysis of the model is presented in Section 3 along with the algorithmic procedures for computing the steady state probabilities. In Section 4 some key system performance measures describing the queueing model are presented along with their formulas. The stationary waiting time distribution of an admitted customer is shown to be of phase type in Section 5 and some interesting numerical examples are presented in Section 6.

2 THE MATHEMATICAL MODEL

We consider a multiserver queueing system with a finite buffer of size K in which customers arrive according to a batch Markovian arrival process ($BMAP$). The service facility consists of a maximum of c and a minimum of $r, 1 \leq r < c$, identical servers with parameter μ. The number of servers is increased by one (from r up to the maximum limit of c) upon completion of a *search* process and decreased by one (up to the minimum level of r) upon completion of a *release* process. Any batch arriving when the waiting room is full is considered lost. However, when an arriving batch finds at least one empty space in the buffer, the customers in the batch are admitted to the extent of the available space. We assume that the service is offered to groups of varying size $i, L \leq i \leq K$. An arriving batch of size at least L finding an idle server enters into service immediately. If a service has to be initiated through an arrival of a batch (for example, i, $i < L$, customers are in the queue with at least one server idle, and the arriving batch has at least $L - i$ customers), then only one idle server can be activated for a group of size at least L but at most K; that is to say that two idle servers cannot become busy simultaneously. However, our analysis can be carried out by modifying this assumption by allowing more than K customers to be admitted and possibly activating more than one server at a time. The details are omitted.

When a group of L or more customers finds all servers busy and when the number of busy servers is less than c, a search for a free server is initiated. The searching time is assumed to be exponential with parameter θ. If during the search process, one of the busy servers becomes free, the search is cancelled instantaneously and the freed server will initiate a new service for the waiting group. Upon completion of a service, a free server not able to offer a new service for lack of customers will be released if the number of busy servers at that epoch is at least r. The release process is assumed to be exponentially distributed with parameter η. If during the release process enough customers accumulate to form a group ready for a service, then the server on release will be recalled instantaneously to initiate a service for this group. Note that at any given time there can be at most one search running and the number of servers on release can be between 0 to $c - r$. We assume that the arrival process, the service times, the search process and the release process are all mutually independent.

The $BMAP$, a special class of tractable Markov renewal process, is a rich class of point processes that includes many well-known processes such as Poisson, PH-renewal processes, and Markov-modulated Poisson process. One of the most significant features of the $BMAP$ is the underlying Markovian structure and fits ideally in the context of matrix-analytic solutions to stochastic models. Matrix-analytic methods were first introduced and studied by Neuts [20-21]. As is well known, Poisson processes are the simplest and most tractable ones used extensively in stochastic modelling. The idea of the $BMAP$ is to significantly generalize the Poisson processes and still keep the tractability for modelling purposes. Furthermore, in many practical applications, notably in communications engineering, production and manufacturing engineering, the arrivals do not usually form a renewal process. So, $BMAP$ is a convenient tool to model both renewal and non-renewal arrivals. While $BMAP$ is defined for both discrete and continuous times, here we will need only the continuous time case.

The $BMAP$ in continuous time is described as follows. Let the underlying Markov chain be irreducible and let $Q^* = (q_{i,j})$ be the generator of this Markov chain. At the end of a sojourn time in state i, that is exponentially distributed with parameter $\lambda_i \geq -q_{i,i}$, one of the following two events could occur: with probability $p_{ij}(k)$ the transition corresponds to an arrival of group size $k \geq 1$, and the underlying Markov chain is in state j with $1 \leq i, j \leq m$; with probability $p_{i,j}(0)$ the transition corresponds to no arrival and the state of the Markov chain is j, $j \neq i$. Note that the Markov chain can go from state i to state i only through an arrival. Also, we have

$$\sum_{k=1}^{\infty}\sum_{j=1}^{m} p_{i,j}(k) + \sum_{j=1, j\neq i}^{m} p_{i,j}(0) = 1, 1 \leq i \leq m.$$

For $k \geq 0$, define matrices $D_k = (d_{ij}(k))$ such that $d_{i,i}(0) = -\lambda_i$, $1 \leq i, j \leq m;$, $d_{ij}(0) = \lambda_i p_{ij}(0)$, for $j \neq i$, $1 \leq i$, $j \leq m$, and $d_{ij}(k) = \lambda_i p_{ij}(k)$. By assuming D_0 to be a nonsingular matrix, the interarrival times will be finite with probability one and the arrival process does not terminate. Hence, we see that D_0 is a stable matrix. The generator Q^* is then given by $Q^* = \sum_{k=0}^{\infty} D_k$.

Thus, the $BMAP$ is described by the matrices $\{D_k\}$ with D_0 governing the transitions corresponding to **no** arrival and D_k governing those corresponding to arrivals of group size $k, k \geq 1$. It can be shown that $BMAP$ is equivalent to Neuts' versatile Markovian point process. The point process described by the $BMAP$ is a special class of semi-Markov processes with transition probability matrix given by

$$\int_0^x e^{D_0 t} dt D_k = [I - e^{D_0 x}](-D_0)^{-1} D_k, fork \geq 1.$$

For use in sequel, let $\mathbf{e}(n)$, $\mathbf{e}_j(n)$ and I_n denote, respectively, the (column)

vector of dimension n consisting of 1's, column vector of dimension n with 1 in the jth position and 0 elsewhere, and an identity matrix of dimension n. When there is no need to emphasize the dimension of these vectors we will suppress the suffix. Thus, $\mathbf{e}$ will denote a column vector of 1's of appropriate dimension. The notation "$\prime$" appearing in a matrix will stand for the matrix transpose. The notation $\otimes$ will stand for the Kronecker product of two matrices. Thus, if A is a matrix of order $m \times n$ and if B is a matrix of order $p \times q$, then $A \otimes B$ will denote a matrix of order $mp \times nq$ whose $(i,j)^{th}$ block matrix is given by $a_{ij}B$. For more details on Kronecker products, we refer the reader to Bellman[3].
Let δ be the stationary probability vector of the Markov process with generator Q^*. That is, δ is the unique (positive) probability vector satisfying.

$$\delta Q^* = \mathbf{0}, \delta \mathbf{e} = 1. \tag{1}$$

Let α be the initial probability vector of the underlying Markov chain governing the $BMAP$. Then, by choosing α appropriately we can model the time origin to be (a) an arbitrary arrival point; (b) the end of an interval during which there are at least k arrivals; (c) the point at which the system is in specific state such as the busy period ends or busy period begins. The most interesting case is the one where we get the stationary version of the $BMAP$ by $\alpha = \delta$. The constant $\lambda = \delta \sum_{k=1}^{\infty} kD_k e$, referred to as the **fundamental rate** gives the expected number of arrivals per unit of time in the stationary version of the MAP. The quantity $\lambda_g = \delta(-D_0)\mathbf{e}$, gives the arrival rate of groups. Note that for a $BMAP$ with single arrivals $\lambda = \lambda_g$.

In this paper we will assume that $D_k = 0$ for $k > K$ so that the maximum batch size of arrivals is K. Often, in model comparisons, it is convenient to select the time scale of the $BMAP$ so that λ_g has a certain value. That is accomplished, in the continuous $BMAP$ case, by multiplying the coefficient matrices $D_k, k \geq 0$, by the appropriate common constant. For further details on $BMAP$ and their usefulness in Stochastic modelling, we refer to Lucantoni[19], Neuts [21]and for a review and recent work on $BMAP$ we refer the reader to Chakravarthy [14].

For economic reasons, it is better to have a minimum number of customers to form a group before they are processed. The maximum number of customers that can be processed at a time is the size of the service capacity, which is taken as K. Larger values of L will result in more waiting time for the customers and smaller values of L will result in frequent services with smaller group sizes. It would be of interest to see the influence of L, r, and c, as well as the correlation of the interarrival times on the behavior of the system performance measures. Also, a number of optimization problems of practical interest can be handled using our algorithmic procedures.

3 THE STEADY STATE PROBABILITY VECTOR AT AN ARBITRARY EPOCH

The model described above can be modelled as a continuous-time Markov chain with state space given by

$$\Omega = \{\mathbf{0}^*, \cdots, (\boldsymbol{L}{-}1)^*, \mathbf{1}', \cdots, (\boldsymbol{r}{-}1)', \boldsymbol{r}', \cdots, (\boldsymbol{c}{-}1)', \mathbf{0}, \cdots, \boldsymbol{K}, \mathbf{1}^{**}, \cdots, (\boldsymbol{c}{-}\boldsymbol{r})^{**}\}$$

where the set of states

$$\boldsymbol{i}^* = \{(i,k) : 1 \le k \le m\}, 0 \le i \le L-1,$$

of dimension m corresponds to the case where all r servers are idle with i customers waiting in the queue; the set of states

$$\boldsymbol{i}' = \{(i,j,k) : 0 \le j \le L-1, 1 \le k \le m\}, 1 \le i \le r-1,$$

of dimension Lm corresponds to the case where i servers are busy with j customers waiting in the queue; the set of states

$$\boldsymbol{i}' = \{(i,j,k) : 0 \le j \le K, 1 \le k \le m\}, r \le i \le c-1,$$

of dimension $(K+1)m$ corresponds to the case where i servers are busy with j customers waiting in the queue; the set of states

$$\boldsymbol{i} = \{(i,k) : 1 \le k \le m\}, 0 \le i \le K,$$

of dimension m corresponds to the case where all c servers are busy with i customers waiting in the queue; and the set of states

$$\boldsymbol{i}^{**} = \{(i,l,j,k) : 0 \le l \le c-i, 0 \le j \le L-1, 1 \le k \le m\}, 1 \le i \le c-r,$$

of dimension $(c+1-i)Lm$ corresponds to the case where i servers are on release, l servers are busy and j customers are waiting in the queue.

The above sets of states are written in lexicographic order. Define the following auxiliary matrices for use in the sequel.

$$B_0 = \begin{pmatrix} D_0 & D_1 & \cdots & D_{L-1} \\ 0 & D_0 & \cdots & D_{L-2} \\ \vdots & \vdots & \cdots & \vdots \\ 0 & 0 & \cdots & D_0 \end{pmatrix}, \tag{2}$$

$$\tilde{D}_i = \sum_{j=i}^{K} D_i, 1 \le i \le K, \tilde{D} = \begin{pmatrix} \tilde{D}_L \\ \tilde{D}_{L-1} \\ \vdots \\ \tilde{D}_1 \end{pmatrix}, \tag{3}$$

$$B_1 = e_1'(L) \otimes \tilde{D}, B_2 = e_1'(K+1) \otimes \tilde{D}, \tag{4}$$

for $r \le j \le c-1$,

$$A_0(j) = \begin{pmatrix} D_0 & D_1 & \cdots & D_{L-1} & D_L & \cdots & \tilde{D}_K \\ 0 & D_0 & \cdots & D_{L-2} & D_{L-1} & \cdots & \tilde{D}_{K-1} \\ \vdots & \vdots & \cdots & \vdots & \vdots & \cdots & \vdots \\ 0 & 0 & \cdots & D_0 & D_1 & \cdots & \tilde{D}_{K-L+1} \\ j\mu I & 0 & \cdots & 0 & D_0 - \theta I & \cdots & \tilde{D}_{K-L} \\ \vdots & \vdots & \cdots & \vdots & \vdots & \cdots & \vdots \\ j\mu I & 0 & \cdots & 0 & 0 & \cdots & Q^* - \theta I \end{pmatrix} - j\mu I, \tag{5}$$

$$A_1 = \theta \sum_{j=L+1}^{K+1} e_j(K+1) \otimes e_1'(K+1) \otimes I, \tag{6}$$

$$A_2 = r\mu \sum_{j=1}^{L} e_j(K+1) \otimes e_j'(L) \otimes I, \tag{7}$$

$$A_3 = \theta \sum_{j=L+1}^{K+1} e_j(K+1) \otimes I, \tag{8}$$

for $r+1 \le j \le c-1$,

$$A_4(j) = j\mu \begin{pmatrix} 0_{Lm\times(j-1)Lm} & I_{Lm\times Lm} & 0_{Lm\times(c-j)Lm} \\ 0_{(K-L+1)m\times(j-1)Lm} & 0_{(K-L+1)m\times Lm} & 0_{(K-L+1)m\times(c-j)Lm} \end{pmatrix}, \tag{9}$$

for $1 \le i \le L$,

$$F_i = c\mu \left(e_c'(c) \otimes e_i'(L) \otimes I \right), \tag{10}$$

for $r \le i \le r-1$,

$$G_i = \eta e_{i+1}(c) \otimes I, \tag{11}$$

for $r \le i \le c-1$,

$$G_i = e_i(c) \otimes e_i'(K+1) \otimes \tilde{D} + \eta e_{i+1}(c) \otimes [I_{Lm}, 0_{Lm\times(K-L+1)m}], \tag{12}$$

$$G_c = e_c(c) \otimes \tilde{D}, \tag{13}$$

for $r \le j \le c-1$,

$$H_0(j) = \begin{pmatrix} B_0 & e_1'(L)\otimes\tilde{D} & 0 & \cdots & 0 & 0 & \cdots & 0 \\ \mu I & B_0-\mu I & e_1'(L)\otimes\tilde{D} & \cdots & 0 & 0 & \cdots & 0 \\ 0 & 2\mu I & B_0-2\mu I & \cdots & 0 & 0 & \cdots & 0 \\ \vdots & \vdots & \vdots & \cdots & \vdots & \vdots & \cdots & \vdots \\ 0 & 0 & 0 & \cdots & r\mu I & B_0-r\mu I & \cdots & 0 \\ \vdots & \vdots & \vdots & \cdots & \vdots & \vdots & \cdots & \vdots \\ 0 & 0 & 0 & \cdots & 0 & 0 & \cdots & B_0-j\mu I \end{pmatrix} - (c-j)\eta I, \tag{14}$$

for $r+1 \le j \le c-1$,

$$H_1(j) = \begin{pmatrix} 0 & \cdots & 0 & 0 & 0 & \cdots & 0 \\ \vdots & \cdots & \vdots & \vdots & \vdots & \cdots & \vdots \\ 0 & \cdots & 0 & 0 & 0 & \cdots & 0 \\ 0 & \cdots & 0 & (r+1)\mu I & 0 & \cdots & 0 \\ 0 & \cdots & 0 & 0 & (r+2)\mu I & \cdots & 0 \\ \vdots & \cdots & \vdots & \vdots & \vdots & \cdots & \vdots \\ 0 & \cdots & 0 & 0 & 0 & \cdots & j\mu I \end{pmatrix}, \tag{15}$$

for $r \le j \le c-2$,

$$H_2(j) = \begin{pmatrix} (c-j)\eta I & 0 & \cdots & 0 & 0 & \cdots & 0 & 0 \\ 0 & (c-j)\eta I & \cdots & 0 & 0 & \cdots & 0 & 0 \\ \vdots & \vdots & \cdots & \vdots & \vdots & \cdots & \vdots & \vdots \\ 0 & 0 & \cdots & (c-j)\eta I & e_1'(L)\otimes\tilde{D} & \cdots & 0 & 0 \\ \vdots & \vdots & \cdots & \vdots & \vdots & \cdots & \vdots & \vdots \\ 0 & 0 & \cdots & 0 & 0 & \cdots & (c-j)\eta I & e_1'(L)\otimes\tilde{D} \end{pmatrix}, \tag{16}$$

$$Q_{11} = \begin{pmatrix} B_0 & B_1 & 0 & 0 & \cdots & 0 & 0 & 0 & \cdots & 0 \\ \mu I & B_0-\mu I & B_1 & 0 & \cdots & 0 & 0 & 0 & \cdots & 0 \\ 0 & 2\mu I & B_0-2\mu I & B_1 & \cdots & 0 & 0 & 0 & \cdots & 0 \\ \vdots & \vdots & \vdots & \vdots & \cdots & \vdots & \vdots & \vdots & \cdots & \vdots \\ 0 & 0 & 0 & 0 & \cdots & B_0-(r-1)\mu I & B_2 & 0 & \cdots & 0 \\ 0 & 0 & 0 & 0 & \cdots & A_2 & A_0(r) & A_1 & \cdots & 0 \\ \vdots & \vdots & \vdots & \vdots & \cdots & \vdots & \vdots & \vdots & \cdots & \vdots \\ 0 & 0 & 0 & 0 & \cdots & 0 & 0 & 0 & \cdots & A_0(c-1) \end{pmatrix}, \tag{17}$$

$$Q_{12} = \begin{pmatrix} 0 & 0 & \cdots & 0 & 0 \\ 0 & 0 & \cdots & 0 & 0 \\ \vdots & \vdots & \cdots & \vdots & \vdots \\ A_3 & 0 & \cdots & 0 & 0 \end{pmatrix}, \tag{18}$$

$$Q_{13} = \begin{pmatrix} 0 & 0 & \cdots & 0 \\ \vdots & \vdots & \cdots & \vdots \\ 0 & 0 & \cdots & 0 \\ A_4(r+1) & 0 & \cdots & 0 \\ \vdots & \vdots & \cdots & \vdots \\ A_4(c-1) & 0 & \cdots & 0 \end{pmatrix}, \tag{19}$$

$$Q_{22} = \begin{pmatrix} D_0 - c\mu I & D_1 & D_2 & \cdots & D_{L-1} & D_L & \cdots & D_{K-1} & \tilde{D}_K \\ 0 & D_0 - c\mu I & D_1 & \cdots & D_{L-2} & D_{L-1} & \cdots & D_{K-2} & \tilde{D}_{K-1} \\ \vdots & \vdots & \vdots & \cdots & \vdots & \vdots & \cdots & \vdots & \vdots \\ 0 & 0 & 0 & \cdots & D_0 - c\mu I & D_1 & \cdots & D_{K-L} & \tilde{D}_{K-L+1} \\ c\mu I & 0 & 0 & \cdots & 0 & D_0 - c\mu I & \cdots & D_{K-L-1} & \tilde{D}_{K-L} \\ \vdots & \vdots & \vdots & \cdots & \vdots & \vdots & \cdots & \vdots & \vdots \\ c\mu I & 0 & 0 & \cdots & 0 & 0 & \cdots & 0 & Q^* - c\mu I \end{pmatrix}, \tag{20}$$

$$Q_{23} = \begin{pmatrix} F_1 & 0 & \cdots & 0 \\ F_2 & 0 & \cdots & 0 \\ \vdots & \vdots & \cdots & \vdots \\ F_L & 0 & \cdots & 0 \\ 0 & 0 & \cdots & 0 \\ \vdots & \vdots & \cdots & \vdots \\ 0 & 0 & \cdots & 0 \end{pmatrix}, \tag{21}$$

$$Q_{31} = \begin{pmatrix} \eta e_1(c) \otimes I & G_1 & G_2 & \cdots & G_{c-1} \\ 0 & 0 & 0 & \cdots & 0 \\ \vdots & \vdots & \vdots & \cdots & \vdots \\ 0 & 0 & 0 & \cdots & 0 \end{pmatrix}, \tag{22}$$

$$Q_{32} = \begin{pmatrix} G_c & 0 & \cdots & 0 \\ 0 & 0 & \cdots & 0 \\ \vdots & \vdots & \cdots & \vdots \\ 0 & 0 & \cdots & 0 \end{pmatrix}, \tag{23}$$

$$Q_{33} = \begin{pmatrix} H_0(c-1) & H_1(c-1) & 0 & 0 \cdots & 0 & 0 & 0 \\ H_2(c-2) & H_0(c-2) & H_1(c-2) & 0 \cdots & 0 & 0 & 0 \\ \vdots & \vdots & \vdots & \vdots \cdots & \vdots & \vdots & \vdots \\ 0 & 0 & 0 & 0 \cdots & H_2(r+1) & H_0(r+1) & H_1(r+1) \\ 0 & 0 & 0 & 0 \cdots & 0 & H_2(r) & H_0(r) \end{pmatrix}. \tag{24}$$

The Markov process describing the model under study has the generator Q, in partitioned form, given by

$$Q = \begin{pmatrix} Q_{11} & Q_{12} & Q_{13} \\ 0 & Q_{22} & Q_{23} \\ Q_{31} & Q_{32} & Q_{33} \end{pmatrix}, \tag{25}$$

where the coefficient matrices appearing in Q are as given in (17-24).

Let $\boldsymbol{x}$, partitioned as $\boldsymbol{x} = (\boldsymbol{u}, \boldsymbol{v}, \boldsymbol{w}, \boldsymbol{z})$, denote the steady-state probability vector of Q. That is,$\boldsymbol{x}$ satisfies

$$\boldsymbol{x}Q = \mathbf{0}, \boldsymbol{x}\mathbf{e} = 1. \tag{26}$$

We further partition the vectors, $\boldsymbol{u}$ of dimension Lm, $\boldsymbol{v}$ of dimension $(r-1)Lm + (c-r)(K+1)m$, $\boldsymbol{w}$ of dimension $(K+1)m$, and $\boldsymbol{z}$ of dimension $0.5(c-r)(c+r+1)Lm$, into vectors of dimension m as follows.

$$\boldsymbol{u} = (\boldsymbol{u}_0, \cdots, \boldsymbol{u}_{L-1})$$

$$\boldsymbol{v} = (\boldsymbol{v}_0(1), \cdots, \boldsymbol{v}_{L-1}(1), \cdots, \boldsymbol{v}_0(r-1), \cdots, \boldsymbol{v}_{L-1}(r-1),$$

$$\boldsymbol{v}_0(r), \cdots, \boldsymbol{v}_K(r), \cdots, \boldsymbol{v}_0(c-1), \cdots, \boldsymbol{v}_K(c-1))$$

$$\boldsymbol{w} = (\boldsymbol{w}_0, \cdots, \boldsymbol{w}_K)$$

$$\boldsymbol{z} = (\boldsymbol{z}_0(1,0), \cdots, \boldsymbol{z}_{L-1}(1,0), \cdots, \boldsymbol{z}_0(1,c-1), \cdots, \boldsymbol{z}_{L-1}(1,c-1),$$

$$\boldsymbol{z}_0(2,0), \cdots, \boldsymbol{z}_{L-1}(2,0), \cdots, \boldsymbol{z}_0(2,c-2), \cdots, \boldsymbol{z}_{L-1}(2,c-2), \cdots$$

$$\boldsymbol{z}_0(c-r,0), \cdots, \boldsymbol{z}_{L-1}(c-r,0), \cdots, \boldsymbol{z}_0(c-r,r), \cdots, \boldsymbol{z}_{L-1}(c-r,r))$$

3.1. Computation of the steady-state probability vector Due to the special structure of the matrix Q as given in (25) and the coefficient matrices (2-24), the vector $\boldsymbol{x}$ is computed efficiently using vectors of dimension m. For example, the vector $\boldsymbol{u}$ is computed as follows.

$$\boldsymbol{u}_0 = [\mu \boldsymbol{v}_0(1) + \eta \boldsymbol{z}_0(1,0)]\,(-D_0^{-1}),$$

$$u_j = \left[\sum_{k=0}^{L-1} u_k \tilde{D}_{L-k} + \mu v_j(1) + \eta z_j(1,0)\right] (-D_0^{-1}), 1 \le j \le L-1.$$

The other equations are similarly written and the details are omitted.

3.2. The steady-state probability vector at an arrival epoch of a batch: Suppose that $\hat{x}$ partitioned as $\hat{x} = (\hat{u}, \hat{v}, \hat{w}, \hat{z})$, denote the steady-state probability vector that the system will be in various states *immediately after* an arrival of a batch. It can easily be verified that the components of $\hat{u}$, $\hat{v}$, $\hat{w}$ and $\hat{z}$ partitioned in an obvious manner are given by

$$\hat{u}_i = \frac{1}{\lambda_g} \sum_{l=0}^{i-1} u_l D_{i-l}, 1 \le i \le L-1, \tag{27}$$

$$\hat{v}_0(j) = \begin{cases} \frac{1}{\lambda_g} \sum_{l=0}^{L-1} u_l \tilde{D}_{L-l}, j = 1, \\ \frac{1}{\lambda_g} \sum_{l=0}^{L-1} v_l(j-1) \tilde{D}_{L-l}, 2 \le j \le r, \\ \frac{1}{\lambda_g} \sum_{l=0}^{L-1} z_l(1, j-1) \tilde{D}_{L-l}, r+1 \le j \le c-1, \end{cases} \tag{28}$$

$$\hat{v}_i(j) = \begin{cases} \frac{1}{\lambda_g} \sum_{l=0}^{i-1} v_l(j) D_{i-l}, 1 \le i \le L-1, 1 \le j \le r-1, \\ \frac{1}{\lambda_g} \sum_{l=0}^{i-1} v_l(j) D_{i-l}, 1 \le i \le K-1, r \le j \le c-1, \\ \frac{1}{\lambda_g} \left[\sum_{l=0}^{K-1} v_l(j) \tilde{D}_{K-l} + v_K(j) \tilde{D}_1\right], i = K, r \le j \le c-1, \end{cases} \tag{29}$$

$$\hat{w}_i = \begin{cases} \frac{1}{\lambda_g} \sum_{l=0}^{L-1} z_l(1, c-1) \tilde{D}_{L-l}, i = 0, \\ \frac{1}{\lambda_g} \sum_{l=0}^{i-1} w_l D_{i-l}, 1 \le i \le K-1, \\ \frac{1}{\lambda_g} \left[\sum_{l=0}^{K-1} w_l \tilde{D}_{K-l} + w_K \tilde{D}_1\right], i = K, \end{cases} \tag{30}$$

$$\hat{z}_i(j,k) = \frac{1}{\lambda_g} \sum_{l=0}^{i-1} z_l(j,k) D_{i-l}, 1 \le i \le L-1, 1 \le j \le c-1, 0 \le k \le c-j. \tag{31}$$

3.3. The steady-state probability vector at an arrival of an arbitrary admitted customer in a batch: Suppose that $\bar{x}$ partitioned as $\bar{x} = (\bar{u}, \bar{v}, \bar{w}, \bar{z})$, denote the steady-state probability vector that the system will be in various states *immediately after* an arrival of an arbitrary admitted customer in a batch. Due to the random size of the batch, this vector is different from the one calculated in section 3.2. In the sequel, let

$$\hat{D}_{L-l} = (K-l)\tilde{D}_{K-l+1} + \sum_{k=L-l}^{K-l} kD_k,$$

the probability of customer loss be

$$\frac{1}{\lambda}\left\{\sum_{l=1}^{L-1}\left[\left(u_l + w_l + \sum_{j=1}^{c-1} v_l(j) + \sum_{j=1}^{c-r}\sum_{p=0}^{c-j} z_l(j,p)\right)\sum_{k=K-l+1}^{K}(k+l-K)D_k\right]\right.$$

$$\left.+\sum_{l=L}^{K}\left(w_l + \sum_{j=r}^{c-1} v_l(j)\right)\sum_{k=K-l+1}^{K}(k+l-K)D_k\right\} e \qquad (32)$$

and $\lambda_e = \lambda[1 - P(\text{customer loss})]$.
It can easily be verified that the components of $\bar{u}$, $\bar{v}$, $\bar{w}$ and $\bar{z}$ partitioned in an obvious manner are given by

$$\bar{u}_i = \frac{1}{\lambda_e}\sum_{l=0}^{i-1}(i-l)u_l D_{i-l}, 1 \le i \le L-1, \qquad (33)$$

$$\bar{v}_0(j) = \begin{cases} \frac{1}{\lambda_e}\sum_{l=0}^{L-1} u_l \hat{D}_{L-l}, j = 1, \\ \frac{1}{\lambda_e}\sum_{l=0}^{L-1} v_l(j-1)\hat{D}_{L-l}, 2 \le j \le r, \\ \frac{1}{\lambda_e}\sum_{l=0}^{L-1} z_l(1,j-1)\hat{D}_{L-l}, r+1 \le j \le c-1, \end{cases} \qquad (34)$$

$$\bar{v}_i(j) = \begin{cases} \frac{1}{\lambda_e}\sum_{l=0}^{i-1}(i-l)v_l(j)D_{i-l}, 1 \le i \le L-1, 1 \le j \le r-1, \\ \frac{1}{\lambda_e}\sum_{l=0}^{i-1}(i-l)v_l(j)D_{i-l}, 1 \le i \le K-1, r \le j \le c-1, \\ \frac{1}{\lambda_e}\left[\sum_{l=0}^{K-1}(K-l)v_l(j)\tilde{D}_{K-l}\right], i = K, r \le j \le c-1, \end{cases} \qquad (35)$$

$$\bar{w}_i = \begin{cases} \frac{1}{\lambda_e}\sum_{l=0}^{L-1} z_l(1,c-1)\hat{D}_{L-l}, i = 0, \\ \frac{1}{\lambda_e}\sum_{l=0}^{i-1}(i-l)w_l D_{i-l}, 1 \le i \le K-1, \\ \frac{1}{\lambda_e}\left[\sum_{l=0}^{K-1}(K-l)w_l\tilde{D}_{K-l}\right], i = K, \end{cases} \qquad (36)$$

$$\bar{z}_i(j,k) = \frac{1}{\lambda_e}\sum_{l=0}^{i-1}(i-l)z_l(j,k)D_{i-l}, 1 \le i \le L-1, 1 \le j \le c-1, 0 \le k \le c-j. \quad (37)$$

4 System Performance Measures

In this section we will list some important performance measures along with their formulas. These measures are used to bring out the qualitative behavior of the queueing model under study.

a. The Probability that the system is idle. The probability that at an arbitrary time the system is idle is given by

$$P(idle) = \boldsymbol{ue} + \sum_{i=1}^{c-r}\sum_{j=0}^{L-1} \boldsymbol{z}_j(i,0)\boldsymbol{e}. \quad (38)$$

b. The probability mass function of the number of servers on release. The probability that there are i servers on release at an arbitrary time is given by

$$a_i^{release} = \begin{cases} \sum_{k=0}^{c-i}\sum_{j=0}^{L-1} \boldsymbol{z}_j(i,k)\boldsymbol{e}, 1 \le i \le c-r, \\ 1 - \sum_{i=1}^{c-r}\sum_{k=0}^{c-i}\sum_{j=0}^{L-1} \boldsymbol{z}_j(i,k)\boldsymbol{e}, i = 0. \end{cases} \quad (39)$$

c. The probability mass function of the number of busy servers. The probability that i servers are busy at an arbitrary time is given by

$$a_i^{busy} = \begin{cases} \boldsymbol{ue} + \sum_{j=1}^{c-r}\sum_{k=0}^{L-1} \boldsymbol{z}_k(j,0)\boldsymbol{e}, i = 0, \\ \sum_{k=0}^{L-1} \boldsymbol{v}_k(i)\boldsymbol{e} + \sum_{j=1}^{c-i}\sum_{k=0}^{L-1} \boldsymbol{z}_k(j,i)\boldsymbol{e}, 1 \le i \le r-1, \\ \sum_{k=0}^{K} \boldsymbol{v}_k(i)\boldsymbol{e} + \sum_{j=1}^{c-i}\sum_{k=0}^{L-1} \boldsymbol{z}_k(j,i)\boldsymbol{e}, 1 \le r \le c-1, \\ \boldsymbol{we}, i = c. \end{cases} \quad (40)$$

d. The probability mass function of the number of available servers. The probability that at an arbitrary time there are i servers available in the system is given by

$$a_{r+i}^{available} = \begin{cases} ue + \sum_{j=1}^{r} v(j)e, i = 0, \\ v(r+i)e + \sum_{j=0}^{r} z(i,j)e + \sum_{l=1}^{i-1} z(l, r+i-1)e, 1 \le i \le c-r-1, \\ we + \sum_{j=0}^{r} z(c-r,j)e + \sum_{l=1}^{c-r-1} z(l, c-l)e, i = c-r. \end{cases} \tag{41}$$

e. The probability mass function of the number of customers waiting in the queue. The probability that at an arbitrary time i customers are waiting in the queue is given by

$$a_i^{queue} = \begin{cases} u_i e + w_i e + \sum_{j=1}^{c-1} v_i(j)e + \sum_{j=1}^{c-r}\sum_{k=0}^{c-j} z_i(j,k)e, 0 \le i \le L-1, \\ w_i e + \sum_{j=r}^{c-1} v_i(j)e, L \le i \le K. \end{cases} \tag{42}$$

f. The probability mass function of the number of lost customers. The probability that exactly i customers will be lost when an arriving batch finds insufficient buffer space for all its customers is given by (note that we assume that the batch size cannot exceed K, the maximum buffer size)

$$a_i^{lost} = \begin{cases} \frac{1}{\lambda_g} \sum_{l=i}^{L-1} \left[u_l + w_l + \sum_{j=1}^{c-1} v_l(j) + \sum_{j=1}^{c-r}\sum_{k=0}^{c-j} z_l(j,k) \right] D_{K-l+i} e, \\ \qquad 0 \le i \le L-1, \\ \frac{1}{\lambda_g} \sum_{l=i}^{K} \left[w_l + \sum_{j=r}^{c-1} v_l(j) \right] D_{K-l+i} e, L \le i \le K. \end{cases} \tag{43}$$

g. The throughput of the system. The throughput of the system,γ, defined as the number of customers per unit of time that leave the system, is easily calculated as

$$\gamma = (\lambda - \lambda_g \mu_{NL}), \tag{44}$$

where μ_{NL}, the expected number of lost jobs per batch, is obtained from the probability mass function given in (e) above.

h. The probability of a batch loss. The probability that an arriving batch will be *completely* lost is given by

$$P(loss) = \frac{1}{\lambda_g}\left[\sum_{j=r}^{c-1} \boldsymbol{v}_K(j) + \boldsymbol{w}_K\right]\tilde{D}_1. \tag{45}$$

i. The probability of a customer loss. The probability that an arriving customer in a batch will be lost is as given in (32).

j. The server utilization. The server utilization, ζ, is defined as the average fraction of busy servers in the system and is calculated as

$$\zeta = \frac{\mu_{BS}}{\mu_{AS}}, \tag{46}$$

where μ_{BS} and μ_{AS} are, respectively, the mean number of busy servers and the mean number of available servers and are obtained from their respective probability functions as given in (c) and (d).

Note that once the probability functions of various random variables are known, the corresponding means and variances can be calculated.

5 The Stationary Waiting Time Distribution

First note that due to the type of service mechanism considered in this paper, all customers in an admitted batch (either fully or partially up to the buffer capacity) will have the same waiting time distribution. Hence, we will refer to an admitted (full or partial) batch of customers as an admitted batch in the sequel. In this section we show that that distribution of the stationary waiting time, Y, of an admitted batch in the queue is of phase type. Before we prove this result, we need the following auxiliary matrices.

$$\hat{B}_0 = \begin{pmatrix} D_0 & D_1 & \cdots & D_{L-2} \\ 0 & D_0 & \cdots & D_{L-3} \\ \vdots & \vdots & \cdots & \vdots \\ 0 & 0 & \cdots & D_0 \end{pmatrix}, \tag{47}$$

for $r \le j \le c-1$,

$$\hat{A}_0(j) = \begin{pmatrix} D_0 & D_1 & \cdots & D_{L-1} & D_L & D_{L+1} & \cdots & \tilde{D}_{K-1} \\ 0 & D_0 & \cdots & D_{L-2} & D_{L-1} & D_L & \cdots & \tilde{D}_{K-2} \\ \vdots & \vdots & \cdots & \vdots & \vdots & \vdots & \cdots & \vdots \\ 0 & 0 & \cdots & D_0 & D_1 & D_2 & \cdots & \tilde{D}_{K-L+1} \\ 0 & 0 & \cdots & 0 & D_0-\theta I & D_1 & \cdots & \tilde{D}_{K-L} \\ 0 & 0 & \cdots & 0 & 0 & D_0-\theta I & \cdots & \tilde{D}_{K-L-1} \\ \vdots & \vdots & \cdots & \vdots & \vdots & \vdots & \cdots & \vdots \\ 0 & 0 & \cdots & 0 & 0 & 0 & \cdots & Q^{*}-\theta I \end{pmatrix} - j\mu I, \tag{48}$$

$$\hat{A}_0(c) = \begin{pmatrix} D_0 - c\mu I & D_1 & \cdots & D_{L-1} & D_L & D_{L+1} & \cdots & \tilde{D}_{K-1} \\ 0 & D_0 - c\mu I & \cdots & D_{L-2} & D_{L-1} & D_L & \cdots & \tilde{D}_{K-2} \\ \vdots & \vdots & \cdots & \vdots & \vdots & \vdots & \cdots & \vdots \\ 0 & 0 & \cdots & D_0 - c\mu I & D_1 & D_2 & \cdots & \tilde{D}_{K-L+1} \\ 0 & 0 & \cdots & 0 & D_0 - c\mu I & D_1 & \cdots & \tilde{D}_{K-L} \\ 0 & 0 & \cdots & 0 & 0 & D_0 - c\mu I & \cdots & \tilde{D}_{K-L-1} \\ \vdots & \vdots & \cdots & \vdots & \vdots & \vdots & \cdots & \vdots \\ 0 & 0 & \cdots & 0 & 0 & 0 & \cdots & Q^* - c\mu I \end{pmatrix}, \tag{49}$$

$$\hat{A}_2 = \begin{pmatrix} r\mu I_{(L-1)m} \\ 0_{(K-L+1)m \times (L-1)m} \end{pmatrix}, \tag{50}$$

for $0 \le i \le r-1$,

$$\hat{G}_i = \eta e_{i+1}(c) \otimes I, \tag{51}$$

for $r \le i \le c-1$,

$$\hat{G}_i = \eta e_{i+1}(c) \otimes [I_{(L-1)m}, 0_{(L-1)m \times (K-L+1)m}], \tag{52}$$

for $r+1 \le j \le c-1$,

$$\hat{A}_4(j) = j\mu \begin{pmatrix} 0_{(L-1)m \times (j-1)(L-1)m} & I_{(L-1)m \times (L-1)m} & 0_{(L-1)m \times (c-j)(L-1)m} \\ 0_{(K-L+1)m \times (j-1)(L-1)m} & 0_{(K-L+1)m \times (L-1)m} & 0_{(K-L+1)m \times (c-j)(L-1)m} \end{pmatrix}, \tag{53}$$

for $r \le j \le c-1$,

$$\hat{H}_0(j) = \begin{pmatrix} \hat{B}_0 & 0 & \cdots & 0 & 0 & 0 \cdots & 0 \\ \mu I & \hat{B}_0 - \mu I & \cdots & 0 & 0 & 0 \cdots & 0 \\ \vdots & \vdots & \vdots & \vdots & \vdots & \vdots \cdots & \vdots \\ 0 & 0 & \cdots & r\mu I & \hat{B}_0 - r\mu I & 0 \cdots & 0 \\ 0 & 0 & \cdots & 0 & B_0 - (r+1)\mu I & 0 \cdots & 0 \\ \vdots & \vdots & \cdots & \vdots & \vdots & \vdots \cdots & \vdots \\ 0 & 0 & \cdots & 0 & 0 & 0 \cdots & \hat{B}_0 - j\mu I \end{pmatrix} - (c-j)\eta I, \tag{54}$$

for $r+1 \le j \le c-1$,

$$\hat{H}_1(j) = \begin{pmatrix} 0 \cdots 0 & 0 & 0 & \cdots & 0 \\ \vdots \cdots \vdots & \vdots & \vdots & \cdots & \vdots \\ 0 \cdots 0 & 0 & 0 & \cdots & 0 \\ 0 \cdots 0 & (r+1)\mu I & 0 & \cdots & 0 \\ 0 \cdots 0 & 0 & (r+2)\mu I & \cdots & 0 \\ \vdots \cdots \vdots & \vdots & \vdots & \cdots & \vdots \\ 0 \cdots 0 & 0 & 0 & \cdots & j\mu I \end{pmatrix}, \tag{55}$$

for $r \le j \le c-2$,

$$\hat{H}_2(j) = \begin{pmatrix} (c-j)\eta I & 0 & 0 \cdots & 0 & 0 & 0 \cdots & 0 & 0 \\ 0 & (c-j)\eta I & 0 \cdots & 0 & 0 & 0 \cdots & 0 & 0 \\ \vdots & \vdots & \vdots \cdots & \vdots & \vdots & \vdots \cdots & \vdots & \vdots \\ 0 & 0 & 0 \cdots & (c-j)\eta I & 0 & 0 \cdots & 0 & 0 \\ 0 & 0 & 0 \cdots & 0 & (c-j)\eta I & 0 \cdots & 0 & 0 \\ \vdots & \vdots & \vdots \cdots & \vdots & \vdots & \vdots \cdots & \vdots & \vdots \\ 0 & 0 & 0 \cdots & 0 & 0 & 0 \cdots & (c-j)\eta I & 0 \end{pmatrix}, \quad (56)$$

$$M_{11} = \begin{pmatrix} \hat{B}_0 & 0 & 0 & \cdots & 0 & 0 & 0 & \cdots & 0 \\ \mu I & \hat{B}_0 - \mu I & 0 & \cdots & 0 & 0 & 0 & \cdots & 0 \\ 0 & 2\mu I & \hat{B}_0 - 2\mu I & \cdots & 0 & 0 & 0 & \cdots & 0 \\ \vdots & \vdots & \vdots & \cdots & \vdots & \vdots & \vdots & \cdots & \vdots \\ 0 & 0 & 0 & \cdots & \hat{B}_0 - (r-1)\mu I & 0 & 0 & \cdots & 0 \\ 0 & 0 & 0 & \cdots & \hat{A}_2 & \hat{A}_0(r) & 0 & \cdots & 0 \\ 0 & 0 & 0 & \cdots & 0 & 0 & \hat{A}_0(r+1) & \cdots & 0 \\ \vdots & \vdots & \vdots & \cdots & \vdots & \vdots & \vdots & \cdots & \vdots \\ 0 & 0 & 0 & \cdots & 0 & 0 & 0 & \cdots & \hat{A}_0(c) \end{pmatrix}, \quad (57)$$

$$M_{12} = \begin{pmatrix} 0 & 0 \cdots 0 \\ \vdots & \vdots \cdots \vdots \\ 0 & 0 \cdots 0 \\ \hat{A}_4(r+1) & 0 \cdots 0 \\ \vdots & \vdots \cdots \vdots \\ \hat{A}_4(c-1) & 0 \cdots 0 \end{pmatrix}, \quad (58)$$

$$M_{21} = \begin{pmatrix} \hat{G}_0 & \hat{G}_1 & \hat{G}_2 & \cdots & \hat{G}_{c-1} & 0 \\ 0 & 0 & 0 & \cdots & 0 & \\ \vdots & \vdots & \vdots & \cdots & \vdots & \vdots \\ 0 & 0 & 0 & \cdots & 0 & 0 \end{pmatrix}, \quad (59)$$

$$M_{22} = \begin{pmatrix} \hat{H}_0(c-1) & \hat{H}_1(c-1) & 0 & 0 \cdots & 0 & 0 & 0 \\ \hat{H}_2(c-2) & \hat{H}_2(c-2) & \hat{H}_1(c-2) & 0 \cdots & 0 & 0 & 0 \\ \vdots & \vdots & \vdots & \vdots \cdots & \vdots & \vdots & \vdots \\ 0 & 0 & 0 & 0 \cdots & \hat{H}_2(r+1) & \hat{H}_0(r+1) & \hat{H}_1(r+1) \\ 0 & 0 & 0 & 0 \cdots & 0 & \hat{H}_2(r) & \hat{H}_0(r) \end{pmatrix}. \quad (60)$$

Now we state the main result of this section.

Theorem 1: The distribution of the stationary waiting time, Y, of an admitted batch in the queue is of phase type with representation (ϕ, M) of order $[(L-1)r+(c-r+1)K+0.5(L-1)(c-r)(c+r+1)]m$, where ϕ and M are given by

$$\phi = \frac{1}{1-P(loss)}(\hat{\boldsymbol{u}}_1, \cdots, \hat{\boldsymbol{u}}_{L-1}, \hat{\boldsymbol{v}}_1(1), \cdots, \hat{\boldsymbol{v}}_{L-1}(1), \cdots, \hat{\boldsymbol{v}}_1(r-1), \cdots, \hat{\boldsymbol{v}}_{L-1}(r-1),$$

$$\hat{\boldsymbol{v}}_1(r), \cdots, \tilde{\boldsymbol{v}}_K(r), \cdots, \hat{\boldsymbol{v}}_1(c-1), \cdots, \tilde{\boldsymbol{v}}_K(c-1), \hat{\boldsymbol{w}}_1, \cdots, \tilde{\boldsymbol{w}}_K,$$

$$\hat{\boldsymbol{z}}_1(1,0), \cdots, \hat{\boldsymbol{z}}_{L-1}(1,0), \cdots, \hat{\boldsymbol{z}}_1(1,c-1), \cdots, \hat{\boldsymbol{z}}_{L-1}(1,c-1),$$

$$\hat{\boldsymbol{z}}_1(2,0), \cdots, \hat{\boldsymbol{z}}_{L-1}(2,0), \cdots, \hat{\boldsymbol{z}}_1(2,c-2), \cdots, \hat{\boldsymbol{z}}_{L-1}(2,c-2), \cdots$$

$$\hat{\boldsymbol{z}}_1(c-r,0), \cdots, \hat{\boldsymbol{z}}_{L-1}(c-r,0), \cdots, \hat{\boldsymbol{z}}_1(c-r,r), \cdots, \hat{\boldsymbol{z}}_{L-1}(c-r,r)),$$

and M is of the form

$$M = \begin{pmatrix} M_{11} & M_{12} \\ M_{21} & M_{22} \end{pmatrix}, \tag{61}$$

where the elements of ϕ are as given in (27-31) and the coefficient matrices appearing in M are as given in (57-60), and $\tilde{\boldsymbol{v}}_K(j), r \le j \le c-1$, and $\tilde{\boldsymbol{w}}_K$ are given by

$$\tilde{\boldsymbol{v}}_K(j) = \hat{\boldsymbol{v}}_K(j) - \frac{1}{\lambda_g}\boldsymbol{v}_K(j)\tilde{D}_1, r \le j \le c-1$$

$$\tilde{\boldsymbol{w}}_K = \hat{\boldsymbol{w}}_K - \frac{1}{\lambda_g}\boldsymbol{w}_K\tilde{D}_1.$$

Note: Note that the probability that an admitted customer will enter into service immediately is given by

$$P(Y=0) = 1 - \phi \boldsymbol{e} = \frac{1}{1-P(loss)}\left[\sum_{j=1}^{c-1} \hat{\boldsymbol{v}}_0(j) + \hat{\boldsymbol{w}}_0\right] \boldsymbol{e}.$$

Proof: Follows immediately by considering various scenarios that an *admitted* customer will see and the details are omitted.

Remark. Suppose that we are interested in the stationary waiting time, $\bar{Y}$, of an admitted arbitrary customer in a batch. This is also of phase type with representation $(\bar{\phi}, M)$ where the matrix M is as given in (61) and the vector $\bar{\phi}$ in terms of quantities defined in (33-37), is defined by

$$\bar{\phi} = (\bar{\boldsymbol{u}}_1, \cdots, \bar{\boldsymbol{u}}_{L-1}, \bar{\boldsymbol{v}}_1(1), \cdots, \bar{\boldsymbol{v}}_{L-1}(1), \cdots, \bar{\boldsymbol{v}}_1(r-1), \cdots, \bar{\boldsymbol{v}}_{L-1}(r-1),$$

$$\bar{\boldsymbol{v}}_1(r), \cdots, \bar{\boldsymbol{v}}_K(r), \cdots, \bar{\boldsymbol{v}}_1(c-1), \cdots, \bar{\boldsymbol{v}}_K(c-1), \bar{\boldsymbol{w}}_1, \cdots, \bar{\boldsymbol{w}}_K,$$

$$\bar{\boldsymbol{z}}_1(1,0), \cdots, \bar{\boldsymbol{z}}_{L-1}(1,0), \cdots, \bar{\boldsymbol{z}}_1(1,c-1), \cdots, \bar{\boldsymbol{z}}_{L-1}(1,c-1),$$

$$\bar{\boldsymbol{z}}_1(2,0), \cdots, \bar{\boldsymbol{z}}_{L-1}(2,0), \cdots, \bar{\boldsymbol{z}}_1(2,c-2), \cdots, \bar{\boldsymbol{z}}_{L-1}(2,c-2), \cdots$$

$$\bar{\boldsymbol{z}}_1(c-r,0), \cdots, \bar{\boldsymbol{z}}_{L-1}(c-r,0), \cdots, \bar{\boldsymbol{z}}_1(c-r,r), \cdots, \bar{\boldsymbol{z}}_{L-1}(c-r,r)).$$

The probability that an admitted arbitrary customer in a batch will enter into service immediately is given by

$$P(\bar{Y}=0) = \frac{1}{\lambda_e} \sum_{l=0}^{L-1} \left[\boldsymbol{u}_l + \sum_{p=1}^{r-1} \boldsymbol{v}_l(p) + \sum_{j=1}^{c-r} \sum_{p=0}^{c-j} \boldsymbol{z}_l(j,p) \right] \hat{D}_{L-l} \boldsymbol{e}$$

6 Numerical Examples

In this section we discuss some interesting numerical examples that qualitatively describe the performance of the queuing model under study. For the arrival process, we consider the following special class of $BMAP$. Let $D_k = Da_k$, for $1 \le k \le K$, where $\{a_k\}$ is a legitimate probability mass function. Thus, for this special $BMAP$ we need to specify the matrices D_0 and D of order m, and the probability function $\{a_k\}$. We consider the following five sets of values for D_0 and D.

1. Erlang (ERL):

$$D_0 = \begin{pmatrix} -2 & 2 \\ 0 & -2 \end{pmatrix}, \qquad D = \begin{pmatrix} 0 & 0 \\ 2 & 0 \end{pmatrix}$$

2. Exponential (EXP):

$$D_0 = (-1), \qquad D = (1)$$

3. Hyperexponential (HEX):

$$D_0 = \begin{pmatrix} -10 & 0 \\ 0 & -1 \end{pmatrix}, \qquad D = \begin{pmatrix} 9 & 1 \\ 0.9 & 0.1 \end{pmatrix}$$

4. BMAP with negative correlation (BN1):

$$D_0 = \begin{pmatrix} -2 & 2 & 0 \\ 0 & -2 & 0 \\ 0 & 0 & -450.50 \end{pmatrix}, \qquad D = \begin{pmatrix} 0 & 0 & 0 \\ 0.02 & 0 & 1.98 \\ 445.995 & 0 & 4.505 \end{pmatrix}$$

5. BMAP with positive correlation (BP1):

$$D_0 = \begin{pmatrix} -2 & 2 & 0 \\ 0 & -2 & 0 \\ 0 & 0 & -450.50 \end{pmatrix}, \qquad D = \begin{pmatrix} 0 & 0 & 0 \\ 1.98 & 0 & 0.02 \\ 4.505 & 0 & 445.995 \end{pmatrix}$$

6. BMAP with negative correlation (BN2):

$$D_0 = \begin{pmatrix} -45.5 & 45.5 & 0 & \cdots & 0 & 0 \\ 0 & -45.5 & 45.5 & \cdots & 0 & 0 \\ \vdots & \vdots & \vdots & \cdots & \vdots & \vdots \\ 0 & 0 & 0 & \cdots & -45.5 & 0 \\ 0 & 0 & 0 & \cdots & 0 & -4505 \end{pmatrix},$$

$$D = \begin{pmatrix} 0 & 0 \cdots 0 & 0 \\ 0 & 0 \cdots 0 & 0 \\ \vdots & \vdots \cdots \vdots & \vdots \\ 0.4505 & 0 \cdots 0 & 44.5995 \\ 4459.9500 & 0 \cdots 0 & 45.0500 \end{pmatrix}$$

7. BMAP with positive correlation (BP2):

$$D_0 = \begin{pmatrix} -45.5 & 45.5 & 0 & \cdots & 0 & 0 \\ 0 & -45.5 & 45.5 & \cdots & 0 & 0 \\ \vdots & \vdots & \vdots & \cdots & \vdots & \vdots \\ 0 & 0 & 0 & \cdots & -45.5 & 0 \\ 0 & 0 & 0 & \cdots & 0 & -4505 \end{pmatrix},$$

$$D = \begin{pmatrix} 0 & 0 \cdots 0 & 0 \\ 0 & 0 \cdots 0 & 0 \\ \vdots & \vdots \cdots \vdots & \vdots \\ 44.5995 & 0 \cdots 0 & 0.4505 \\ 45.0500 & 0 \cdots 0 & 4459.9500 \end{pmatrix}$$

All these seven $BMAP$ processes will be normalized so as to have a specific arrival rate λ in the numerical examples discussed below. However, these are qualitatively different in that they have different variance and correlation structure. The first three arrival processes correspond to renewal processes and so the correlation is 0. The arrival process labelled $BN1$ has correlated arrivals with a correlation value of -0.48891, and the arrivals corresponding to the process labelled $BP1$ has a positive correlation with a value of 0.48891. The arrival process labelled $BN2$ has correlated arrivals with a correlation value of -0.8015, and the arrivals corresponding to the process labelled $BP2$ has a positive correlation with a value of 0.8015. Since these arrival processes are normalized later on, the ratios of the variances of the six arrival processes, labelled 2 through 7 above, with respect to the Erlang process are, 1.4142, 3.1745, 1.9934, 1.9934, 0.0172, and 0.0172 respectively.

We take $a_k = d/K$, for $1 \leq k \leq K$ with d being the normalizing constant such that $\{a_k\}$ is a legitimate probability function.

Before we discuss some numerical examples, note that when either θ approaches ∞ or when η approaches 0, the performance measures for the current system will approach the corresponding performance measures of the system in which there is always a fixed number of servers, namely, c, available at any time.

Denote by, η^*(or θ^*), the values of η(or θ), such that for a given value of $\epsilon, 0 < \epsilon < 1$, the throughput as a function of η (or θ) is guaranteed to have at least (1-ϵ) of the throughput of the corresponding system in which all c servers are available at any given time. That is,

$$\gamma(\eta^*) \geq \gamma(0)\ (1 - \epsilon),$$
$$\gamma(\theta^*) \geq \gamma(\infty)\ (1 - \epsilon).$$

Thus, η^* gives the maximum release rate for which the throughput is guaranteed to achieve a given level of the throughput of the corresponding system in which all c servers are available at any time. Similarly, θ^* gives the minimum search rate for which the throughput is guaranteed to achieve a given level of the throughput of the corresponding system in which all c servers are available at any time.

Let ζ^* denote the server utilization at this optimal point and let ζ_c denote the server utilization of the system in which all c servers are available at any given time. Define the *server utilization improvement factor*, ζ_{SUI}, as

$$\zeta_{SUI} = \frac{\zeta^* - \zeta_c}{\zeta_c}.$$

Thus, larger positive values of ζ_{SUI} will indicate that variable server system is much better than the system in which all c servers are available at any given time.

In the following we take $K = 50$, $c = 15$, $L = 5$,$\mu = 0.1$, and $\lambda = 10$.

First we examine the behavior of ζ_{SUI} as a function of r, for the five arrival processes listed above by varying θ. We find the values of η^* such that the throughput for the system with a minimum of r servers is guaranteed to be at least 99% of the throughput of the system with 15 servers. The values of η^* are listed in Table 1. Next we vary η and find the values of θ^* such that the throughput for the system with a minimum of r servers is guaranteed to be at least 99% of the throughput of the system with 15 servers. The values of θ^* are listed in Table 2.

An examination of these tables reveals the following observations.

1. As is to be expected, for a fixed r, η^* appears to increase as θ is increased for all seven arrival processes.

2. Looking at the behavior for the first three arrival processes we notice that η^* appears to decrease with increasing variability of the process for all values of r and θ.

3. While for Erlang arrival process, η^* appears to increase as r increases for all values of θ, a different behavior is seen for the other arrival processes. For example, in the case of $BP1$ when $\theta = 2$ we notice that η^* decreases as r increases. While this seems counter intuitive, a possible explanation is that the server utilization decreases from 0.3134 to 0.2011 as r increases from 1 to 5. This might indicate that it is better to have a relatively larger release rate when the servers are not properly utilized.

4. In the case of Erlang arrivals, it is very interesting to note that in some cases ($\theta = 1.5, r = 5, \theta = 2.0, r = 4$ and $r = 5$) the server can be released immediately. For the same set of values, all other arrival processes have very low release rate.

5. For the $BP1$ arrival process the values of η^* appear to be much smaller compared to the other processes for all values of r and θ. Note that this arrival process has less variance compared to hyperexponential case.

6. For the $BP2$ arrival process the values of η^*, for all values of r considered, appear to be insensitive to θ. Furthermore, only for $r = 5$, η^* is larger compared to that of Erlang process. Note that the variance of $BP2$ process is much smaller than that of Erlang. This seems to indicate the dominance of the variance over correlation for large r.

7. Comparing $BN1$ and $BN2$ processes, there seems to be a trend w.r.t to η^* as a function of r. For example, r up to 4, η^* appears to be larger for

$BN1$ and for $r = 5$, it is reversed. This indicates that for larger value of r, correlation effect may not be that significant compared to the variance.

8. As is to be expected, for a fixed r, θ^* appears to decrease as η is increased for all seven arrival processes.

9. For all seven arrival processes, θ^* appears to be a monotonic (nonincreasing) function of r.

10. For all seven arrival processes, θ^* appears to converge (to values that depend on r as well as the arrival process)as η approaches ∞. The rate of convergence very much depends on r as well as the type of arrival processes. For example, the convergence seems to be faster for the $BP1$ case and slower for the $BN1$ case for all values of r. When comparing the arrival processes that have independent inter-arrival times (namely the first three arrival processes), θ^* appears to increase with increasing variance for all values of r and θ.

11. The interesting observation is that these values are very large for $BN1$ and $BN2$ arrival process compared to the other processes. It is worth mentioning that $BN2$ has the smallest variance, yet the performance measure behavior is much different from that of Erlang.

12. The observations listed in (3),(5),(6),(7),(10) and (11) indicate a significant role played by correlation in the arrival processes. Further careful analysis will shed more light on this largely neglected area in the literature.

We now examine the influence of the arrival processes, r, and θ on ζ_{SUI} by fixing $\eta = \eta^*$, where η^* is such that the throughput for the system with a minimum of r servers is guaranteed to be at least 99% of the throughput of the system with 15 servers. The data for ζ_{SUI} are given in Table 3. An examination of this table reveals that

- Depending on the value of r, there seems to be a cut-off point for θ, say, θ_0 such that for all values of $\theta > \theta_0$, Erlang arrivals tend have a larger value for ζ_{SUI} among the seven arrival processes. This value θ_0 appears to increase with r.

- For all values of r and for the first five arrival processes, ζ_{SUI}, appears to increase as θ increases. However, the situation is different for the other two arrival processes. Note that these two arrival processes have very small variance and a larger correlation compared to the other five processes. This further emphasizes the key role played by the correlation.

- For the range of θ values considered, it appears that Erlang arrivals yield consistently higher improvement factor compared to $BP1$ arrivals.
- The rate of improvement (as a function of θ) appears to be larger for $BP1$ arrivals compared to that of Erlang arrivals, for all r.

The following observation was noted in running several examples.

Observation: For any arrival process, there is a value of θ such that ζ_{SUI} is monotonically nonincreasing in r. This can be intuitively explained as follows. If the search rate is large enough then having a smaller number of servers will definitely increase the server utilization factor. However the value of θ depends on the arrival process. For example, in the case of the first three arrival processes (renewal processes), the value of θ increases with increasing variance.

Table 1: Relationship between η^*, r, θ, and the arrival processes

θ	r	ERL	EXP	HEX	BN1	BP1	BN2	BP2
0.5	1	0.0086	0.0038	0.0030	0.0028	0.0001	0.008	0.002
	2	0.0086	0.0038	0.0031	0.0028	0.0002	0.008	0.002
	3	0.0096	0.0041	0.0032	0.0029	0.0002	0.008	0.002
	4	0.0140	0.0050	0.0033	0.0040	0.0002	0.011	0.011
	5	0.0390	0.0075	0.0033	0.0050	0.0002	0.101	0.101
1.0	1	0.0292	0.0090	0.0071	0.0052	0.0004	0.013	0.002
	2	0.0302	0.0091	0.0071	0.0053	0.0006	0.013	0.002
	3	0.0373	0.0098	0.0072	0.0054	0.0004	0.015	0.002
	4	0.0750	0.0130	0.0072	0.0070	0.0004	0.021	0.011
	5	3.2830	0.0230	0.0072	0.0090	0.0004	0.101	0.101
1.5	1	0.0800	0.0200	0.0110	0.0080	0.0040	0.016	0.002
	2	0.0900	0.0200	0.0110	0.0080	0.0010	0.016	0.002
	3	0.1200	0.0200	0.0100	0.0080	0.0010	0.018	0.002
	4	0.4900	0.0300	0.0100	0.0090	0.0008	0.026	0.011
	5	∞	0.0650	0.0110	0.0140	0.0008	0.101	0.101
2.0	1	0.1800	0.0250	0.0160	0.0090	0.0068	0.017	0.002
	2	0.2100	0.0260	0.0145	0.0090	0.0024	0.017	0.002
	3	0.4000	0.0300	0.0141	0.0090	0.0017	0.020	0.002
	4	∞	0.0460	0.0141	0.0110	0.0014	0.029	0.011
	5	∞	0.1910	0.0147	0.0170	0.0013	0.112	0.101

Table 2: Relationship between θ, r, η^*, and the arrival processes

η	r	ERL	EXP	HEX	BN1	BP1	BN2	BP2
0.50	1	2.700	8.50	14.30	750.20	11	1482	18.06
	2	2.500	7.70	13.10	637.00	11	1275	17.31
	3	2.100	6.20	11.40	463.70	10	875	16.03
	4	1.508	4.25	9.33	278.30	9	424	14.50
	5	0.835	2.45	7.09	123.81	9	86	13.00
1.00	1	2.97	9.99	16.7	981	11.1	2030	18.3
	2	2.78	8.98	15.3	827	10.6	1720	17.5
	3	2.33	7.13	13.3	598	9.9	1160	16.1
	4	1.66	4.85	10.9	358	9.0	559	14.6
	5	0.91	2.75	8.2	164	8.1	131	13.0
1.50	1	3.13	10.74	17.0	1088	11.1	2269	19
	2	2.92	9.63	16.4	917	10.6	1914	18
	3	2.44	7.61	14.3	662	9.9	1284	17
	4	1.74	5.16	11.7	397	9.0	618	15
	5	0.95	2.90	8.8	184	8.1	150	13
100000	1	3.60	13.00	22.3	1360	11.2	2800	19
	2	3.30	11.70	20.5	1150	10.7	2350	18
	3	2.80	9.20	17.9	840	9.9	1560	17
	4	1.95	6.12	14.6	510	9.0	750	15
	5	1.07	3.41	11.0	240	8.1	200	13

Table 3: Relationship between θ, r, ζ_{SUI}, and the arrival processes

θ	r	ERL	EXP	HEX	BN1	BP1	BN2	BP2
0.50	1	1.441	1.043	0.391	0.927	0.117	1.467	1.238
	2	1.473	1.063	0.407	0.953	0.199	1.502	1.441
	3	1.507	1.098	0.440	0.986	0.203	1.506	1.592
	4	1.553	1.158	0.474	1.078	0.188	1.535	2.098
	5	1.615	1.180	0.509	1.091	0.170	1.738	1.724
1.00	1	1.798	1.223	0.488	1.014	0.291	1.545	1.098
	2	1.826	1.253	0.522	1.042	0.356	1.577	1.224
	3	1.857	1.290	0.562	1.079	0.265	1.612	1.316
	4	1.918	1.340	0.630	1.149	0.242	1.630	2.005
	5	1.846	1.378	0.671	1.154	0.220	1.716	1.725
1.50	1	2.161	1.473	0.608	1.101	0.969	1.583	1.017
	2	2.187	1.507	0.650	1.141	0.481	1.612	1.103
	3	2.191	1.496	0.683	1.165	0.454	1.634	1.157
	4	2.199	1.550	0.732	1.185	0.370	1.657	1.939
	5	1.863	1.555	0.789	1.225	0.336	1.705	1.728
2.0	1	2.433	1.547	0.726	1.121	1.255	1.586	0.957
	2	2.447	1.590	0.777	1.161	0.768	1.614	1.019
	3	2.467	1.618	0.813	1.182	0.614	1.648	1.049
	4	2.323	1.655	0.853	1.223	0.512	1.669	1.884
	5	1.869	1.683	0.883	1.253	0.447	1.708	1.731

References

1. Andrews, B.H., and Parsons, H.L., Establishing telephone agent staffing levels through economic optimization, Interfaces, **23**(1993), 14-20.
2. Bell, C.E., Optimal Operation of an M/M/2 queue with removable servers. Operations Research, **28**(1980), 1189-1204.
3. Bellman, R.E., Introduction to Matrix Analysis, McGraw Hill, New York, 1960.
4. Chakravarthy, S., A finite capacity $GI/PH/1$ queue with group services. Naval Research

Logistics Quarterly, **39**(1992), 345-357.

5. Chakravarthy, S., Analysis of a finite $MAP/G/1$ queue with group services. Queueing Systems, **13**(1993), 385-407.
6. Chakravarthy, S., A finite capacity queueing network with single and multiple processing nodes. In R.F. Onvural and I.F. Akyildiz, editors, Queueing Network with Finite Capacity: North-Holland, Netherlands, (1993), 197-211.
7. Chakravarthy, S., Two finite queues in series with nonrenewal input and group services. In Proceedings of the Seventh International Symposium on Applied Stochastic Models and Data Analysis (1995), 78-87.
8. Chakravarthy, S., Analysis of the $MAP/PH/1/K$ queue with service control. Applied Stochastic Models and Data Analysis, **12**(1996), 179-191.
9. Chakravarthy, S.R., Analysis of a priority polling system with group services. Commun. Statist. Stochastic Models, **14**(1998), 25-49.
10. Chakravarthy, S.R., Analysis of a multi-server queue with batch Markovian arrivals and group services. Engineering Simulation, **18**(2000), 51-66.
11. Chakravarthy, S., and A.S. Alfa., A finite capacity queue with Markovian arrivals and two servers with group services. J. of Appl. Math. and Stochastic Analysis, **7**(1994), 161-178.
12. Chakravarthy, S., and L. Bin., A finite capacity queue with nonrenewal input and exponential dynamic group services. INFORMS Journal on Computing, **9**(1997), 276-287.
13. Chakravarthy, S., and S.Y. Lee., An optimization problem in a finite capacity $PH/PH/1$ queue with group services. In et al G.V. Krishna Reddy, editor, Stochastic Models, Optimization Techniques and Computer Applications, (1994), 3-13.
14. Chakravarthy, S.R., The batch Markovian arrival process: A review and future work. To appear in A. Krishnamoorthy, et al. Advances in Probability Theory and Stochastic Processes, 2001.
15. Dudin,A.N., and E. Khalaf., Optimizing the dynamic reorganization of multichannel data bases in computer networks. Automatic Control and Computer Science, **3**(1992),50-55.
16. Grassman, W.K., Finding the right number of servers in real-world queueing systems, Interfaces, **18**(1988), 94-104.
17. Jennings, O.B., Mandelbaum, A., Massey, W.A., and Whitt, W., Server staffing to meet time-varying demand. The second INFORMS Telecommunication Conference, Boca Raton, Florida, USA (1995), 24-26.
18. Li, H., and Yang, T., Queues with a variable number of servers, European J. Operational Research, **124**(2000), 615-628.
19. Lucantoni,D.M., New results on the single server queue with a batch markovian arrival process, Communications in Statistics-Stochastic Models, **7**(1991), 1-46.
20. Neuts, M.F., Matrix-Geometric Solutions in Stochastic Models - An Algorithmic Approach. Dover Publications, 1995 (originally published by Johns Hopkins University Press, 1981).
21. Neuts, M.F., Structured Stochastic Matrices of M/G/1 type and their applications. Marcel Dekker, 1989.
22. Okamura, H., Dohi, T., and Osaki, S. Optimal policies for a controlled queueing system with removable server under a random vacation circumstance, Computers and Mathematics with Applications, **39**(2000), 215-227.
23. Wang, K.H., and Hsieh, W.F., Optimal control of a removable and non-reliable server in a Markovian queueing system with finite capacity, Microelectronics and Reliability, **35**(1995), 189-196.
24. Wang, Kuo-Hsiung., Optimal operation of a Markovian queueing system with a removable and non-reliable server, Microelectronics and Reliability, **35**(1995),1131-1136.
25. Wang, Kuo-Hsiung., Chang, Kuan-Wen., and Sivazlian, B.D., Optimal control of a removable and non-reliable server in an infinite and a finite M/H2/1 queueing system, Applied Mathematical Modelling, **23**(1999), 651-666.
26. Wang, Kuo-Hsiung., and Ke, Jau-Chuan., Recursive method to the optimal control of an M/G/1 queueing system with finite capacity and infinite capacity, Applied Mathematical Modelling, **24**(2000), 899-914.
27. Zhang, R., Phillis, Y.A., and Zhu, X., Fuzzy control of queueing systems with removable servers, Proceedings of the 1998 IEEE International Conference on Systems, Man, and Cybernetics, 2160-2165.

FURTHER RESULTS ON THE SIMILARITY BETWEEN FLUID QUEUES AND QBDS

ANA DA SILVA SOARES AND GUY LATOUCHE
Université Libre de Bruxelles, Département d'Informatique, Boulevard du Triomphe, CP 212, 1050 Bruxelles, Belgium
E-mail: andasilv@ulb.ac.be, Guy.Latouche@ulb.ac.be

We consider an infinite buffer fluid queue driven by a Markovian environment and we analyze it by matrix analytic methods. We extend the recent work of Ramaswami by giving a more direct algorithm for the computation of the stationary distribution, as well as a clear probabilistic interpretation of this algorithm. We conclude the paper by a brief presentation of some numerical examples.

1 Introduction

We consider a Markov modulated fluid queue, that is, a two-dimensional continuous-time Markov process $\{(X(t), \varphi(t)) : t \in \mathbb{R}^+\}$ where $X(t)$ takes values in $\mathbb{R}^+$ and $\varphi(t)$ in $\mathcal{S}$, a finite set. The component $X(\cdot)$ is called the *level* and $\varphi(\cdot)$ is called the *phase*. The level is subordinated to the phase in the following way. The phase process $\{\varphi(t) : t \in \mathbb{R}^+\}$ is an irreducible Markovian process with transition generator T. During those intervals of time in which the phase is constant, say equal to i, the level increases or decreases at a constant rate which depends on i, or it remains constant. If $X(t) = 0$ and if the rate at time t is negative, then the level remains at 0.

Ramaswami [7] shows that $\{X(t)\}$ has a phase-type stationary distribution: first, he proves by a ladder-height argument that the distribution has a matrix exponential form, then, he uses the dual process of $\{(X(t), \varphi(t))\}$ to show that the distribution actually is PH. Most importantly, he also constructs a very efficient computational procedure, based on the logarithmic-reduction algorithm of Latouche and Ramaswami [2,3] for discrete-level QBDs; he thereby reduces a complex continuous time, continuous state space problem to a familiar, simple discrete time, discrete state space system.

The use of the dual process in Ramaswami [7] is motivated by a property which relates the stationary distribution of the original process $\{(X(t), \varphi(t))\}$ to first passage probabilities at level 0 in the dual process. Our purpose in the present paper is to reinforce the similarities with QBDs by showing that one may actually *directly* use these first passage probabilities in the original process. We also give another probabilistic interpretation of Ramaswami's computational procedure.

We do not extensively review the literature but we mention two references, Rogers [8] and Asmussen,[1] which are particularly relevant in our context. Among other things, Rogers shows that the stationary distribution is matrix-exponential but he does not explore algorithmic issues; we shall return to this at the end of Section 4. Asmussen also shows that $\{X(t)\}$ has a phase-type stationary distribution. His argument is based on the dual process of the Markov modulated random walk $\{(Y(t), \varphi(t)) : t \in \mathbb{R}^+\}$ governed by the same dynamics as $\{(X(t), \varphi(t))\}$, with the difference that $Y(t)$ is allowed to be negative. He also gives an algorithm to compute a representation of that distribution. The reader will find interesting references in these two papers, as well as in Ramaswami [7] and Sericola and Tuffin.[9]

The paper is organized as follows. We precisely define the stochastic process in the next section and we recall some basic properties from the literature. In Section 3 we show how the stationary distribution of the process is related to first passage probabilities to the level 0 for the *same* process. We adapt in Section 4 the computational procedure from Ramaswami [7] and we give it a probabilistic interpretation. We show in Section 5 that the stationary distribution of the level is PH and we give a few numerical examples.

2 Background

The level $X(t)$ represents the content at time t of an infinite capacity fluid buffer for which the input and output rates are determined by the environmental phase process. The fluid level increases or decreases linearly, or remains constant, as long as the phase remains constant.

The phase process $\{\varphi(t) : t \in \mathbb{R}^+\}$ is assumed to be irreducible, with infinitesimal generator T and stationary probability vector $\boldsymbol{\xi}$; by this, we mean that $\boldsymbol{\xi}$ is the unique solution of the system $\boldsymbol{\xi}T = \mathbf{0}$, $\boldsymbol{\xi}\mathbf{1} = 1$, where $\mathbf{0}$ and $\mathbf{1}$ respectively denote a vector of zeros and a vector of ones. The context will indicate whether these are row or column vectors.

During an interval of time when $\varphi(\cdot)$ is constant and equal to i, the net rate of input to the fluid buffer is given by r_i, with

$$
\begin{aligned}
r_i &= 0 \text{ if } i \in \mathcal{S}_0, \\
r_i &> 0 \text{ if } i \in \mathcal{S}_1, \\
r_i &< 0 \text{ if } i \in \mathcal{S}_2,
\end{aligned}
$$

where $\mathcal{S} = \mathcal{S}_0 \cup \mathcal{S}_1 \cup \mathcal{S}_2$ is the finite discrete state space of the phase process. We denote by $\mathcal{S}_\bullet = \mathcal{S}_1 \cup \mathcal{S}_2$ the subset of phases for which the input rate is not zero. The sizes of $\mathcal{S}$, $\mathcal{S}_0$, $\mathcal{S}_1$ and $\mathcal{S}_2$ are denoted by s, s_0, s_1 and s_2, respectively.

The fluid model is clearly a two-dimensional Markov process $\{(X(t), \varphi(t)) : t \in \mathbb{R}^+\}$ on the state space $\{(x,i) : x \in \mathbb{R}^+, i \in \mathcal{S}\}$. For fixed t, denote by $f_j(x;t)$ the density of $(X(t), \varphi(t))$ evaluated at (x,j); thus, $\mathrm{P}[X(t) \in dx, \varphi(t) = j] = f_j(x;t)dx$, for $j \in \mathcal{S}$, and $x \geq 0$. Further, denote by $\boldsymbol{\pi}(x) = (\pi_i(x) : i \in \mathcal{S})$, $x \in \mathbb{R}^+$ the stationary density vector defined by $\pi_i(x) = \lim_{t\to\infty} f_i(x;t)$. It exists if and only if the stationary net rate of input is negative, that is, if and only if $\boldsymbol{\xi r} < 0$, where $\boldsymbol{r}$ is the column vector with entries r_i for $i \in \mathcal{S}$, as we assume from now on.

The density functions satisfy the system of partial differential equations (see Ramaswami,[7] for example)

$$\frac{\partial}{\partial t} f_j(x;t) = \sum_{i\in\mathcal{S}} f_i(x;t)T_{ij} - r_j \frac{\partial}{\partial x} f_j(x;t),$$

for $x > 0$ and for all $j \in \mathcal{S}$. Letting $t \to +\infty$, we obtain the steady state equations

$$-r_j \frac{d}{dx}\pi_j(x) + \sum_{i\in\mathcal{S}} \pi_i(x)T_{ij} = 0, \tag{1}$$

for $x > 0$ and for all $j \in \mathcal{S}$.

We separate the phases in $\mathcal{S}_0$ from the others and partition the infinitesimal generator T and the stationary density vector $\boldsymbol{\pi}(x)$ in the obvious manner:

$$T = \begin{bmatrix} T_{\bullet\bullet} & T_{\bullet 0} \\ T_{0\bullet} & T_{00} \end{bmatrix}$$

and

$$\boldsymbol{\pi}(x) = [\boldsymbol{\pi}_\bullet(x), \boldsymbol{\pi}_0(x)].$$

The equations (1) become

$$\frac{d}{dx}\boldsymbol{\pi}_\bullet(x)C = \boldsymbol{\pi}_\bullet(x)T_{\bullet\bullet} + \boldsymbol{\pi}_0(x)T_{0\bullet}$$

$$0 = \boldsymbol{\pi}_\bullet(x)T_{\bullet 0} + \boldsymbol{\pi}_0(x)T_{00},$$

where C is the diagonal matrix $\mathrm{diag}(r_i : i \in \mathcal{S}_\bullet)$. Since T is irreducible, T_{00} is nonsingular and we immediately have the following property.

Proposition 2.1 *For $x > 0$, the stationary density vector $\boldsymbol{\pi}(x)$ for the buffer content of the fluid model is a solution of the equations*

$$\frac{d}{dx}\boldsymbol{\pi}_\bullet(x)C = \boldsymbol{\pi}_\bullet(x)T^*, \tag{2}$$

$$\boldsymbol{\pi}_0(x) = \boldsymbol{\pi}_\bullet(x)T_{\bullet 0}(-T_{00})^{-1}, \tag{3}$$

where

$$T^* = T_{\bullet\bullet} + T_{\bullet 0}(-T_{00})^{-1}T_{0\bullet}.$$

□

Observe that T^* is the generator of the process $\{\varphi(t) : t \in \mathbb{R}^+\}$ embedded at visits to phases in $\mathcal{S}_\bullet$.

We give in the next theorem an expression for the steady state density of the buffer content, it essentially consists of a restatement of Theorem 2.1, Corollary 2.2 in Ramaswami.[7] Our expression (4) below is slightly different because it is assumed there that $\mathcal{S}_0$ is empty, that is, there are no phases with zero net input rates. Here, we need to factor in the normalizing constant $\boldsymbol{\xi}_\bullet \mathbf{1}$. See also Sericola and Tuffin.[9]

Theorem 2.2 *The stationary distribution of the system has a mass at the level zero and a continuous density for strictly positive values.*

There exist a matrix K of order s_1 and a matrix Ψ with dimensions $s_1 \times s_2$ such that the stationary density vector is given by

$$\boldsymbol{\pi}_\bullet(x) = (\boldsymbol{\xi}_\bullet \mathbf{1})(-\boldsymbol{\xi}_1 K)[\exp(Kx), \exp(Kx)\Psi], \qquad \text{for } x > 0, \qquad (4)$$

where the vectors $\boldsymbol{\xi}_\bullet$, $\boldsymbol{\xi}_1$ and $\boldsymbol{\xi}_2$ are the row vectors respectively containing the components ξ_j for $j \in \mathcal{S}_\bullet$, $j \in \mathcal{S}_1$ and $j \in \mathcal{S}_2$. The expressions for K and Ψ will be given later.

Denote by $\boldsymbol{\beta}_\bullet = (\beta_i : i \in \mathcal{S}_\bullet)$ the stationary probability mass vector of the empty buffer. It is given by

$$\boldsymbol{\beta}_\bullet = (\boldsymbol{\xi}_\bullet \mathbf{1})(\mathbf{0}, \boldsymbol{\xi}_2 - \boldsymbol{\xi}_1 \Psi).$$

□

Ramaswami proves this by following a level-crossing argument first given for discrete-level QBDs in Ramaswami [6] and also presented in Section 5.2 of Latouche and Ramaswami.[3] The starting point is to calculate the probability of being at some level $x + y$ at time t by conditioning on the last epoch of visit to the level x; the details are to be found in Ramaswami.[7]

It is well known (Neuts,[4] Latouche and Ramaswami [3]) that for a continuous-time QBD on the *discrete* state space $\mathbb{N} \times \mathcal{S}$, the stationary density vectors $\boldsymbol{\pi}_n = (\pi_{n,i} : i \in \mathcal{S})$ are of the form $\boldsymbol{\pi}_n = \boldsymbol{\pi}_0 R^n$, for $n \geq 0$. As pointed out in Ramaswami,[7] the matrices $\exp(Kx)$ and $\exp(Kx)\Psi$ in (4) play the role of the matrix sequence $\{R^n : n \geq 1\}$ in the matrix-geometric solution, because the (i,j)th entry of the matrix $\exp(Kx)$ gives the average number of visits to state (x, j), before returning to level 0, given that the initial state is $(0, i)$, with i and j in $\mathcal{S}_1$ and similarly, the (i,j)th entry of the matrix

$\exp(Kx)\Psi$ records the same expected number of visits, for i in $\mathcal{S}_1$ and j in $\mathcal{S}_2$.

For continuous-time QBDs, the matrix G of first passage probabilities from the level 1 to the level 0 also plays a prominent role. The rate matrix R is related to G in two different ways. On the one hand, we have that

$$R = A_0(-A_1 - A_0 G)^{-1} \tag{5}$$

(equations (6.9, 6.11) in Latouche and Ramaswami [3]). On the other hand, it is shown in Ramaswami [5] that

$$R = \Delta^{-1}\tilde{G}'\Delta, \tag{6}$$

where $\tilde{G}$ is the first passage probability matrix of the dual process obtained by time-reversal, $\Delta = \text{diag}(\boldsymbol{\alpha})$ and $\boldsymbol{\alpha}$ is the stationary probability vector of the stochastic matrix $A = A_0 + A_1 + A_2$, and we denote by M' the transpose of M.

The equations developed in Ramaswami [7] are the analog of (6) for fluid queues and they relate $\exp(Kx)$ to the first passage probability matrix $G(\cdot)$ of the dual process. By contrast, our equations (7, 9) below form the analog of (5) and deal with the given process only.

3 First Passage Probabilities

If the net rates of input into the buffer are different from zero, that is, if $\mathcal{S}_0$ is empty, then the differential equations (1) may be written as

$$\frac{d}{dx}\pi_j(x)|r_j|\delta_j + \sum_{i\in\mathcal{S}} \pi_i(x) r_i r_i^{-1} T_{ij} = 0$$

for all $j \in \mathcal{S}$, $x > 0$, where δ_j is equal to -1 if $j \in \mathcal{S}_1$, and to $+1$ if $j \in \mathcal{S}_2$. Denote by $|C|$ the diagonal matrix $\text{diag}(|r_i| : i \in \mathcal{S})$. Clearly, the vector $\boldsymbol{\pi}(x)|C|$ is proportional to the stationary density vector for the fluid queue with infinitesimal generator $|C|^{-1}T$ and net input rates equal to 1 or -1.

We restrict ourselves to this special model, and temporarily assume that the r_i's are all equal to 1 or -1. As shown above, this is not a restrictive assumption and it will make it easier for us to develop our argumentation. We show in Section 5 how to return to the general setting. Note that the same trick is used in Rogers.[8]

If the r_i's are all equal to 1 or -1, then $T = T_{\bullet\bullet} = T^*$ and, decomposing the rate matrix in a manner conformant to the partition $\mathcal{S}_\bullet = \mathcal{S}_1 \cup \mathcal{S}_2$, we write that

$$T = T^* = \begin{bmatrix} T_{11} & T_{12} \\ T_{21} & T_{22} \end{bmatrix}.$$

By Ramaswami,[7] the matrices K and Ψ are given by

$$K = T_{11} + \Psi T_{21} \tag{7}$$

and

$$\Psi = \int_0^\infty \exp(Ky) T_{12} \exp(T_{22} y) dy. \tag{8}$$

Define $G(x)$ as the matrix of first passage probabilities from the level x to the level 0. More precisely, with $\theta = \inf\{t > 0 : X(t) = 0\}$ being the first return time to the level 0, we define

$$G_{ij}(x) = \mathrm{P}[\theta < \infty \text{ and } \varphi(\theta) = j | X(0) = x, \varphi(0) = i].$$

Clearly, $\varphi(\theta)$ necessarily belongs to $\mathcal{S}_2$ and $G(x)$ has the following structure:

$$G(x) = \begin{bmatrix} 0 & G_{12}(x) \\ 0 & G_{22}(x) \end{bmatrix}, \qquad \text{for } x \geq 0.$$

We now state our main result which, together with (7), establishes a direct connection between the matrix K and the matrices of first passage probabilities to the level 0. By contrast, the relation in Ramaswami [7] is between $\exp(Kx)$ and the first passage probability matrices of the *dual* fluid queue obtained by time reversal.

Theorem 3.1 *The matrix Ψ is given by*

$$\Psi = \int_0^\infty \exp(T_{11} y) T_{12} G_{22}(y) dy; \tag{9}$$

it is equal to $\lim_{x \to 0} G_{12}(x)$, and for i in $\mathcal{S}_1$ and j in $\mathcal{S}_2$, Ψ_{ij} is the probability that, starting from $(0, i)$ at time 0, the fluid queue returns to the level 0 at some time θ, with $0 < \theta < \infty$, and does so in phase j.

Proof We proceed in a manner similar to the approach in Ramaswami [7] albeit with a slight difference in focus. We assume throughout that $X(0) = 0$.

For $x > 0$ and j in $\mathcal{S}_2$, we have that $(X(t), \varphi(t)) = (x, j)$ if and only if at time t the fluid queue crosses the level x from above, at phase j. For this event to occur, it is necessary that there exist some $\tau < t$ and i in $\mathcal{S}_1$ such that

- at time $t - \tau$ the fluid queue is in state (x, i), and

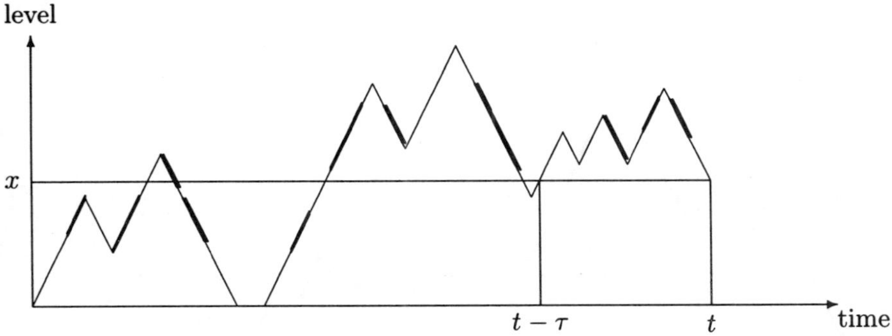

Figure 1. The last epoch at which the process crosses the level x before t.

- in $(t-\tau, t)$, it continuously remains above the level x.

We illustrate this on Figure 1 which is to be interpreted as follows. The piecewise linear curve shows how the level evolves in time; we assume that there are four phases in all, with $\mathcal{S}_1 = \{1,2\}$ and $\mathcal{S}_2 = \{3,4\}$. The graph is drawn with a thin line when the phase is 1 or 3, with a thick line otherwise. We see that the process is in phase 1 at time 0, then it jumps to phase 2, then to phase 3, at which time the level begins to decrease, etc.

Hence, we may write that

$$f_j(x;t) = \int_0^t \sum_{i \in \mathcal{S}_1} f_i(x;t-\tau)\Gamma_{ij}(x;d\tau) \tag{10}$$

where $\Gamma_{ij}(x;t)$ is a probability of first return to the level x at or before time t; precisely, $\Gamma_{ij}(x;t)$ is the conditional probability that there exists t' with $0 < t' \le t$ such that $X(h) > x$ for $0 < h < t'$ and $(X(t'), \varphi(t')) = (x, j)$, given that $(X(0), \varphi(0)) = (x, i)$.

When the level of the buffer is strictly positive, the rate of increase or decrease is independent of its specific value, thus, the probabilities $\Gamma_{ij}(x;t)$ do not depend on x for $x > 0$ and we omit that parameter, writing (10) in vector notation as

$$\boldsymbol{f}_2(x;t) = \int_0^t \boldsymbol{f}_1(x;t-\tau)\Gamma(d\tau), \qquad \text{for } x > 0. \tag{11}$$

For fixed i and x, $f_i(x;t)$ is a continuous function of t, converging to $\pi_i(x)$ as t goes to infinity. It is, therefore, uniformly bounded and we may take in (11) the limit as $t \to \infty$, to find that

$$\boldsymbol{\pi}_2(x) = \boldsymbol{\pi}_1(x)\Gamma,$$

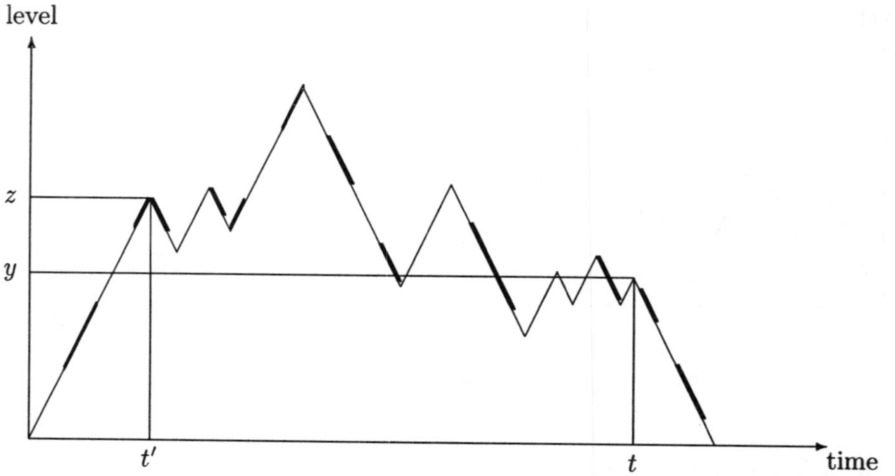

Figure 2. Conditioning on the beginning of the last downturn or on the end of the first slope upward.

where $\Gamma_{ij} = \int_0^\infty \Gamma(d\tau) = \lim_{t\to\infty} \Gamma_{ij}(x,t)$, independently of $x > 0$.

A moment reflection shows that $\Gamma = \lim_{x\to 0} G_{12}(x)$. In order to prove the theorem, we need to show that $\Gamma = \Psi$ and also that Γ is given by (9).

Starting in $(0, \mathcal{S}_1)$ at time 0, the queue returns to the level 0 at a time which is positive and finite if and only if the following event holds (see Figure 2 for an illustration): there is a time t and a level y such that

- $X(h) > 0$ for $0 < h < t$,
- $X(t) = y$ and $\varphi(t)$ is in $\mathcal{S}_1$,
- at time t the phase changes to $\mathcal{S}_2$,
- after time t, the phase remains in $\mathcal{S}_2$ at least until the level returns to 0.

Since $r_i = -1$ for all i in $\mathcal{S}_2$, it takes exactly y units of time for the fluid queue to become empty if it starts in $(y, \mathcal{S}_2)$ and continuously remains in $\mathcal{S}_2$. Therefore, we find that

$$\Gamma = \int_0^\infty [\int_0^\infty F_{11}^*(y;t)dt] T_{12} \exp(T_{22}y) dy,$$

where $(F_{11}^*(y;t))_{ij}$ $(i, j \in \mathcal{S}_1)$ is for fixed t the conditional density of $(X(t), \varphi(t))$ evaluated at (y, j), given that the initial state is $(0, i)$ and that the process avoids the level 0 in the interval $(0, t)$.

The integral $\int_0^\infty F_{11}^*(y;t)dt$ is, therefore, the expected number of visits to (y,j) before returning to the level 0 and is equal to $\exp(Ky)$ by Ramaswami,[7] equation (2.5), so that

$$\Gamma = \int_0^\infty \exp(Ky) T_{12} \exp(T_{22} y) dy$$

which, by equation (2.10) in Ramaswami,[7] proves that $\Gamma = \Psi$.

To complete the proof, we need to show that Γ is given by (9). We condition on the first transition from $\mathcal{S}_1$ to $\mathcal{S}_2$, instead of the last, as we did earlier. Starting in $(0, \mathcal{S}_1)$, the queue returns to the level 0 in a finite time if and only if there exist a time t' and a level z such that

- $\varphi(h)$ is in $\mathcal{S}_1$ for $0 < h < t'$,
- $X(t') = z$,
- at time t' the phase changes to $\mathcal{S}_2$, and
- the queue returns to the level 0 in a finite time afterwards

(see Figure 2 again).

Since $r_i = 1$ for all i in $\mathcal{S}_1$, necessarily t' and z are equal, and

$$\Gamma = \int_0^\infty \exp(T_{11} z) T_{12} G_{22}(z) dz$$

as claimed. □

A straightforward adaptation of Theorem 3.2 in Ramaswami [7] shows that $G_{22}(x) = \exp(Ux)$, where $U = T_{22} + T_{21}\Psi$. The interpretation of U is as follows.

Consider the random walk $\{(Y(t), \varphi(t))\}$ where the level is allowed to be negative. Take t_0 arbitrary, $\varphi(t_0)$ in $\mathcal{S}_2$ and $Y(t_0) = y$ arbitrary. Define the sequence $\{y_k, d_k, t_k : k \in \mathbb{N}\}$ as follows: $y_0 = y$, $d_k = \inf\{t > t_k : \varphi(t) \in \mathcal{S}_1\}$, $y_{k+1} = Y(d_k)$, and $t_{k+1} = \inf\{t > d_k : Y(t) = y_{k+1}, \varphi(t) \in \mathcal{S}_2\}$. During the intervals (t_k, d_k), the fluid is steadily decreasing; at time d_k, a temporary record low level y_{k+1} is reached and the fluid begins to build up; during the interval (d_k, t_{k+1}), the fluid goes up and down until time t_{k+1} when it reaches its previous record y_{k+1}. We illustrate this in Figure 3 by removing from the trajectory in Figure 2 all indications about the phases outside of the intervals $[t_i, d_i]$

The matrix U is the generator of the phase process obtained if one excises the intervals (d_k, t_{k+1}) and only keeps track of the phase during the intervals

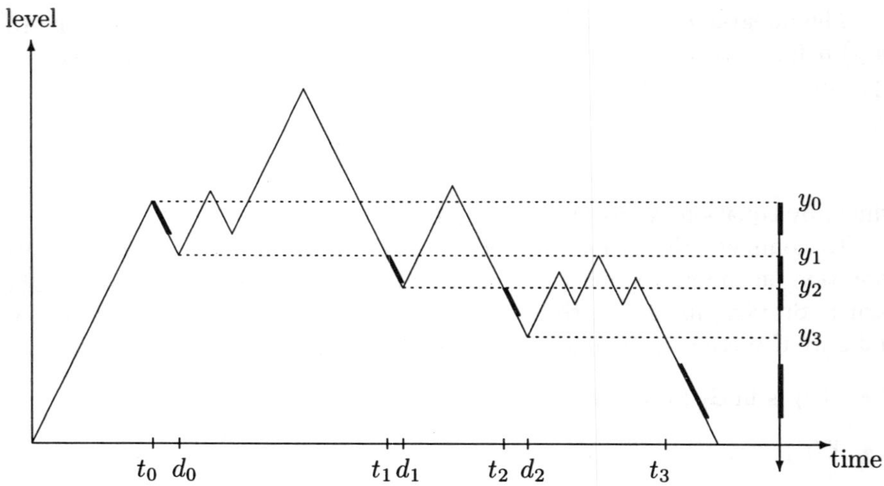

Figure 3. Illustration of the process of downward records.

(t_k, d_k); we shall write that this is the process of *downward records.* To illustrate this in Figure 3, we project the phases on the vertical line on the right, marked with an arrow to indicate the direction of the flow.

4 Discretization and Interpretation

We shall now give a probabilistic interpretation for the computational procedure proposed in Ramaswami,[7] Section 4.

The equation (9) may also be written as

$$\Psi = \int_0^\infty \exp(T_{11}y)T_{12}\exp(Uy)dy \tag{12}$$

with

$$U = T_{22} + T_{21}\Psi. \tag{13}$$

We uniformize the phase process and define $P = I + 1/\mu T$, where $\mu \geq \max_{i \in \mathcal{S}} |T_{ii}|$. The matrix P is decomposed in a manner conformant to the partition of T. With these, we have that

$$\exp(T_{11}y) = \sum_{k \geq 0} e^{-\mu y} \frac{(\mu y)^k}{k!} P_{11}^k.$$

Since $U \geq T_{22}$, we may use the same parameter μ to discretize the process of downward records and write that

$$\exp(Uy) = \sum_{n\geq 0} e^{-\mu y} \frac{(\mu y)^n}{n!} V^n$$

where

$$V = P_{22} + P_{21}\Psi. \tag{14}$$

We shall write that V is the transition matrix of the *discretized* process of downward records.

The equation (12) becomes

$$\Psi = \int_0^\infty \sum_{k\geq 0} e^{-\mu y} \frac{(\mu y)^k}{k!} \mu \sum_{n\geq 0} e^{-\mu y} \frac{(\mu y)^n}{n!} P_{11}^k P_{12} V^n dy \tag{15}$$

and the right-hand side is a discretized version of the fluid/phase process, which we interpret as follows. One considers the epochs of a Poisson process with rate μ, and a phase process which starts in $\mathcal{S}_1$. The equation (15) states that Ψ is equal to the probability matrix of the following event: there exist y, k and n such that

- an epoch occurs at time y, k epochs occur in $(0, y)$ and n epochs occur in $(y, 2y)$;
- the epoch at time y is the first at which the phase enters $\mathcal{S}_2$;
- at each epoch in $(0, y)$, a transition occurs from $\mathcal{S}_1$ to $\mathcal{S}_1$ with transition matrix P_{11};
- at each epoch in $(y, 2y)$, a transition occurs from $\mathcal{S}_2$ to $\mathcal{S}_2$ with transition matrix V.

Next, one writes (15) as

$$\Psi = \sum_{k,n\geq 0} \gamma_{kn} P_{11}^k P_{12} V^n, \tag{16}$$

where

$$\gamma_{kn} = \int_0^\infty e^{-2\mu y} \frac{(\mu y)^{k+n}}{k!n!} \mu dy = \frac{(k+n)!}{k!n!} \left(\frac{1}{2}\right)^{k+n+1}$$

and is the probability of n failures before the $k+1$st success in a Bernoulli sequence with probability 1/2 of success.

This, in turn, may be interpreted as follows. Denote by $\{t_1, t_2, \ldots\}$ the Poisson epochs before the first passage to $\mathcal{S}_2$ and by $\{T_1, T_2, \ldots\}$ the Poisson epochs afterwards. Since they occupy the non overlapping intervals $(0, y)$ and $(y, 2y)$, they are independent. Therefore, one may replace the Poisson process over two disjoint intervals by two independent processes over the same interval and consider $\mathcal{P}_t = \{t_i : i \geq 0\}$ and $\mathcal{P}_T = \{T_i : i \geq 0\}$, both with intensity μ and with $t_0 = T_0 = 0$.

The superposition $\mathcal{P}_t \cup \mathcal{P}_T = \{\theta_i : i \geq 0\}$ characterizes a Poisson process with intensity 2μ and each epoch θ_i belongs to $\mathcal{P}_t$ or $\mathcal{P}_T$ with probability $1/2$, independently of the others.

In (16), we count the number n of epochs of $\mathcal{P}_T$ which occur before the epoch t_{k+1} which marks the first passage to $\mathcal{S}_2$.

The second transformation consists in completely disconnecting the discretized process from any reference to the fluid buffer. Here, we write (16) as

$$\Psi = \sum_{n \geq 0} \{\sum_{k \geq 0} \gamma_{kn} P_{11}^k\} P_{12} V^n$$

and the nth term in the right-hand side is interpreted as follows. We consider a Bernoulli process with probability $1/2$ of success, we start with a phase in $\mathcal{S}_1$ and a counter D initialized to 1, and we perform the operations described below:

- in the case of a failure, we increase the counter D by 1 and do not change the phase;
- in the case of a success, either we make a transition to $\mathcal{S}_1$ with the probability matrix P_{11} and we keep D constant,
- or we make a transition to $\mathcal{S}_2$ with the probability matrix P_{12} and we decrement D by 1;
- once the phase has moved to $\mathcal{S}_2$, we stop the Bernoulli process, we systematically apply the transition matrix V and we decrement D by 1 at each step until it becomes zero.

The counter and the phase now evolve like in a discrete time QBD with transition matrices

$$A_0 = \begin{bmatrix} \frac{1}{2}I & 0 \\ 0 & 0 \end{bmatrix}, \qquad A_1 = \begin{bmatrix} \frac{1}{2}P_{11} & 0 \\ 0 & 0 \end{bmatrix} \quad \text{and} \quad A_2 = \begin{bmatrix} 0 & \frac{1}{2}P_{12} \\ 0 & V \end{bmatrix}.$$

Here, Ψ is a matrix of first passage probabilities to lower levels; specifically, Ψ_{ij} is the conditional probability of eventually reaching $(0, j)$, before any other state in $(0, \mathcal{S})$, given that the process starts from $(1, i)$ at time 0, with i in $\mathcal{S}_1$ and j in $\mathcal{S}_2$.

Of course, we do not know V, so that we need to pursue the matter a little further. In view of the interpretation we have given to Ψ, (14) tells us that there are two ways to reduce D by one, starting from $\mathcal{S}_2$:

- either one does it directly, using the transition probabilities of P_{22},
- or a transition is made to $\mathcal{S}_1$, with probabilities in P_{21}, in which case one must recursively apply the same procedure in order to eventually reduce D by one, with probability matrix Ψ.

Thus, we finally interpret Ψ as the matrix of first passage probabilities from $(1, \mathcal{S}_1)$ to $(0, \mathcal{S}_2)$ for the QBD with transition matrices

$$A_0 = \begin{bmatrix} \frac{1}{2}I & 0 \\ 0 & 0 \end{bmatrix}, \quad A_1 = \begin{bmatrix} \frac{1}{2}P_{11} & 0 \\ P_{21} & 0 \end{bmatrix} \quad \text{and} \quad A_2 = \begin{bmatrix} 0 & \frac{1}{2}P_{12} \\ 0 & P_{22} \end{bmatrix} \tag{17}$$

and we have that

$$G = \begin{bmatrix} 0 & \Psi \\ 0 & V \end{bmatrix},$$

where G is the matrix of first passage probabilities to lower levels for the QBD process defined by (17), for which simple and efficient computational algorithms abound.

Rogers [8] analyses the Wiener-Hopf factorization of the matrix $C^{-1}T$ where $C = \text{diag}(r_i : i \in \mathcal{S})$ has a very simple form in the present case. In our notations, it is shown that a Wiener-Hopf factorization is given by

$$C^{-1}T \begin{bmatrix} I & \Psi \\ \widehat{\Psi} & I \end{bmatrix} = \begin{bmatrix} I & \Psi \\ \widehat{\Psi} & I \end{bmatrix} \begin{bmatrix} \widehat{U} & 0 \\ 0 & -U \end{bmatrix},$$

where the matrices $\widehat{\Psi}$ and $\widehat{U}$ are the matrices of return probability to zero, and the generator of the process of downward records, for the *level reversed* fluid queue, that is, for the queue with the same infinitesimal generator T, but with r_i negative in $\mathcal{S}_1$ and positive in $\mathcal{S}_2$.

A thorough discussion of the relation between the results in Asmussen,[1] Ramaswami,[7] Rogers,[8] and here is beyond the scope of the present paper. We shall just mention that Theorem 2 in Rogers [8] states, mutatis mutandis, that Ψ is a solution of

$$T_{11}\Psi + \Psi T_{22} + T_{12} + \Psi T_{21}\Psi = 0$$

which, together with (13) is equivalent to the statement that G is a solution of $A_2 + A_1 G + A_0 G^2 = G$.

5 Phase-type Representation

We now return to the general fluid model of Section 2 where the net rates of input r_i can take any value, and we denote again by T the generator of the general environmental process. We define $\tilde{T} = |C|^{-1}T$; this is the generator of the phase process for the restricted model where all the r_i's are equal to 1 or -1. Furthermore, we use $\tilde{K}$ and $\tilde{\Psi}$ for the matrices defined by (7, 9) with T replaced by $\tilde{T}$.

To connect the two models, one needs to relate K and Ψ of the general case to $\tilde{K}$ and $\tilde{\Psi}$ of the restricted model. The relationship between these matrices is given by

$$K = C_1\tilde{K}C_1^{-1} \qquad \text{and} \qquad \Psi = C_1\tilde{\Psi}|C_2|^{-1}$$

where C_1 and $|C_2|$ are the diagonal matrices diag($r_i : i \in \mathcal{S}_1$) and diag($|r_i| : i \in \mathcal{S}_2$), respectively. To show this, we recall that $\tilde{K}$ and $\tilde{\Psi}$ satisfy (7, 9) and find that the matrices K and Ψ defined above satisfy

$$K = (T_{11} + \Psi T_{21})C_1^{-1}$$

and

$$\Psi = \int_0^{\infty} \exp(Ky)T_{12}|C_2|^{-1}\exp(T_{22}|C_2|^{-1}y)dy$$

which are the same equations as in Ramaswami.[7]

There, a phase-type representation is given for the stationary buffer content of the fluid queue in the case where there are no r_i's equal to zero. We show that one still has a PH distribution in the general case considered in the present paper. The phase-type characterization of the steady state buffer content allows one to use the machinery available for these distributions and to perform numerical computations with great accuracy.

Theorem 5.1 *The stationary distribution of the fluid queue is phase-type with representation* $(\boldsymbol{\omega}, W)$ *of order* s_1*, with*

$$\boldsymbol{\omega} = (\boldsymbol{\xi}_{\bullet}\mathbf{1})\{\Delta_1[I,\Psi][\mathbf{1} + \boldsymbol{T}_{\bullet 0}(-T_{00})^{-1}\mathbf{1}]\}'$$

and

$$W = \Delta_1^{-1}K'\Delta_1,$$

where $\Delta_1 = diag(\boldsymbol{\xi}_1)$.

Proof From Proposition 2.1 and Theorem 2.2, the steady state fluid density is given by

$$\begin{aligned} f(x) &= \boldsymbol{\pi}_{\bullet}(x)\mathbf{1} + \boldsymbol{\pi}_0(x)\mathbf{1} \\ &= -(\boldsymbol{\xi}_{\bullet}\mathbf{1})\boldsymbol{\xi}_1 K\exp(Kx)[I,\Psi][\mathbf{1} + \boldsymbol{T}_{\bullet 0}(-T_{00})^{-1}\mathbf{1}] \end{aligned}$$

for $x > 0$. By transposing both sides of this equation, we obtain that

$$\begin{aligned} f(x) &= -(\boldsymbol{\xi}_{\bullet}\mathbf{1})[\mathbf{1} + T_{\bullet 0}(-T_{00})^{-1}\mathbf{1}]'[I, \Psi]' \exp(K'x) K' \boldsymbol{\xi}_1' \\ &= -(\boldsymbol{\xi}_{\bullet}\mathbf{1})[\mathbf{1} + T_{\bullet 0}(-T_{00})^{-1}\mathbf{1}]'[I, \Psi]' \Delta_1 \\ &\quad \Delta_1^{-1} \exp(K'x) \Delta_1 \Delta_1^{-1} K' \Delta_1 \Delta_1^{-1} \boldsymbol{\xi}_1' \\ &= -(\boldsymbol{\xi}_{\bullet}\mathbf{1})[\mathbf{1} + T_{\bullet 0}(-T_{00})^{-1}\mathbf{1}]'[I, \Psi]' \Delta_1 \exp(Wx) W \mathbf{1} \\ &= -\boldsymbol{\omega} \exp(Wx) W \mathbf{1} \end{aligned}$$

which is the announced result. □

We give some illustrative examples to conclude this section. We consider a random environment which cycles through three periods: one during which the fluid builds up at the constant rate c, followed by one where the fluid level remains constant and finally the third period during which the fluid decreases at the constant rate 0.5. After the third period, the cycle repeats. The first period lasts 1 unit of time, on average, and the second and third periods last 2 units of time each, on average. The traffic intensity, that is, the ratio of the amount of fluid going in the buffer to the amount going out, is therefore equal to c and one clearly sees that the queue is stable for $c < 1$.

The generator T has the following structure

$$T = \begin{bmatrix} -\lambda_1 & \lambda_1 & & & \\ & -\lambda_2 & \lambda_2 & & \\ & & \ddots & \ddots & \\ & & & -\lambda_{s-1} & \lambda_{s-1} \\ \lambda_s & & & & -\lambda_s \end{bmatrix}$$

and the λ_i's and r_i's are given in the following table:

$$\begin{aligned} &\lambda_i = s_1, \quad r_i = c, \qquad 1 \le i \le s_1, \\ &\lambda_i = s_0/2, \; r_i = 0, \qquad s_1 + 1 \le i \le s_1 + s_0, \\ &\lambda_i = s_2/2, \; r_i = -0.5, \; s_1 + s_0 + 1 \le i \le s_1 + s_0 + s_2, \end{aligned}$$

so that the system is fully parametrized by s_0, s_1, s_2 and c.

We show on Figure 4 the steady-state distribution function $F(x) = \lim_{t\to\infty} \mathrm{P}[X(t) \le x]$ for the fluid queue with $s_1 = 2$ and $s_0 = s_2 = 4$ in four different cases: $c = 0.5$, $c = 0.75$, $c = 0.9$ and $c = 0.95$. We observe three effects resulting from increasing the rate c: the probability mass moves to the right and is spread over a larger interval, both resulting from the fact that the fluid reaches higher values at the end of the first period; furthermore, the probability $F(0)$ of an empty buffer decreases, because of shorter intervals at the end of each cycle where the fluid has returned to zero.

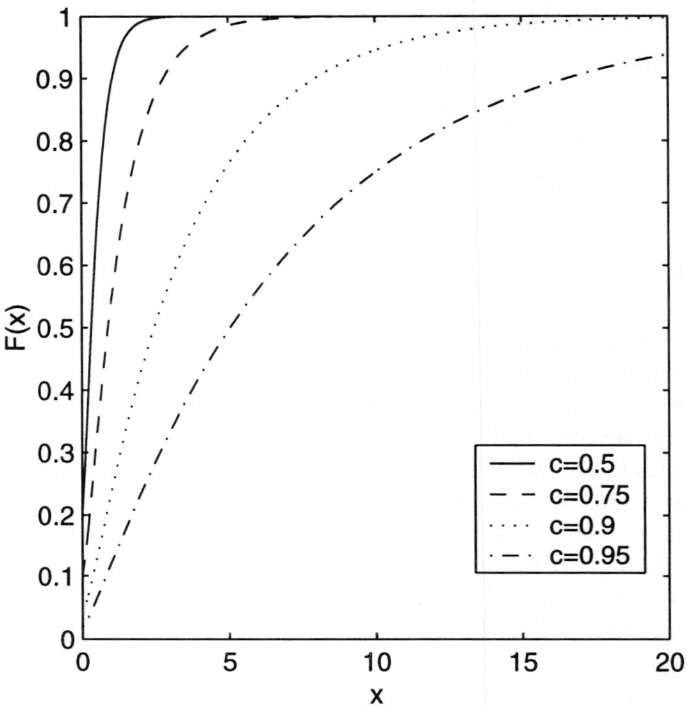

Figure 4. Distribution function of the stationary buffer content for a fluid queue driven by a Markov process with parameters $s_1 = 2$ and $s_0 = s_2 = 4$. The traffic intensity c varies from 0.5 to 0.95.

The first two moments are given below:

c	0.50	0.75	0.90	0.95
Mean	0.44	1.21	3.48	7.23
Variance	0.21	1.44	11.79	51.49

For the examples in Figure 5, we fix $c = 0.9$ and analyze four fluid queues for which we vary the number of phases:

case	a	b	c	d
s_1	1	2	2	4
s_0	1	2	4	4
s_2	1	2	4	4

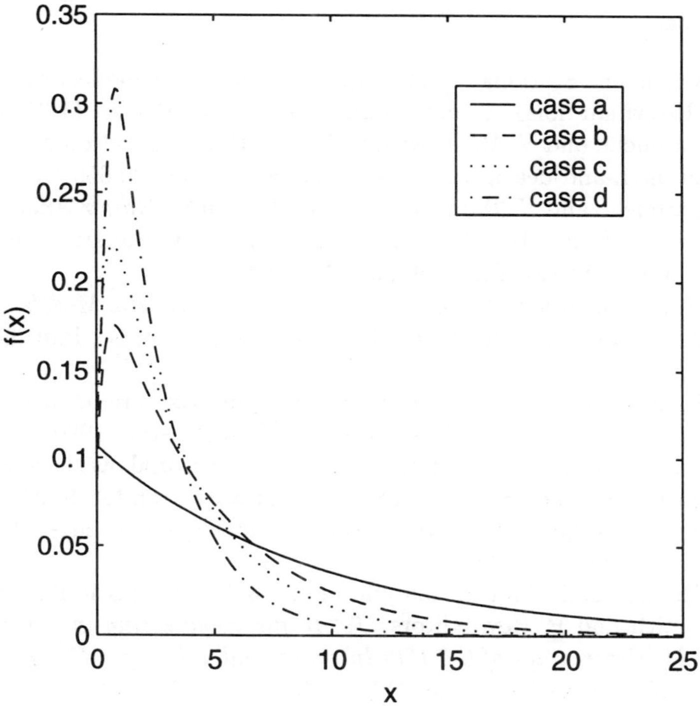

Figure 5. Density function of the stationary buffer content for fluid queues driven by Markov processes with increasingly regular cycles. The traffic intensity c is equal to 0.9.

By increasing the s_i's while keeping the average lengths of the three periods constant, we make them more regular (their probability distribution is more concentrated around the mean). We observe that the effect is to make the fluid density more concentrated around its mean as well. The first two moments are given below:

case	a	b	c	d
Mean	8.64	4.50	3.48	2.47
Variance	80.87	20.37	11.79	5.21

Acknowledgments

We are grateful to the referees for having made very interesting comments and, in particular, for drawing our attention to Rogers.[8]

References

1. S. Asmussen. Stationary distributions for fluid flow models with or without Brownian noise. *Comm. Statist. Stochastic Models*, 11:21–49, 1995.
2. G. Latouche and V. Ramaswami. A logarithmic reduction algorithm for quasi-birth-and-death processes. *J. Appl. Probab.*, 30:650–674, 1993.
3. G. Latouche and V. Ramaswami. *Introduction to Matrix Analytic Methods in Stochastic Modeling.* ASA-SIAM Series on Statistics and Applied Probability. SIAM, Philadelphia PA, 1999.
4. M. F. Neuts. *Matrix-Geometric Solutions in Stochastic Models. An Algorithmic Approach.* The Johns Hopkins University Press, Baltimore, MD, 1981.
5. V. Ramaswami. A duality theorem for the matrix paradigms in queueing theory. *Comm. Statist. Stochastic Models*, 6:151–161, 1990.
6. V. Ramaswami. Matrix analytic methods: A tutorial overview with some extensions and new results. In S. R. Chakravarthy and A. S. Alfa, editors, *Matrix-Analytic Methods in Stochastic Models*, pages 261–295. Marcel Dekker, New York, 1996.
7. V. Ramaswami. Matrix analytic methods for stochastic fluid flows. In D. Smith and P. Hey, editors, *Teletraffic Engineering in a Competitive World (Proceedings of the 16th International Teletraffic Congress)*, pages 1019–1030. Elsevier Science B.V., Edinburgh, UK, 1999.
8. L. C. G. Rogers. Fluid models in queueing theory and Wiener-Hopf factorization of Markov chains. *The Annals of Applied Probability*, 4:390–413, 1994.
9. B. Sericola and B. Tuffin. A fluid queue driven by a Markovian queue. *Queueing Systems Theory Appl.*, 16:253–264, 1999.

PENALISED MAXIMUM LIKELIHOOD ESTIMATION OF THE PARAMETERS IN A COXIAN PHASE-TYPE DISTRIBUTION

M. J. FADDY

School of Mathematics and Statistics, The University of Birmingham Edgbaston, Birmingham B15 2TT, U.K.
E-mail: mfaddy@for.mat.bham.ac.uk

It has been noted that when fitting Coxian phase-type distributions to observed and simulated data by maximum likelihood, often a maximum corresponding to equality of two or more eigenvalues of the matrix of transition rates was found. Such equality of eigenvalues would contribute to the smoothness of the resulting probability density function. It is proposed to adjust the log-likelihood of the data by subtracting a quantity which penalises configurations that have disparate eigenvalues. The resulting penalised maximum likelihood estimation of the parameters specifying the transition rate matrix is discussed with reference to two example data-sets.

Key Words: Phase-type distributions; maximum likelihood; smooth density estimation; penalty function; penalised likelihood.

1 Introduction

Phase-type distributions have been popularised by Neuts[12] and others in the context of applied probability modelling to allow for sojourn time distributions other than exponential, while retaining some analytical tractability. More recently Aalen[1] and Faddy and McClean[5] have argued for phase-type distributions in a more statistical context as a data-analytical tool. Such applications require procedures for fitting phase-type distributions to data from a variety of observational circumstances. Johnson and Taafe[8,9] used moment matching where moments of observed data were matched to those of a phase-type distribution. Asmussen *et al.*[2] commented that such moment matching was not entirely satisfactory from a statistical point of view, and in any case this cannot be used if the data contain censored values or if any other features of the observed data prevent the calculation of moments. Asmussen *et al.*[2] also pointed out that, due to their denseness, phase-type modelling of data can be viewed as semi-parametric density estimation, with the number of phases determining the degree of smoothness. This is a rather crude measure of smoothness, and in Section 3 there is an example data-set where fitted phase-type distributions of low order can have multi-modal shapes with one mode at zero. Green[6], in the context of semi-parametric regression modelling, has suggested including a roughness penalty in the fitting criterion which penalises rough forms of the function being fitted.

Faddy[4] noted that, when fitting Coxian distributions to observed and simulated data by maximum likelihood, often a maximum corresponding to equality of two or more eigenvalues of the matrix of transition rates was found. Although such solutions might not be global maxima, the reduced number of parameters in the phase-type formulation would contribute to the smoothness of the fitted distribution. This leads to the suggestion of subtracting a quantity from the log-likelihood which penalises phase-type configurations that have eigenvalues which are very disparate. It is the purpose of this paper to discuss such an approach to fitting phase-type distributions, using two example data-sets to illustrate the methodology.

2 The Distributions

The so-called Coxian[3] phase-type distributions have probability density function taking the form:

$$f(t) = \mathbf{p}\exp\{\mathbf{Q}t\}\mathbf{q}, \tag{1}$$

where $\mathbf{p}$ is a probability vector:

$$\mathbf{p} = (1\ 0\ 0 \cdots 0\ 0), \tag{2}$$

$\mathbf{Q}$ a matrix of transition rates:

$$\mathbf{Q} = \begin{pmatrix} -(\lambda_1+\mu_1) & \lambda_1 & 0 \cdots & 0 & 0 \\ 0 & -(\lambda_2+\mu_2) & \lambda_2 \cdots & 0 & 0 \\ \vdots & \vdots & \vdots & \vdots & \vdots \\ 0 & 0 & 0 \cdots & -(\lambda_{n-1}+\mu_{n-1}) & \lambda_{n-1} \\ 0 & 0 & 0 \cdots & 0 & -\mu_n \end{pmatrix} \tag{3}$$

and $\mathbf{q}$ the vector of absorption rates:

$$\mathbf{q} = (\mu_1\ \mu_2\ \mu_3 \cdots \mu_{n-1}\ \mu_n)^{\mathrm{T}}. \tag{4}$$

Fitting these distributions to data $t_1, t_2, \cdots, t_m$ can be done by estimating the parameters $\lambda_1, \lambda_2, \cdots, \lambda_{n-1}$ and $\mu_1, \mu_2, \cdots, \mu_n$ by maximising the likelihood (Kotz and Johnson[10] pp 639–644) $\prod_{i=1}^{m} f(t_i)$ or equivalently the log-likelihood $\sum_{i=1}^{m} \log\{f(t_i)\}$. Such maximisation can be carried out quickly and efficiently using MATLAB[11] routines to calculate the matrix exponential in (1) and to perform the optimisation. If some of the data refer to censored values (*i.e.*, observation greater than the recorded value) which is quite common in survival analysis (Aalen[1]) then for such data the probability density component in the likelihood is replaced by the survivor function:

$$\mathcal{F}(t) = \mathbf{p}\exp\{\mathbf{Q}t\}\mathbf{1}. \tag{5}$$

Faddy and McClean[5] described fitting such distributions to data on durations of treatment in hospital of geriatric patients. They chose a distribution of order $n = 4$ in preference to one of order $n = 5$ because the latter distribution was multi-modal. Such multi-modality was a consequence of the eigenvalues of the matrix $\mathbf{Q}$ (3) being too disparate. If a function that penalises disparate values of the eigenvalues $-(\lambda_i + \mu_i)$ ($i = 1, 2, \cdots, n$ with $\lambda_n = 0$) such as:

$$p \times \frac{1}{n-1}\sum_{i=1}^{n}(\gamma_i - \bar{\gamma})^2, \tag{6}$$

where $\gamma_i = \lambda_i + \mu_i$ and $\bar{\gamma} = \frac{1}{n}\sum_{i=1}^{n}\gamma_i$, is subtracted from the log-likelihood (*e.g.*, Green[6]) then configurations with disparate eigenvalues will be progressively discounted for increasing values of the multiple p in (6). To this end, it is proposed to fit phase-type distributions (1) by maximising the penalised log-likelihood:

$$\sum_{i=1}^{m}\log\{f(t_i)\} - p \times \frac{1}{n-1}\sum_{i=1}^{n}(\gamma_i - \bar{\gamma})^2. \tag{7}$$

3 Examples

The first data-set, from Faddy and McClean[5], refers to the lengths of occupancy of geriatric beds by 2090 male patients from a number of London hospitals over the period 1969-85. These data, shown in Figure 2, appear to have a distribution with a single mode away from zero and a very long upper tail. Coxian phase-type distributions of increasing order were fitted to these data, and illustrated in Figure 1 is the five phase fit referred to in Faddy and McClean[5]. This fit corresponded to a maximum of the log-likelihood with $\mu_2 = 0$, which was not a global maximum as the log-likelihood function is in fact unbounded. The distribution is bi-modal with a sharp mode at zero, and as such might be considered an inimical description of the data. Penalising the log-likelihood by using (7) reduces the effect of any mode at zero, and shown in Table 1 are the maximised penalised log-likelihoods (7) from fitting phase-type distributions (1) with increasing numbers of phases n.

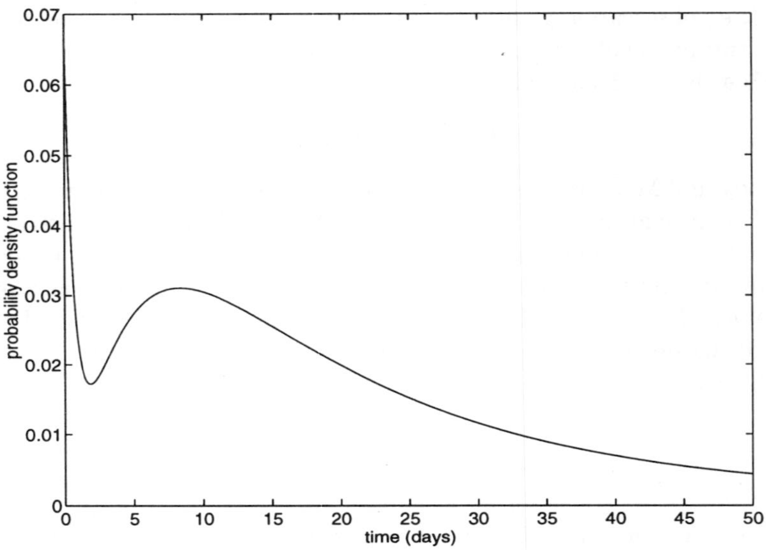

Figure 1. Example bi-modal 5-phase fitted distribution

Table 1: penalised log-likelihood values

n	p				
	0	100	1000	5000	10000
1	-9648.2	-9648.2	-9648.2	-9648.2	-9648.2
2	-9370.1	-9370.1	-9370.7	-9373.0	-9376.0
3	-9357.7	-9357.7	-9358.2	-9360.1	-9362.4
4	-9332.5	-9332.8	-9334.8	-9343.0	-9352.1
5	∞	-9319.8	-9329.2	-9339.5	-9345.6

The penalised log-likelihood values for the five phase fits show some improvement over the four phase fits, with the mode of the fitted distribution at zero disappearing by $p = 5000$. Shown in Figure 2 are the four and five phase fitted distributions corresponding to $p = 5000$, the four phase fit here being very similar to that when $p = 0$ whereas there were more appreciable differences at $p = 10000$. There is little difference apparent between these fitted distributions and they both show quite good agreement with the observed distribution. Thus Faddy and McClean's[5] choice of the four phase fit would seem reasonable.

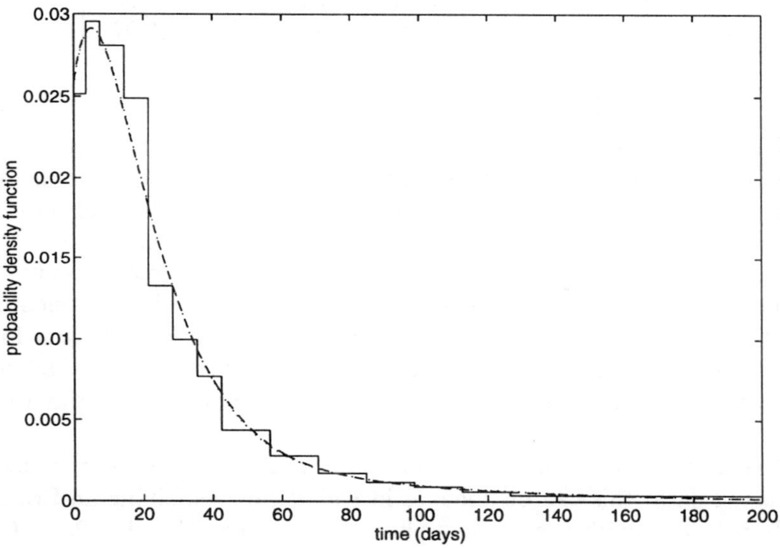

Figure 2. Observed (——), and fitted 4-phase (- - -) and 5-phase (· · · ·) distributions

The second data-set is taken from Jiang and Murthy[7] and refers to the failures of throttles from a number of general purpose vehicles. The failure "times" here are in kilometres and 25 of the 50 observations were censored; that is, the throttles were still working at the observed number of kilometres and $\mathcal{F}(\cdot)$ from (5) is used in the construction of the log-likelihood in (7) for these censored data while $f(\cdot)$ from (1) is used for the uncensored data. None of the observations is less than 478 kilometres so, as in Faddy[4], this quantity is first subtracted from the data and phase-type distributions of increasing order using (3) are fitted to these adjusted data. Shown in Table 2 are values of the maximised penalised log-likelihood (7) from these fits.

Table 2: penalised log-likelihood values

n	p			
	0	100	1000	10000
1	-248.16	-248.16	-248.16	-248.16
2	∞	-246.25	-246.26	-246.30
3	∞	-245.82	-245.94	-245.97

Here, the un-penalised log-likelihood functions ($p = 0$) for the $n \geq 2$ phase distributions are unbounded, and as the value of p increases the mode of the fitted distributions at zero becomes less sharp. The penalised log-likelihood values in Table 2 give a more reliable assessment of the two and three phase fits to these adjusted data, and show that a three phase fit offers little improvement over one with two phases. However, the improvement from using two phases over a single phase (exponential) is apparent from Figure 3, which shows the one and two phase ($p = 10000$) fitted survivor functions along with the Kaplan-Meier (Kotz and Johnson[10] pp 346–352) estimate from the data.

It remains to construct an appropriate phase-type distribution for the unadjusted data: this can be done, as in Faddy[4], by combining an Erlang(k) distributed delay with the above 2-phase distribution; *i.e.*, $(k+2) \times (k+2)$ matrix (3) with parameters $\lambda_1 = \lambda_2 = \cdots = \lambda_k$, $\mu_1 = \mu_2 = \cdots = \mu_k = 0$, λ_{k+1}, μ_{k+1} and μ_{k+2}. Fitting such a distribution (1) and (5) gave a maximised un-penalised ($p = 0$) log-likelihood of -246.25 when $k = 19$ (and a penalised, $p = 10000$, log-likelihood of -246.57 when $k = 17$).

4 Concluding Remarks

The notion of penalising a log-likelihood to discount rough forms of a fitted function is common in a regression context (*e.g.*, Green[6]). Here it has been used to discount unsatisfactory features of phase-type densities (1) when fitting them to data. The examples have illustrated how useful comparisons can be made between phase-type distributions of increasing order using a penalised log-likelihood function. The penalty increased as the eigenvalues of $\mathbf{Q}$ (3) became more disparate, and the fitted phase-type distributions did have some equal eigenvalues: $\lambda_1 + \mu_1 = \lambda_2 + \mu_2$ in the first example, and $\lambda_k + \mu_k = \lambda_{k+1} + \mu_{k+1}$ in the second. So a recommendation would be to use the penalised log-likelihood function (7) in comparing the fits of phase-type distributions of increasing order, and then to exploit any eigenvalue equalities

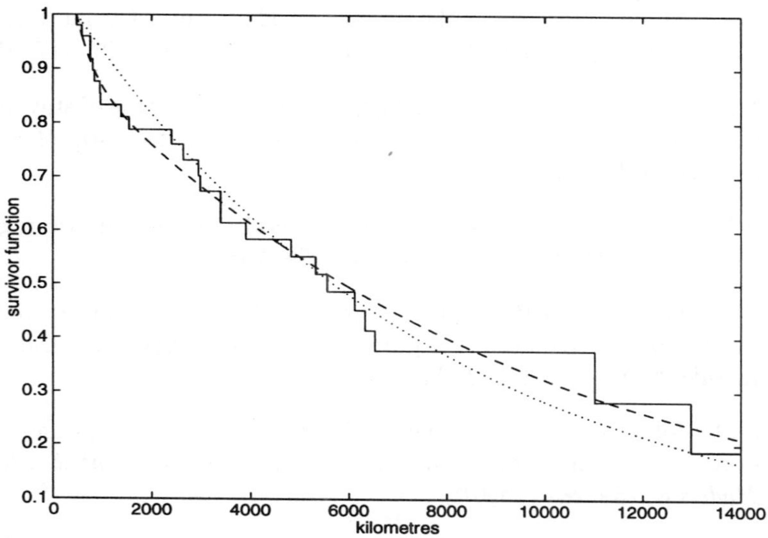

Figure 3. Kaplan-Meier estimate (——) of the survivor function, and fitted 1-phase (····) and 2-phase (- - -) forms

in the final estimation of the parameters of a phase-type distribution of chosen order.

References

1. O.O. Aalen, Phase type distributions in survival analysis, *Scandanavian Journal of Statistics* **22**, 447–463 (1995).

2. S. Asmussen, O. Nerman and S. Olsson, Fitting phase-type distributions via the EM algorithm, *Scandanavian Journal of Statistics* **23**, 419–441 (1996).

3. D.R. Cox, A use of complex probabilities in the theory of stochastic processes, *Proceedings of the Cambridge Philosophical Society* **51**, 313–319 (1955).

4. M.J. Faddy, On inferring the number of phases in a Coxian phase-type distribution, *Commun. Statist. Stochastic Models* **14**, 407–417 (1998).

5. M.J. Faddy and S.I. McClean, Analysing data on lengths of stay of hospital patients using phase-type distributions, *Appl. Stochastic Models Bus. Ind.* **15**, 311–317 (1999).

6. P.J. Green, Penalised likelihood for general semi-parametric regression models, *Int. Statist. Review* **55**, 245–259 (1987).

7. R. Jiang and D.N.P. Murthy, Modelling failure-data by a mixture of two Weibull distributions: a graphical approach, *IEEE Transactions on Reliability* **44**, 477–488 (1995).

8. M.A. Johnson and M.R. Taaffe, Matching moments to phase distributions: non-linear programming approaches, *Commun. Statist. Stochastic Models* **6**, 259–281 (1990).

9. M.A. Johnson and M.R. Taaffe, Matching moments to phase distributions: density function shapes, *Commun. Statist. Stochastic Models* **6**, 283–306 (1990).

10. S. Kotz and N.L. Johnson (Eds.), *Encyclopedia of Statistical Sciences* **4**, John Wiley and Sons Inc., New York, 1983.

11. MATLAB, *Using MATLAB*, The MathWorks Inc., Natick, Massachussetts, 1996.

12. M.F. Neuts, *Matrix Geometric Solutions in Stochastic Models*, Johns Hopkins University Press, Baltimore, Maryland, 1981.

MAP/PH/1 QUEUES WITH LEVEL-DEPENDENT FEEDBACK AND THEIR DEPARTURE PROCESSES

DAVID GREEN

Department of Applied Mathematics, The University of Adelaide, South Australia, 5005.

E-mail: dgreen@maths.adelaide.edu.au

A family of approximations to the departure process from the *MAP/PH/1* queue was proposed and demonstrated in Bean, Green and Taylor [1]. In this paper we extend these models to the departure process of a *MAP/PH/1* feedback queue, where the probability of feedback may be level-dependent.

1 Introduction

In communications networks where data transmissions need to be guaranteed error free to within some specified probability, feedback schemes are used to request retransmission of packets that are lost or received in a corrupt form. A specific example is given in Green [2], where the feedback queue is used to model the automatic repeat request (ARQ) protocol in a high frequency (HF) communications network.

Feedback queues are an important feature of many networks and network models such as those of Jackson [3]. The Jackson networks however, are restricted by the use of Poisson arrival models. A more general arrival process model is found in the *MAP*, which has been widely used in modelling bursty arrival processes, and has even been shown capable of modelling self similar traffic over an arbitrary time frame. Some numerical examples of the approximations to a *MAP/PH/1* feedback queue with level dependent feedback are given.

An important feature of the family of approximations demonstrated in Bean, Green and Taylor [1] is that the kth approximation exactly captures the first $k-1$ lag-correlation coefficients of the departure process (see Green [4]). This is also true for the *MAP* approximations to the departure process from the *MAP/PH/1* feedback queue if the probability of feedback is constant for all levels. Some numerical examples of these approximations are given.

2 Notation for the *MAP/PH/1* feedback queue

Let D_0 and $D_1 \geq 0$ be the $m \times m$ matrix descriptors of the *MAP* and (β, S) be the *PH*-type service distribution, where the matrix S is $n \times n$. The probability of feedback is level-dependent and has a description given by some probability relation $f(i)$. For $i > 0$, a customer who completes service leaving $i-1$ in the queue will immediately rejoin the queue with probability $f(i)$ or with probability $(1-f(i))$ leaves the system. Here, the variable i represents the size or level of the queue. Define $\boldsymbol{S}^0 = -S\boldsymbol{e}$. The *MAP/PH/1* level-dependent feedback queue then has the following level-dependent quasi birth and death process (*LDQBD*, see Bright and Taylor [5]) description given by

$$Q = \begin{bmatrix} B_1 & B_0 & & & \\ (1-f(1))B_2 & A_1 + f(1)A_2 & A_0 & & \\ & (1-f(2))A_2 & A_1 + f(2)A_2 & A_0 & \\ & & (1-f(3))A_2 & A_1 + f(3)A_2 & A_0 \\ & & & \ddots & \ddots \quad \ddots \end{bmatrix}, \quad (1)$$

where

$$B_0 = D_1 \otimes \boldsymbol{\beta}, \quad (2)$$
$$B_1 = D_0, \quad (3)$$
$$B_2 = I_m \otimes \boldsymbol{S}^0, \quad (4)$$
$$A_0 = D_1 \otimes I_n, \quad (5)$$
$$A_1 = D_0 \oplus S, \quad (6)$$
$$A_2 = I_m \otimes \boldsymbol{S}^0 \boldsymbol{\beta}. \quad (7)$$

Here, $\otimes$ and $\oplus$ are the Kronecker product and sum respectively. Assume Q defines an irreducible, regular Markov chain, so that it has at most one stationary distribution $\boldsymbol{\Psi}$ such that $\boldsymbol{\Psi}Q = 0$. If it exists, this stationary distribution is given by the following matrix form

$$\boldsymbol{\Psi} = \boldsymbol{\pi}_0[I, R_0, R_0R_1, R_0R_1R_2, \ldots], \quad (8)$$

where for $i \geq 1$, the $mn \times mn$ matrices R_i and the $m \times mn$ matrix R_0 are the minimal non-negative solutions to the system of equations

$$R_iR_{i+1}\Big(1 - f(i+2)\Big)A_2 + R_i\Big(A_1 + (1 - f(i+1))A_2\Big) + A_0 = 0$$
$$B_0 + R_0\Big(A_1 + f(1)A_2 + R_1A_2(1 - f(2))\Big) = 0 \quad (9)$$

and the vector $\boldsymbol{\pi}_0$ is the unique positive solution to the system of equations

$$\boldsymbol{\pi}_0(B_1 + R_0B_2) = \mathbf{0} \quad \text{and} \quad \boldsymbol{\pi}_0 \sum_{j=0}^{\infty} \prod_{i=0}^{j-1} R_i\boldsymbol{e} = 1. \quad (10)$$

For later use we re-write

$$\Psi = [\pi_0, \pi_1, \pi_2, \ldots], \tag{11}$$

where for $i \geq 1$, $\pi_i = \pi_0 \prod_{j=0}^{i-1} R_j$. For further discussion on the equilibrium distribution of an *LDQBD* see Bright and Taylor [5].

3 Approximating the departure process from a *MAP/PH/1* feedback queue

In this section we consider the departure process from a *MAP/PH/1* feedback queue, where the probability of feedback is given by a possibly level-dependent probability $f(i)$. The departure process from the *MAP/PH/1* feedback queue which includes those departures which rejoin the queue can be observed using the following filtration matrices Q_0^* and Q_1^*, where $Q = Q_0^* + Q_1^*$.

$$Q_0^* = \begin{bmatrix} B_1 & B_0 & & & & \\ & A_1 & A_0 & & & \\ & & A_1 & A_0 & & \\ & & & A_1 & A_0 & \\ & & & & \ddots & \ddots \end{bmatrix} \quad \text{and}$$

$$Q_1^* = \begin{bmatrix} 0 & & & & & \\ B_2(1-f(1)) & f(1)A_2 & & & & \\ & A_2(1-f(2)) & f(2)A_2 & & & \\ & & A_2(1-f(3)) & f(3)A_2 & & \\ & & & \ddots & \ddots & \end{bmatrix}, \tag{12}$$

where the A_i and B_i have the same interpretation as those given in equation (7). Note that a measure $f(i)$ of the "departures" at level i is fed back to level i. These "departures" do not leave the system, but immediately return to service. This effectively maintains the number i of customers present at the server and associated queue. The matrix $Q = Q_0^* + Q_1^*$ given by (12) is of an *LDQBD*. As we are concerned with those departures which actually leave the system, we partition this particular Q matrix into the following filtration matrices Q_0 and Q_1, so as to capture the actual process of departures leaving the system (hereafter referred to as the departure process).

$$Q_0 = \begin{bmatrix} B_1 & B_0 & & & \\ & A_1 + f(1)A_2 & A_0 & & \\ & & A_1 + f(2)A_2 & A_0 & \\ & & & A_1 + f(3)A_2 & A_0 \\ & & & & \ddots & \ddots \end{bmatrix} \text{ and}$$

$$Q_1 = \begin{bmatrix} 0 & & & & \\ B_2(1-f(1)) & 0 & & & \\ & A_2(1-f(2)) & 0 & & \\ & & A_2(1-f(3)) & 0 & \\ & & & \ddots & \ddots \end{bmatrix}, \quad (13)$$

The observed transitions recorded by Q_1 are departure transitions and the observed process is the departure process. However, the matrices Q_0 and Q_1 do not provide a *MAP* representation for the departure process because there are infinitely many states.

The resultant filtration matrices in (13) when $f(i) = \rho$ for all $i \geq 1$ are in fact representative of a *MAP/PH/1* queue without feedback, where the arrival process is the same but with a modified service time distribution. We assume here that this *MAP/PH/1* feedback queue has a stationary distribution.

Similarly to Bean,Green and Taylor [1], we construct a family of approximations indexed by a parameter k, where the accuracy of the approximation increases with the value of the parameter k. The kth approximation assumes that

1. the phase of the arrival process when the *QBD* moves from level k to level $k-1$ is given by its correct marginal distribution, and

2. the number of services during a sojourn at level k and above is geometrically distributed with the parameter chosen such that the sojourn at level k and above has the correct mean.

Thus the $k = 1$ approximation assumes that

1. the phase of the arrival process when a busy period ends has the correct marginal distribution, and

2. the number of services during a busy period is geometrically distributed with the mean chosen such that the busy period has the correct mean.

Physically, the kth approximation amalgamates levels k and above into a super level $\bar{k}$, approximates the distribution of the sojourn in level $\bar{k}$ by a geometric mixture of convolutions of *PH*-type distributions, and also approximates the phase on return to level $k-1$ by its correct marginal distribution. What is lost in this approximation is the exact distribution of the sojourn at and above level k and correlations between the return phases and sojourn times.

Intuitively the stationary rate of departures from a *MAP/PH/1* feedback queue must be equivalent to the stationary rate of arrivals given by $\boldsymbol{\nu} D_1 \boldsymbol{e}$, where $\boldsymbol{\nu}$ is the stationary probability vector of the *MAP* satisfying $\boldsymbol{\nu}(D_0 + D_1) = 0$. We will use this result in the construction of the distribution of the *QBD* at level $k-1$, conditional on a departure having just occurred.

The distribution of the *QBD* given in (13) at level $k-1$, conditional on a departure having just occurred, can be calculated from its stationary distribution by (see Neuts [6])

$$\boldsymbol{x}_{k-1} = \begin{cases} \boldsymbol{\pi}_0 R_0 (1 - f(1)) B_2 (\boldsymbol{\nu} D_1 \boldsymbol{e})^{-1} & \text{for } k = 1 \\ \\ \boldsymbol{\pi}_0 \prod_{i=0}^{k-1} R_i (1 - f(k-1)) A_2 (\boldsymbol{\nu} D_1 \boldsymbol{e})^{-1} & \text{for } k > 1. \end{cases} \tag{14}$$

In our approximation, the probability of return to level $k-1$ after each service in the super-level $\bar{k}$ is

$$\frac{\boldsymbol{x}_{k-1}\boldsymbol{e}}{\left(\sum_{j=k-1}^{\infty} \boldsymbol{x}_j \boldsymbol{e} \right)},$$

and the distribution of the return phase given that a return occurs is given by

$$\frac{\boldsymbol{x}_{k-1}}{\boldsymbol{x}_{k-1}\boldsymbol{e}}.$$

Thus the unconditional distribution of return phase at level $k-1$ is given by

$$\boldsymbol{y}_{k-1} = \frac{\boldsymbol{x}_{k-1}}{\left(\sum_{j=k-1}^{\infty} \boldsymbol{x}_j \boldsymbol{e} \right)}. \tag{15}$$

The kth *MAP* approximation for the departure process of the *MAP/PH/1* feedback queue is given by the $\left(m + n + (k-1)mn \right) \times \left(m + n + (k-1)mn \right)$

matrices

$$Q_0 = \begin{bmatrix} B_1 & B_0 & & & & \\ & A_1 + f(1)A_2 & A_0 & & & \\ & & A_1 + f(2)A_2 & \ddots & & \\ & & & \ddots & A_0 & \\ & & & & A_1 + f(k-1)A_2 & E_0 \\ & & & & & E_1 \end{bmatrix} \text{ and}$$

$$Q_1 = \begin{bmatrix} 0 & & & & & \\ B_2(1-f(1)) & 0 & & & & \\ & A_2(1-f(2)) & \ddots & & & \\ & & \ddots & 0 & & \\ & & & A_2(1-f(k-1)) & 0 & \\ & & & & E_2 & E_3 \end{bmatrix}. \quad (16)$$

Here the sub-matrices are as defined previously with

$$\begin{aligned} E_0 &= D_1 \boldsymbol{e} \otimes I_n, \\ E_1 &= S + \gamma_k \boldsymbol{S}^0 \boldsymbol{\beta}, \\ E_2 &= (1-\gamma_k)\boldsymbol{S}^0 \boldsymbol{y}_{k-1}, \\ E_3 &= (1-\boldsymbol{y}_{k-1}\boldsymbol{e})(1-\gamma_k)\boldsymbol{S}^0 \boldsymbol{\beta}, \end{aligned}$$

where γ_k is a scaling factor for the service time distribution which reflects the level-dependent feedback mechanism in operation above the super level k. The scaling factor γ_k is given by

$$\gamma_k = \sum_{i=k}^{\infty} f(i) \frac{(\pi_i \boldsymbol{e})}{(1 - \sum_{j=0}^{k-1} \pi_j \boldsymbol{e})}, \quad (17)$$

where recall from Equation (11) that $\boldsymbol{\pi}_i = \boldsymbol{\pi}_0 \prod_{j=0}^{i-1} R_j$ for all $i \geq 1$. The stationary distribution of this *MAP* approximation can then be shown to be given by

$$\boldsymbol{\nu}(k) = \boldsymbol{\pi}_0 \left[I, R_0, R_0 R_1, \ldots, \prod_{i=0}^{k-2} R_i, \left(\sum_{j=k-1}^{\infty} \prod_{i=0}^{j} R_i \right) (\boldsymbol{e}_m \otimes I_n) \right]. \quad (18)$$

Note that when $f(i) = \rho \in [0,1)$ for all $i \geq 1$, we have $\gamma_k = \rho$ for all $k \geq 1$.

For the special case of $k = 1$, the *MAP* approximation to the departure process of the *MAP/PH/1* feedback queue reduces to a *PH*-renewal process.

For $k = 1$, the *MAP* approximation is given by

$$Q_0(1) = \begin{pmatrix} D_0 & D_1 e\beta \\ 0 & \gamma_1 \boldsymbol{S}^0 \boldsymbol{\beta} + S \end{pmatrix} \text{ and}$$

$$Q_1(1) = \begin{pmatrix} 0 & 0 \\ (1-\gamma_1)\boldsymbol{S}^0 \boldsymbol{x}_0 & (1-\boldsymbol{x}_0 e)(1-\gamma_1)\boldsymbol{S}^0 \boldsymbol{\beta} \end{pmatrix}.$$

This can also be represented as a *PH*-distribution $(\boldsymbol{\alpha}, Q_0(1))$, where $Q_0(1)$ is as above and

$$\boldsymbol{\alpha} = (\boldsymbol{x}_0, (1-\boldsymbol{x}_0 \boldsymbol{e})\boldsymbol{\beta}). \tag{19}$$

Here, $\boldsymbol{x}_0$ is the distribution of phases of the arrival process immediately after a departure that leaves the queue empty, calculated using (14). The stationary distribution of this *MAP* approximation is given by

$$\boldsymbol{\nu}(1) = \left[\boldsymbol{\pi}_0, \boldsymbol{\pi}_0 \left(\sum_{j=0}^{\infty} \prod_{i=0}^{j} R_i \right) (\boldsymbol{e}_m \otimes I_n) \right].$$

This approximation closely captures the distribution of the inter-departure times but ignores any correlation structures between these times.

4 Some numerical examples

For a *MAP/PH/1* feedback queue with constant feedback, the results given in Bean, Green and Taylor [1] are essentially sufficient to demonstrate the accuracy of the *MAP* approximations. This is because the *MAP/PH/1* queue with constant feedback is effectively another *MAP/PH/1* queue. However for completeness, we give some numerical examples in Section 4.1, which also demonstrate that Poisson and or negative-exponential assumptions are inappropriate in many instances.

For comparison, since the stationary distributions for the actual departure process and the approximations to the departure process are identical, we need a measure which addresses the difference in correlation structure between departures. The difference in behaviour when the approximations to the departure process and the actual departure process are applied to a second queue will give such a measure. Ideally, it would be better to establish some sort of measure on the difference between the distributions directly. However, this is a difficult task.

We use a tandem queueing system, comprising firstly a feedback queue and then a non-feedback queue, which can be represented by a *QBD* as outlined in Bean, Taylor and Li [7]. We do this by setting the number in the first and second queues to be part of the phase description and the level respectively. Under this regime, the size of the first queue must be truncated at a "sufficiently large" value that will not affect the calculation of the stationary distribution of the queue length. We refer to the results calculated for this *QBD* model as "exact" throughout this paper.

We calculate the probability distributions of the stationary second queue length so that any queue length probability being less than 10^{-14} is considered as 0. This distribution is then used to calculate the mean and variance for the stationary queue length.

Two different functional forms of the feedback probability $f(i)$ are numerically demonstrated.

- Regime 1 in Section 4.1 has $f(i) = \rho \in [0, 1)$ for all $i \geq 1$.
- Regime 2 in Section 4.2 has $f(i) = \rho^i$ for all $i \geq 1$, with $\rho \in [0, 1)$.

Note that other forms of $f(i)$ are easily implementable, with the proviso that the queue remain stable. The stability of a queue is easily assured for a decaying or constant feedback mechanism, whereas it is not a trivial problem for other forms of feedback.

4.1 Regime 1: Level-independent feedback.

Here we have $f(i) = \rho \in [0, 1)$ for all $i \geq 1$.

We note here that the same special properties as demonstrated in Bean, Green and Taylor [1] for the non-feedback queue approximations are also applicable here. We will now demonstrate the approximations and in the process show that Poisson and or negative-exponential assumptions are inappropriate in many instances. We fix the feedback rate and the traffic intensities at queue 1 and queue 2, and maintain a hyper-exponential server of Appendix A.2 at queue one.

In Table 1, the "exact" first two central moments for the second queue length are presented for a Poisson arrival stream against an Erlang renewal arrival process and a bursty *MMPP*. We also compare the actual results against our $k = 10$ approximation for these tandems in Table 2. The moments for the Poisson arrivals with the H_2 server can be considered as the results for a Poisson approximation to both the Erlang renewal and *MMPP* arrival processes. This Poisson approximation ($\frac{\sigma^2}{\mu^2} = 1$) over-estimates the stationary second queue length for the Erlang renewal arrival process ($\frac{\sigma^2}{\mu^2} = 0.5$) and greatly under-estimates for the *MMPP* ($\frac{\sigma^2}{\mu^2} = 4.9721$). Hence the case for using processes other than the Poisson process is obviated.

Table 1. Poisson and negative-exponential assumptions comparison.

Feedback Parameter	*Traffic Intensities*		*Two moments of the stationary second queue length for different arrival processes to the tandem (first server also given)*							
			Erlang H_2 server		Poisson H_2 server		MMPP H_2 server		Poisson neg-exp server	
ρ	η_1	η_2	$E[\mu]$	$E[\sigma^2]$	$E[\mu]$	$E[\sigma^2]$	$E[\mu]$	$E[\sigma^2]$	$E[\mu]$	$E[\sigma^2]$
0.25	0.25	0.25	0.302	0.266	0.336	0.369	0.593	1.719	0.333	0.361
0.25	0.25	0.5	0.839	0.897	1.006	1.52	3.244	26.326	1.000	1.500
0.25	0.25	0.75	2.368	5.415	3.008	9.777	12.631	232.060	3.000	9.750
0.25	0.5	0.25	0.317	0.312	0.338	0.378	0.434	0.763	0.333	0.361
0.25	0.5	0.5	0.898	1.086	1.02	1.580	2.122	10.446	1.000	1.500
0.25	0.5	0.75	2.494	5.882	3.041	9.959	10.764	189.796	3.000	9.750
0.25	0.75	0.25	0.330	0.351	0.340	0.382	0.369	0.487	0.333	0.361
0.25	0.75	0.5	0.969	1.360	1.033	1.640	1.329	3.201	1.000	1.500
0.25	0.75	0.75	2.741	7.354	3.107	10.475	6.383	63.321	3.000	9.750
0.5	0.25	0.25	0.301	0.265	0.335	0.367	0.590	1.702	0.333	0.361
0.5	0.25	0.5	0.837	0.892	1.004	1.512	3.243	26.314	1.000	1.500
0.5	0.25	0.75	2.364	5.404	3.005	9.766	12.632	232.081	3.000	9.750
0.5	0.5	0.25	0.317	0.31	0.337	0.374	0.432	0.753	0.333	0.361
0.5	0.5	0.5	0.893	1.064	1.014	1.554	2.107	10.291	1.000	1.500
0.5	0.5	0.75	2.479	5.808	3.026	9.873	10.751	189.614	3.000	9.750
0.5	0.75	0.25	0.329	0.348	0.339	0.379	0.368	0.482	0.333	0.361
0.5	0.75	0.5	0.961	1.325	1.025	1.602	1.316	3.122	1.000	1.500
0.5	0.75	0.75	2.706	7.128	3.071	10.214	6.323	62.198	3.000	9.750
0.75	0.25	0.25	0.301	0.263	0.334	0.364	0.587	1.682	0.333	0.361
0.75	0.25	0.5	0.835	0.886	1.002	1.506	3.242	26.308	1.000	1.500
0.75	0.25	0.75	2.361	5.396	3.002	9.757	12.633	232.109	3.000	9.750
0.75	0.5	0.25	0.315	0.305	0.336	0.369	0.429	0.738	0.333	0.361
0.75	0.5	0.5	0.887	1.043	1.007	1.527	2.091	10.125	1.000	1.500
0.75	0.5	0.75	2.464	5.747	3.012	9.803	10.74	189.466	3.000	9.750
0.75	0.75	0.25	0.328	0.343	0.337	0.373	0.366	0.474	0.333	0.361
0.75	0.75	0.5	0.951	1.285	1.014	1.557	1.299	3.020	1.000	1.500
0.75	0.75	0.75	2.671	6.923	3.035	9.971	6.260	61.015	3.000	9.750

Table 2. The $k = 10$ approximation comparison.

Feedback Parameter	*Traffic Intensities*		*Two moments of the stationary second queue length for different arrival processes to the tandem*							
			Erlang				*MMPP*			
			exact		$k = 10$		*exact*		$k = 10$	
ρ	η_1	η_2	$E[\mu]$	$E[\sigma^2]$	$E[\mu]$	$E[\sigma^2]$	$E[\mu]$	$E[\sigma^2]$	$E[\mu]$	$E[\sigma^2]$
0.25	0.25	0.25	0.302	0.266	0.302	0.266	0.593	1.719	0.593	1.719
0.25	0.25	0.5	0.839	0.897	0.839	0.897	3.244	26.326	3.242	26.236
0.25	0.25	0.75	2.368	5.415	2.368	5.415	12.631	232.060	12.620	231.179
0.25	0.5	0.25	0.317	0.312	0.317	0.312	0.434	0.763	0.434	0.763
0.25	0.5	0.5	0.898	1.086	0.898	1.086	2.122	10.446	2.104	10.035
0.25	0.5	0.75	2.494	5.882	2.495	5.883	10.764	189.796	10.238	156.993
0.25	0.75	0.25	0.330	0.351	0.330	0.351	0.369	0.487	0.369	0.487
0.25	0.75	0.5	0.969	1.360	0.969	1.360	1.329	3.201	1.329	3.192
0.25	0.75	0.75	2.741	7.354	2.742	7.364	6.383	63.321	5.907	47.481
0.5	0.25	0.25	0.301	0.265	0.301	0.265	0.590	1.702	0.590	1.701
0.5	0.25	0.5	0.837	0.892	0.837	0.892	3.243	26.314	3.241	26.222
0.5	0.25	0.75	2.364	5.404	2.364	5.404	12.632	232.081	12.621	231.212
0.5	0.5	0.25	0.317	0.31	0.317	0.309	0.432	0.753	0.432	0.753
0.5	0.5	0.5	0.893	1.064	0.893	1.064	2.107	10.291	2.089	9.875
0.5	0.5	0.75	2.479	5.808	2.479	5.808	10.751	189.614	10.218	156.403
0.5	0.75	0.25	0.329	0.348	0.329	0.348	0.368	0.482	0.368	0.482
0.5	0.75	0.5	0.961	1.325	0.961	1.325	1.316	3.122	1.315	3.113
0.5	0.75	0.75	2.706	7.128	2.707	7.137	6.323	62.198	5.843	46.415
0.75	0.25	0.25	0.301	0.263	0.301	0.263	0.587	1.682	0.587	1.682
0.75	0.25	0.5	0.835	0.886	0.835	0.887	3.242	26.308	3.240	26.217
0.75	0.25	0.75	2.361	5.396	2.361	5.396	12.633	232.109	12.622	231.269
0.75	0.5	0.25	0.315	0.305	0.315	0.305	0.429	0.738	0.429	0.738
0.75	0.5	0.5	0.887	1.043	0.887	1.043	2.091	10.125	2.073	9.704
0.75	0.5	0.75	2.464	5.747	2.464	5.747	10.740	189.466	10.200	155.930
0.75	0.75	0.25	0.328	0.343	0.328	0.343	0.366	0.474	0.366	0.474
0.75	0.75	0.5	0.951	1.285	0.951	1.285	1.299	3.020	1.298	3.012
0.75	0.75	0.75	2.671	6.923	2.672	6.930	6.260	61.015	5.777	45.317

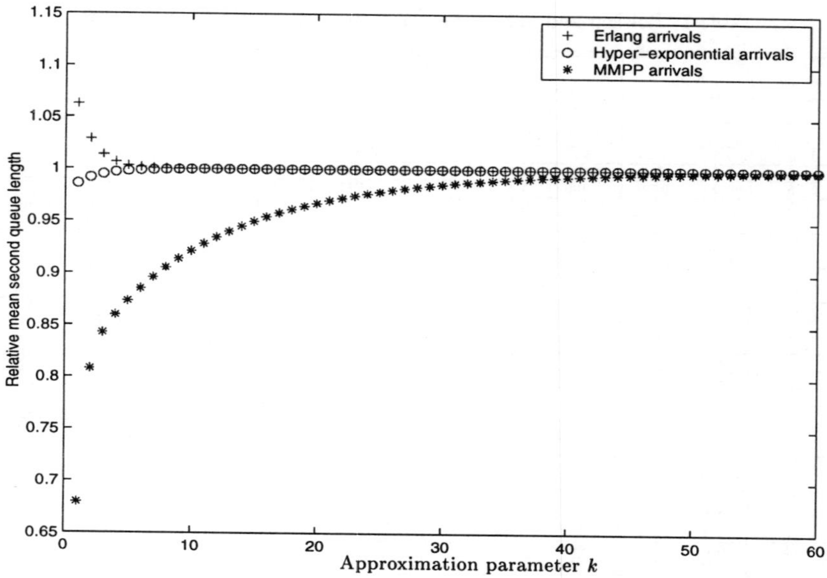

Figure 1. Relative convergence of the mean second queue length for the kth approximation to the *MAP/M/1* departure process with the Erlang, Hyper-exponential and *MMPP* arrival processes.

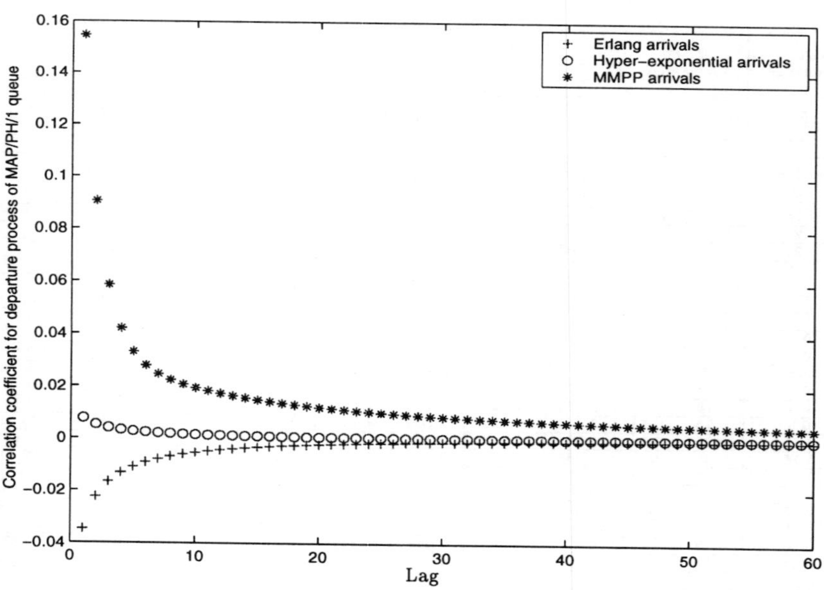

Figure 2. Correlation coefficients for the *MAP/M/1* departure process for the Erlang, Hyper-exponential and *MMPP* arrival processes.

The use of a negative exponential approximation for the server was also considered while maintaining the actual arrival process, but the results were not particularly sensitive to this modification and it was not presented.

We show the relative convergence of the approximations to the exact result in graphical form in Figure 1 for three arrival processes with tandem queue parameters $\rho = 0.25, \eta_1 = 0.75$ and $\eta_2 = 0.75$, by plotting the ratio (approximate mean)/(exact mean) for the second queue length against k. Note that the parameter set corresponds to the case when the *MAP* approximations to both the *MAP/M/1* and *MAP/PH/1* feedback queue realised their worst performance with the *MMPP* arrival process.

As an indicator of the correlation structures for the departure processes of *MAP/M/1* feedback queues using the parameter set for the first queue $\rho = 0.25, \eta_1 = 0.75$, the correlation coefficients are plotted against the actual lag in Figure 2. By comparing Figures 1 and 2, the convergence of the mean second queue length using the *MAP* approximations for increasing k towards the exact result is shown to be very dependent on the lag-correlation structure of the actual departure process of the first queue. That is, if the lag-correlation coefficients for the departure processes of the *MAP/M/1* feedback queue tend towards zero rapidly for an increasing number of lags, then the absolute difference between the mean second queue length calculated using the kth approximation *MAP* and the exact mean second queue length also tends towards zero rapidly for increasing k.

A long tailed correlation structure is seen for the *MAP/M/1* feedback queue with *MMPP* arrivals in Figure 2, yet the approximations still give good results albeit for a larger value of k.

4.2 Regime 2: Level-dependent feedback.

In this section, the level-dependent feedback has a geometrically decaying form given by

$$f(i) = \rho^i \quad \text{for all } i \geq 1, \text{ with } \rho \in [0, 1).$$

This regime seems to be sensible from the perspective that a customer wanting to rejoin a queue and finding a large number already present, is less likely to rejoin it.

Although increasing the level k of the *MAP* approximation gathers more information about the actual departure process of the *MAP/PH/1* queue with level-dependent feedback, the same special properties of the *MAP* approximations given in the previous section do not hold. The correlation structure of the departure process of the *MAP/PH/1* queue with level-dependent feedback is not exactly captured by the *MAP* approximations as it is in the non-level-dependent case. The physical reason for this can be seen by considering the proof given in Green [8,?] for the lag-correlation structure in the case of zero feedback. Alternatively a numerical calculation of the correlation coefficients for a sequence of approximation *MAP*s for $k = 2, 3, 4$ will reveal a numerical difference, albeit marginal. The lag-correlation coefficients for increasingly large k do however appear to converge to some number which is assumed to be the actual departure process lag-correlation coefficients.

Three different arrival processes as given in Appendix A.1 were used to feed the tandem feedback queueing system, including a bursty Hyper-exponential renewal process, an Erlang renewal process and a positively correlated, bursty Markov modulated Poisson process or *MMPP*. As an indicator of the relative burstiness of the processes, a sample path of one hundred arrivals is displayed in the appendix immediately after the description of the respective arrival process. A negative-exponential server with feedback mechanism as described was used at the first queue, as the process does not appear to be that sensitive to the form of the service time distribution. A negative exponential server was also used at the second or reference queue.

In Appendix A.1, the squared coefficient of variation $\frac{\sigma^2}{\mu^2}$ is given for each of the arrival processes. For the *MMPP*, which has a non-zero lag-correlation structure, an indication of the level of this structure is given by the first two lag correlation coefficients c_1 and c_2, calculated by (see Neuts [9])

$$c_i = \frac{\frac{1}{\boldsymbol{\nu} D_1 \boldsymbol{e}} \boldsymbol{\nu}[(-D_0)^{-1} D_1]^i (-D_0)^{-1} \boldsymbol{e} - \frac{1}{(\boldsymbol{\nu} D_1 \boldsymbol{e})^2}}{\frac{2}{\boldsymbol{\nu} D_1 \boldsymbol{e}} \boldsymbol{\nu}(-D_0)^{-1} \boldsymbol{e} - \frac{1}{(\boldsymbol{\nu} D_1 \boldsymbol{e})^2}}, \tag{20}$$

where $\boldsymbol{\nu}$ is the stationary probability vector of phase for the arrival process. Note that all *PH*-renewal arrival processes have a zero lag-correlation structure by their very nature.

The two servers have an infinite buffer with the traffic intensity at the second or reference queue being one of $\{0.25, 0.5, 0.75\}$. The traffic intensity at the first queue is a little non-descript as this now has level dependency. The service time distribution was configured as if the feedback rate was constant as in the previous level-independent case, as the code was set up initially to deal with constant feedback. The figure quoted in the tables of results indicates the traffic intensity had the queue been subject to a constant feedback rate. Note that this form of queue is stable as long as the queue without feedback is stable. This is irrespective of whether or not in the lower states the total arrival rate (exogenous and endogenous feedback) exceeds the service rate.

Table 3. The $k = 2$ and $k = 5$ approximation results for Erlang and Hyper-exponential arrivals to a feedback queue with level-dependent geometrically decaying feedback.

Feedback Parameter	*Traffic Intensities**		*Absolute difference of the kth approximation to the exact stationary second queue length for*							
			Erlang arrivals				*Hyper-exponential arrivals*			
			$k = 2$		$k = 5$		$k = 2$		$k = 5$	
ρ	η_1^*	η_2	E[μ]	E[σ^2]	E[μ]	E[σ^2]	E[μ]	E[σ^2]	E[μ]	E[σ^2]
0.25	0.25*	0.25	0.025	0.144	0.000	0.000	0.005	0.028	0.000	0.000
0.25	0.25*	0.5	0.060	0.311	0.000	0.000	0.012	0.055	0.000	0.000
0.25	0.25*	0.75	0.070	0.331	0.000	0.000	0.013	0.055	0.000	0.000
0.25	0.5*	0.25	0.125	0.731	0.000	0.000	0.027	0.133	0.000	0.001
0.25	0.5*	0.5	0.509	2.785	0.002	0.017	0.114	0.530	0.003	0.022
0.25	0.5*	0.75	0.753	3.788	0.008	0.062	0.169	0.743	0.008	0.050
0.25	0.75*	0.25	0.258	1.028	0.000	0.001	0.057	0.198	0.000	0.002
0.25	0.75*	0.5	1.540	6.310	0.015	0.098	0.355	1.246	0.017	0.091
0.25	0.75*	0.75	3.252	13.610	0.152	0.942	0.836	2.968	0.115	0.578
0.5	0.25*	0.25	0.029	0.167	0.000	0.000	0.006	0.033	0.000	0.000
0.5	0.25*	0.5	0.111	0.561	0.000	0.000	0.023	0.102	0.000	0.001
0.5	0.25*	0.75	0.195	0.898	0.000	0.002	0.038	0.162	0.000	0.002
0.5	0.5*	0.25	0.107	0.612	0.000	0.000	0.022	0.102	0.000	0.000
0.5	0.5*	0.5	0.546	2.893	0.001	0.007	0.112	0.493	0.001	0.009
0.5	0.5*	0.75	1.085	5.386	0.007	0.053	0.234	0.989	0.007	0.045
0.5	0.75*	0.25	0.196	0.762	0.000	0.000	0.039	0.131	0.000	0.000
0.5	0.75*	0.5	1.252	4.867	0.003	0.019	0.247	0.818	0.004	0.020
0.5	0.75*	0.75	2.988	11.711	0.040	0.244	0.641	2.111	0.032	0.159
0.75	0.25*	0.25	0.030	0.168	0.000	0.000	0.006	0.032	0.000	0.000
0.75	0.25*	0.5	0.144	0.714	0.000	0.000	0.028	0.125	0.000	0.001
0.75	0.25*	0.75	0.300	1.366	0.000	0.003	0.057	0.241	0.000	0.003
0.75	0.5*	0.25	0.091	0.510	0.000	0.000	0.018	0.080	0.000	0.000
0.75	0.5*	0.5	0.524	2.702	0.000	0.002	0.099	0.415	0.000	0.003
0.75	0.5*	0.75	1.181	5.738	0.002	0.020	0.232	0.940	0.003	0.018
0.75	0.75*	0.25	0.154	0.588	0.000	0.000	0.029	0.093	0.000	0.000
0.75	0.75*	0.5	1.023	3.837	0.000	0.003	0.180	0.569	0.001	0.004
0.75	0.75*	0.75	2.559	9.531	0.007	0.043	0.474	1.474	0.006	0.031

Table 4. The $k = 2$, $k = 5$ and $k = 10$ approximation results for *MMPP* arrivals to a feedback queue with level-dependent geometrically decaying feedback.

Feedback Parameter	*Traffic Intensities**		*Absolute difference of the kth approximation to the exact stationary second queue length for MMPP arrivals*					
			$k = 2$		$k = 5$		$k = 10$	
ρ	η_1^*	η_2	E[μ]	E[σ^2]	E[μ]	E[σ^2]	E[μ]	E[σ^2]
0.25	0.25*	0.25	1.826	6.290	0.051	0.328	0.001	0.007
0.25	0.25*	0.5	1.338	4.677	0.044	0.224	0.001	0.008
0.25	0.25*	0.75	0.438	1.621	0.006	0.028	0.000	0.001
0.25	0.5*	0.25	3.589	10.451	0.152	0.783	0.006	0.035
0.25	0.5*	0.5	9.788	28.198	2.633	10.555	0.715	3.563
0.25	0.5*	0.75	6.690	20.307	1.953	7.668	0.629	2.907
0.25	0.75*	0.25	3.838	9.318	0.186	0.644	0.007	0.030
0.25	0.75*	0.5	14.024	33.600	4.795	14.882	1.461	5.621
0.25	0.75*	0.75	25.063	56.098	14.212	39.303	7.620	25.470
0.5	0.25*	0.25	1.962	6.306	0.022	0.141	0.000	0.001
0.5	0.25*	0.5	4.067	12.973	0.217	1.085	0.005	0.034
0.5	0.25*	0.75	3.049	9.969	0.159	0.718	0.003	0.017
0.5	0.5*	0.25	3.129	8.405	0.029	0.159	0.000	0.001
0.5	0.5*	0.5	9.210	24.960	0.943	4.247	0.089	0.512
0.5	0.5*	0.75	14.623	38.085	3.425	13.214	0.856	4.213
0.5	0.75*	0.25	3.209	7.447	0.032	0.112	0.000	0.001
0.5	0.75*	0.5	9.473	22.217	1.165	3.837	0.125	0.509
0.5	0.75*	0.75	18.536	41.494	5.688	17.184	1.694	6.399
0.75	0.25*	0.25	1.864	5.700	0.007	0.038	0.000	0.000
0.75	0.25*	0.5	5.408	16.136	0.178	0.874	0.001	0.006
0.75	0.25*	0.75	8.393	24.560	0.742	3.248	0.013	0.089
0.75	0.5*	0.25	2.872	7.342	0.005	0.026	0.000	1.042
0.75	0.5*	0.5	7.823	19.871	0.182	0.846	0.001	0.008
0.75	0.5*	0.75	14.345	34.798	1.114	4.633	0.036	0.221
0.75	0.75*	0.25	2.931	6.633	0.005	0.017	0.000	0.000
0.75	0.75*	0.5	7.321	16.508	0.171	0.564	0.001	0.006
0.75	0.75*	0.75	13.176	28.864	1.048	3.320	0.043	0.173

The infinite sum in Equation (17) has to be truncated at some point in order to find an approximation to the scaling factor γ_k for the construction of the kth approximation *MAP*. For the results presented here, this was accomplished by truncating the sum in Equation (17) at the value i for which $\pi_i e < 10^{-24}$ and hence $\rho^i \pi_i e << 10^{-24}$.

The results presented in Tables 3 and 4 are for the absolute percentage difference between the exact first/second central moment for the second queue length and the first/second central moment for the second queue length as calculated from the kth approximation. That is, the results given are for

$$\frac{100\ |\text{exact- approximation}|}{\text{exact}},$$

for the first and second central moments of the queue length of the stationary second queue.

Once again it should be noted that the correlation structure of the departure process from the *MMPP/M/1* feedback queue with level-dependent feedback has a long tail for the parameter set at the first queue of $\rho = 0.25, \eta_1 = 0.75$. This requires a substantial level of k for an adequate approximation result. By observing the convergence of the calculated lag-correlation coefficients for the approximation *MAP*s for higher k, we can approximate the actual lag-correlation coefficients for the departure process from the *MMPP/M/1* feedback queue with level-dependent feedback. A plot of the relative convergence of the lag-correlation coefficients for lags $1 - 30$ is given in Figure 3 for the approximations with parameters $k = 2, 3, \ldots, 30$.

The mesh of Figure 3 makes it appear that the nice property of the approximation *MAP*s in the previous section is replicated in the case of level-dependent feedback but upon closer inspection of the calculated values they are marginally different. Each line of the mesh corresponds to either one particular lag-correlation coefficient or a particular value of the approximation parameter k according to the labelled axis.

We again plot an approximate set of correlation coefficients for the departure process against the actual lag in Figure 4 and then plot the relative convergence for the approximation to the mean second queue length against parameter k in Figure 5. By comparison of the two plots the dependency again appears to be that a larger value of k is required for good accuracy if the lag-correlation coefficient is significant for higher lags (a nice visual correspondence between plots).

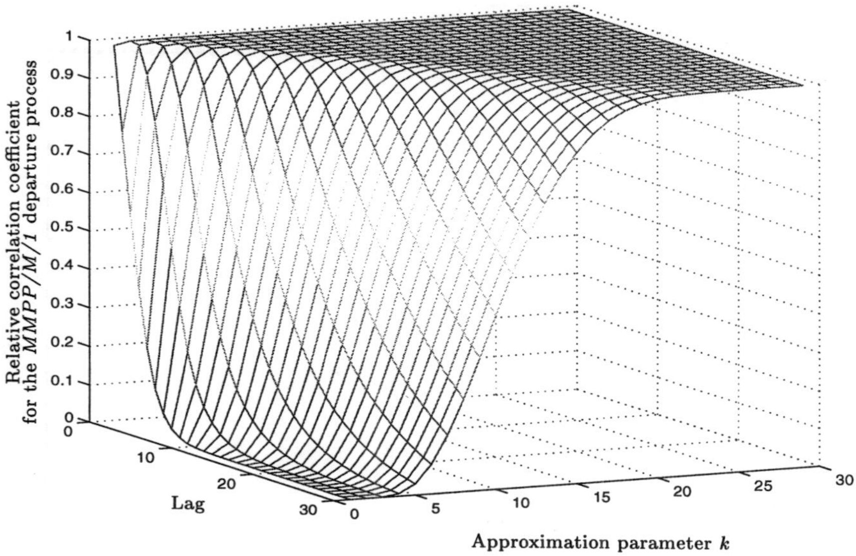

Figure 3. Relative convergence of the lag-correlation coefficients for the kth approximation to the *MMPP/M/1* departure process.

5 Summary

The *MAP* approximations to the departure process from the two forms of *MAP/PH/1* feedback queue have been shown to be very effective, with the level of accuracy being determined by the parameter k. The special properties of the *MAP* approximations to the departure process from a *MAP/PH/1* feedback queue with constant or level-independent feedback are that the correlation coefficients are exactly matched up to the $k - 1^{st}$ for the kth approximation. This special property is not shared for *MAP* approximations to the departure process for *MAP/PH/1* feedback queues with level-dependent feedback. Although this special property is not exactly replicated, it appears that in the case of level-dependent feedback the difference to an exact match is only marginal.

A high dependency between the level required of the parameter k to achieve good accuracy and the significance of the lag correlation coefficient at higher lags is evident. That is, if the lag-correlation coefficients for the departure processes of the *MAP/M/1* feedback queue tend towards zero rapidly for an increasing number of lags, then the absolute difference between the mean second queue length calculated using the kth approximation *MAP* and the exact mean second queue length also tends towards zero rapidly for increasing k.

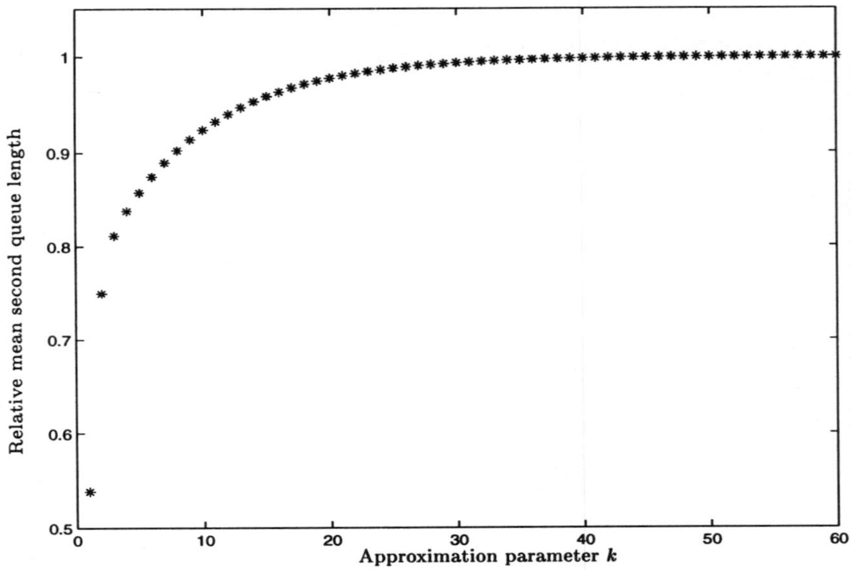

Figure 4. Relative convergence of the mean second queue length for the kth approximation to the departure process of the *MMPP/M/1* feedback queue.

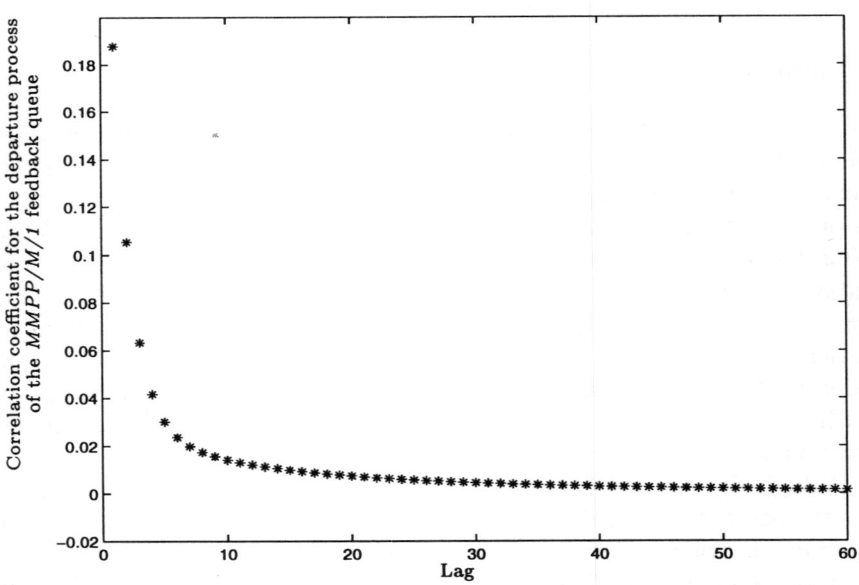

Figure 5. Correlation coefficients for the departure process of the *MMPP/M/1* feedback queue.

The feedback node is used in Jackson networks where of course the exogenous arrivals are Poisson. From the perspective of Jackson like networks with non-Poisson arrivals this type of analysis is promising. The problems associated with dimensionality however are restrictive if the approximations are directly used in a pedestrian manner in a network situation. That is, the systematic analysis of nodes in isolation, using previous departure approximations without modification. This will be addressed in future.

Appendix

A Tandem queue processes

A.1 The arrival processes

1. Erlang (E_2), $\frac{\sigma^2}{\mu^2} = 0.5000$.

2. Hyper-exponential, $D_0 = \begin{pmatrix} -2 & 0 \\ 0 & -1 \end{pmatrix}$, $\boldsymbol{\alpha} = (0.75, 0.25)$, $\frac{\sigma^2}{\mu^2} = 1.2400$.

3. Markov Modulated Poisson process *MMPP* (Bursty)

$$D_0 = \begin{pmatrix} -10.0 & 1.0 \\ 0.4 & -0.8 \end{pmatrix}, D_1 = \begin{pmatrix} 9.0 & 0.0 \\ 0.0 & 0.4 \end{pmatrix},$$

$$\frac{\sigma^2}{\mu^2} = 4.9721, c_1 = 0.1892, c_2 = 0.0896.$$

A.2 The first server

Hyper-exponential

$$S = \begin{pmatrix} -6 & 0 \\ 0 & -12 \end{pmatrix}, \beta = (0.2, 0.8), \frac{\sigma^2}{\mu^2} = 1.2222.$$

References

1. N.G. Bean, D.A. Green, and P.G. Taylor. Approximations to the Output Process of *MAP/PH/1* Queues. In *Advances in Matrix Analytic Methods for Stochastic Models - Proceedings of the 2nd International Conference on Matrix Analytic Methods*, pages 151–169. Notable Publications Inc., NJ, 1998.
2. D.A. Green. A (*MAP/PH/1*) Feedback Queueing Model for a HF Channel using the ARQ protocol . *Submitted to JSAC Wireless Communications Series*, 2001.
3. J.R. Jackson. Networks of Waiting Lines. *Operations Research*, 5:518–521, 1957.
4. D.A. Green. Lag Correlations of Approximating Departure Processes for *MAP/PH/1* Queues. In *Proceedings of the third International Conference on Matrix-Analytic Methods in Stochastic Models*, pages 135–151, NJ, 2000. Notable Publications Inc.
5. L.W. Bright and P.G. Taylor. Equilibrium Distributions for Level-Dependent Quasi-Birth-and Death Processes. In *Matrix Analytic Methods in Stochastic Models (Flint, MI)*. Dekker, New York, 1997.
6. M.F. Neuts. *Matrix-geometric Solutions in Stochastic Models : An Algorithmic Approach.* The John Hopkins University Press, Baltimore, 1981.
7. N.G. Bean, J. Li, and P.G. Taylor. Some Asymptotic Properties of Two-stage Tandem Networks of *PH/PH/1* Queues. In *Advances in Matrix Analytic Methods for Stochastic Models - Proceedings of the 2nd International Conference on Matrix Analytic Methods*, pages 171–193. Notable Publications Inc., NJ, 1998.
8. D.A. Green. *Departure Processes from MAP/PH/1 Queues.* PhD thesis, Department of Applied Mathematics, The University of Adelaide, 1998.
9. M.F. Neuts. *Algorithmic Probability: A collection of problems.* Chapman and Hall, London, 1995.

A MATRIX ANALYTIC MODEL FOR MACHINE MAINTENANCE

DAVID GREEN

Department of Applied Mathematics, The University of Adelaide, South Australia, 5005.

E-mail: dgreen@maths.adelaide.edu.au

ANDREW V. METCALFE

Department of Applied Mathematics, The University of Adelaide, South Australia, 5005.

E-mail: ametcalf@maths.adelaide.edu.au

DAVID C. SWAILES

Department of Engineering Mathematics, Newcastle University, United Kingdom

E-mail: D.C.Swailes@newcastle.ac.uk

In this paper, we consider a production line consisting of machines working in series, at the same speed with independent exponential times before failure and times for repair. It can be shown that this production line has an exponential time before failure with rate equal to the sum of the individual machine failure rates. The repair time for the line is distributed as a mixture of exponentials (hyper-exponential). We compare an analysis using non-linear cost functions, based on the hyper-exponential distribution of repair times for the line with an approximation that assumes an exponential repair time. The approximate exponential repair time has a mean equal to a weighted average of the individual repair times, with weights proportional to the failure rates. Non-linear cost functions allow for differences between costs of: overtime, extra shifts, and failure to meet deadlines; or for deterioration of a product if it is left standing on a production line. We use a realistic example to demonstrate that the approximate analysis can underestimate costs by over 10%. We also present models for two production lines with one or two repair crews. In the instance of a single repair crew, we distinguish between a zero priority and a priority for line one case, and show, in contrast to a single line, that the stationary distributions are slightly different from approximations which assume the line repair times are exponentially distributed.

Keywords: *Matrix analytic methods; phase-type Markov models; hyper-exponential distribution; maintenance strategies; reliability; maintenance costs.*

1 Introduction

Many manufactured goods are made on production lines that consist of machines arranged in series. Typical examples are car manufacture and the assembly of printed circuit boards. The capital cost of setting up modern production lines, especially those which make extensive use of robots, is high, and it is important to keep them working in order to justify the investment. An inherent drawback of series systems is that if any machine breaks down the entire line will be stopped for the duration of the repair. Having duplicate machines in parallel will greatly improve the reliability of the line, but this is usually prohibitively expensive. A more practical alternative is to attempt to prevent machine breakdowns, and there has been a move towards preventative maintenance which can often be undertaken by the employees who are responsible for operating individual machines , or of monitoring the performance of a completely automated line. Some companies choose to formalise this approach as "total productive maintenance" (TPM), which was defined by the Japan Institute for Plant Maintenance (JIPM) [4] in 1971 as "a system of maintenance which covers the entire life of the equipment in every division including planning, manufacturing and maintenance". The aim of TPM is to achieve, unrelentingly, "zero-defect" and "zero-breakdown" (Nakajima 1988, 1991 [5,6]) in a plant, and thereby provide support for manufacturing strategies such as Just-In-Time (JIT) and Six Sigma (e.g. Bendell 2000 [3]) in a company's strive for business excellence (Al-Hassan *et al.* 2000a [1]).

However, many managers in Europe still have to be convinced of the usefulness of TPM and many more are perhaps unsure of the losses which can accrue as a consequence of not having an appropriate maintenance programme implemented in their organisations. In a recent paper (Al-Hassan *et al.* 2000b [2]) a simple Markov model (sMm) was proposed as a technique for identifying the prime costs involved in production line downtimes, and for assessing the benefits of the replacement of unreliable equipment and improved maintenance strategies. This work assumed that a fault on a single machine would stop a production line, that a factory consisted of several identical production lines, that monetary loss was proportional to downtime, and that lost production on one line per day could be made up by overtime on another. The purpose of the sMm was to support decisions about equipment maintenance and replacements.

In principle it is straightforward to modify the sMm to accommodate different production lines, but as the number of states increases factorially the algebra becomes awkward for more than three different lines.

The sMm assumed that the times between individual machine breakdowns, on a single line, are independent exponential random variables and that repair times are also independent exponential random variables. A single line was then treated as a fictitious single machine with a failure rate equal to the sum of the failure rates of the constituent machines, and a repair rate equal to a weighted average of the constituent machine repair rates. It was also assumed that the times between failures, and repair times, of the fictitious machine are exponentially distributed. While the assumption of an exponential distribution of times between failures of the fictitious machine can be justified, the repair time is not precisely exponential. In this paper we use matrix analytic methods to model the repair time distribution as a mixture of exponentials. This is achieved by supplementing the two state, working and under repair, Markov process for the fictitious machine with a random environment which models the changes in repair time distributions. We investigate the sensitivity of the sMm to the simplifying assumption that it is a single exponential distribution.

2 Matrix analytic models for one and two production lines

2.1 Model for One Production Line

We consider a line consisting of N machines arranged in series. These machines have failure rates λ_j, and hence mean time between failures (MTBF) $1/\lambda_j$, $j = 1, ..., n$. Now, if any one machine fails the line fails, and so if W_j denotes the time before failure for machine j, and W the time until the line fails, then

$$\Pr(W > w) \quad = \quad \Pr(W_1 > w \ , \ \dots \ , \ W_N > w) \ . \tag{1}$$

With the W_j independent and exponentially distributed it follows that W has an exponential distribution with rate
$\lambda = \lambda_1 + \cdots + \lambda_N$, and hence the MTBF for the line is $1/\lambda$.

Now define $\boldsymbol{\beta}$ as the vector of proportions of line failures, which are due to machine j for $j = 1, \dots N$. That is

$$\boldsymbol{\beta} = \frac{1}{\lambda}(\lambda_1, \lambda_2, \dots, \lambda_N) \, .$$

Let μ_j be the repair rate for machine j. Then upon breakdown, the time to repair (T) is hyper-exponentially distributed as $(\boldsymbol{\beta}, S)$ (see Neuts 1981 [7]),

where

$$S = \begin{bmatrix} -\mu_1 & 0 & \cdots & 0 \\ 0 & -\mu_2 & & \vdots \\ \vdots & 0 & \ddots & 0 \\ 0 & \cdots & 0 & -\mu_N \end{bmatrix}.$$

Then the expected time to repair is given by

$$E[T] = \frac{1}{\mu} = -\beta S^{-1} e$$
$$= \frac{1}{\lambda} \sum_j \frac{\lambda_j}{\mu_j}.$$

We note that $E[T]$ is a weighted average of the mean repair times for the individual machines, the weights being proportional to the failure rates. The variance σ^2 of T is given by

$$\sigma^2 = \frac{2}{\lambda} \sum_j \frac{\lambda_j}{\mu_j^2} - \left(\frac{1}{\lambda} \sum_j \frac{\lambda_j}{\mu_j} \right)^2 .$$

The approximation used in the sMm assumes repair times have an exponential distribution with the same mean $1/\mu$, as the hyper-exponential distribution. In general, the variance of an exponential variable will be less than that of a hyper-exponential variable with the same mean. It follows that the fictitious machine approximation will underestimate the costs, if for example, the cost per minute of downtime increases with the length of time for which the line has stopped. The size of this error is investigated in Section 3.

The hyper-exponential model also distinguishes different machines under repair as different states. This would be useful if repair costs are incorporated in the model and differ for different machines. The $(N+1) \times (N+1)$ rate matrix is given by

$$Q = \begin{bmatrix} -\lambda & \lambda\beta \\ \boldsymbol{S}^0 & S \end{bmatrix},$$

where it is convenient to label the $N+1$ states using the natural partition by 0 and $(1, j)$ for $j \in \{1, \ldots, N\}$. Here state 0 indicates that the production line is operational and state $(1, j)$ corresponds to the j rows of the matrices $\boldsymbol{S}^0$ and S, and indicates that the production line is stopped with machine j under repair. We refer to states 0 and 1 with state 1 having associated phases (indicating which machine is under repair) $j \in \{1, \ldots .\}$.

2.2 Models For Multiple Lines

In this section we present models for two lines with: two repair crews; one repair crew with no priority; and one repair crew with priority for line 1. The rate matrices for these three cases are followed by approximations which assume, as discussed in Section 2.1, that the repair rates for each line are exponential with a mean equal to a weighted average of the constituent machine repair rates. Unlike the single line case, the stationary distributions for the models with the one repair crew, with states combined for comparability with their approximations, are not identical to their approximations. However, the discrepancy is typically rather small and a numerical example is given in Section 3.

The notation is similar to that of Section 2.1, with subscripts 1 and 2 added to denote lines 1 and 2 respectively. Also m and n will be used for the number of machines in line 1 and line 2 respectively.

(i) Two repair crews

The rate matrix of order $\Big(1+(m+n)+(mn)\Big)$ is again naturally partitioned to have states 0, $(1,j)$ and $(2,k)$, which correspond to: both production lines operating; a single production line stopped with one machine under repair; and two production lines stopped with two machines concurrently under repair. We assume that there is no cooperation between repair crews. The indices j and k may be interpreted as follows.

$$j = \begin{cases} 1,\ldots,m & \text{machine } j \text{ on line one is under repair} \\ m+1,\ldots,m+n & \text{machine } j-m \text{ on line two is under repair} \end{cases}$$

$$k = \begin{cases} 1,\ldots,n & \begin{array}{l}\text{machine 1 on line one is under repair and} \\ \text{machine } k \text{ on line two is under repair}\end{array} \\ n+1,\ldots,2n & \begin{array}{l}\text{machine 2 on line one is under repair and} \\ \text{machine } k-n \text{ on line two is under repair}\end{array} \\ \vdots & \vdots \\ (m-1)n+1,\ldots,mn & \begin{array}{l}\text{machine } m \text{ on line one is under repair and} \\ \text{machine } k-(m-1)n \text{ on line two is under repair.}\end{array} \end{cases}$$

$$Q=\begin{bmatrix} (-\lambda_1-\lambda_2) & (\ \lambda_1\boldsymbol{\beta}_1 \quad \lambda_2\boldsymbol{\beta}_2\ ) & (\ 0\ ) \\ \begin{pmatrix} \boldsymbol{S}_1^0 \\ \boldsymbol{S}_2^0 \end{pmatrix} & \begin{pmatrix} S_1-\lambda_2 I_1 & 0 \\ 0 & S_2-\lambda_1 I_2 \end{pmatrix} & \begin{pmatrix} I_1\otimes\lambda_2\boldsymbol{\beta}_2 \\ \lambda_1\boldsymbol{\beta}_1\otimes I_2 \end{pmatrix} \\ (\ 0\ ) & (I_1\otimes\boldsymbol{S}_2^0 \quad \boldsymbol{S}_1^0\otimes I_2) & (S_1\oplus S_2) \end{bmatrix}$$

where $S_1 \oplus S_2 = S_1 \otimes I_2 + I_1 \otimes S_2$.

(ii)One crew no priority

The partitioned rate matrix of order $\Big(1+(m+n)+(m+n)\Big)$ again has states 0, $(1,j)$ and $(2,k)$, where states 0 and $(1,j)$ have the same interpretation as for two repair crews. Here, state $(2,k)$ corresponds to the situation where two production lines are broken down with only one machine under repair. When a single line is under repair and the other line fails, we do not need to establish which machine has failed on this line until immediately after the repairs of the initial breakdown are complete. This leads to a considerable saving in the order of the rate matrix needed to model the situation. The index k of state $(2,k)$ in this situation has the following interpretation.

$$k=\begin{cases} 1,\ldots,m & \text{machine } k \text{ on line one is under repair} \\ m+1,\ldots,m+n & \text{machine } k-m \text{ on line two is under repair.} \end{cases}$$

$$Q=\begin{bmatrix} (-\lambda_1-\lambda_2) & (\ \lambda_1\boldsymbol{\beta}_1 \quad \lambda_2\boldsymbol{\beta}_2\ ) & (\ 0\ ) \\ \begin{pmatrix} \boldsymbol{S}_1^0 \\ \boldsymbol{S}_2^0 \end{pmatrix} & \begin{pmatrix} S_1-\lambda_2 I_1 & 0 \\ 0 & S_2-\lambda_1 I_2 \end{pmatrix} & \begin{pmatrix} \lambda_2 I_1 & 0 \\ 0 & \lambda_1 I_2 \end{pmatrix} \\ \begin{pmatrix} 0 \\ 0 \end{pmatrix} & \begin{pmatrix} 0 & \boldsymbol{S}_1^0\otimes\boldsymbol{\beta}_2 \\ \boldsymbol{\beta}_1\otimes\boldsymbol{S}_2^0 & 0 \end{pmatrix} & \begin{pmatrix} S_1 & 0 \\ 0 & S_2 \end{pmatrix} \end{bmatrix}$$

(iii)One crew priority line 1

The rate matrix now is of order $\Big(1+(m+n)+(m+nm)\Big)$ and again using the natural partitioning has states 0, $(1,j)$ and $(2,k)$, with states 0 and $(1,j)$ having the same interpretation as for two repair crews. In state $(2,k)$ there are two possible scenarios which must be dealt with distinctly. When line one is under repair and line two fails, the repair crew does not have to discover which machine has failed on line two until immediately after line one is repaired. On the other hand, if line two is under repair when line one fails, the repair crew immediately leave line two and repair line one. The repair crew in this instance must remember which machine on line two to return to when line one is repaired. This incurs a higher order for the rate matrix. Note, no allowance is made for any reduction in the repair time for the machine on line 2 on account of the repair having been started (Exponential repair times). The interpretation for the index k of state $(2, k)$ here is given by

$$k = \begin{cases} 1, \ldots, m & \text{machine } k \text{ on line one is under repair} \\ m+1, \ldots, 2m & \text{machine } k-m \text{ on line one is under repair and} \\ & \text{crew must return to machine 1 on line two} \\ \vdots & \vdots \\ mn+1, \ldots, (m+1)n & \text{machine } k-mn \text{ on line one is under repair and} \\ & \text{crew must return to machine } n \text{ on line two.} \end{cases}$$

$$Q = \begin{bmatrix} (-\lambda_1 - \lambda_2) & (\ \lambda_1 \boldsymbol{\beta}_1 \quad \lambda_2 \boldsymbol{\beta}_2\ ) & (\ 0 \quad 0\ ) \\ \begin{pmatrix} \boldsymbol{S}_1^0 \\ \boldsymbol{S}_2^0 \end{pmatrix} & \begin{pmatrix} S_1 - \lambda_2 I_1 & 0 \\ 0 & S_2 - \lambda_1 I_2 \end{pmatrix} & \begin{pmatrix} \lambda_2 I_1 & 0 \\ 0 & I_2 \otimes \lambda_1 \boldsymbol{\beta}_1 \end{pmatrix} \\ \begin{pmatrix} 0 \\ 0 \end{pmatrix} & \begin{pmatrix} 0 & \boldsymbol{S}_1^0 \otimes \boldsymbol{\beta}_2 \\ 0 & I_2 \otimes \boldsymbol{S}_1^0 \end{pmatrix} & \begin{pmatrix} S_1 & 0 \\ 0 & I_2 \otimes S_1 \end{pmatrix} \end{bmatrix}$$

(iv)Approximation for two repair crews

There are now simply four states, indicating whether each of the two lines is operational or under repair: both lines operating; line 1 under repair and line 2 operating; line 1 operating and line 2 under repair; both lines under repair.

$$Q = \begin{bmatrix} (-\lambda_1 - \lambda_2) & (\ \lambda_1 \quad \lambda_2\ ) & (\ 0\ ) \\ \begin{pmatrix} \mu_1 \\ \mu_2 \end{pmatrix} & \begin{pmatrix} -\mu_1 - \lambda_2 & 0 \\ 0 & -\mu_2 - \lambda_1 \end{pmatrix} & \begin{pmatrix} \lambda_2 \\ \lambda_1 \end{pmatrix} \\ (\ 0\ ) & (\ \mu_2 \quad \mu_1\ ) & (-\mu_1 - \mu_2) \end{bmatrix}$$

(v)Approximation for one crew no priority

The states are now defined as: both lines operating; line 1 under repair and line 2 operating; line 1 operating and line 2 under repair; line 1 under repair and line 2 failed and awaiting repair; line 2 under repair and line 1 failed and awaiting repair.

$$Q = \begin{bmatrix} (-\lambda_1 - \lambda_2) & \begin{pmatrix} \lambda_1 & \lambda_2 \end{pmatrix} & \begin{pmatrix} 0 & 0 \end{pmatrix} \\ \begin{pmatrix} \mu_1 \\ \mu_2 \end{pmatrix} & \begin{pmatrix} -\mu_1 - \lambda_2 & 0 \\ 0 & -\mu_2 - \lambda_1 \end{pmatrix} & \begin{pmatrix} \lambda_2 & 0 \\ 0 & \lambda_1 \end{pmatrix} \\ \begin{pmatrix} 0 \\ 0 \end{pmatrix} & \begin{pmatrix} 0 & \mu_1 \\ \mu_2 & 0 \end{pmatrix} & \begin{pmatrix} -\mu_1 & 0 \\ 0 & -\mu_2 \end{pmatrix} \end{bmatrix}$$

(vi)Approximation for one repair crew priority line 1

The states are: both lines operating; line 1 under repair and line 2 operating; line 2 under repair and line 1 operating; line 1 under repair and line 2 failed awaiting repair.

$$Q = \begin{bmatrix} (-\lambda_1 - \lambda_2) & \begin{pmatrix} \lambda_1 & \lambda_2 \end{pmatrix} & \begin{pmatrix} 0 \end{pmatrix} \\ \begin{pmatrix} \mu_1 \\ \mu_2 \end{pmatrix} & \begin{pmatrix} -\mu_1 - \lambda_2 & 0 \\ 0 & -\mu_2 - \lambda_1 \end{pmatrix} & \begin{pmatrix} \lambda_2 \\ \lambda_1 \end{pmatrix} \\ \begin{pmatrix} 0 \end{pmatrix} & \begin{pmatrix} 0 & \mu_1 \end{pmatrix} & (-\mu_1) \end{bmatrix}$$

The adequacy of the approximations for specific, but realistic, numerical examples is investigated in the next section.

3 Numerical examples

3.1 One production line

As a simple example of a single line production system we consider the assembly and packaging of audio cassettes. These are typically produced in large quantities from fully automated assembly lines. Each line consists of a sequence of multi-tasking machines, with each machine performing a number of different operations in the assembly process before delivering the unit to the next machine. For example

Machine 1:

Position lower half of cassette housing; locate and fix transparent plastic window; locate and fix metal foil; locate guide rollers.

Machine 2:

Wind audio tape onto spools; locate spools onto lower cassette housing and thread tape through integral feeder guides and around guide rollers.

Machine 3:

Locate and fix window and metal foil to upper half of cassette housing; position onto lower half of housing and screw two halves together.

Machine 4:

Test functionality of unit.

Machine 5:

Add labels to sides of cassette.

Machine 6:

Assemble parts of cassette box; place cassette in box.

The various machines are fully synchronised and rely on the immediate supply of units from the previous machine. There is therefore no storage of units between machines, and should any function of any one of the machines fail the production process is halted.

Table 1 gives illustrative values for the failure and repair rates, λ_j and μ_j respectively, of the various machines.

Machine No.	A1	A2	A3	A4	A5	A6
Failure rate (per hour)	0.0133	0.0179	0.0086	0.0053	0.0102	0.0057
Repair rate (per hour)	0.5000	0.2500	0.3333	0.2000	0.3333	1.0000

Table 1: Failure and repair rates for audio cassette assembly line.

From these figures we obtain the line failure rate $\lambda = 0.0611/hour$, and the sMm approximation for the line repair rate $\mu = 0.3266/hour$.

The cost incurred when the line fails depends crucially on the magnitude of the down-time of the line, and the significance of lost production: The cost per time is not a constant but increases with down-time. This is because whilst it is relatively easy to compensate for short stoppages, longer stoppages incur additional overhead costs as well overtime payments to machine supervisors. Further, serious faults requiring extensive maintenance time can result in penalty payments for failure to deliver contracted goods on time.

The precise dependence of the cost per unit time, $\dot{C}$, on the line repair time, T, will clearly depend on the particulars of a given manufacturer. In this paper we consider two simple models; (a) $\dot{C}$ varies linearly with T, and (b) $\dot{C}$ piecewise constant.

(a) $\dot{C}$ and linear function of T.

We model the cost per unit time in the form $\dot{C} = kT$, k constant, so that the cost function $C(T)$ is given by

$$C(T) \quad = \quad \frac{1}{2}kT^2 \, .$$

This gives the following expression for the estimated mean cost per line failure, based on the sMm (exponential) approximation for T with line failure rate μ

$$\overline{C}_\mu \quad = \quad \int_0^\infty \frac{1}{2}kt^2 e^{-\mu t}\, dt \quad = \quad \frac{k}{\mu^2} \, .$$

The mean cost based on the correct, hyper-exponential, distribution for T is given by

$$\overline{C} \;=\; \int_0^\infty \frac{1}{2}kt^2 \beta e^{St} \boldsymbol{S}^0 \, dt \;=\; k\beta S^{-2} \boldsymbol{e} \;= \frac{1}{\lambda}\sum_j \lambda_j \frac{k}{\mu_j^2} \;=\; \frac{1}{\lambda}\sum_j \lambda_j \overline{C}_{\mu_j} \, ,$$

where $\overline{C}_{\mu_j}$ denotes the sMm approximation with $\mu = \mu_j$.

It follows that $\overline{C}_\mu = 2\big(\overline{C} - \; k\sigma^2/2\big)$, indicating that the approximation $\overline{C}_\mu$ will underestimate the true value $\overline{C}$ significantly as the variance of T increases. Using the data given in Table 1, and taking $k = 1$ (for simplicity), we obtain the approximation $\overline{C}_\mu = 9.3762$ which significantly underestimates the correct value $\overline{C} = 10.6075$, with a percentage error of approximately 11.6%

(b) $\dot{C}$ a piecewise constant function of T.

Here we consider a model for $\dot{C}$ in the form

$$\dot{C} = \begin{cases} \gamma_1 & 0 < t < t_1 \\ \gamma_2 & t_1 < t < t_2 \\ \gamma_3 & t > t_2 \end{cases}$$

where the γ_i are constants, with $\gamma_1 < \gamma_2 < \gamma_3$. This gives

$$\overline{C}_\mu = \frac{1}{\mu}\Big(e^{-\mu t_1}(\gamma_2 - \gamma_1) + e^{-\mu t_2}\big((\gamma_2 - \gamma_1)t_1 + (\gamma_3 - \gamma_2)\big) + \gamma_1\Big)$$

and

$$\overline{C} = \frac{1}{\lambda}\sum_j \lambda_j \overline{C}_{\mu_j} .$$

Assuming that up to 3 hours lost production can be made up relatively cheaply ($t_1 = 3$, $\gamma_1 = 1$), that between 3 and 20 hours lost production can be recovered realistically by use of overtime ($t_2 = 20$, $\gamma_2 = 4$), and that more severe penalty payments are imposed for production losses over 20 hours ($\gamma_3 = 25$) we obtain the following results based on the values given in Table 1.

$$\overline{C}_\mu = 6.6445 \qquad \text{and} \qquad \overline{C} = 7.2796 ,$$

so that, in this case, $\overline{C}_\mu$ would underestimate $\overline{C}$ by approximately 8.7%.

3.2 Two production lines

In this section we compare results obtained from the detailed rate matrix formulations of section 2.2 with those obtained from the corresponding sMm rate matrix approximations. Specifically, we compare the resulting stationary distributions; these give the distribution of the system over the different states, *i.e.* the proportion of the time the system spends in each state. Given the rate matrix, Q, the stationary state distribution vector $\boldsymbol{\pi}$ satisfies $Q^{\mathsf{T}}\boldsymbol{\pi} = \mathbf{0}$.

The two-production line system considered here comprises the audio cassette production line presented in section 3.1, together with a second line engaged in the assembly and packaging of cassettes containing photographic film. This second line, like the first, is fully automated with a sequence of machines performing the following operations:

Machine 1: Wind photographic film onto spool.

Machine 2: Seal spool and film into light-proof cassette; place cassette into plastic container.

Machine 3: Assemble film box; place plastic container into box; seal box.

Table 2 gives illustrative values for the failure and repair rates, λ_j and μ_j respectively, of the various machines on this second line.

Machine No.	B1	B2	B3
Failure rate (per hour)	0.0169	0.0137	0.0083
Repair rate (per hour)	0.4000	0.2857	0.6666

Table 2: Failure and repair rates for photo cassette assembly line.

From these figures we obtain the line failure rate $\lambda = 0.0390/hour$, and the sMm approximation for the line repair rate $\mu = 0.3791/hour$.

For a single repair crew, with and without priority, the stationary state distributions obtained from both the detailed (exact) model and the sMm approximation are given in Table 3. Comparison of these figures indicates that, at least in this case, the sMm approximation provides a reasonable approximation to the stationary state distribution. In both cases the approximation slightly underestimates the proportion of the time that both lines are out of action.

		no priority		priority (line 1)	
Line 1	Line 2	exact	approx	exact	approx
operating	operating	0.7533	0.7530	0.7526	0.7525
failed	operating	0.1370	0.1389	0.1240	0.1256
operating	failed	0.0778	0.0789	0.0900	0.0901
failed	waiting repair	0.0183	0.0165	0.0334	0.0318
waiting repair	failed	0.0136	0.0127	—	—

Table 3: Stationary distributions; single crew, with and without priority.

4 Discussion and Conclusions

For a single production line, the approximation provides the exact stationary distribution. Furthermore, with only two global states, any distributional assumptions will do, provided that the expected sojourn time in each state is correctly evaluated. However, the variance of repair times is underestimated by the approximation and this will have an effect if the cost function is non-linear. In practice non-linear cost functions will often be more realistic than linear cost functions. For example, a manufacture of car seats on the same site as a motor manufacturer supplies seats to the car production line on a JIT basis. However, the seat manufacturer may carry a small stock of seats in case its own production facility fails. The cost of downtime to the seat manufacturer will be relatively low if it can continue to supply seats to the car production line. In contrast, the costs will be very high if the car manufacturer has to stop the line until seat production resumes. Other cases that would be better modelled with non-linear cost functions are processes in which the product deteriorates if it is left standing on the production line. Dairy products are an example. Non-linear cost functions can also allow for differences in overtime, extra shifts, and penalty clauses for failure to meet deadlines. If the cost function is non-linear, the sMm may underestimate costs by around 10%. This is quite substantial. However, the sMm will give the same result as the use of the hyper-exponential distribution if the cost function is assumed linear. This is because the phase-type model, with states combined in the appropriate manner, and sMm have the same stationary distribution.

In the case of two production lines with two repair crews the approximation again gives an exact result for the stationary distribution provided the random variables are assumed independent. But, if there is a single repair crew the stationary distributions of the phase-type models are not identical to their approximations. Nevertheless, in the numerical example we considered the difference was slight and would be of little practical importance if linear cost functions were being used. As in the single line case, use of the the phase-type models will be worthwhile if non-linear cost functions are considered, or if the specific machine that fails is an important detail. This will be the case if machine maintenance costs differ.

References

1. Al-Hassan, K, Chan, J.F., Metcalfe, A.V. (2000a) The Role of TPM in Business Excellence, *Total Quality Management*, **11** (4,5,6), S596-S601.
2. Al-Hassan, K, Chan, J.F., Metcalfe, A.V. (2000b) Markov Models For

Promoting Total Productive Maintenance, *In Proceedings of Industrial Statistics In Action (Newcastle University, UK)*, **1** (10), 1–12. ISBN 0-70-170092-0.

3. Bendell, T. (2000) What Is Six Sigma?, **Quality World**, **26** (1), 14–17.
4. Japan Institute for Plant Maintenance (as at July 2000), *http://www.jipm.or.jp/en/index.html; http://wwww.tpm.co.jpm/*
5. Nakajima, S. (1988) Introduction to TPM: Total Productive Management, *Productivity Press*, USA. ISBN 0-915299-23-2.
6. Nakajima, S. (1991) TPM Development Program – Implementing Total Productive Maintenance, *Productivity Press*, USA, ISBN 0-915299-37-2.
7. Neuts, M.F. (1981) *Matrix Geometric Solutions in Stochastic Models*, John Hopkins, Baltimore.

A LINEAR PROGRAM APPROACH TO ERGODICITY OF *M/G/*1 TYPE MARKOV CHAINS WITH A TREE STRUCTURE

QI-MING HE

Department of Industrial Engineering, Dalhousie University
Halifax, Nova Scotia, Canada B3J 2X4
E-mail: Qi-Ming.HE@dal.ca

HUI LI

Department of Mathematics, Mount Saint Vincent University
Halifax, Nova Scotia, Canada B3M 2J6
E-mail: hli@msvu1.msvu.ca

It has been shown recently that the Perron-Frobenius eigenvalue of a nonnegative matrix provides information for a complete classification of *M/G/*1 type Markov chains with a tree structure. The use of that ergodicity condition depends largely on the computation of a set of nonnegative matrices, which can be quite challenging. In this paper, without using a set of nonnegative matrices, we develop two linear programs whose solutions provide sufficient conditions for ergodicity of the Markov chains of interest. We also introduce a simple approximation to the ergodicity problem. Numerical examples demonstrate that the linear program approach, as well as the approximation approach, can be quite useful.

1 Introduction

Markov chains with a tree structure, introduced by Takine, Sengupta, and Yeung [12], have broad applications in stochastic modeling, especially in queueing theory. For instance, the queueing processes of a number of queueing systems with a last-come-first-served (LCFS) service discipline can be formulated into Markov chains with a tree structure (see HE and Alfa [5] and Takine, Sengupta, and Yeung [12], and references therein). In Van Houdt and Blondia [7], the data transmission process of a random access system is formulated as a Markov chain with a tree structure. As a result, the stability of these stochastic systems is closely related to the ergodicity of the corresponding Markov chains with a tree structure.

The ergodicity of Markov chains with a tree structure has attracted considerable attention recently. In HE [3, 4], it has been shown that the Perron-Frobenius eigenvalue of a nonnegative matrix provides information for a complete classification of *M/G/*1 type Markov chains with a tree structure. Unfortunately, the ergodicity condition is based on a set of nonnegative matrices that are the fixed points of certain matrix equations. When the number of phases involved is large, the computations required for calculating those matrices are quite demanding and, in some cases, impossible to implement because of computer space limitations (e.g., the random access memory of a computer). Therefore, other simpler conditions (sufficient or necessary) can be quite useful in practice.

In HE and Li [6], a linear program approach is used to find sufficient conditions for stability of a queueing system with multiple types of customers and a last-come-first-served preemptive repeat service discipline. In this paper, we generalize this linear program approach to *M/G/*1 type Markov chains with a tree structure. We develop two linear programs whose solutions provide information about ergodicity of the Markov chain of interest. The two linear programs are formulated using only original system parameters. Since efficient algorithms have been developed for solving linear programs, information for ergodicity can be obtained efficiently even when the number of phases is large. This is the main contribution of this paper. In addition, we also introduce a simple (approximation) condition for ergodicity.

In queueing theory and queueing networks, stability has been an important issue. Various approaches have been explored (Chen and Zhang [1], Kumar and Meyn [8]). In fact, the linear program approach has been used to find stability conditions for queueing networks with reentry (Kumar and Meyn [8]). Our work shows that the ergodicity problem of complicated Markov chains can be transformed into a linear program, if the Markov chains possess a certain structure.

Our work is based on matrix analytic methods and Foster's criteria for Markov chains. Latouche and Ramaswami [9] and Neuts [10, 11] provide an introduction to matrix analytic methods. Fayolle, et al. [2] gives an introduction to the classification of Markov chains, including Foster's criteria.

The rest of the paper is organized as follows. In Section 2, we introduce *M/G/*1 type Markov chains with a tree structure. In Section 3, we introduce three existing approaches to the ergodicity problem. In Section 4, we present two linear programming formulations whose solutions give sufficient conditions for ergodicity. In Section 5, we give some details about the implementation of numerical algorithms. In Section 6, we present some numerical examples to gain insight into the methods introduced in this paper and to draw general conclusions about the usefulness of the methods.

2 Markov Chain of Matrix *M/G/*1 Type with a Tree Structure

The following discrete time Markov process of matrix *M/G/*1 type with a tree structure was first introduced in Takine, Sengupta, and Yeung [12]. Consider a discrete time two-dimensional Markov chain $\{(C_n, \eta_n), n \geq 0\}$ in which the values of C_n are represented by the nodes of a K-ary tree, and η_n takes integer values between 1 and m, where m is a positive integer. C_n is referred to as the node variable and η_n the auxiliary (phase) variable of the Markov chain at time n.

The K-ary tree of interest is a tree for which each node has a parent and K children, except the root node of the tree. The root node is denoted as 0. Strings of integers between 1 and K are used to represent nodes of the tree. For instance, the

kth child of the root node is represented by k, the lth child of node k is represented by kl, and so on.

Let $\aleph = \{J\colon\ J=k_1k_2\cdots k_n,\ 1\le k_i \le K,\ 1\le i \le n,\ n>0\}\cup\{0\}$. Any string $J\in\aleph$ represents a node in the K-ary tree. The length of a string J is defined as the number of integers in the string and is denoted by $|J|$. When $J = 0$, $|J| = 0$. The *addition operation* and the *subtraction operation* for strings in $\aleph$ are defined as follows: if $J = k_1\cdots k_n \in \aleph$, $J\neq 0$, $H = h_1\cdots h_i \in \aleph$, and $H\neq 0$, then $J+H = k_1\cdots k_n h_1\cdots h_i \in \aleph$; if $J\in\aleph$, then $J+0 = 0+J = J$; if $J = k_1\cdots k_n \in \aleph$ and $H = k_i\cdots k_n \in \aleph$, $i>0$, then $J-H = k_1\cdots k_{i-1} \in \aleph$.

The Markov chain $\{(C_n, \eta_n), n\ge 0\}$ takes values in $\aleph\times\{1, 2, \cdots, m\}$. To be called a homogenous Markov chain of matrix *M/G/*1 type with a tree structure, (C_n, η_n) transits at each step either to its parent node or to a descendent of its parent node. Assuming that $(C_n, \eta_n) = (H+k, i)$ for $k>0$ and $1\le i, i'\le m$, then $(C_{n+1}, \eta_{n+1}) = (H+J, i')$ with probability $a_{(i,i')}(k, J)$ for $J\in\aleph$. If $(C_n, \eta_n) = (0, i)$, then $(C_{n+1}, \eta_{n+1}) = (J, i')$ with probability $b_{(i,i')}(J)$ for $J\in\aleph$.

Note that transition probabilities depend only on the last integer in the string representing the current node $H+k$. If we call the node $J+k$ a type k node, it is clear that all type k nodes have the same transition probabilities, $1\le k\le K$. In matrix form, transition probabilities are represented as:

> $A(k, J)$ is an $m\times m$ matrix with elements $a_{(i,i')}(k, J)$, $1\le k \le K$, for $J\in\aleph$;
> $B(J)$ is an $m\times m$ matrix with elements $b_{(i,i')}(J)$ for $J\in\aleph$.

Let

$$\begin{aligned} A(k) &= \sum_{J\in\aleph} A(k,J), \quad 1\le k\le K; \\ B^{*(1)}(k) &= \sum_{J\in\aleph} B(J)N(J,k), \ 1\le k\le K, \end{aligned} \tag{2.1}$$

where $N(J, k)$ is the number of appearances of integer k in the string J. By the law of total probability, we must have $A(k)\mathbf{e} = \mathbf{e}$, $1\le k\le K$, and $\left(\sum_{J\in\aleph} B(J)\right)\mathbf{e}=\mathbf{e}$, where $\mathbf{e}$ is the column vector with all components being one.

3 Three Existing Approaches to Ergodicity

In this section, we introduce three approaches to establish ergodicity conditions of the Markov chain $\{(C_n, \eta_n), n\ge 0\}$ defined in Section 2.

3.1 The Perron-Frobenius Eigenvalue (PFE) Approach

Let $\mathbf{X} = \{X_1, \cdots, X_K\}$, where $X_1, X_2, \cdots$, and X_K are $m\times m$ stochastic matrices, i.e., X_k is nonnegative and $X_k\mathbf{e} = \mathbf{e}$, $1\leq k\leq K$. Let $\mathfrak{R}$ be a set of all $\mathbf{X}$ for which $X_1, \cdots$, and X_K are stochastic matrices and satisfy the following equations, for $1\leq k\leq K$, ($J = k_1\cdots k_{|J|}$)

$$X_k = A(k,0) + \sum_{J\in\aleph,\, J\neq 0} A(k,J) X_{k_{|J|}} X_{k_{|J|-1}} \cdots X_{k_1}. \tag{3.1}$$

By the well-known Brouwer's fixed point theorem, it was shown in HE [4] that the set $\mathfrak{R}$ is nonempty. For any fixed point $\mathbf{X} = \{X_1, \cdots, X_K\}\in\mathfrak{R}$, define $X^{(J)} = X_{k_{|J|}} X_{k_{|J|-1}} \cdots X_{k_1}$ for all $J=k_1\cdots k_{|J|}\in\aleph$ and define the following $m\times m$ matrices, for $J=k_1\cdots k_{|J|}\in\aleph$,

$$\begin{aligned} &N(0,j,\mathbf{X}) = 0, \quad 1\leq j\leq K; \\ &N(J,j,\mathbf{X}) = I\delta(k_{|J|}, j) + \sum_{n=1}^{|J|-1} X_{k_{|J|}} X_{k_{|J|-1}} \cdots X_{k_{n+1}} \delta(k_n, j), \ 1\leq j\leq K; \\ &p(k,j,\mathbf{X}) = \sum_{J\in\aleph} A(k,J)\, N(J,j,\mathbf{X}), \quad 1\leq k,\ j\leq K, \end{aligned} \tag{3.2}$$

where $\delta(k, j) = 1$, if $k=j$; 0, otherwise, and I is the identity matrix. Note that the matrix $N(J, j, \mathbf{X})$ counts the number of appearances of integer j in the string J and keeps track of the phase changes in the transition process. The matrix $p(k, j, \mathbf{X})$ can be interpreted as the average number of appearances of integer j in the next transition, given that the Markov chain is currently in node $H+k$ for $H\in\aleph$. Define an $mK\times mK$ matrix $P(\mathbf{X})$ by

$$P(\mathbf{X}) = \begin{pmatrix} p(1,1,\mathbf{X}) & \cdots & p(1,K,\mathbf{X}) \\ \vdots & \vdots & \vdots \\ p(K,1,\mathbf{X}) & \cdots & p(K,K,\mathbf{X}) \end{pmatrix}. \tag{3.3}$$

Let $sp(P(\mathbf{X}))$ be the Perron-Frobinus eigenvalue of the matrix $P(\mathbf{X})$ (i.e., the eigenvalue with the largest real part).

Theorem 3.1 (Theorem 3.2, HE [4]) Assume that the Markov chain $\{(C_n, \eta_n), n\geq 0\}$ is irreducible and aperiodic and that $B^{*(1)}(k)$ is finite, $1\leq k\leq K$. For any element $\mathbf{X}\in\mathfrak{R}$, if the matrix $P(\mathbf{X})$ is irreducible, then the Markov chain of matrix *M/G/1* type with a tree structure $\{(C_n, \eta_n), n\geq 0\}$ is

1) positive recurrent if and only if $sp(P(\mathbf{X})) < 1$;
2) null recurrent if and only if $sp(P(\mathbf{X})) = 1$;
3) transient if and only if $sp(P(\mathbf{X})) > 1$. □

If m=1, **X** is reduced to $\mathbf{X} = \{1, 1, \ldots, 1\}$. Then Theorem 3.1 gives an explicit ergodicity condition. If m>1, since the matrix set **X** has to be calculated in order to construct the matrix $P(\mathbf{X})$, the usefulness of Theorem 3.1 is compromised. Thus, there is a need to find ergodicity conditions without the presence of **X**.

Remark: Let $\mathbf{G} = \{G_1, \cdots, G_K\}$ be the minimal nonnegative solutions to equation (3.1). Then G_k is the (matrix) probability of the first passage from a node $J+k$ to its parent node J for any $J \in \aleph$ and $1 \le k \le K$ (see Takine, et al. [12]). It has been shown in HE [4] that $sp(P(\mathbf{G})) \le 1$, a fact that is quite useful in accuracy check and validating algorithms.

3.2 *The Perron-Frobenius Eigenvalue sp(Q)*

In this subsection, we introduce a descriptor for ergodicity without using any fixed point **X** in $\Re$. It is easy to calculate the descriptor, though it may not provide correct information about the ergodicity of the Markov chain of interest.

Let $\theta(k)$ be the left invariant vector of the stochastic matrix $A(k)$, where $\theta(k)$ is nonnegative and is normalized by $\theta(k)\mathbf{e} = 1$, $1 \le k \le K$. Let

$$q(k, j) = \theta(k) \sum_{J \in \aleph} A(k, J) N(J, j) \mathbf{e}, \quad 1 \le k, j \le K. \tag{3.4}$$

Let Q be a $K \times K$ matrix with the (k, j)th element being $q(k, j)$. Denote by $sp(Q)$ the Perron-Frobenius eigenvalue of the matrix Q.

Intuitively, $sp(Q)$, similar to $sp(P(\mathbf{X}))$, measures the average magnitude of an one-step movement of the Markov chain $\{(C_n, \eta_n), n \ge 0\}$. Thus, $sp(Q)$ should have a close relationship with ergodicity of the Markov chain. Furthermore, the computations of the matrix Q and of $sp(Q)$ are straightforward. The size of the matrix Q is smaller than that of the matrix $P(\mathbf{X})$. Therefore, it would be ideal if $sp(Q)$ could replace $sp(P(\mathbf{X}))$ for ergodicity (i.e., if $sp(Q) < 1$, the Markov chain is positive recurrent; if $sp(Q) > 1$, the Markov chain is transient.) Unfortunately, $sp(Q)$ may not provide correct information for ergodicity of the Markov chain. The change of the phase variable η_n depends on the type of node C_n. Therefore, $\theta(k)$ may not provide accurate information about the steady state distribution of the phase η_n. Consequently, $sp(Q)$ may not accurately measure the average magnitude of the one-step movement of the Markov chain.

Nonetheless, our numerical examples show that $sp(Q)$ is close to $sp(P(\mathbf{X}))$ and can be useful in practice since its computation is much easier than that of $sp(P(\mathbf{X}))$. In Section 6, we shall present a large number of examples to show the relationship between $sp(Q)$ and $sp(P(\mathbf{X}))$.

3.3 *Sufficient Conditions for Ergodicity*

The following sufficient conditions for ergodicity have been obtained in HE [4]. Denote by R_+ the set of nonnegative real numbers. Let $\mathbf{z} = (z_1, \cdots, z_K) \in R_+^K$ and define, for $1 \le k \le K$,

$$A^*(k,\mathbf{z}) = \sum_{J\in\aleph} \mathbf{z}^{(J)} A(k,J), \quad B^*(\mathbf{z}) = \sum_{J\in\aleph} \mathbf{z}^{(J)} B(J), \tag{3.5}$$

where $\mathbf{z}^{(J)} = z_{j_{|J|}} \cdots z_{j_1}$ if $J \in \aleph$ and $J \neq 0$, and $\mathbf{z}^{(J)} = 1$ if $J=0$.

Lemma 3.2 (Lemma 6.1, HE [4]) Assume that the Markov chain $\{(C_n, \eta_n), n \ge 0\}$ is irreducible and aperiodic and that $B^{*(1)}(k)$ is finite, $1 \le k \le K$. If there exists an $m \times 1$ positive vector $\mathbf{u}$ such that $A^*(k, \mathbf{z})\mathbf{u} < z_k\mathbf{u}$, (i.e., every element of $A^*(k, \mathbf{z})\mathbf{u}$ is strictly smaller than its counterpart in $z_k\mathbf{u}$,) and $B^*(\mathbf{z})\mathbf{u} < \infty$ for some $\mathbf{z}$ satisfying $1 < z_k < \infty$, $1 \le k \le K$, then the Markov chain $\{(C_n, \eta_n), n \ge 0\}$ is positive recurrent. □

Lemma 3.3 (Lemma 6.2, HE [4]) Assume that the Markov chain $\{(C_n, \eta_n), n \ge 0\}$ is irreducible and aperiodic and that $B^{*(1)}(k)$ is finite, $1 \le k \le K$. If there exists an $m \times 1$ positive vector $\mathbf{u}$ such that $A^*(k, \mathbf{z})\mathbf{u} \le z_k\mathbf{u}$ for some $\mathbf{z}$ satisfying $z_k > 0$, $1 \le k \le K$, and $0 < z_k < 1$ for at least one k ($1 \le k \le K$), then the Markov chain $\{(C_n, \eta_n), n \ge 0\}$ is transient. □

These two sufficient conditions do not make use of any fixed point $\mathbf{X} = \{X_1, \cdots, X_K\}$ in $\Re$ and they provide correct information about ergodicity. However, it is not straightforward to verify the conditions. In Section 4, based on Lemmas 3.2 and 3.3, we develop two linear programs for the ergodicity problem.

4 Linear Programs for Ergodicity Conditions

Let $\delta = (\delta_1, \delta_2, \ldots, \delta_K)^T$ and $\mathbf{v} = (v_1, v_2, \ldots, v_m)^T$, where superscript "T" represents matrix transpose. Define a linear system with variables $(\delta, \mathbf{v}, \varepsilon)$ as follows:

$$\sum_{j=1}^{K} \delta_j \mathbf{d}(k, j) + \big(A(k) - I\big)\mathbf{v} + \varepsilon\mathbf{e} \le 0, \quad 1 \le k \le K, \tag{4.1}$$

where $\delta \ge 0$, $-\infty < v_j < \infty$, $1 \le j \le m$, $\varepsilon \ge 0$, and

$$\mathbf{d}(k, j) = \begin{cases} \sum_{J\in\aleph} A(k,J)N(J,j)\mathbf{e}, & 1 \le j \le K,\ j \ne k; \\ \sum_{J\in\aleph} A(k,J)N(J,j)\mathbf{e} - \mathbf{e}, & j = k. \end{cases} \tag{4.2}$$

Lemma 4.1 Consider the Markov chain $\{(C_n, \eta_n), n\ge0\}$ defined in Section 2. We assume that $B^{*(1)}(k)$ is finite, $1\le k\le K$, and $B^*(\mathbf{z}) < \infty$ for some $\mathbf{z}$ satisfying $1 < z_k < \infty$, $1\le k\le K$. If the linear system (4.1) has a solution $(\delta, \mathbf{v}, \varepsilon)$ with positive δ_k, $1\le k\le K$, and positive ε, then the Markov chain $\{(C_n, \eta_n), n\ge0\}$ is positive recurrent.

Proof. We use Lemma 3.2 to prove Lemma 4.1. The idea is to choose a direction $\delta = (\delta_1, \delta_2, \ldots, \delta_K)^T$ with $\delta_k \ge 0$, $1\le k\le K$, such that, when $z_k(t) = 1 + \delta_k t$, $1\le k\le K$, (i.e., $\mathbf{z}(t) = \mathbf{e} + \delta t$) and $\mathbf{u}(t) = \mathbf{e} + t\mathbf{v}$, we have, for $1\le k\le K$,

$$A^*(k, \mathbf{z}(t))\mathbf{u}(t) \le z_k(t)\mathbf{u}(t) - t\varepsilon\mathbf{e}, \tag{4.3}$$

with positive $\mathbf{u}(t)$ and $\delta\mathbf{e} = 1$ for some positive ε and positive t. By using the Taylor expansion of $A^*(k, \mathbf{z}(t))$ with respect to the variable t, the problem can be transformed into the linear system (4.1) in the following way:

$$\begin{aligned} A^*(k,\mathbf{z}(t))\mathbf{u}(t) &= \left(\sum_{J\in\aleph} A(k,J)(\mathbf{z}(t))^{(J)}\right)\mathbf{u}(t) \\ &= \left[\sum_{J\in\aleph} A(k,J) + \left(\sum_{J\in\aleph}\sum_{j=1}^{K} A(k,J)N(J,j)\delta_j\right)t + o(t)\right](\mathbf{e}+t\mathbf{v}) \\ &= \left(\sum_{J\in\aleph} A(k,J)\right)\mathbf{e} + t\left[\sum_{J\in\aleph} A(k,J)\mathbf{v} + \sum_{j=1}^{K}\delta_j \sum_{J\in\aleph} A(k,J)N(J,j)\mathbf{e}\right] + o(t). \end{aligned} \tag{4.4}$$

Then, for $1\le k\le K$, inequality (4.3) becomes,

$$A(k)\mathbf{e}+t\left[A(k)\mathbf{v}+\sum_{j=1}^{K}\delta_j\sum_{J\in\aleph}A(k,J)N(J,j)\mathbf{e}\right]+o(t).$$
$$\leq(1+\delta_k t)(\mathbf{e}+t\mathbf{v})-t\varepsilon\mathbf{e}=\mathbf{e}+t(\delta_k\mathbf{e}+\mathbf{v}-\varepsilon\mathbf{e})+o(t). \tag{4.5}$$

Canceling the vector **e** and letting $t \to 0$ on both sides of the inequality in equation (4.5), we obtain

$$\sum_{j=1}^{K}\delta_j\left(\sum_{J\in\aleph}A(k,J)N(J,j)\right)\mathbf{e}+A(k)\mathbf{v}\leq\delta_k\mathbf{e}+\mathbf{v}-\varepsilon\mathbf{e}$$
$$\Leftrightarrow\quad\sum_{j=1}^{K}\delta_j\left(\sum_{J\in\aleph}A(k,J)N(J,j)\right)\mathbf{e}-\delta_k\mathbf{e}+\left(A(k)-I\right)\mathbf{v}+\varepsilon\mathbf{e}\leq 0. \tag{4.6}$$

It is easy to see that inequality (4.6) is equivalent to the linear system (4.1). If the linear system (4.1) has a solution (δ, **v**, ε) with positive δ_k, $1\leq k\leq K$, and positive ε, then equation (4.3) holds for small enough positive t and some positive $\varepsilon' \leq \varepsilon$. Note that we choose positive t small enough to ensure that 1) inequality (4.3) is satisfied, 2) $z_k(t)>1$, $1\leq k\leq K$, 3) $\mathbf{u}(t)$ is positive, and 4) $B^*(\mathbf{z}(t))$ is finite. By Lemma 3.2, the Markov chain is positive recurrent. □

To find a condition for transient Markov chains, we define the following linear system for (δ, **v**, ε):

$$-\sum_{j=1}^{K}\delta_j\mathbf{d}(k,j)+\left(A(k)-I\right)\mathbf{v}+\varepsilon\mathbf{e}\leq 0,\quad 1\leq k\leq K. \tag{4.7}$$

with $\delta \geq 0$, $-\infty < v_j < \infty$, $1 \leq j \leq m$, and $\varepsilon \geq 0$.

Lemma 4.2 Consider the Markov chain $\{(C_n, \eta_n), n\geq 0\}$ defined in Section 3. If the linear system (4.7) has a solution (δ, **v**, ε) with nonzero nonnegative vector δ and positive ε, then the Markov chain $\{(C_n, \eta_n), n\geq 0\}$ is transient.

Proof. The proof is similar to that of Lemma 4.1. Choose $z_k(t) = 1-\delta_k t$, $1\leq k\leq K$, i.e., $\mathbf{z}(t) = \mathbf{e} - \delta t$, and $\mathbf{u}(t) = \mathbf{e} + t\mathbf{v}$. By Lemma 3.3, we need to find δ and **v** such that $A^*(k, \mathbf{z}(t))\mathbf{u}(t) \leq z_k(t)\mathbf{u}(t)$ holds for some positive t. As in the proof of Lemma 4.1, we expand $A^*(k, \mathbf{z}(t))\mathbf{u}(t) \leq z_k(t)\mathbf{u}(t)$ and evaluate the expanded expressions in the following way:

$$
\begin{aligned}
&A^*(k,\mathbf{z}(t))\mathbf{u}(t) \\
&= A(k)\mathbf{e} + t\left[A(k)\mathbf{v} - \sum_{j=1}^{K}\delta_j \sum_{J\in\aleph} A(k,J)N(J,j)\mathbf{e}\right] + o(t) \qquad (4.8)\\
&\le \mathbf{e} + t(\mathbf{v} - \delta_k\mathbf{e}) + o(t).
\end{aligned}
$$

Canceling the vector **e** and letting $t \to 0$ on both sides of the inequality in equation (4.8), we obtain

$$
-\sum_{j=1}^{K}\delta_j\left(\sum_{J\in\aleph} A(k,J)N(J,j)\right)\mathbf{e} + \delta_k\mathbf{e} + \big(A(k) - I\big)\mathbf{v} \le 0. \qquad (4.9)
$$

We add the vector $\varepsilon\mathbf{e}$ to the left hand side of equation (4.9) to obtain equation (4.7). If the linear system (4.7) has a solution (δ, **v**, ε) with a nonzero nonnegative vector δ and positive ε, inequality (4.9) is satisfied in strict sense. That implies that inequality (4.8) holds for small enough positive t. Thus the conditions given in Lemma 3.3 are satisfied for small enough positive t. Therefore, the Markov chain is transient. □

It is easy to see that the key step in the application of Lemmas 4.1 and 4.2 is to show the existence of the required solutions to the linear systems (4.1) and (4.7). For that purpose, we introduce the following two linear programs. First, we define a linear program from linear system (4.1) to get a sufficient condition for an ergodic Markov chain:

$$
\begin{aligned}
&\xi_1 = \max\{\varepsilon\} \\
&s.t. \begin{pmatrix}
\mathbf{d}(1,1) & \cdots & \mathbf{d}(1,K) & A(1)-I & \mathbf{e} \\
\mathbf{d}(2,1) & \cdots & \mathbf{d}(2,K) & A(2)-I & \mathbf{e} \\
\vdots & \vdots & \vdots & \vdots & \vdots \\
\mathbf{d}(K,1) & \cdots & \mathbf{d}(K,K) & A(K)-I & \mathbf{e} \\
1 & \cdots & 1 & 0 \;\cdots\; 0 & 0
\end{pmatrix}
\begin{pmatrix} \delta \\ \mathbf{v} \\ \varepsilon \end{pmatrix}
\le
\begin{pmatrix} 0 \\ 0 \\ \vdots \\ 0 \\ 1 \end{pmatrix}; \qquad (4.10)\\
&\qquad \delta \ge 0,\;\; \varepsilon \ge 0, \quad -\infty < v_j < \infty, \;\; 1 \le j \le m.
\end{aligned}
$$

Note that the constraints of the above linear program are from linear system (4.1), except that the constraint $\delta\mathbf{e} = 1$ is added to ensure a finite optimal solution (which may not be unique). Next, we define a linear program from linear system (4.7) to get a sufficient condition for a transient Markov chain:

$$\xi_2 = \max\{\varepsilon\}$$

$$s.t. \quad \begin{pmatrix} -\mathbf{d}(1,1) & \cdots & -\mathbf{d}(1,K) & A(1)-I & \mathbf{e} \\ -\mathbf{d}(2,1) & \cdots & -\mathbf{d}(2,K) & A(2)-I & \mathbf{e} \\ \vdots & \vdots & \vdots & \vdots & \vdots \\ -\mathbf{d}(K,1) & \cdots & -\mathbf{d}(K,K) & A(K)-I & \mathbf{e} \\ 1 & \cdots & 1 & 0 \ \cdots \ 0 & 0 \end{pmatrix} \begin{pmatrix} \delta \\ \mathbf{v} \\ \varepsilon \end{pmatrix} \le \begin{pmatrix} 0 \\ 0 \\ \vdots \\ 0 \\ 1 \end{pmatrix} \tag{4.11}$$

$$\delta \ge 0, \quad \varepsilon \ge 0, \quad -\infty < v_j < \infty, \quad 1 \le j \le m.$$

Now, we are ready to present the main theorem of this paper.

Theorem 4.3 Consider the Markov chain $\{(C_n, \eta_n), n\geq 0\}$ defined in Section 2. The linear system (4.1) has a solution with positive (δ, ε) if and only if $\xi_1 > 0$. The linear system (4.7) has a solution with nonzero δ and positive ε if and only if $\xi_2 > 0$. Consequently, if $\xi_1 > 0$, the Markov chain is positive recurrent (provided that other assumptions in Lemma 4.1 are satisfied); if $\xi_2 > 0$, the Markov chain is transient.

Proof. First, we note that both (4.10) and (4.11) have a feasible solution $(\delta, \mathbf{v}, \varepsilon) = (0, \ldots, 0)$. Therefore, optimal solutions exist for both problems and $\xi_1, \xi_2 \ge 0$.

If the linear system (4.1) has a solution with positive (δ, ε), then we have $\xi_1 > \varepsilon > 0$. On the other hand, if the objective function of the optimal solution of (4.10) is positive, then the linear system (4.1) has a solution $(\delta, \mathbf{v}, \varepsilon)$ with a positive ε. Next, we show that δ is also positive. If $\delta=0$, then the constraints in equation (4.10) (except the last line) become $(A(k)-I)\mathbf{v} + \varepsilon\mathbf{e} \le 0$, $1\le k\le K$. Multiplying both sides of these inequalities by the nonnegative and nonzero vector $\theta(k)$ yields $\varepsilon\theta(k)\mathbf{e} = \varepsilon \le 0$, $1\le k\le K$, which is a contradiction. (Note that $\theta(k)$ is the left invariant vector of $A(k)$ defined in Subsection 3.2.) Therefore, δ is nonzero. If the vector δ is not positive, i.e., if there exists k such that $\delta_k=0$, then we multiply $\theta(k)$ on both sides of the kth inequality in (4.10) to obtain

$$\sum_{j=1, j\neq k}^{K} \delta_j \theta(k)\mathbf{d}(k,j) + \varepsilon\theta(k)\mathbf{e} \le 0. \tag{4.12}$$

Since all of the components of the above vectors are nonnegative and some of them are positive ($\varepsilon>0$), the inequality cannot hold. Therefore, the vector δ is positive. According to Lemma 4.1, the Markov chain is positive recurrent.

The second part of the theorem about equation (4.11) can be proved similarly, except that the vector δ only has to be nonzero for this case. Details are omitted. □

We note that, if m=1, information provided by the solutions of (4.10) and (4.11) for ergodicity of Markov chains is sufficient and necessary (provided that the other assumptions in Lemma 4.1 are satisfied). That is, the Markov chain is positive recurrent if and only if (4.10) has a positive optimal objective value; the Markov chain is transient if and only if (4.11) has a positive optimal objective value. Consequently, if neither (4.10) nor (4.11) has a positive optimal objective value, then the Markov chain of interest is null recurrent. Unfortunately, it is not easy to check how accurate the information provided by equations (4.10) and (4.11) is if m>1. In Section 6, a numerical analysis will be carried out to analyze the usefulness of Theorem 4.3.

Remark: The sufficient conditions given by Theorem 4.3 are closely related to $P(\mathbf{X})$. To see the relationship, let $\mathbf{v}$=0. Then the constraints of equation (4.10) becomes $P(\mathbf{X})\Delta - \Delta + \varepsilon\mathbf{e} \leq 0$, where the vector $\Delta = (\delta_1\mathbf{e}^T, \delta_2\mathbf{e}^T, \ldots, \delta_K\mathbf{e}^T)^T$. Thus, finding a solution to (4.10) is equivalent to finding a special type of subinvariant measure of $P(\mathbf{X})$. However, $P(\mathbf{X})$ may not have such a subinvariant vector. Thus, Theorem 4.3 may fail to provide information for ergodicity.

It can be shown that (4.10) and (4.11) provide consistent information about ergodicity. If $\xi_1 > 0$, then $\xi_2 = 0$ and the Markov chain is positive recurrent. On the other hand, if $\xi_2 > 0$, then $\xi_1 = 0$ and the Markov chain is transient. It is possible that $\xi_1 = \xi_2 = 0$. For this case, ξ_1 and ξ_2 provide no information about the ergodicity of the Markov chain. These results can be proved in a way similar to that of Property 5.5 in HE and Li [6]. Details are omitted.

To end this section, we outline a computational scheme to check whether or not the Markov chain defined in Section 2 is ergodic.

Step 1. Calculate $\{A(k), 1 \leq k \leq K\}$.
Step 2. Calculate $\{\mathbf{d}(k, j), 1 \leq k, j \leq K\}$ by equation (4.2).
Step 3. Solve linear programs (4.10) and (4.11).
Step 4. If neither (4.10) nor (4.11) provides information about system stability, use the PFE method given in Section 3.1.

Compared to the methods introduced in Section 3.1, the linear program approach has a larger matrix (the constraints of equations (4.10) and (4.11)) to deal with. In fact, the space complexity of the PFE approach is $O(Km^2)$ and the space complexity of the linear program approach is $O(K+m+1)(Km+1) = O(K^2m+Km^2)$. If m is much larger than K, then the space complexity of the two methods is more or less the same. On the other hand, the matrix iterations for $\mathbf{X} = \{X_1, \ldots, X_K\}$ that are necessary for the PFE approach are avoided for the linear program approach so that the time complexity of the linear program approach is low and numerical precision can be ensured. Furthermore, there are well-developed algorithms and software that can solve linear programs efficiently, even when the number of phases is large. Therefore, the linear program approach has its advantages over the other approach.

5 Computational Details

In order to use the methods introduced in Sections 3 and 4, we have to compute summations of matrices over string $J \in \aleph$. Examples of such summations can be found in equations (2.1), (3.1), (3.2), (3.4), and (4.2). However, the actual implementation of such a summation is not straightforward. We introduce the following transformations that transform the summations over $J \in \aleph$ into summations over two indices. The latter can be implemented easily in computation.

For any string $J=k_1k_2\ldots k_{|J|} \in \aleph$ and $J \neq 0$, we introduce a pair of integers (n, t) as follows:

$$n = |J|, \quad t = \sum_{i=1}^{n} (k_i - 1)K^{n-i}. \tag{5.1}$$

On the other hand, for any pair of integers (n, t) with $n>0$ and $0 \leq t \leq K^n - 1$, we introduce a string $J = k_1 \ldots k_n$ as follows:

$$k_1 = \left\lfloor \frac{t}{K^{n-1}} \right\rfloor, \quad k_2 = \left\lfloor \frac{t - k_1 K^{n-1}}{K^{n-2}} \right\rfloor, \quad \ldots, \quad k_n = \left\lfloor \frac{t - \sum_{i=1}^{n-1} k_i K^{n-i}}{K^{n-n}} \right\rfloor, \tag{5.2}$$

where $\lfloor x \rfloor$ represents the largest integer that is smaller than or equal to x.

Lemma 5.1. Assume that the string $J=0$ corresponds to the pair $(n, t) = (0, 0)$. Then the transformations defined by equations (5.1) and (5.2) are two one-to-one transforms between the two sets $\aleph$ and $\{(n, t)\colon 0 \leq t \leq K^n - 1, n \geq 0\}$. Furthermore, define $\aleph_n = \{J\colon J \in \aleph \text{ and } |J| = n\}$ for $n \geq 0$. Then the transformations defined by equations (5.1) and (5.2) are one-to-one transformations between the two sets $\aleph_n$ and $\{(n, t)\colon 0 \leq t \leq K^n - 1\}$ for all $n \geq 0$.

Proof. The conclusion is obtained from the fact that the transformation defined by equation (5.1) is the inverse of the transformation defined by equation (5.2) and vice versa. □

6 Numerical Examples

To study the usefulness of the linear program approach and $sp(Q)$, we have run and analyzed a large number of numerical examples. In this section, we present the results of our numerical analysis. First, we show in Example 6.1 that the PFE,

$sp(Q)$, and linear program methods may provide different information for ergodicity. Second, in Example 6.2, we show how good $sp(Q)$ and the linear program approach can be by summarizing the results for a large number of randomly chosen examples.

Example 6.1 Consider an *M/G/*1 type Markov chain with a tree structure with the following transition blocks: $K=m=2$,

$$A(1,0)=\begin{pmatrix}0 & \mu\\ \mu & 0\end{pmatrix},\ A(1,1)=\begin{pmatrix}0.1 & 0\\ 0.1 & 0\end{pmatrix},\ A(1,2)=\begin{pmatrix}0 & 0.1\\ 0.1 & 0\end{pmatrix},\ A(1,11)=\begin{pmatrix}0 & 0\\ 0.1 & 0\end{pmatrix},$$

$$A(1,12)=\begin{pmatrix}0 & 0.1\\ 0 & 0\end{pmatrix},\ A(1,21)=\begin{pmatrix}0.1 & 0\\ 0.35-\mu/2 & 0\end{pmatrix},\ A(1,22)=\begin{pmatrix}0 & 0.6-\mu\\ 0 & 0.35-\mu/2\end{pmatrix},$$

$$A(2,0)=\begin{pmatrix}\frac{\mu}{2} & \frac{\mu}{2}\\ 0 & \mu\end{pmatrix},\ A(2,1)=\begin{pmatrix}0 & 0\\ 0.1 & 0\end{pmatrix},\ A(2,2)=\begin{pmatrix}0.2 & 0.1\\ 0.2 & 0\end{pmatrix},\ A(2,11)=\begin{pmatrix}0 & 0\\ 0 & 0\end{pmatrix},$$

$$A(2,12)=\begin{pmatrix}0 & 0\\ 0 & 0\end{pmatrix},\ A(2,21)=\begin{pmatrix}0 & 0.1\\ 0.1 & 0\end{pmatrix},\ A(2,22)=\begin{pmatrix}0 & 0.6-\mu\\ 0.6-\mu & 0\end{pmatrix},$$

where $0<\mu<0.6$. For μ goes from 0.2 to 0.5, the values of $sp(P(\mathbf{X}))$, $sp(Q)$, ξ_1, and ξ_2 are shown in Table 6.1. Note that, in Table 6.1, $sp(P(\mathbf{X}))$ is the only measurement that always provides correct information about ergodicity.

Table 6.1 Values of $sp(P(\mathbf{X}))$, $sp(Q)$, ξ_1, and ξ_2 for Example 6.1

μ	0.2	0.3	0.35	0.3597	0.37	0.4	0.5
$sp(P(\mathbf{X}))$	1.31422	1.1170	1.0188	0.999863	0.9883	0.9211	0.7278
$sp(Q)$	1.31421	1.1171	1.0190	1.000084	0.9809	0.9214	0.7285
ξ_1	0	0	0	0	0.0098	0.0370	0.1317
ξ_2	0.1530	0.0558	0.0089	0	0	0	0

As shown in Table 6.1, if $sp(P(\mathbf{X})) \approx 1$ ($\mu \approx 0.36$), ξ_1 and ξ_2 do not provide information about ergodicity since they are both zero. If $\mu < 0.35$, ξ_2 is positive so that the Markov chain is transient. If $\mu > 0.37$, ξ_1 is positive so that the Markov chain is positive recurrent. If $\mu = 0.3597$, $sp(P(\mathbf{X})) = 0.999863 < 1 < 1.000084 = sp(Q)$. For this case, $sp(Q)$ does not provide correct information about ergodicity. Nonetheless, if $\mu < 0.359$ or $\mu > 0.36$, information about ergodicity provided by $sp(Q)$ is consistent with that provided by $sp(P(\mathbf{X}))$.

Example 6.2 In this example, we plot $sp(P(\mathbf{X}))$ against $sp(Q)$, ξ_1, and ξ_2 respectively for a large number of randomly chosen examples. In Figure 6.1, we plot $(sp(P(\mathbf{X})), sp(Q))$. In Figure 6.2, we plot $(sp(P(\mathbf{X})), \xi_1)$ and $(sp(P(\mathbf{X})), \xi_2)$.

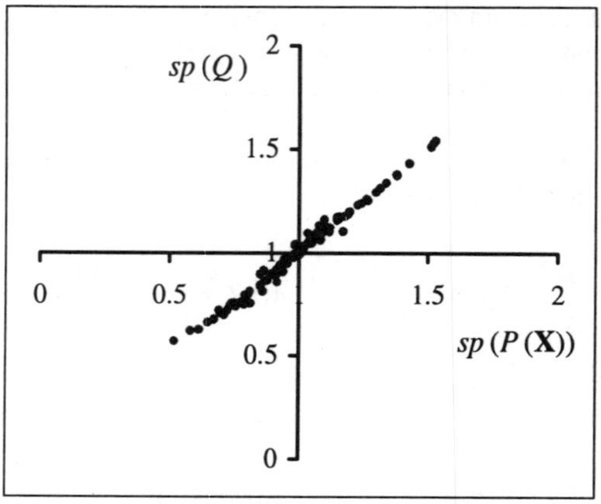

Figure 6.1 Plot of $(sp(P(\mathbf{X})), sp(Q))$.

Figure 6.1 demonstrates clearly that $sp(Q)$ and $sp(P(\mathbf{X}))$ are very close for almost all the examples, though outliers do exist. Thus, $sp(Q)$ can be a useful indicator of ergodicity of the Markov chain.

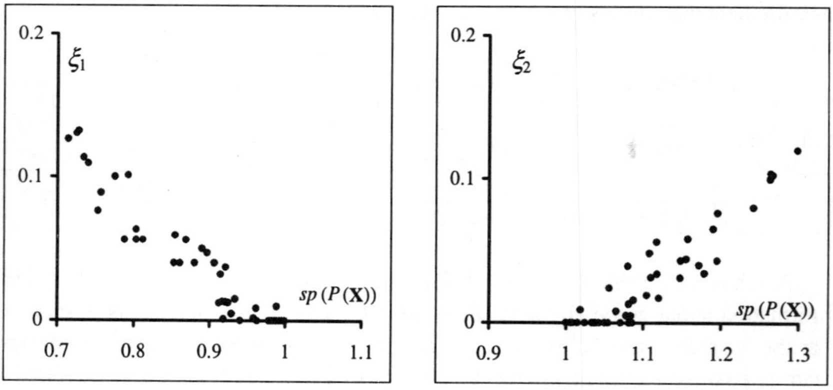

Figure 6.2 Plots of $(sp(P(\mathbf{X})), \xi_1)$ and $(sp(P(\mathbf{X})), \xi_2)$

By Theorem 4.3, ξ_1 may be useful if the Markov chain is positive recurrent and ξ_2 may be useful if the Markov chain is transient. Therefore, we plot $(sp(P(\mathbf{X})), \xi_1)$ only for $sp(P(\mathbf{X})) < 1$ and $(sp(P(\mathbf{X})), \xi_2)$ only for $sp(P(\mathbf{X})) > 1$. Figure 6.2 shows that both ξ_1 and ξ_2 may be zero and so fail to provide information about ergodicity if

$sp(P(\mathbf{X})) \approx 1$ ($0.9 < sp(P(\mathbf{X})) < 1.1$). Nonetheless, Figure 6.2 does show that ξ_1 starts to provide information if $sp(P(\mathbf{X}))$ goes below 0.9 and ξ_2 starts to provide information if $sp(P(\mathbf{X}))$ goes beyond 1.1. Figure 6.2 also shows that ξ_1 and ξ_2 may provide useful information even if $sp(P(\mathbf{X})) \approx 1$ ($0.9 < sp(P(\mathbf{X})) < 1.1$). It is reasonable to conclude that if neither (4.10) nor (4.11) provides information about ergodicity, i.e., $\xi_1 = \xi_2 = 0$, then the Markov chain is "*close*" to null recurrent. More numerical examples for the relationship between $sp(P(\mathbf{X}))$, ξ_1, and ξ_2 can be found in HE and Li [6].

In summary, $sp(Q)$ is close to $sp(P(\mathbf{X}))$ and may provide correct information about ergodicity if $sp(P(\mathbf{X}))$ is not close to one. The linear program approach provides correct information about ergodicity if one of the two linear programs has a positive optimal value, which is usually true if the Markov chain is not on the border of being transient or recurrent. Without using $sp(P(\mathbf{X}))$, the following scheme can be used for the ergodicity problem: 1) If $\xi_1 > 0$, the Markov chain is positive recurrent; 2) If $\xi_2 > 0$, the Markov chain is transient; 3) If $\xi_1 = \xi_2 = 0$ and $sp(Q)$ is close to one, the Markov chain is on the border of being transient or recurrent; 4) Otherwise, the ergodicity of the Markov chain is unsure.

Acknowledgements This research is supported by the National Science and Engineering Research Council of Canada. The authors would like to thank two anonymous referees for their valuable comments on the paper. The authors would also like to thank Dr. S. Seager for proofreading this paper.

References

1. Chen, H. and H. Zhang, Stabiliy of multiclass queueing networks under priority service disciplines, *Operations Research*, **Vol 48** (2000), 26-37.
2. Fayolle, G., V.A Malyshev, and M.V. Menshikov, *Topics in the Constructive Theory of Countable Markov Chains*, Cambridge University Press (1995).
3. HE, Q-M, Classification of Markov processes of *M/G/*1 type with a tree structure and its applications to queueing systems, *O.R. Letters*, **Vol 26** (2000), 67-80.
4. HE, Q-M, Classification of Markov processes of matrix *M/G/*1 type with a tree structure and its applications to the *MMAP*[*K*]/*PH*[*K*]/1 queue, *Stochastic Models*, **Vol 15** (2000), 407-433.
5. HE, Q-M and A.S. Alfa, The discrete time *MMAP*[*K*]/*PH*[*K*]/1/LCFS-GPR queues and its variants, *Advances in Algorithmic Methods for Stochastic Models – Proceedings of the 3rd International conference on Matrix analytic Methods*, G. Latouche and P.G. Taylor (Editors), 2000, Notable Publications Inc., NJ., 167-190 (2000).

6. HE, Q-M and Hui Li, Stability conditions of the *MMAP*[*K*]/*G*[*K*]/1/LCFS preemptive repeat queue, Working paper #01-02, Department of Industrial Engineering, Dalhousie University (2001) (Submitted for publication).
7. Van Houdt, B. and C. Blondia, Stability and performance of stack algorithms for random access communication modeled as a tree structured QBD Markov chain, *Stochastic Models*, **Vol 17** (2001).
8. Kumar, P.R. and S.P. Meyn, Duality and linear programs for stability and performance analysis of queueing networks and scheduling policies, *IEEE Transactions on Automatic Control*, **Vol 41/1** (1996), 4-17.
9. Latouche, G. and V. Ramaswami, *Introduction to Matrix Analytic Methods in Stochastic Modelling*, SIAM, Philadelphia, Pennsylvania (1999).
10. Neuts, M.F., *Matrix-Geometric Solutions in Stochastic Models: An algorithmic Approach*, The Johns Hopkins University Press, Baltimore (1981).
11. Neuts, M.F., *Structured Stochastic Matrices of M/G/1 type and Their Applications*, Marcel Dekker, New York (1989).
12. Takine, T., B. Sengupta, and R.W. Yeung, A generalization of the matrix *M/G/1* paradigm for Markov chains with a tree structure, *Stochastic Models*, **Vol 11** (1995), 411-421.

MATRIX GEOMETRIC SOLUTION OF FLUID STOCHASTIC PETRI NETS

ANDRÁS HORVÁTH, MARCO GRIBAUDO
Dipartimento di Informatica, Università di Torino
Corso Svizzera 185, 10149 Torino, Italy,
horvath@di.unito.it, marcog@di.unito.it

Fluid (or Hybrid) Petri Nets with flush-out arcs are Petri net based models with two classes of places: discrete places that carry a natural number of distinct objects (tokens), and fluid places that hold a positive amount of fluid, represented by a real number. For this kind of formalism, equations can be automatically derived from the model. Such equations, however, are often too complex to be solved analytically and simple discretization techniques usually can be successfully applied only to simple cases. In this paper we present a numerical technique for steady state solution that makes use of known matrix geometric techniques.
Keywords: Non-Markovian Models, Fluid Stochastic Petri Nets, Numerical Techniques

1 Introduction

Fluid Stochastic Petri Nets (FSPN) or Hybrid Petri Nets (HPN) are Petri net based models in which some places may hold a discrete number of tokens, and some places a continuous quantity represented by a non-negative real number. Places that hold continuous quantities are referred to as *fluid* or continuous places, and the non-negative real number is said to represent the fluid level in the place. Discrete tokens move along discrete arcs with the enabling and firing rules of standard Petri Nets (PN), while the fluid moves along special continuous (or fluid) arcs according to an assigned instantaneous flow rate.

Several different versions of FSPNs have been defined in the literature (see for example [1,2,3,4,5]), and for many of them a method to derive the equations that describe the underlying stochastic process has been provided. In general, the solution of these equations is not a trivial task, and this problem has been directly addressed in many papers. In particular, steady state solution for the case of FSPN in which there is no dependency of the fluid places has been considered in [4]. In that paper a solution technique which requires *spectral decomposition* of a matrix has been presented. Transient analysis has also been considered in the same paper, proposing a technique called *upwind semi-discretization.* A more complex discretization technique for transient analysis of second order differential equations has been proposed in [5]. This technique uses an *implicit discretization* scheme which requires the solution of a linear system at every time step.

Even if the proposed techniques are quite general, none of them is really appropriate to overcome the difficulties of the numerical solution of the equations describing a FSPN. Spectral decomposition, for example, may cause excessive roundoff errors if the matrix is large and thus could be successfully applied only to systems characterized by a small number of states. Pure upwind semi-discretization can instead lead to system with a very high number of unknown variables, limiting the applicability of the algorithm only to systems with a small number of discrete states and/or fluid places. Implicit techniques instead are not stable for systems which involves only first order differential equations such as the one describing common FSPNs.

In this paper we present a numerical solution technique for steady state analysis. The technique can be applied only to a special subclass of FSPNs and its purpose is to solve equations faster than conventional discretization techniques. The technique consists in applying matrix geometric techniques to the block matrix that arises when the differential equations of the underlying stochastic process are solved with upwind semidiscretization. Simple upwind discretization can give very accurate results when sufficiently small discretization steps can be applied. However since the number of discretization points increases when the discretization step becomes smaller, it is not always possible to achieve the desired accuracy.

The main contribution of this work is thus the application of known matrix geometric techniques to the the particular matrix structures that appears when dealing with Fluid Stochastic Petri Nets. By using matrix geometric techniques, the complexity of the solution grows only linearly with the number of discretization points, making possible to use very large (even infinite) fluid levels and very small discretization step.

The technique proposed in this paper has already been used successfully in [6] to compute the performance indexes of a GPRS system. The technique, however, was not described in that paper due to space constraints.

The rest of the paper is organized as follows. Section 2 introduces the considered FSPN formalism and the notations. Section 3 gives the set of equations that describe the evolution of the stochastic behavior of the FSPN. Section 4 present the solution technique. In Section 5 the proposed technique is used to analyze a simple producer/consumer system as a numerical example.

2 Definitions and Notations

The definition of the FSPN is derived from [2] with standard notation inherited from [7]. A FSPN is a tuple $\langle \mathcal{P}, \mathcal{T}, \mathcal{A}, B, F, W, R, M_0 \rangle$, where

- $\mathcal{P}$ is the set of places partitioned into a set of discrete places $\mathcal{P}_d = \{p_1, \ldots, p_{|\mathcal{P}_d|}\}$, and a set of continuous places $\mathcal{P}_c = \{c_1, \ldots, c_{|\mathcal{P}_c|}\}$ (with $\mathcal{P}_d \cap \mathcal{P}_c = \emptyset$ and $\mathcal{P}_d \cup \mathcal{P}_c = \mathcal{P}$). The discrete places may contain a natural number of tokens, while the marking of a continuous place is a non negative real number. In the graphical representation a discrete place is drawn as a single circle while a continuous place is drawn with two concentric circles. The complete state (marking) of a FSPN is described by a pair of vectors $M = (\boldsymbol{m}, \boldsymbol{x})$, where the vector $\boldsymbol{m}$, of dimension $|\mathcal{P}_d|$, is the marking of the discrete part of the FSPN and the vector $\boldsymbol{x}$, of dimension $|\mathcal{P}_c|$, represents the fluid levels in the continuous places (with $x_l \geq 0$ for any $c_l \in \mathcal{P}_c$). We use $\mathcal{S}$ to denote the partially discrete and partially continuous state space. In the following we denote by $\mathcal{S}_d$ and $\mathcal{S}_c$ the discrete and the continuous component of the state space respectively. Time is denoted by τ, and the stochastic marking process is by $\mathcal{M}(\tau) = \{(\boldsymbol{m}(\tau), \boldsymbol{x}(\tau)), \tau \geq 0\}$.
- $\mathcal{T}$ is the set of transitions partitioned into a set of stochastically timed transitions $\mathcal{T}_e$ and a set of immediate transitions $\mathcal{T}_i$ (with $\mathcal{T}_e \cap \mathcal{T}_i = \emptyset$ and $\mathcal{T}_e \cup \mathcal{T}_i = \mathcal{T}$). A timed transition $T_j \in \mathcal{T}_e$ is drawn as a rectangle and has an instantaneous firing rate associated to it. An immediate transition $t_h \in \mathcal{T}_i$ is drawn with a thin bar and has a constant zero firing time. We denote the timed transitions with uppercase letters and the immediate transitions with lowercase letters.
- $\mathcal{A}$ is the set of arcs partitioned into four subsets: $\mathcal{A}_d$, $\mathcal{A}_h$, $\mathcal{A}_c$, and $\mathcal{A}_f$. The subset $\mathcal{A}_d$ contains the discrete arcs which can be seen as a function $A_d : ((\mathcal{P}_d \times \mathcal{T}) \cup (\mathcal{T} \times \mathcal{P}_d)) \to I\!N$. The arcs $\mathcal{A}_d$ are drawn as single arrows. The subset $\mathcal{A}_h$ contains the inhibitor arcs, $A_h : (\mathcal{P}_d \times \mathcal{T}) \to I\!N$. These arcs are drawn with a small circle at the end. The subset $\mathcal{A}_c$ defines the continuous arcs. These arcs are drawn as double arrows to suggest a pipe. $\mathcal{A}_c$ is a subset of $(\mathcal{P}_c \times \mathcal{T}_e) \cup (\mathcal{T}_e \times \mathcal{P}_c)$, i.e., a continuous arc can connect a fluid place to a timed transition or it can connect a timed transition to a fluid place. The subset $\mathcal{A}_f$ contains the *flush-out* arcs. $\mathcal{A}_f$ is a subset of $(\mathcal{P}_c \times \mathcal{T}_e)$. These arcs connect continuous places to timed transitions, and describe the capability of a transition to empty in zero time the existing fluid from a continuous place when it fires. The arcs $\mathcal{A}_f$ are drawn as thick single arrows.
- The function $B : \mathcal{P}_c \to I\!R^+ \cup \{\infty\}$ describes the fluid upper bounds on each continuous place. This bound has no effect when it is set to infinity. Each fluid place has an implicit lower bound at level 0.
- The firing rate function F is defined for timed transitions $\mathcal{T}_e$ so that $F : \mathcal{T}_e \times \mathcal{S} \to I\!R^+$. Therefore, a timed transition T_j enabled at time τ in a discrete marking $\boldsymbol{m}(\tau)$ with fluid level $\boldsymbol{x}(\tau)$, may fire with rate

$F(T_j, \boldsymbol{m}(\tau), \boldsymbol{x}(\tau))$, that is:

$$\lim_{\Delta\tau \to 0} Pr\{T_j \text{ fires in } (\tau, \tau + \Delta\tau) |$$

$$\mathcal{M}(\tau) = (\boldsymbol{m}(\tau), \boldsymbol{x}(\tau))\} = F(T_j, \boldsymbol{m}(\tau), \boldsymbol{x}(\tau))\Delta\tau$$

- The weight function W for immediate transitions $\mathcal{T}_i$ ($W : \mathcal{T}_i \times \mathcal{S}_d \rightarrow I\!R^+$) has the usual meaning and it may depend only on the discrete part of the marking [7].
- The function $R : \mathcal{A}_c \times \mathcal{S} \rightarrow I\!R^+ \cup \{0\}$ is called the *flow rate function* and describes the marking dependent flow of fluid across continuous arcs.
- The initial state of the FSPN is denoted by the pair $M_0 = (\boldsymbol{m_0}, \boldsymbol{x_0})$.

In this paper we consider only FSPN with a single fluid place c_1. The level of c_1 is denoted by x. For a more detailed explanation of the previous sets and of the dynamics of a FSPN with flush-out arcs, see [2].

3 Analysis

In this section, we derive the equations for the joint process $\mathcal{M}(\tau) = (\boldsymbol{m}(\tau), x(\tau))$ that describes the dynamic behavior of the FSPN model as a function of the time. The derivation of the equations is based on the inclusion of a supplementary variable [8].

3.1 The infinitesimal generators

In order to derive the complete equations we start investigating the behavior of the discrete part of the system. Since fluid arcs and flush-out arcs do not change the enabling condition of a transition, standard analysis techniques can be applied to the discrete marking process $\boldsymbol{m}(\tau)$ [7]. These techniques split the discrete state space into two disjoint subsets, called respectively the *tangible marking* set and the *vanishing marking* set. Since the process spends no time in vanishing markings, they can be removed and their effect can be included in the transitions between tangible markings. From this point on, we will consider only tangible markings.

In a FSPN with flush-out arcs the marking process $\mathcal{M}(\tau)$ can be characterized by two matrices $\boldsymbol{Q}$ and $\boldsymbol{Q}'$, both of size $\mathcal{S}_d \times \mathcal{S}_d$, that we call *infinitesimal generators*. Both matrices $\boldsymbol{Q}$ and $\boldsymbol{Q}'$ depend on the fluid part of the marking, since the firing rate of the various transitions may be fluid dependent.

The matrix $\boldsymbol{Q}$ accounts for the transition rates among tangible states when no flush-out occurs, and $\boldsymbol{Q}'$ accounts for the transition rates among tangible states when flush-out of fluid place c_1 does occur. An entry of $\boldsymbol{Q}(x)$

$q_{ij}(x), i \neq j$ represents the transition rate from state $\boldsymbol{m_i}$ to state $\boldsymbol{m_j}$ when the level of the fluid is x and the considered transition does not flush out the fluid place, that is:

$$q_{ij}(x) = \sum_{\substack{T_k \in \mathcal{E}(\boldsymbol{m_i}) \mid \\ \boldsymbol{m}_i \xrightarrow{T_k} \boldsymbol{m}_j \wedge \\ (c_1, T_k) \notin \mathcal{A}_f}} F(T_k, \boldsymbol{m}_i, x), \tag{1}$$

where $\mathcal{E}(\boldsymbol{m_i})$ represents the set of enabled transitions in state $\boldsymbol{m_i}$.

An entry of $\boldsymbol{Q}'(x)$ $q'_{ij}(x)$ represents the transition rate from state $\boldsymbol{m_i}$ to state $\boldsymbol{m_j}$ when the level of the fluid is x and the considered transition flushes out the fluid place:

$$q'_{ij}(x) = \sum_{\substack{T_k \in \mathcal{E}(\boldsymbol{m_i}) \mid \\ \boldsymbol{m}_i \xrightarrow{T_k} \boldsymbol{m}_j \wedge \\ (c_1, T_k) \in \mathcal{A}_f}} F(T_k, \boldsymbol{m}_i, x). \tag{2}$$

Note that flush-out may occur without change in the discrete part of the marking.

The summations (1) and (2) consider the transition rate of all the transitions T_k that have concession in marking $\boldsymbol{m}_i$ and bring the net from state $\boldsymbol{m}_i$ to $\boldsymbol{m}_j$ (denoted by $\boldsymbol{m}_i \xrightarrow{T_k} \boldsymbol{m}_j$), respectively flush out and do not flush out place c_1. In the standard equations that describe a CTMC, the terms of the diagonal of the infinitesimal generator accounts for the possibility of exiting from a state. Here we have to consider not only standard transitions, but also changes of state that cause a flush-out. We denote by

$$q_i(x) = \sum_{\boldsymbol{m}_j \in \mathcal{S}_d, \boldsymbol{m_j} \neq \boldsymbol{m_i}} q_{ij}(x) + \sum_{\boldsymbol{m}_j \in \mathcal{S}_d} q'_{ij}(x) \tag{3}$$

the total exit rate from state $\boldsymbol{m_i}$, when the fluid level is x. This function takes into account the sum of the rates from state $\boldsymbol{m}_i$ to any state $\boldsymbol{m_j}$, with or without flush-outs, The diagonal element defined in (3) is included in the matrix $\boldsymbol{Q}$ and hence

$$q_{ii}(x) = -q_i(x). \tag{4}$$

The sum of above defined matrices $\boldsymbol{Q}(x) + \boldsymbol{Q}'(x)$ of dimensions $|\mathcal{S}_d| \times |\mathcal{S}_d|$ is a proper infinitesimal generator.

3.2 Equations of the model

We describe the fluid flow by collecting all the possible flow rates in a diagonal matrix $\boldsymbol{R}$. The element $r_{jj}(x)$ of $\boldsymbol{R}(x)$, represents the fluid flow rate of continuous place c_1, in discrete state $\boldsymbol{m}_j$, conditioned to the fluid level x. We define the potential flow rate as:

$$r_j^{\mathcal{P}} = \sum_{T_j \in \mathcal{E}(\boldsymbol{m_j})} R((T_j, c_1), \boldsymbol{m_j}) - R((c_1, T_j), \boldsymbol{m_j}),$$

where $R((T_j, c_1), \boldsymbol{m_j})$ denotes the fluid rate entering the place, while $R((c_1, T_j), \boldsymbol{m_j})$ exiting it. Following [2], we include the boundary condition in the definition of matrix $\boldsymbol{R}$ by making it dependent on the fluid level of model. In particular we define $\boldsymbol{R}(x) = \mathrm{diag}\,(r_{jj}(x))$, with:

$$r_{jj}(x) = \begin{cases} 0 & \text{if } x = 0 \text{ and } r_j^{\mathcal{P}}(0) < 0 \\ 0 & \text{if } x = B(c_1) \text{ and } r_j^{\mathcal{P}}(B(c_1)) > 0 \\ r_j^{\mathcal{P}} & \text{otherwise} \end{cases} .$$

Matrices $\boldsymbol{Q}$ and $\boldsymbol{Q}'$, together with matrix $\boldsymbol{R}$, describe completely the stochastic process. With this definition, if we call $\boldsymbol{\pi}(x, \tau)$ a vector whose component $\pi_i(x, \tau)$ represents the probability at time τ of being in discrete state $\boldsymbol{m_i}$ with the fluid level of place c_1 in the interval $(x, x + \Delta x)$, then the equation that completely describes the underlying stochastic process becomes:

$$\frac{\partial \boldsymbol{\pi}(x,\tau)}{\partial \tau} + \frac{\partial \left(\boldsymbol{\pi}(x,\tau)\boldsymbol{R}(x)\right)}{\partial x} = \boldsymbol{\pi}(x,\tau)\boldsymbol{Q}(x) + \delta(x)\int_0^B \boldsymbol{\pi}(y,\tau)\boldsymbol{Q}'(y)dy, \quad (5)$$

where $\delta(x)$ represent the conventional Dirac's delta function. The steady state solution is obtained by taking the limit $\tau \to \infty$. In this case (5) becomes

$$\frac{d\left(\boldsymbol{\pi}(x)\boldsymbol{R}(x)\right)}{dx} = \boldsymbol{\pi}(x)\boldsymbol{Q}(x) + \delta(x)\int_0^B \boldsymbol{\pi}(y)\boldsymbol{Q}'(y)dy. \qquad (6)$$

For a complete derivation of the previous equation in a more general setting, the interested reader may refer to [2].

4 Matrix geometric solution techniques

4.1 Discretization

Consider a FSPN model with a single unbounded fluid place c_1, without flush-out arcs, and in which neither the flow rates associated to fluid arcs nor the firing rates associated to timed transition depend on the continuous part of the net (i.e. they do not depend on the level of place c_1). The equation that describes the steady state solution $\boldsymbol{\pi}(x)$ (away from the boundary) is

$$\frac{d\boldsymbol{\pi}(x)}{dx}\boldsymbol{R} = \boldsymbol{\pi}(x)\boldsymbol{Q}, \tag{7}$$

where, according to our assumptions, $\boldsymbol{R}$ and $\boldsymbol{Q}$ do not depend on the level x of fluid place c_1. Equation (7) has been obtained from (6) introducing the considered simplifications.

Numerical solution of (7) can be obtained using the method called *upwind discretization* [4] with which the derivative along the fluid level x is approximated by the "upwind" finite difference. The fluid level is discretized with step-size Δx. Let us use the notation $\boldsymbol{\pi}^{(i)} = \boldsymbol{\pi}(i\Delta x)$. If $r_j > 0$ (the actual flow rate corresponding to place c_1 in discrete state $\boldsymbol{m}_j$), then

$$\frac{d\pi_j(i\Delta x)}{dx} \approx \frac{\pi_j^{(i)} - \pi_j^{(i-1)}}{\Delta x}. \tag{8}$$

Instead, if $r_j < 0$, the derivative is approximated by

$$\frac{d\pi_j(i\Delta x)}{dx} \approx \frac{\pi_j^{(i+1)} - \pi_j^{(i)}}{\Delta x}. \tag{9}$$

The method is called "upwind" because it tries to follow the flux described by the equation by appropriately choosing the discretization points used to approximate the derivative. This way the discretized version of (7) becomes

$$\frac{\boldsymbol{\pi}^{(i)} - \boldsymbol{\pi}^{(i-1)}}{\Delta x}\boldsymbol{R}^+ + \frac{\boldsymbol{\pi}^{(i+1)} - \boldsymbol{\pi}^{(i)}}{\Delta x}\boldsymbol{R}^- = \boldsymbol{\pi}^{(i)}\boldsymbol{Q}, \tag{10}$$

where the diagonal matrices $\boldsymbol{R}^+$ and $\boldsymbol{R}^-$ are defined as

$$r_{ii}^+ = \begin{cases} r_{ii}, \text{ if } r_{ii} > 0, \\ 0, \quad \text{otherwise} \end{cases} \qquad r_{ii}^- = \begin{cases} r_{ii}, \text{ if } r_{ii} < 0, \\ 0, \quad \text{otherwise.} \end{cases} \tag{11}$$

Using the notation

$$\boldsymbol{X}^+ = \frac{1}{\Delta x}\boldsymbol{R}^+, \quad \boldsymbol{X}^- = -\frac{1}{\Delta x}\boldsymbol{R}^-,$$

(10) can be rewritten as

$$-\boldsymbol{\pi}^{(i-1)}\boldsymbol{X}^+ + \boldsymbol{\pi}^{(i)}[\boldsymbol{X}^+ + \boldsymbol{X}^-] - \boldsymbol{\pi}^{(i+1)}\boldsymbol{X}^- = \boldsymbol{\pi}^{(i)}\boldsymbol{Q}. \tag{12}$$

Denoting $\boldsymbol{Q} - \boldsymbol{X}^+ - \boldsymbol{X}^-$ by $\boldsymbol{C}$ (12) becomes

$$\boldsymbol{\pi}^{(i-1)}\boldsymbol{X}^+ + \boldsymbol{\pi}^{(i)}\boldsymbol{C} + \boldsymbol{\pi}^{(i+1)}\boldsymbol{X}^- = \boldsymbol{0}. \tag{13}$$

The equations at the boundary are slightly different. Using the notation $\boldsymbol{C}^+ = \boldsymbol{Q} - \boldsymbol{X}^+$, the boundary condition is

$$\boldsymbol{\pi}^{(0)}\boldsymbol{C}^+ + \boldsymbol{\pi}^{(1)}\boldsymbol{X}^- = \boldsymbol{0}. \tag{14}$$

Equations (13) and (14) form a linear system. This system of equations is characterized by a matrix having the structure represented in Figure 1.

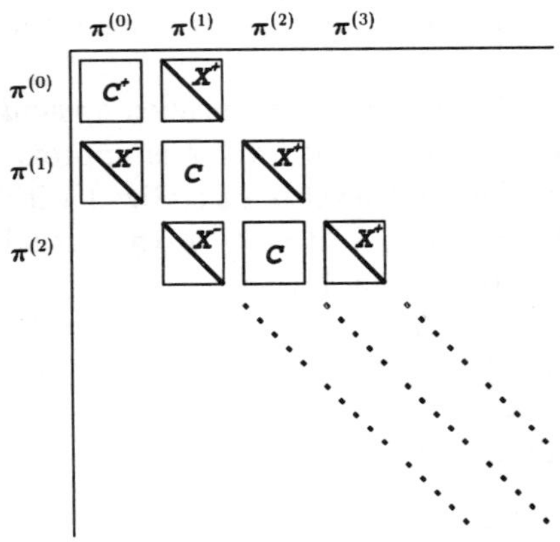

Figure 1. Matrix of the unbounded case

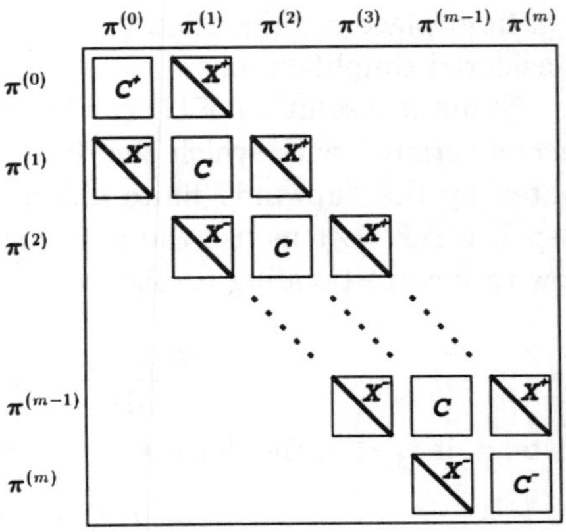

Figure 2. Matrix of the bounded case

4.2 Matrix Geometric Solution

Such linear system as described in (13) and (14) can be solved applying matrix geometric techniques. By $\boldsymbol{\Lambda}$ let us denote the solution of the matrix equation

$$\boldsymbol{X}^+ + \boldsymbol{\Lambda}\boldsymbol{C} + \boldsymbol{\Lambda}^2\boldsymbol{X}^- = \boldsymbol{0}, \tag{15}$$

then, the solution of the linear system can be obtained as

$$\boldsymbol{\pi}^{(0)} = \boldsymbol{u_0}, \quad \boldsymbol{\pi}^{(i)} = \boldsymbol{u_1}\boldsymbol{\Lambda}^{(i-1)}, \; i \geq 1, \tag{16}$$

with

$$\left| \boldsymbol{u_0} \; \boldsymbol{u_1} \right| \begin{vmatrix} \boldsymbol{C}^+ & \boldsymbol{X}^+ \\ \boldsymbol{X}^- & \boldsymbol{C} + \boldsymbol{\Lambda}\boldsymbol{X}^- \end{vmatrix} = \boldsymbol{0}, \tag{17}$$

under the normalization condition

$$\left[\sum_{i=0}^{\infty} \boldsymbol{\pi}^{(i)}\right] \boldsymbol{1} = \left[\boldsymbol{u_0} + \boldsymbol{u_1} \sum_{i=0}^{\infty} \boldsymbol{\Lambda}^i\right] \boldsymbol{1} = \tag{18}$$

$$= \left[\boldsymbol{u_0} + \boldsymbol{u_1} \left(\boldsymbol{I} - \boldsymbol{\Lambda}\right)^{-1}\right] \boldsymbol{1} = 1.$$

Equation (15) can be solved in a very efficient manner using the technique of Latouche and Ramaswami [9]. Matrix Λ can be determined by the iterative procedure called "logarithmic reduction", given in Table 1, that has logarithmic convergence.

```
Procedure_Logarithmic_Reduction
  B0 := -C^{-1} X^+; B2 := -C^{-1} X^-; G := B2; T := B0
  while ||1^T - G 1^T|| > ε do
     D := B0 B2 + B2 B0
     B0 := (I - D)^{-1} B0^2
     B2 := (I - D)^{-1} B2^2
     G := G + T B2
     T := T B0
  end while
  U := C + X^+ G
  Λ := -X^+ U^{-1}
End Procedure_Logarithmic_Reduction
```

```
Procedure_Unbounded_System
  Compute Λ.
  Compute a solution |u0 u1| of (17).
  Normalize |u0 u1| using (18).
  Compute the discretized solution π^(i) using (16).
End Procedure_Unbounded_System
```

```
Procedure_Bounded_System
  Compute Λ.
  Compute Γ.
  Compute Λ^m, Γ^m, Σ_{i=0}^m Λ^i and Σ_{i=0}^m Γ^i
  Compute a solution |u0 um| of (23).
  Normalize |u0 u1| using (24).
  Compute the discretized solution π^(i) using (22).
End Procedure_Bounded_System
```

```
Procedure_Fluid_Dependent
  Compute Λ_m using (34).
  for i := m - 1 downto 1 do
     Compute Λ_i using (33).
  end for
  Compute π^(0) using (35).
  Compute π^(i) using (32).
End Procedure_Fluid_Dependent
```

Table 1. Solution algorithms

Each iteration involves a matrix inversion, but generally very few iterations (about 15) are required to reach a precision $\epsilon = 10^{-10}$. The complete solution algorithm is given in Table 1 (called "unbounded system").

4.3 Bounded systems

Suppose now that the single fluid place c_1 has an upper bound B. The matrix geometric technique previously described can be extended also to such systems.

The equations that describe the model are identical to the ones describing the unbounded case with the addition of another boundary condition at $x = B$. If we discretize these equations using upwind discretization, we obtain (13) and (14), with the additional boundary condition

$$\boldsymbol{\pi}^{(m)}\boldsymbol{C}^{-} + \boldsymbol{\pi}^{(m-1)}\boldsymbol{X}^{+} = \mathbf{0}, \tag{19}$$

where $\boldsymbol{C}^{-} = \boldsymbol{Q} - \boldsymbol{X}^{-}$, and $m = \lceil B/\Delta x \rceil$ gives the number of discretization points ($\lceil y \rceil$ denotes the smallest integer larger than y). This leads to the linear system represented by the matrix shown in Figure 2. A slightly different matrix procedure should be followed in this case. Denoting by Λ and Γ the solution of the two matrix equations

$$\boldsymbol{X}^{+} + \boldsymbol{\Lambda}\boldsymbol{C} + \boldsymbol{\Lambda}^{2}\boldsymbol{X}^{-} = \mathbf{0}, \tag{20}$$

$$\boldsymbol{X}^{-} + \boldsymbol{\Gamma}\boldsymbol{C} + \boldsymbol{\Gamma}^{2}\boldsymbol{X}^{+} = \mathbf{0}, \tag{21}$$

the solution is obtained by

$$\boldsymbol{\pi}^{(i)} = \boldsymbol{u}_0\boldsymbol{\Lambda}^{i} + \boldsymbol{u}_m\boldsymbol{\Gamma}^{m-i}, \tag{22}$$

with

$$\left| \boldsymbol{u}_0 \ \boldsymbol{u}_m \right| \boldsymbol{J} = \mathbf{0}, \tag{23}$$

where

$$\boldsymbol{J} = \begin{vmatrix} \boldsymbol{C}^{+} + \boldsymbol{\Lambda}\boldsymbol{X}^{-} & \boldsymbol{\Lambda}^{m-1}\boldsymbol{X}^{+} + \boldsymbol{\Lambda}^{m}\boldsymbol{C}^{-} \\ \boldsymbol{\Gamma}^{m-1}\boldsymbol{X}^{-} + \boldsymbol{\Gamma}^{m}\boldsymbol{C}^{+} & \boldsymbol{C}^{-} + \boldsymbol{\Gamma}\boldsymbol{X}^{+} \end{vmatrix}$$

under the normalization condition

$$\left[\sum_{i=0}^{\infty} \boldsymbol{\pi}^{(i)}\right] \mathbf{1} = \left[\boldsymbol{u}_0 \sum_{i=0}^{m} \Lambda^{i} + \boldsymbol{u}_m \sum_{i=0}^{m} \Gamma^{i}\right] \mathbf{1} = 1. \tag{24}$$

In this case, the logarithmic reduction algorithm (presented in Table 1) must be performed two times in order to compute both Λ and Γ. The computation of Λ^m, Γ^m, $\sum_{i=0}^{m} \Lambda^i$ and $\sum_{i=0}^{m} \Gamma^i$ can also be carried out in logarithmic time. The complete solution algorithm is given in Table 1 (called "bounded system"). We should note that other solution techniques can also be applied in this case, see the comparison provided in [10].

4.4 Systems with flush-outs

When we consider a system with constant firing rates and constant flow rates but with flush-outs, equations becomes a little more complex. In particular, the equations that describe the system are

$$\frac{d\boldsymbol{\pi}(x)}{dx}\boldsymbol{R} = \boldsymbol{\pi}(x)\boldsymbol{Q}, \tag{25}$$

$$-\boldsymbol{\pi}(0)\boldsymbol{R} = \boldsymbol{\pi}(0)\boldsymbol{Q} + \int_0^B \boldsymbol{\pi}(x)dx\boldsymbol{Q}', \tag{26}$$

where $B = \infty$ for the unbounded case, and $B < \infty$ for the bounded case. Equations (25) and (26) have been obtained from (6) by writing explicitly the boundary conditions. Using the notation $\boldsymbol{C} = \boldsymbol{Q} - \boldsymbol{X}^+ - \boldsymbol{X}^-$ and $\boldsymbol{D} = \boldsymbol{Q}'$, then the discretized version of (25) and (26) become

$$\boldsymbol{\pi}^{(i-1)}\boldsymbol{X}^+ + \boldsymbol{\pi}^{(i)}\boldsymbol{C} + \boldsymbol{\pi}^{(i+1)}\boldsymbol{X}^- = \boldsymbol{0} \tag{27}$$

$$\boldsymbol{\pi}^{(0)}\boldsymbol{C}^+ + \boldsymbol{\pi}^{(1)}\boldsymbol{X}^- + \sum_{i=0}^{B}\boldsymbol{\pi}^{(i)}\boldsymbol{D} = \boldsymbol{0}, \tag{28}$$

Figure 3a) and Figure 3b) represent the matrices of the discretized system for the unbounded and bounded case, respectively.

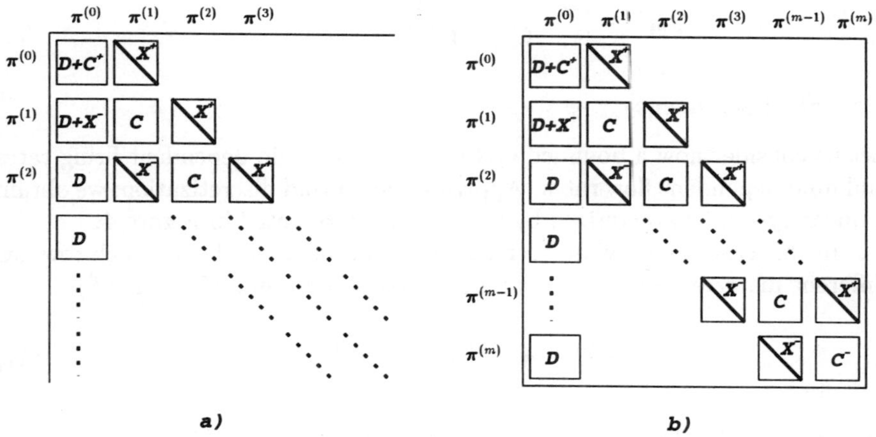

Figure 3. Matrices of the flush-out model: a) bounded b) unbounded

Matrix geometric solution can be applied in both situations by appropriately changing the computation of u_0, u_1 and u_m. The new system that has to be solved in the unbounded case is

$$\left| \boldsymbol{u_0} \; \boldsymbol{u_1} \right| \boldsymbol{J} = \boldsymbol{0}, \tag{29}$$

where

$$\boldsymbol{J} = \left| \begin{array}{cc} \boldsymbol{C}^{+} + \boldsymbol{D} & \boldsymbol{X}^{+} \\ \boldsymbol{X}^{-} + \sum_{i=0}^{\infty} \boldsymbol{\Lambda}^{i} \boldsymbol{D} & \boldsymbol{C} + \boldsymbol{\Lambda X}^{-} \end{array} \right| =$$

$$= \left| \begin{array}{cc} \boldsymbol{C}^{+} + \boldsymbol{D} & \boldsymbol{X}^{+} \\ \boldsymbol{X}^{-} + (\boldsymbol{I} - \boldsymbol{\Lambda})^{-1} \boldsymbol{D} & \boldsymbol{C} + \boldsymbol{\Lambda X}^{-} \end{array} \right|.$$

While for the bounded case

$$\left| \boldsymbol{u_0} \; \boldsymbol{u_m} \right| \left| \boldsymbol{j_0}^T \; \boldsymbol{j_m} T \right| = \boldsymbol{0}, \tag{30}$$

where

$$\boldsymbol{j_0}^T = \left| \begin{array}{c} \boldsymbol{C}^{+} + \boldsymbol{\Lambda X}^{-} + \sum_{i=0}^{m} \boldsymbol{\Lambda}^{i} \boldsymbol{D} \\ \boldsymbol{\Gamma}^{m-1} \boldsymbol{X}^{-} + \boldsymbol{\Gamma}^{m} \boldsymbol{C}^{+} + \sum_{i=0}^{m} \boldsymbol{\Gamma}^{i} \boldsymbol{D} \end{array} \right|,$$

$$\boldsymbol{j_m}^T = \left| \begin{array}{c} \boldsymbol{\Lambda}^{m-1} \boldsymbol{X}^{+} + \boldsymbol{\Lambda}^{m} \boldsymbol{C}^{-} \\ \boldsymbol{C}^{-} + \boldsymbol{\Gamma X}^{+} \end{array} \right|.$$

4.5 Fluid dependency

Let us consider now a bounded system with both fluid dependent firing rates and fluid dependent flow rates. Applying the upwind discretization, we obtain a linear system characterized by the matrix represented in Figure 4.

In this case the previous techniques cannot be applied since each row has different matrixes. Instead, it can be solved by imposing (following [9])

$$\boldsymbol{\pi}^{(i)} = \boldsymbol{\pi}^{(i-1)} \boldsymbol{\Lambda_i} = \boldsymbol{\pi}^{(0)} \prod_{j=0}^{i} \boldsymbol{\Lambda_j}. \tag{31}$$

Substituting the definition of $\boldsymbol{\pi}^{(i)}$ in i-th equation, we obtain

$$\boldsymbol{\pi}^{(i-1)} \boldsymbol{X}^{+}{}_{i-1} + \boldsymbol{\pi}^{(i)} \boldsymbol{C_i} + \boldsymbol{\pi}^{(i+1)} \boldsymbol{X}^{-}{}_{i+1} = \boldsymbol{0},$$
$$\boldsymbol{\pi}^{(i-1)} \boldsymbol{X}^{+}{}_{i-1} + \boldsymbol{\pi}^{(i)} \left(\boldsymbol{C_i} + \boldsymbol{\Lambda_{i+1}} \boldsymbol{X}^{-}{}_{i+1} \right) = \boldsymbol{0},$$

and

$$\boldsymbol{\pi}^{(i)} = -\boldsymbol{\pi}^{(i-1)} \boldsymbol{X}^{+}{}_{i-1} \left(\boldsymbol{C_i} + \boldsymbol{\Lambda_{i+1}} \boldsymbol{X}^{-}{}_{i+1} \right)^{-1}, \tag{32}$$

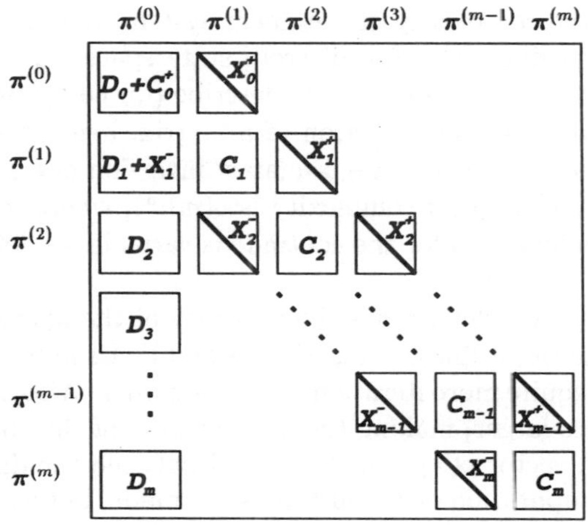

Figure 4. Matrix of the bounded fluid dependent case

Comparing (31) and (32) we obtain

$$\Lambda_i = -X^{+}{}_{i-1}\left(C_i + \Lambda_{i+1} X^{-}{}_{i+1}\right)^{-1}, \tag{33}$$

where Λ_i can be computed by backward substitution starting from

$$\Lambda_m = -X^{+}{}_{m-1}\left(C^{-}{}_m\right)^{-1}. \tag{34}$$

Then $\pi^{(0)}$ can be computed by solving the linear system

$$\begin{aligned} \pi^{(0)}\left[D_0 + C^{+}{}_0 + \Lambda_1 X^{-}{}_1 + \sum_{i=1}^{m}\prod_{j=1}^{i}\Lambda_j D_i\right] &= 0, \\ \pi^{(0)}\left(\sum_{i=1}^{m}\prod_{j=1}^{i}\Lambda_j + I\right)\mathbf{1} &= 1. \end{aligned} \tag{35}$$

The solution algorithm, called "fluid dependent", is given in Table 1.

4.6 Complexity

The complexity of the proposed technique is different for the constant and the fluid dependent case. For the constant case, the most time consuming tasks are the computation of matrices $\mathbf{\Lambda}$ and $\mathbf{\Gamma}$ and the solution of the linear system. Even if the initial matrix is sparse, the coefficient matrix of the linear system is not, so the complexity of the linear system solution is $o(n^3)$, where $n = |S_d|$ is the cardinality of the discrete state space. The complexity of the linear reduction algorithm is hard to describe (for complete discussion see [9]) but each step requires the inversion of a matrix, hence the complexity is $o(n^3)$. For the fluid dependent case, a matrix inversion operation is required at each iteration, so the total complexity is $o(mn^3)$, where m is the number of discretization points used for the continuous variable, as defined in Section 4.3.

The parameter ΔX determines the accuracy of the discretization procedure and influences the stiffness of matrixes Λ and Γ. Usually, as ΔX becomes smaller, Λ and Γ require more iteration to be computed. Also, as ΔX becomes smaller, m becomes larger, making the computation of the coefficient matrix and normalization vector $o(\log m)$. A tradeoff between the discretization accuracy and the computation of Λ and Γ must be made, but as a rule of thumb $\Delta X < 10 \max(|\boldsymbol{R}|)$ is enough to obtain good solutions, where $\max(|\boldsymbol{R}|)$ is the maximum flow rate in absolute value.

4.7 Extensions

The proposed solution technique may be applied as well to *second order fluid stochastic Petri nets* [5]. In second order FSPN, each fluid arc has an associated mean flow rate r, and also a variance v. The equations that arise from these kind of models are second order differential equations. For example the steady state solution of a second order FSPN can be computed by solving the following equation:

$$\frac{d\boldsymbol{\pi}(x)}{dx}\boldsymbol{R} - \frac{1}{2}\frac{d^2\boldsymbol{\pi}(x)}{dx^2}\boldsymbol{V} = \boldsymbol{\pi}(x)\boldsymbol{Q}, \tag{36}$$

where $\boldsymbol{V}$ is a diagonal matrix whose elements represent the variance of the flow rate in a particular state. Note that this is the equivalent of (7) for second order models.

The first order term can be discretized as proposed in (8) and (9). The second order term can be instead discretized as:

$$\frac{d^2\pi_j(i\Delta x)}{dx^2} \approx \frac{\dfrac{\pi_j^{(i+1)} - \pi_j^{(i)}}{\Delta x} - \dfrac{\pi_j^{(i)} - \pi_j^{(i-1)}}{\Delta x}}{\Delta x} = \frac{\pi_j^{(i+1)} - 2\pi_j^{(i)} + \pi_j^{(i-1)}}{\Delta x^2}. \tag{37}$$

We can then define $\boldsymbol{X}^+$ and $\boldsymbol{X}^-$ as:

$$\boldsymbol{X}^+ = \frac{1}{\Delta x}\boldsymbol{R}^+ + \frac{1}{2\Delta x^2}\boldsymbol{E}, \quad \boldsymbol{X}^- = -\frac{1}{\Delta x}\boldsymbol{R}^- + \frac{1}{2\Delta x^2}\boldsymbol{E},$$

and apply the matrix geometric solution technique proposed in Section 4.2.

5 A numerical example

Consider a producer/consumer system with N producers and M consumers and a finite buffer. A FSPN model for the considered problem is shown in Figure 5. The number of tokens contained in place $OFF-P$ represents the number of producers that are not active. Place $ON-P$ represents the active producers, and its marking corresponds to their number. Transition T_{end} represents the end of a production phase for a producer, and transition T_{new} the start of a new phase. Place $ON-c$ models the number of active consumers, and place $OFF-C$ represents the inactive ones. Transition T_{down} models the failure of a consumer, and transition T_{up} its repair. The buffer is modeled by fluid place $BUF-F$. Transition T_{in-buf} pumps fluid into the buffer. The flow rate of the arc that connects this transition to the buffer is proportional to the number of active producers. Transition $T_{out-buf}$ drags fluid out of the buffer. The rate of this flow is proportional to the number of active consumers. Since the buffer has finite capacity, all the fluid that is pumped in after the upper boundary level is reached, is lost. This loss is modeled by transition T_{loss-d}. The flow rate of the associated arc depends on the level of the fluid place. It is zero as long as the level is below the limit (no loss case). It corresponds to the rate of the fluid that is lost due to the finite capacity when the boundary is reached.

Even if the the flow rate of the fluid arc connecting T_{loss-d} to the buffer is fluid level dependent, the technique proposed in Section 4.3 can still be applied since the flow rate changes only on the boundaries.

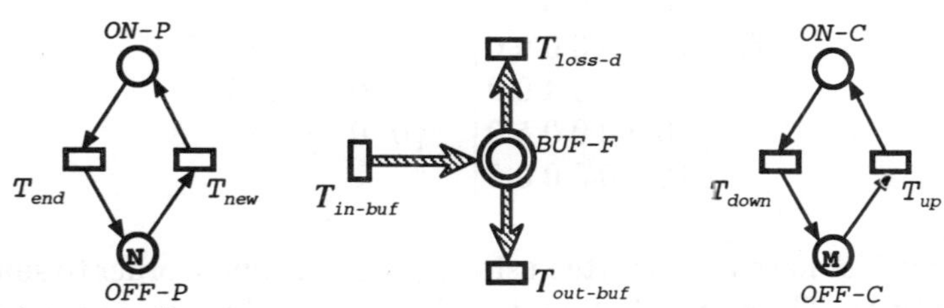

Figure 5. A producer/consumer FSPN model

A producer/consumer model has been studied in [11] using SPNs with reward structures. The main difference between the two approaches is that

FSPN allows the modeler to account for fluid behavior in a simple manner at the description level. Fluid like behavior can be captured with SPNs as well, however, the modeler has to approximate it by using high number of tokens in the model. In other words, discretization appears at the description level using SPNs, while it appears only at the solution level in case of FSPNs and it can be handled in an automatic manner hidden from the modeler. For what concerns computational complexity, the two approaches are similar when applied to model the same phenomena (assuming that the same solution technique is used to solve the discretized FSPN model and the SPN model).

In the following, we consider a system in which all transitions T_i have firing rate equal to 1.0, with 6 producers (that produces at rate $\lambda_p = 1.0$) and 4 consumers (that consumes at rate $\lambda_c = 2.0$). In particular, following the method proposed in [2], matrices $\boldsymbol{Q}$ and $\boldsymbol{R}$ can be computed as:

$$\boldsymbol{Q} = \begin{vmatrix} -6 & 6 & 0 & 0 & 0 & 0 & 0 \\ 1 & -6 & 5 & 0 & 0 & 0 & 0 \\ 0 & 2 & -6 & 4 & 0 & 0 & 0 \\ 0 & 0 & 3 & -6 & 3 & 0 & 0 \\ 0 & 0 & 0 & 4 & -6 & 2 & 0 \\ 0 & 0 & 0 & 0 & 5 & -6 & 1 \\ 0 & 0 & 0 & 0 & 0 & 6 & -6 \end{vmatrix} \oplus \begin{vmatrix} -4 & 4 & 0 & 0 & 0 \\ 1 & -4 & 3 & 0 & 0 \\ 0 & 2 & -4 & 2 & 0 \\ 0 & 0 & 3 & -4 & 1 \\ 0 & 0 & 0 & 4 & -4 \end{vmatrix}$$

$$\boldsymbol{R} = \begin{vmatrix} 0 & 0 & 0 & 0 & 0 & 0 & 0 \\ 0 & 1 & 0 & 0 & 0 & 0 & 0 \\ 0 & 0 & 2 & 0 & 0 & 0 & 0 \\ 0 & 0 & 0 & 3 & 0 & 0 & 0 \\ 0 & 0 & 0 & 0 & 4 & 0 & 0 \\ 0 & 0 & 0 & 0 & 0 & 5 & 0 \\ 0 & 0 & 0 & 0 & 0 & 0 & 6 \end{vmatrix} \oplus \begin{vmatrix} 0 & 0 & 0 & 0 & 0 \\ 0 & -2 & 0 & 0 & 0 \\ 0 & 0 & -4 & 0 & 0 \\ 0 & 0 & 0 & -6 & 0 \\ 0 & 0 & 0 & 0 & -8 \end{vmatrix}$$

(Both matrices have been written using Kronecker sum in order to simplify the presentation). The system has been analyzed using the technique presented in section 4.3 using $\Delta x = 0.1$ and various buffer sizes. Figure 6a) represents the cumulative distribution function (cdf) of the buffer level when the capacity is 15, and Figure 6b) represents the mean buffer occupancy as function of the buffer capacity B (from $B = 10$ to $B = 20$).

Two discontinuities are present in the buffer distribution. These discontinuities correspond respectively to the probabilities of having the buffer empty

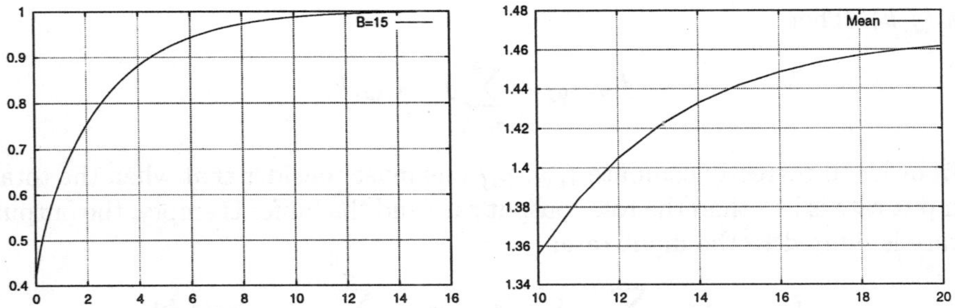

Figure 6. Buffer occupancy: a) cdf in case of $B = 15$ b) mean buffer level as the function of capacity

and of having the buffer full. Figure 7a) and b) represents the probability masses at the two boundaries.

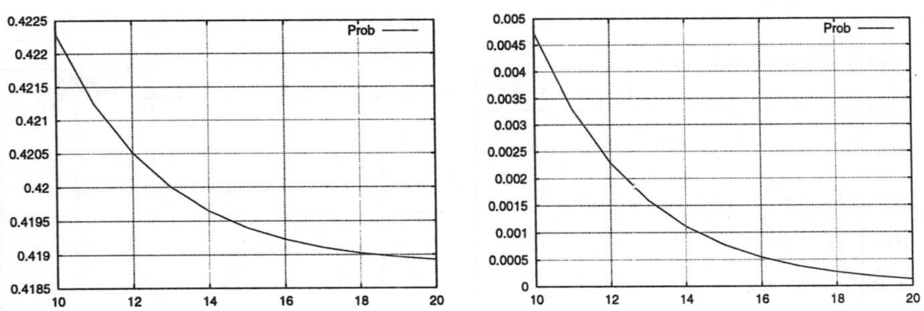

Figure 7. Probability of a) empty and b) full buffer as the function of capacity

Other interesting measures that can be derived from the $FSPN$ description of the model are the mean fluid flow along the fluid arcs connecting transitions T_{in-buf}, $T_{out-buf}$ and T_{loss-d} (which we will call respectively f_{in-buf}, $f_{out-buf}$ and f_{loss-d}). f_{in-buf} can be computed by simply summing the flow rate across the arc in each discrete / continuous state. The probability of having j producers active, k consumers and fluid level $\approx i\Delta x$ is denoted by

$\pi_i(j,k)$. Then

$$f_{in-buf} = \sum_{i,j,k} j\lambda_p \pi_i(j,k).$$

In order to correctly compute $f_{out-buf}$ one must consider that when the total input rate is less than the total output rate and the buffer is empty, the output rate is limited to the input rate:

$$f_{out-buf} = \sum_{i>0,j,k} k\lambda_c \pi_i(j,k) + \sum_{j,k:k\lambda_c > j\lambda_p} j\lambda_p \pi_0(j,k).$$

Figure 8a) represents f_{in-buf} and $f_{out-buf}$ as function of the buffer size. Note that f_{in-buf} does not depend on the buffer size, since losses does not changes the production speed. Also $f_{out-buf}$ approaches f_{in-buf} as the buffer size B becomes bigger. f_{loss-d} can be simply computed as $f_{loss-d} = f_{in-buf} - f_{out-buf}$. Another interesting performance indexes that can be computed is the percentage of fluid produced that is lost p_{loss-d}. This probability can be calculated simply as $p_{loss-d} = f_{loss-d}/f_{in-buf}$. Figure 8b) represents f_{loss-d} and p_{loss-d} as function of the buffer size.

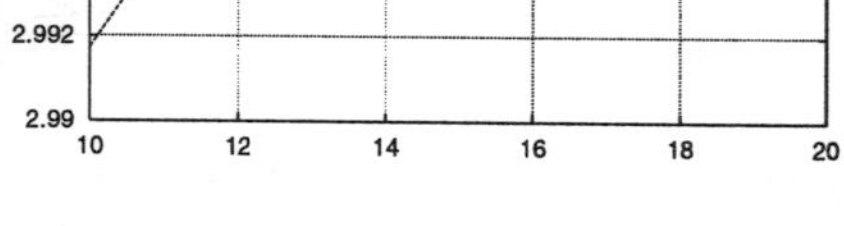

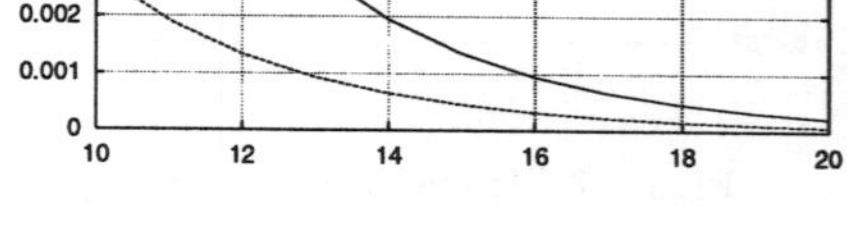

Figure 8. a) input and output throughput b) throughput and probability of loss as the function of the buffer size

Acknowledgments

We acknowledge the support of the Italian Ministry for University and Scientific Research, through the Planet-IP project.

6 Conclusions

In this paper a numerical solution technique for the steady state analysis of Fluid Stochastic Petri Nets with flush-out arcs has been presented. The proposed technique uses known matrix geometric solution techniques to solve the equations that describes the stochastic process underlying a FSPN with flush-out. This technique has already provided good results in [6] (where the technique itself was not described) to analyze the GPRS communication system. In that work it was shown that the matrix geometric technique presented in this paper, applied to the solution of a FSPN model of the GPRS system, can produce results 3 order of magnitude faster than conventional solution techniques for GSPN or DSPN models of the same system. In that paper it was also shown that the obtained solutions have satisfactory accuracy for realistic configurations, even if in some cases provide more optimistic results. In this paper a simple application to a producer/consumer system has been presented. A possible extension to second order models has also been proposed.

References

1. H. Alla and R. David. Continuous and Hybrid Petri Nets. *Journal of Systems Circuits and Computers*, 8(1), Feb 1998.
2. M. Gribaudo, M. Sereno, A. Horváth, and A. Bobbio. Fluid stochastic petri nets augmented with flush-out arcs: Modeling and analysis. *Discrete Event Dynamic Systems*, 11(1/2):97–117, January 2001.
3. G. Horton. Computation of the distribution of accumulated reward with fluid stochastic petri nets. In *Proc. of 2th Inter. Computer Performance & Dependability Symposium (IPDS '96)*, Urbana-Champaigne, USA, 1996.
4. G. Horton, V. G. Kulkarni, D. M. Nicol, and K. S. Trivedi. Fluid stochastic Petri Nets: Theory, Application, and Solution Techniques. *European Journal of Operations Research*, 105(1):184–201, Feb 1998.
5. K. Wolter. Second order fluid stochastic petri nets: an extension of GSPNs for approximate and continuous modeling. In *Proc. of World Congress on System Simulation*, pages 328–332, Singapore, Sep 1997.
6. M. Ajmone Marsan, M. Gribaudo, M. Meo, and M. Sereno. On petri net-based modeling paradigms for the performance analysis of wireless internet accesses. In *Proc. of 9th Intern. Workshop on Petri Nets and Performance Models*, Aachen, Germany, Sep. 2001. IEEE-CS Press. To appear.
7. M. Ajmone Marsan, G. Balbo, G. Conte, S. Donatelli, and G. Franceschi-

nis. *Modeling with Generalized Stochastic Petri Nets.* John Wiley & Sons, 1995.

8. D.R. Cox. The analysis of non-markovian stochastic processes by the inclusion of supplementary variables. *Proceedings of the Cambridge Phylosophical Society*, 51:433–440, 1955.
9. G. Latouche and V. Ramaswami. *Introduction to Matrix Geometric Methods in Stochastic Modeling.* ASA-SIAM Series on Statistics and Applied Probability. SIAM, Philadelphia PA, 1999.
10. B. R. Haverkort and A. Ost. Steady-state analysis of infinite stochastic petri nets: Comparing the spectral expansion and the matrix-geometric method. In *Proc. of 7th Inter. Workshop on Petri Nets And Performance Model (PNPM'97)*, Saint Malo, France, 1997.
11. G. Ciardo, J. K. Muppala, and K. S. Trivedi. Analyzing concurrent and fault-tolerant software using stochastic petri nets. *Journal of Parallel and Distributed Comp.*, 15(3), July 1992.

A MARKOVIAN POINT PROCESS EXHIBITING MULTIFRACTAL BEHAVIOR AND ITS APPLICATION TO TRAFFIC MODELING

ANDRÁS HORVÁTH

Dipartimento di Informatica, Università di Torino
Corso Svizzera 185, 10149 Torino, Italy, horvath@di.unito.it

MIKLÓS TELEK

Department of Telecommunications, Technical University of Budapest
Sztoczek u. 2, 1521 Budapest, Hungary, telek@hit.bme.hu

This paper introduces a set of Markovian Arrival Processes (MAPs) with a special structure exhibiting multifractal behavior. The considered MAP structure is motivated by the unnormalized Haar wavelet transform representation of finite sequences. A parameter fitting method is also proposed to approximate the multifractal behavior of experimental data sets by MAPs of the given structure. The goodness of the fitting method is evaluated via the log-moment diagrams, the partition function, the Legendre transform, and also by comparing the queue length distribution resulting from the measured data set with the one resulted from the approximating MAP.

1 Introduction

The traffic of high-speed communication networks, carrying the data packets of various applications, shows high variability and burstiness over several time scales (references to many measurement studies are provided by Willinger et al.[1]). The statistical analysis of some experimental traffic traces suggested a self-similar behavior over a range of time scales. Since measured data sets are finite (the large ones contain $10^6 - 10^8$ samples) the statistical properties of these data sets could be studied only over a range of time scales and the asymptotic behavior is determined from the range of known time scales.

The importance of the observed self-similar behavior lies in the fact that the queue length and the waiting time distribution of packet queues with self-similar traffic significantly differs from the ones with "regular traffic".

In the early 90's the research was focused on checking the self-similar behavior and the evaluation of the scaling parameter of self-similarity (referred to as Hurst parameter). It has to be noted that the majority of the practically applied statistical tests (e.g., variance-time plot, R/S plot) checks only the second order properties of the data sets and provides information only on the second order self-similarity of the analyzed data set. (Actually, there are self-similar processes, like the fractional Gaussian noise [2], that are completely characterized by their first and second order behavior, but measured data sets are usually far more complex).

The first observations of self-similarity in measured traffic processes resulted in a fertile research for applying complex mathematical models in traffic engineering. The two main goals of these research efforts were to find "solvable" mathematical models with identical (or similar) properties and to create random sequence generators with predefined statistical properties. Some of the considered models are: fractional Gaussian noise [3,2], traditional [4] and fractional ARIMA processes [5], fractal [6] and Markovian models (MMPP, MAP) [7,8,9]. A valuable advantage of Markovian models is that effective numerical methods are available to analyze systems with Markovian arrival processes [10,11]. Furthermore, Markovian models also represent a simple and computationally effective way of generating random data series.

Unfortunately, some of the statistical properties of measured data sets differ from the ones of self-similar processes. The fact that the observed scaling behavior differs from the one of self-similar processes suggested the application of multifractal models to better capture the behavior of measured data sets [12]. A common approach to study multifractal models is wavelet analysis. Riedi et al. proposed a wavelet model to approximate the scaling behavior of measured data sets and based on this model they presented an algorithm to generate random sequences with similar scaling behavior [13]. The proposed method shows a good fit according to several statistical tests, but it is computationally rather expensive and does not allow any numerical analysis of queues.

Based on the mentioned advantages of Markovian models there is a need to approximate multifractal behavior with Markovian models. In this paper we propose Markovian models of a special structure to approximate the multifractal scaling behavior of measured data sets.

The flexibility of Markovian models in exhibiting complex stochastic behavior along the practically interesting time scales is known from previous works [8,7,9]. This paper attempts to extend this set of results with approximating multifractal models. The proposed MAP structure is motivated by the unnormalized Haar wavelet transform representation of finite sequences as it was applied in the multifractal wavelet model of Riedi et al. [13].

The rest of the paper is organized as follows. Section 2 summarizes the basic concepts of multiscale analysis. The proposed MAP structure is introduced in Section 3. Section 4 presents various properties of the proposed MAP structure and an example in fitting with measured data sets. The paper is concluded in Section 5.

2 Multiscale analysis

This section gives an overview of three methods which we will use in this paper to carry out the multiscale analysis of a stationary sequence of numbers $\mathcal{X} = \{X_i, i \geq 1\}$. As far as traffic modeling is concerned, this sequence may represent any kind of important characteristics of the traffic load arriving to the network. That is, $\mathcal{X}$ may be the series of interarrival times, the series of the number of arrivals in successive time-slots or the number of bytes per arrival. (See [14] for an exhaustive study of multifractal properties of time series describing different aspects of TCP traffic.)

In the sequel $\mathcal{X}^{(m)}$ denotes the corresponding aggregated process with level of aggregation m

$$\mathcal{X}^{(m)} = \left\{X_i^{(m)}, i \geq 1\right\} = \left\{\frac{X_1 + \ldots + X_m}{m}, \ldots, \frac{X_{(i-1)m+1} + \ldots + X_{mi}}{m}, \ldots\right\}.$$

This section is organized as follows. First we look at statistical scaling in Section 2.1 that aims at determining the Hurst parameter of the process. In Section 2.2 a method to analyze multifractal scaling, that results in the Legendre spectrum, is described with which we will compare our approximating MAP with the real traffic trace. The fitting procedure is based on another way of examining a finite sequence of numbers, the Haar wavelet transform (Section 2.3), because it allows us to compute the desired properties of our MAP structure analytically.

The introduction to statistical and multifractal scaling given hereinafter is partly based on [15] and [14].

2.1 Statistical scaling

Recently, it has been agreed [16,17,2] that when one studies a traffic trace the most significant parameter to be estimated is the degree of self-similarity, usually given by the so-called Hurst-parameter. The aim of the statistical approach, based on the theory of self-similarity, is to find the Hurst-parameter.

The standard definition of self-similarity is stated for continuous-time processes: $\mathcal{Y} = \{Y(t), t > 0\}$ is self-similar if

$$Y(t) \stackrel{d}{=} a^{-H} Y(at), \forall t > 0, \forall a > 0, 0 < H < 1, \tag{1}$$

where $\stackrel{d}{=}$ denotes equality in the sense of finite-dimensional distributions and H is the Hurst-parameter. The most broadly applied signal model satisfying (1) is the fractional Brownian motion [18,17,2] whose power lies in its simple parameterization. It is fully determined by its mean, variance and the Hurst-parameter.

There are several, different definitions of self-similarity involving stationary sequences $\mathcal{X} = \{X_i, i \geq 1\}$; in the context of traffic modeling these are more appropriate than the one given by (1). A stationary discrete-time stochastic process $\mathcal{X} = \{X_i, i \geq 1\}$ is said to be *exactly self-similar* if

$$\mathcal{X} \stackrel{d}{=} m^{1-H} \mathcal{X}^{(m)} \tag{2}$$

for all aggregation levels m. In other words $\mathcal{X}$ is said to be exactly self-similar if $\mathcal{X}$ and $\mathcal{X}^{(m)}$ are identical within a scale factor in the sense of finite-dimensional distributions. (We note here that if $\mathcal{X}$ is the incremental process of an exactly self-similar continuous-time process it satisfies (2) for all aggregation levels m.) A stationary sequence is said to be *asymptotically self-similar* if (2) holds as $m \to \infty$. A covariance-stationary sequence $\mathcal{X}$ is *exactly second-order self-similar* or *asymptotically second-order self-similar* if $m^{1-H}\mathcal{X}^{(m)}$ has the same variance and auto-correlation as $\mathcal{X}$ for all aggregation level m, or as $m \to \infty$.

As it was proposed in [15] one may perform a test of self-similarity by analyzing the behavior of the absolute moments of the aggregated process. If $\mathcal{X}$ is exactly self-similar

$$log(\mathbf{E}(|X^{(m)}|^q)) = log(\mathbf{E}(|m^{H-1}X|^q)) = q(H-1)log(m) + log(\mathbf{E}(|X|^q)). \tag{3}$$

According to (3), in case of a self-similar process plotting $log(\mathbf{E}(|X^{(m)}|^q))$ against $log(m)$ for a fixed q results in a straight line. The slope of the line is $q(H-1)$. Based on the above observations the test is performed as follows. Having a series of length N, the moments may be estimated as

$$\mathbf{E}(|X^{(m)}|^q) = \frac{1}{\lfloor N/m \rfloor} \sum_{i=1}^{\lfloor N/m \rfloor} |X_i^{(m)}|^q,$$

where $\lfloor x \rfloor$ denotes the largest integer number smaller or equal to x. To test for self-similarity $log(\mathbf{E}(|X^{(m)}|^q))$ is plotted against $log(m)$ and a straight line is fitted to the curve. If the straight line shows good correspondence with the curve, then the process is self-similar and its Hurst-parameter may be calculated by the slope of the straight line.

It is worth pointing out that (2) and stationarity imply that either $\mathbf{E}(X) = 0$, or $\mathbf{E}(X) = \pm\infty$, or $H = 1$. But $H = 1$ implies as well that $X_i = X_j$, $\forall i, j$ almost surely. As a consequence, to test for statistical self-similarity makes sense only having zero-mean data, i.e., the data has to be centered before the analysis. The variance-time plot, which is used widespread to gain evidence of self-similarity, is the special case with $q = 2$. It depicts the behavior of the 2nd moments for the centered data. On the other hand, as we show later, multifractal analysis may be carried out on data with non-zero mean as well.

2.2 Multifractal scaling

As it is described above, statistical tests of self-similarity try to gain evidence through examining the behavior of the absolute moments $\mathbf{E}(|X^{(m)}|^q)$. Multifractal analysis looks at the behavior of the absolute moments as well, but in a different manner which may result in more detailed information on the sequence. While the above described statistical view looks for a single number, the Hurst parameter, that completely describes the behavior of $\mathbf{E}(|X^{(m)}|^q)$ for any q, multifractal analysis results in a *spectrum* to illustrate the behavior of the absolute moments.

As for self-similarity we start the discussion with a continuous-time process $\mathcal{Y} = \{Y(t), t > 0\}$. The scaling of the absolute moments of the increments is observed through the *partition function*

$$T(q) = \lim_{n\to\infty} \frac{1}{-n} \log_2 \mathbf{E}\left[\sum_{k=0}^{2^n-1} |Y((k+1)2^{-n}) - Y(k2^{-n})|^q\right]. \qquad (4)$$

Then, a multifractal spectrum, the so-called *Legendre spectrum* is given as the *Legendre transform* of (4) by

$$f_L(\alpha) = T^*(\alpha) = \inf_q(q\alpha - T(q))$$

Since $T(q)$ is always concave, the Legendre spectrum $f_L(\alpha)$ may be found by simple calculations using

$$T^*(\alpha) = q\alpha - T(q), \text{ and } (T^*)'(\alpha) = q \text{ at } \alpha = T'(q). \qquad (5)$$

Let us mention here that there are also other kinds of fractal spectrum defined in the fractal world (see for example [19]). The Legendre spectrum is the most attractive one from numerical point of view, and even though in some cases it is less informative than, for example, the large deviation spectrum, it provides enough information in the cases considered herein.

In the case of a discrete-time process $\mathcal{X}$ we assume that we are given the increments of a continuous-time process. This way, assuming that the sequence we examine consists of $N = 2^L$ numbers, the sum in (4) becomes

$$S_n(q) = \sum_{k=0}^{N/2^n-1} |X_k^{(2^n)}|^q, \; 0 \le n \le L, \qquad (6)$$

where the expectation is ignored. Ignoring the expectation is accurate for small n, i.e., for the finer resolution levels. In order to estimate $T(q)$, we plot $\log_2(S_n(q))$ against $(L-n)$, $n = 0, 1, ..., L$, then $T(q)$ is found by the slope of the straight line fitted to the plotted points. If the straight line shows good correspondence with the plotted points, that is, if $\log_2(S_n(q))$ scales linearly with $\log(n)$, then the sequence $\mathcal{X}$ can be considered a multifractal process.

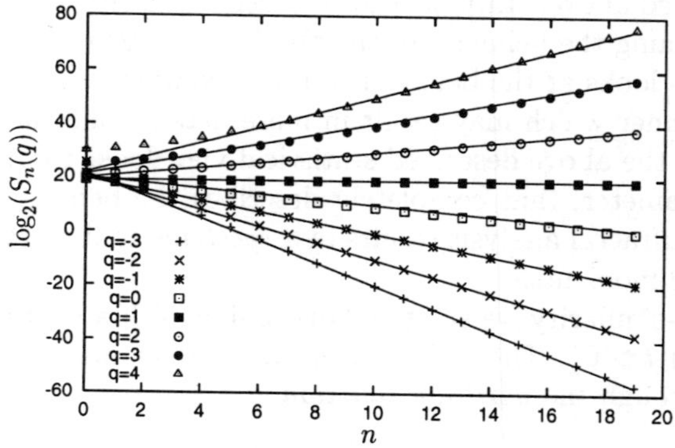

Figure 1. Scaling of log-moments with linear fits for the interarrival times of the Bellcore *pAug* trace

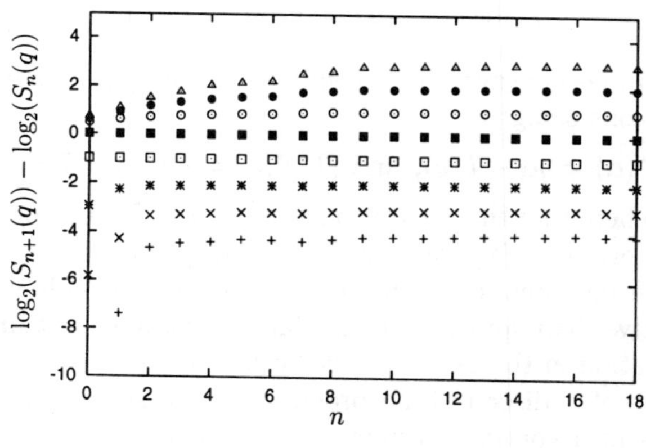

Figure 2. Increments of log-moments for the interarrival times of the Bellcore *pAug* trace

Figure 1, 2 (the same legend applies to this figure as to Figure 1), 3 and 4 illustrate the above described procedure to obtain the Legendre spectrum of the famous Bellcore ***pAug*** traffic trace (the trace may be found at [20]). Figure 1 depicts the scaling behavior of the log moments calculated through (6). With q in the range $[-3, 4]$, excluding the finest resolution levels $n = 0, 1$ the moments show good linear scaling. For values of q outside the range $[-3, 4]$ the curves deviate more and more from linearity. As, for example, in [13] one may look at non-integer values of q as well, but, in general, it does not provide notably more information on the process. To better visualize the deviation from linearity Figure 2 depicts the increments of the log-moment curves of Figure 1. Completely horizontal lines would represent linear log-moment curves.

The partition function $T(q)$ is depicted in Figure 3. The three slightly different curves differ only in the considered range of the log-moments curves, since different ranges result in different linear fitting. The lower bound of the linear fitting is set to 3, 5 and 7, while the upper bound is 18 in each case. (In the rest of this paper the fitting range is 5 - 18 and there are 200 moments evaluated in the range $[-5, +5]$.) Since the partition function varies only a little (its derivative is in the range $[0.8, 1.15]$), it is not as informative as its Legendre transform is (Figure 4). According to (5) the Legendre spectrum is as wide as the range of derivatives of the partition function is. That is, the more the partition function deviates from linearity the wider the Legendre spectrum is. The Legendre transform significantly amplifies the scaling information, but it is also sensitive to the considered range of the log-moments curves.

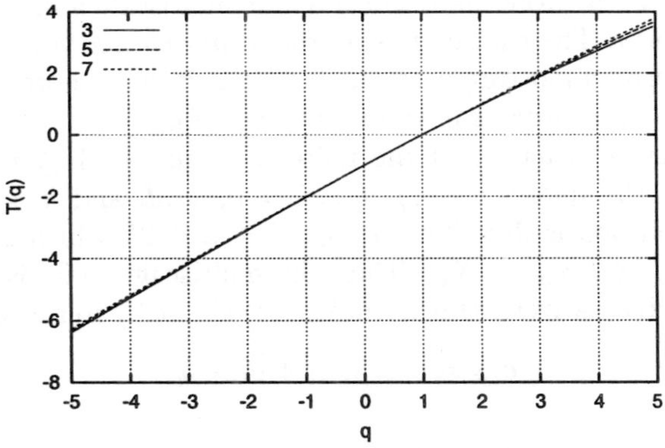

Figure 3. Partition function estimated through the linear fits shown in Figure 1

See [13] for basic principles of interpreting the spectrum. We mention here

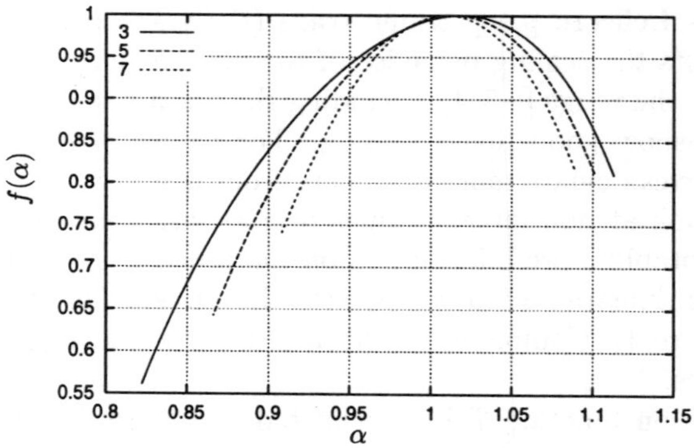

Figure 4. The Legendre transform of the partition function (Figure 3) results in the Legendre spectrum

only that a curve like the one depicted in Figure 4 reveals a ***rich multifractal spectrum.*** On the contrary, as it was shown in [21], the fractional Brownian motion (fBm) has a trivial spectrum. The partition function of the fBm is a straight line which indicates that its spectrum consists of one point, i.e., the behavior of its log-moments is identical for any q.

2.3 The unnormalized Haar wavelet transform

The third way we mention here to carry out multiscale analysis is the Haar wavelet transform. The choice of using the ***unnormalized*** version of the Haar wavelet transform is motivated by the fact that it suits more the analysis of the Markovian point process introduced further on.

The multiscale behavior of the finite sequence $X_i, 1 \leq i \leq 2^L$ will be represented by the quantities $c_{j,k}, d_{j,k}, j = 0, \ldots, L$ and $k = 1, \ldots, 2^L/2^j$. The finest resolution is described by $c_{0,k}, 1 \leq k \leq 2^L$ which gives the finite sequence itself, i.e., $c_{0,k} = X_k$. Then the multiscale analysis based on the unnormalized Haar wavelet transform is carried out by iterating

$$c_{j,k} = c_{j-1,2k-1} + c_{j-1,2k}, \tag{7}$$

$$d_{j,k} = c_{j-1,2k-1} - c_{j-1,2k}, \tag{8}$$

for $j = 1, \ldots, L$ and $k = 1, \ldots, 2^L/2^j$. The quantities $c_{j,k}, d_{j,k}$ are the so-called scaling and wavelet coefficients of the sequence, respectively, at scale j and position k. The procedure to obtain the scaling and wavelet coefficients

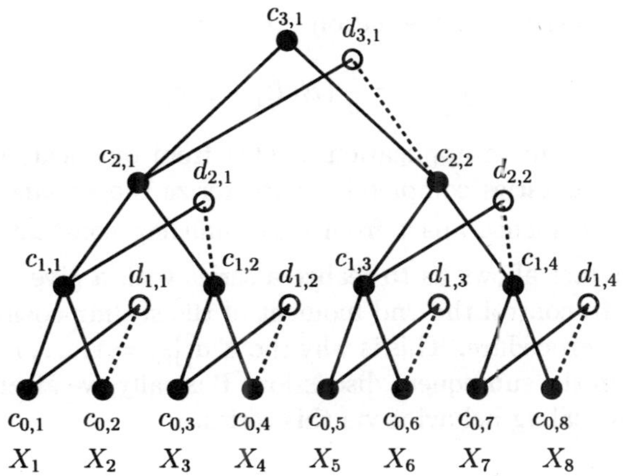

Figure 5. Haar wavelet transform

is depicted in Figure 5 for a series of length 8. In the figure, the scaling (wavelet) coefficients are drawn as filled (empty) circles, and solid (dashed) lines are connected to quantities that are taken into account with positive (negative) sign. At each scale the coefficients are represented by the vectors $\mathbf{c}_j = [c_{j,k}]$ and $\mathbf{d}_j = [d_{j,k}]$ with $k = 1, \ldots, 2^L/2^j$. For what concerns $\mathbf{c}_j$, the higher j the lower the resolution level at which we have information on the sequence. The information that we lose as a result of the step from $\mathbf{c}_{j-1}$ to $\mathbf{c}_j$, is conveyed by the sequence of wavelet coefficients $\mathbf{d}_j$. It is easy to see that $\mathbf{c}_{j-1}$ can be perfectly reconstructed from $\mathbf{c}_j$ and $\mathbf{d}_j$. As a consequence the whole $X_i, 1 \leq i \leq 2^L$ sequence can be constructed (in a top to bottom manner) based on a normalizing constant, $\mathbf{c}_L = c_{L,1} = \sum_{i=1}^{2^L} X_i$, and the $\mathbf{d}_j, j = 1, \ldots, L$ vectors.

By taking the expectation of the square of (7) and (8) we get

$$E[c_{j,k}^2] = E[c_{j-1,2k-1}^2] + 2E[c_{j-1,2k-1}c_{j-1,2k}] + E[c_{j-1,2k}^2], \tag{9}$$

$$E[d_{j,k}^2] = E[c_{j-1,2k-1}^2] - 2E[c_{j-1,2k-1}c_{j-1,2k}] + E[c_{j-1,2k}^2]. \tag{10}$$

Let us assume that the series we analyze is stationary[a]; then, by summing (9)

[a]Clearly, since we are given only a finite trace, we cannot have strong evidence that the series under analysis is obtained from a stationary process. However, we apply this assumption

and (10) and rearranging the equation, we have

$$E[c_{j-1}^2] = \frac{1}{4}\left(E[d_j^2] + E[c_j^2]\right). \qquad (11)$$

Similarly, by consecutive application of (11) from one scale to another the $E[d_j^2], j = 1, \ldots, L$ series completely characterize the variance decay of the $X_i, 1 \leq i \leq 2^L$ sequence apart from a normalizing constant ($\mathbf{c}_L = c_{L,1} = \sum_{i=1}^{2^L} X_i$). This fact allows us to realize a series with a given variance decay if it is possible to control the 2nd moment of the scaling coefficient with the chosen synthesis procedure. This is why the $E[d_j^2], j = 1, \ldots, L$ series plays an important role in the subsequent discussion. Basically, we attempt to capture the multifractal scaling behavior via this series.

3 The proposed MAP structure

To exhibit multifractal behavior we propose to apply a special MAP, a Markov modulated Poisson process (MMPP) whose background CTMC has a symmetric[b] n-dimensional cube structure and the arrival intensities are set according to the variation of the arrival process at the different time scales. We believe that other MAP structures can also exhibit multifractal behavior. Our special choice is motivated by the generation of the Haar wavelet transform. Basically the Haar wavelet transform evaluates the variation of the data set at different aggregation levels (time scales), and similarly, the proposed MAP structure provides different variation of the arrival rate at different time scales.

The composition of the proposed MAP structure follows a very similar pattern as the generation of the Haar wavelet transform. Without loss of generality, we assume that the time unit is such that the long term arrival intensity is one. A MAP of one state with arrival rate 1 represents the arrival process at the largest (considered) time scale.

At the next time scale, $1/\lambda$, an MMPP of two states with generator

$$\begin{bmatrix} -\lambda & \lambda \\ \lambda & -\lambda \end{bmatrix}$$

and with arrival rates $1 - a_1$ and $1 + a_1$ ($-1 \leq a_1 \leq 1$) represents the variation of the arrival process. This composition leaves the long term average arrival rate unchanged.

---

because it simplifies the discussion significantly.

[b]We also investigated the effect of applying asymmetric n-dimensional cubes. According to our experience, the asymmetric models perform similarly to the symmetric ones but have more parameters.

In the rest of the composition we perform the same step. We introduce a new dimension and generate the n-dimensional cube such that the behavior at the already set time scales remains unchanged. E.g., considering also the $1/\gamma\lambda$ ($\gamma > 1$) time scale an MMPP of four states with generator

$$\begin{array}{cccc} \bullet & \lambda & \gamma\lambda & \\ \lambda & \bullet & & \gamma\lambda \\ \gamma\lambda & & \bullet & \lambda \\ & \gamma\lambda & \lambda & \bullet \end{array}$$

and with arrival rates $(1-a_1)(1-a_2)$, $(1+a_1)(1-a_2)$, $(1-a_1)(1+a_2)$ and $(1+a_1)(1+a_2)$ ($-1 \le a_1, a_2 \le 1$) represents the variation of the arrival process. With this MMPP, parameter a_1 (a_2) determines the variance of the arrival process at the $1/\lambda$ ($1/\gamma\lambda$) time scale. If γ is large enough ($>\sim 30$) the process behavior at the $1/\lambda$ time scale is independent of a_2. The proposed model is also applicable with a small γ. In this case, the only difference is that the model parameters and the process behavior of different time scales are dependent.

Finally, Figure 6 introduces the MAP structure with 3 levels. Following the same construction rule one can compose any high level MAP structure of this kind.

$$\begin{array}{cccccccc} \bullet & \lambda & \gamma\lambda & & \gamma^2\lambda & & & \\ \lambda & \bullet & & \gamma\lambda & & \gamma^2\lambda & & \\ \gamma\lambda & & \bullet & \lambda & & & \gamma^2\lambda & \\ & \gamma\lambda & \lambda & \bullet & & & & \gamma^2\lambda \\ \gamma^2\lambda & & & & \bullet & \lambda & \gamma\lambda & \\ & \gamma^2\lambda & & & \lambda & \bullet & & \gamma\lambda \\ & & \gamma^2\lambda & & \gamma\lambda & & \bullet & \lambda \\ & & & \gamma^2\lambda & & \gamma\lambda & \lambda & \bullet \end{array} \qquad \begin{array}{c} (1-a_1)(1-a_2)(1-a_3) \\ (1+a_1)(1-a_2)(1-a_3) \\ (1-a_1)(1+a_2)(1-a_3) \\ (1+a_1)(1+a_2)(1-a_3) \\ (1-a_1)(1-a_2)(1+a_3) \\ (1+a_1)(1-a_2)(1+a_3) \\ (1-a_1)(1+a_2)(1+a_3) \\ (1+a_1)(1+a_2)(1+a_3) \end{array}$$

Figure 6. The generator matrix of the proposed MAP structure with 3 levels and the associated arrival rates

A level n MAP of the proposed structure is composed by 2^n states and has $n+2$ parameters. Parameters γ and λ define the considered time scales, and parameters $a_1, a_2, \ldots, a_n$ determine the variance of the arrival process at the n considered time scales. It can be seen that the ratio of the largest and the smallest considered time scales is γ^n. Having a fixed n (i.e., a fixed cardinality of the MAP), any large ratio of the largest and the smallest considered time scales can be captured by using a sufficiently large γ.

3.1 Analysis of the proposed MAP structure

In the case of an m-phase MMPP with descriptors $\mathbf{D_0}$ and $\mathbf{D_1}$, the distribution of the sum of k consecutive interarrival times can be viewed as PH distribution of order mk whose descriptors are the following. In order to describe the initial probability vector of the PH distribution we need the MAPs stationary probability vector embedded at arrival epochs which can be obtained by

$$\boldsymbol{\tau} = (\boldsymbol{\pi}\mathbf{D_1}\mathbf{e})^{-1}\boldsymbol{\pi}\mathbf{D_1} \tag{12}$$

where $\boldsymbol{\pi}$ is the stationary probability vector of the CTMC with infinitesimal generator $\mathbf{D_0}+\mathbf{D_1}$, and $\mathbf{e}$ is a vector of ones. Using (12) the initial probability vector of length $m \cdot k$ is given by

$$\mathbf{t}_{(k)} = [\boldsymbol{\tau}\ 0 \cdots 0]\,,$$

and the generator matrix $\mathbf{T}$ that describes the state-transitions among the transient states is of size $mk \times mk$ and is given by

$$\mathbf{T}_{(k)} = \begin{bmatrix} \mathbf{D_0} & \mathbf{D_1} & \mathbf{0} & \cdots & & \\ & \mathbf{0} & \mathbf{D_0} & \mathbf{D_1} & \mathbf{0} & \cdots \\ & & \ddots & \ddots & & \\ & & \cdots & \mathbf{0} & \mathbf{D_0} & \mathbf{D_1} \\ & & & \cdots & \mathbf{0} & \mathbf{D_0} \end{bmatrix}.$$

Let us denote by $Z_{(k)}$ a PH distributed random variable with descriptors $\mathbf{t}_{(k)}$ and $\mathbf{T}_{(k)}$. Then, applying (10), the second moment of the wavelet coefficients can be calculated as

$$E[d_j^2] = 2E[Z^2_{(2^j-1)}] - 2\sum_{i=1}^{km} Pr(z_l = i)E[Z_{(2^j-1)}|z_l = i]E[Z_{(2^j-1)}|z_0 = i]$$

where z_l denotes the last transient phase before absorption while z_0 denotes the initial phase. The first term of the right hand side can be obtained by

$$E[Z^2_{(2^j-1)}] = 2\mathbf{t}_{(2^j-1)}\mathbf{T}^{-2}_{(2^j-1)}\mathbf{e},$$

while the second term can be calculated based on

$$Pr(z_l = i)E[Z_{(2^j-1)}|z_l = i] = -\mathbf{t}_{(2^j-1)}\mathbf{T}^{-2}_{(2^j-1)}\mathbf{T}_{(2^j-1)}\mathbf{e_i},$$

where $\mathbf{e_i}$ denotes the vector whose only nonzero entry is 1 at position i, and

$$E[Z_{(2^j-1)}|z_0 = i] = -\mathbf{e_i}\mathbf{T}^{-1}_{(2^j-1)}\mathbf{e}.$$

In order to compute the quantities in the above expressions one has to determine some entries of $\mathbf{T}_{(k)}^{-1}$. Note that this can be done easily by performing computations on matrices of size $m \times m$ since

$$\mathbf{T}_{(k)}^{-1} =$$

$$\begin{bmatrix} \mathbf{D_0}^{-1} & -\mathbf{D_0}^{-1}\mathbf{D_1}\mathbf{D_0} & \mathbf{D_0}^{-1}(\mathbf{D_1}\mathbf{D_0})^2 & \cdots & (-1)^{k-1}\mathbf{D_0}^{-1}(\mathbf{D_1}\mathbf{D_0})^{k-1} \\ \mathbf{0} & \mathbf{D_0}^{-1} & -\mathbf{D_0}^{-1}\mathbf{D_1}\mathbf{D_0} & \cdots & (-1)^{k-2}\mathbf{D_0}^{-1}(\mathbf{D_1}\mathbf{D_0})^{k-2} \\ \mathbf{0} & \mathbf{0} & \mathbf{D_0}^{-1} & \cdots & \\ \mathbf{0} & \cdots & \cdots & \mathbf{0} & \mathbf{D_0}^{-1} \end{bmatrix}.$$

3.2 A parameter fitting method

We apply a simple numerical procedure to fit a MAP of the given structure with a measured data set. Actually, our heuristic approach is composed by "engineering considerations" based on the properties of the measured data set and a parameter fitting method.

First, we fix the value of n. According to our experience a "visible" multiscaling behavior can be obtained from $n = 3 \sim 4$. The computational complexity of the fitting procedure grows exponentially with the dimension of the MAP. The response time with $n = 6$ (MAP of 64 states) is still acceptable (in the order of minutes).

Similar to [13], we set the γ and the λ parameters based on the inspection of the data set. Practically, we define the largest, T_M, and the smallest, T_m, considered time scales and calculate γ and λ from

$$T_M = \frac{1}{\lambda}\ ; \quad T_m = \frac{1}{\gamma^n \lambda},$$

where $\gamma > 1$.

The extreme values of T_M and T_m can be set based on simple practical considerations. For example when the measured data set is composed of N arrival instances, T_M can be chosen to be less than the mean time of $N/4$ arrivals, and T_m can be chosen to be greater than the mean time of 4 arrivals. A similar approach was applied in [13]. These boundary values can be refined based on a detailed statistical test of the data set. For example, if the scaling behavior disappears beyond a given time scale T_M can be set to that value.

Having the γ and the λ parameters we apply a downhill simplex method to find the optimal values of the variability parameters $a_1, a_2, \ldots, a_n$. The goal function that our parameter fitting method minimizes is the sum of the relative errors of the second moment of Haar wavelet coefficients up to a predefined time scale S:

$$\min_{a_1,\dots,a_n} \sum_{j=1}^{S} \frac{|E[d_j^2] - E[\hat{d}_j^2]|}{E[d_j^2]} .$$

4 Numerical analysis

This section presents a collection of numerical analysis results using the proposed Markovian model. The first subsection investigates the multifractal scaling properties of the considered MAP structure. The second subsection presents the comparison of the Bellcore *pAug* trace and its approximating MAP.

4.1 Multifractal scaling properties of the proposed MAP structure

The logarithmic moments of the MAP structure

As mentioned in Section 2 the scaling behavior of the data samples is usually checked using log-moment diagrams. The log-moment diagram plots the logarithm of different moments of the m aggregated process against $log(m)$. Linear curves in the log-moments plot suggest scaling behavior. Figures 7 and 9 show the log-moment plots of the MAPs with the following sets of parameters:

- $n = 5, \lambda = 1/2^{18}, \gamma = 8, a_1 = a_2 = a_3 = a_4 = a_5 = 0.3$
- $n = 5, \lambda = 1/2^{18}, \gamma = 8, a_1 = a_2 = a_3 = a_4 = a_5 = 0.5$

In the range of 4 to 18 the curves are very close to linear, as it can be seen from the increment plots in Figures 8 and 10. Based on Figures 8 and 10, we conclude that the considered MAP exhibits scaling behavior over the time scales from 4 to 18.

Mono-fractal scaling is assumed when the slope of the log-moment curve is linear with q and a non-linear relation suggests multifractal scaling. The slope of the log-moment curves of the two MAPs are collected in Table 1. None of the considered MAPs show a linear relation, but they are not too far from that. The slopes of the MAP with larger variability parameters are further from linearity.

The other test of multifractal scaling considered is the analysis of the partition function and its Legendre transform. The visual inspection of a partition function curve (e.g., in Figure 3) is very hard because the slope of the

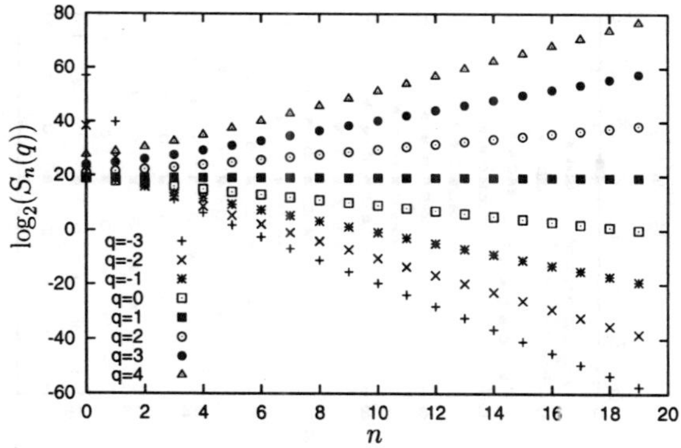

Figure 7. Scaling of log-moments for $a_i = 0.3$

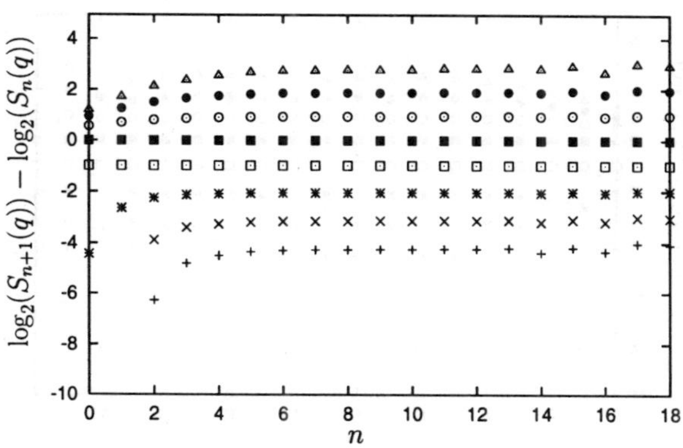

Figure 8. Increments of the log-moment curves for $a_i = 0.3$

curve carries the required information that is very hard to see. The Legendre transform of the partition function somehow amplifies the important information. The differences of the scaling behavior of two multifractal processes are much more perceptible from the Legendre transform. It is analyzed in the following subsection.

The effect of variability parameters on the Legendre spectrum

Legendre transform presents the scaling behavior of a multifractal process in a way which is not closely related to the "physical understanding" of the

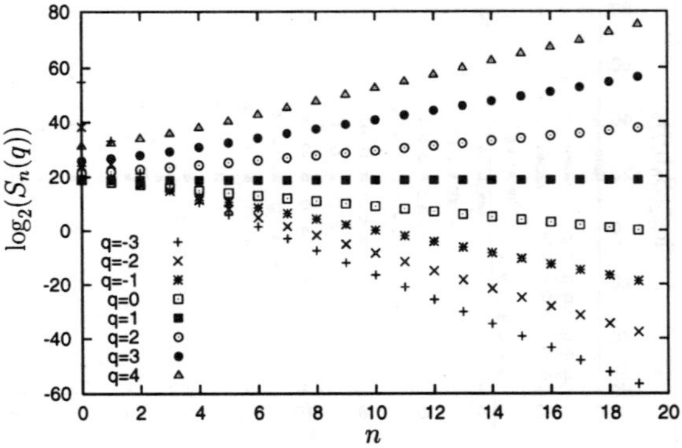

Figure 9. Scaling of log-moments for $a_i = 0.5$

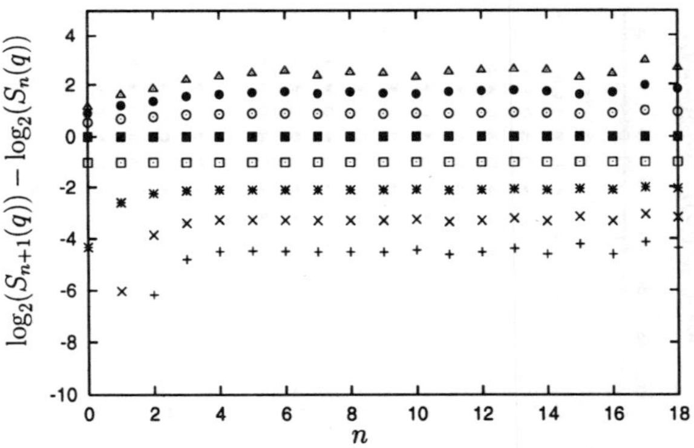

Figure 10. Increments of the log-moment curves for $a_i = 0.5$

process. It is useful in visualizing the scaling behavior, but it is not easy to interpret. We should emphasize again that according to the definition of the partition function (4) and its Legendre transform only infinite data sets can be analyzed. In our experiments we always use finite data sets because the measured traffic samples are finite and the considered MAPs exhibit scaling behavior only through a range of time scales. Using finite data sets we approximate the partition function according to (6). The $n \to \infty$ limiting behavior is approximated via the linear fit of the function over the available time scales. According to our experiment the range of time scales for which the linear line

Table 1. The slope of the log-moments

q	$a_i = 0.3$	$a_i = 0.5$
-5	-6.5319	-6.8912
-4	-5.3821	-5.6784
-3	-4.2457	-4.4719
-2	-3.1298	-3.2766
-1	-2.0446	-2.1076
0	-1.0000	-1.000
1	0	0
2	0.9602	0.8895
3	1.8904	1.7085
4	2.7995	2.4900
5	3.6498	3.2132

is fitted significantly affects the Legendre transform curves (as it is presented in 4). We found that the Legendre transform curves are comparable only when the same range of time scales are considered. Due to this fact in the following set of results the $n = 4, \lambda = 1/2^{15}, \gamma = 8$ parameters, as well as the range of time scales of the linear fit are fixed. We only analyze the effect of the variability parameters, $a_i, i = 1, \ldots, 4$, on the Legendre spectrum.

Figure 11 shows the Legendre transform of the considered MAP structure with uniform variability parameters. It can be seen that low variability, $a_i = 0.1, i = 1, \ldots, 4$, results in a narrow Legendre transform and large variability, $a_i = 0.9, i = 1, \ldots, 4$, results in a wide one. Theoretically, the Legendre transform of a mono-fractal is a single point, since a mono-fractal has the same scaling parameter for each moment. On the other hand, a wide Legendre transform curve represents a "rich multifractal spectrum". Note that our MAP structure results in a Poisson process when $a_i = 0, i = 1, \ldots, 4$, which is not a scaling process (its log-moment curves are not linear).

Figure 12 - 14 display the effect of different variability patterns on the Legendre transform. In Figure 12 the average of the variability parameters is fixed, $\sum_{i=1}^{4} a_i/4 = 0.5$, and the variability parameters form an arithmetic sequence with different increments. In Figure 13 the a_2, a_3, a_4 parameters are fixed to 0.5 and the variability parameter associated with the slowest time scale, a_1, is changed. In Figure 14 $a_1 = a_2 = a_3 = 0.5$ and a_4 varies. From Figures 12 - 14, one can conclude that the variability at the slowest time scale effects the width of the Legendre transform curve. The higher a_1 is the wider the Legendre transform curve is. According to Figure 14 the variability

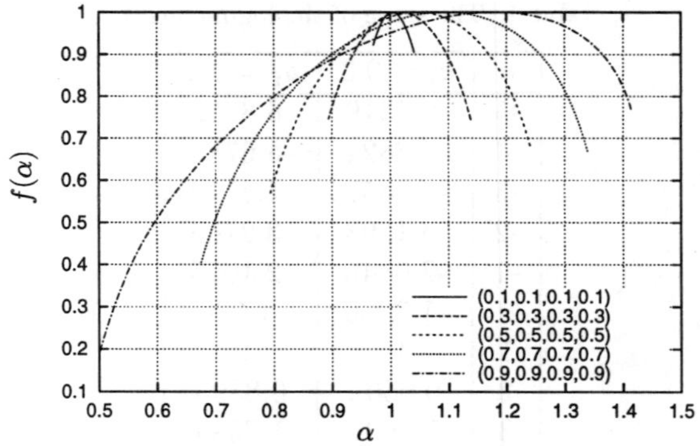

Figure 11. Effect of increasing variability at each time scale

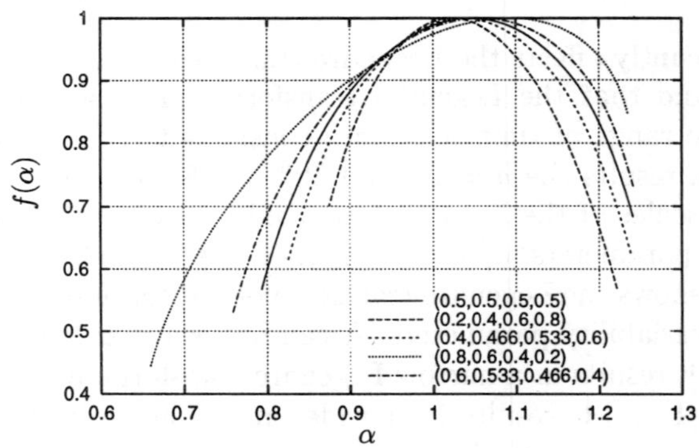

Figure 12. Effect of different variability patterns on the Legendre spectrum

parameter of the fastest time scale does not affect the width of the Legendre transform too much. Instead it turns the curve a bit.

4.2 Approximating the Bellcore pAug *trace*

To test the properties of the proposed MAP structure and the fitting method, we fit a MAP to the famous Bellcore *pAug* trace [20]. This data set is composed of $10^6 \sim 2^{20}$ interarrival times. We applied the fitting method with $n = 5$ and several different predefined setting of γ, λ. We found that the goodness of the fitting is not very sensitive to the predefined parameters around the

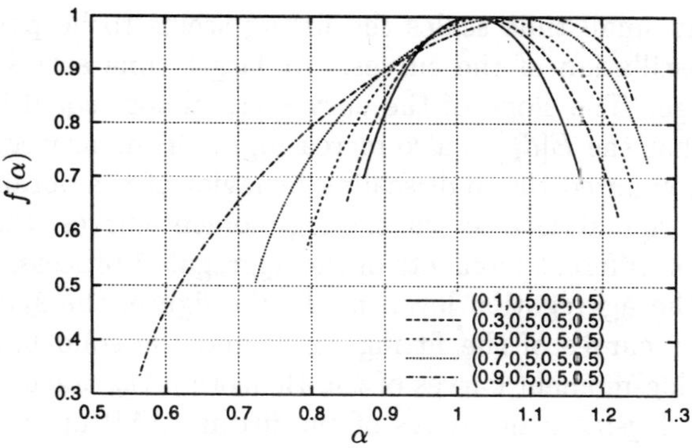

Figure 13. Effect of increasing variability at each time scale

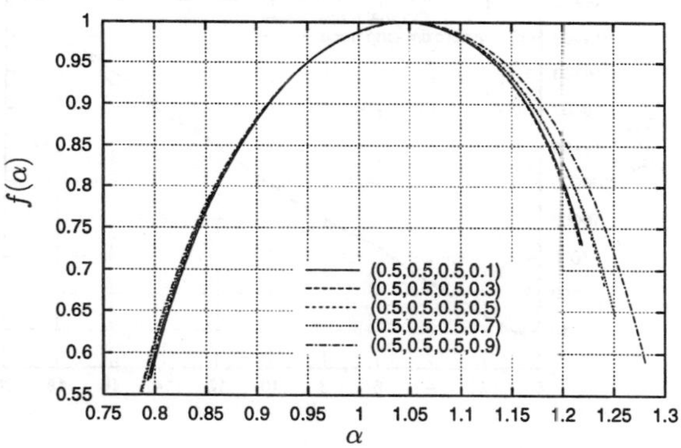

Figure 14. Effect of different variability patterns on the Legendre spectrum

reasonable region. The best "looking" fit is obtained when T_m is the mean time of 16 arrivals and $\gamma = 8$. In this case T_M is the mean time of $16 * 8^5 = 2^{19}$ arrivals which corresponds to the coarsest time scale we can analyze in the case of the Bellcore *pAug* trace. The simplex method minimizing the relative error of the second moment of the Haar wavelet coefficients over $S = 12$ time scales resulted in: $a_1 = 0.144, a_2 = 0.184, a_3 = 0.184, a_4 = 0.306, a_5 = 0.687$. The result of fitting the second moment of the Haar wavelet transform at different aggregation levels is plotted in Figure 15. Since our fitting method intends to minimize a sum of these differences the obtained small differences come from the structural limits of the applied MAP with the given fixed parameters. At small time scales the fitting seems to be perfect. There is only a small oscillation of the curves. At larger time scales the oscillation seems to enlarge. The slope of the curves are almost equal in the depicted range. (Note that the $E[d_k^2]$ is also increasing with the aggregation level.)

First, we compared the multiscaling behavior of the obtained MAP with the one of the original data set via the log-moment curves. Figure 16 depicts the logarithm of different moments of the aggregated process, $\log_2(S_n(q))$, as a function of the aggregation level, n. In the figure, the symbols represent the log-moment curves of the fitting MAP and the solid lines indicate the corresponding log-moment curves of the Bellcore *pAug* trace. In the range of $n \in (3, 19)$ the log-moment curves of the fitting MAP are very close to the ones of the original trace. The log-moment curves of the approximate MAP are also very close to linear in the considered range.

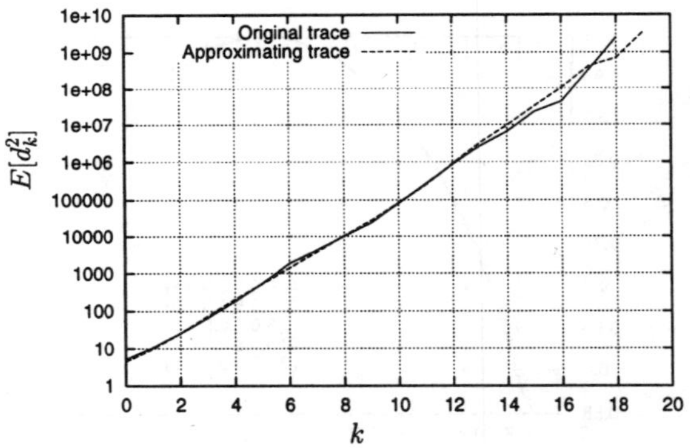

Figure 15. The second moment of the Haar wavelet transform at different aggregation levels

The partition functions of the fitting MAP and the original trace are depicted in Figure 17. As it is mentioned in the previous section, the visual appearance of the partition function is not very informative about the multifractal scaling behavior. Figure 18 depicts the Legendre transform of the partition functions of the original data set and the approximating MAP. The visual appearance of the Legendre transform significantly amplifies the differences of the partition functions. In Figure 18, it can be seen that both processes exhibit multifractal behavior but the original data set has a bit richer multifractal spectrum. The difference of the Legendre transforms comes from the differences of the high negative and high positive moments (< -3 and > 4), which are not provided in Figure 16. The reason why the Legendre

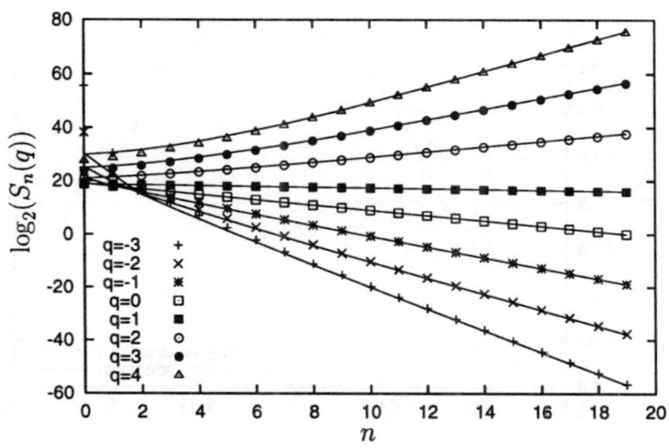

Figure 16. Scaling of log-moments of the original trace and the fitting MAP

transform of the approximating trace seems to be the rotation of the Legendre spectrum of the original trace can be found examining the partition function. One can observe in Figure 17 that the trace generated by the MAP gives higher (lower) values for $T(q)$ for high (low) values of q than the original trace. This difference appears as a rotation in Figure 18.

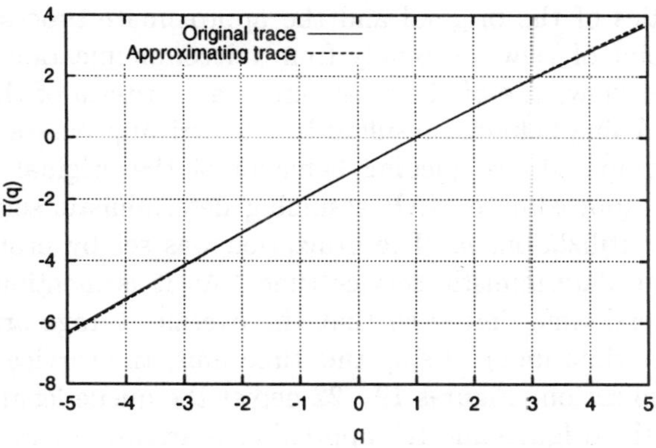

Figure 17. Partition function estimated through the linear fits shown in Figure 16

Actually, Figure 18 shows a rather poor fitting of the multifractal spectrum. Based only on this figure one cannot accept the proposed fitting method. We believe that the differences in the Legendre transforms have to be handled with care. The Legendre transform might overemphasize the

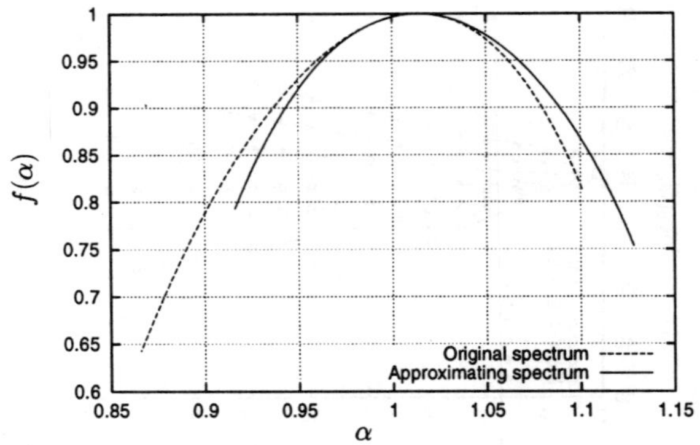

Figure 18. The Legendre transform of the original data set and the one of the approximate MAP

differences of the multifractal spectrum and it is very sensitive to the applied numerical procedure as it was shown in Figure 4. The Legendre spectrum of the approximate multifractal wavelet models proposed in [13] also show significant differences from the one of the original trace (Figure 9 in [13]).

The above tests of the proposed MAP fitting method considered the statistical properties of the original and the approximate processes. From an applications point of view, especially from telecommunication related applications point of view, one of the most important criteria of the goodness of fitting is the queuing behavior resulted by the arrival processes.

We also compared the queuing behavior of the original data set with the one of the approximate MAP assuming deterministic service time and different queue utilization, ρ. The utilization was set by properly choosing the value of the deterministic service time. As it is mentioned above, we apply an artificial time unit such that the overall average arrival rate is 1 (one arrival per time unit). Using this time unit, the service time (< 1) is equal to the utilization. Figures 19 - 22 depict the queue length distribution resulting from the original and the approximate arrival processes. The queue length distribution curves show a quite close fit. The probability of an empty queue, which is not displayed in the figures, is the same for the MAP as for the original trace since the MAP has the same average arrival intensity as the original trace. The fit is better with a higher queue utilization, which might mean that different scaling behaviors play a dominant rule at different utilizations, and the ones that are dominant at high utilization are better approximated by the proposed MAP.

5 Conclusion

The paper presents a MAP structure that is able to exhibit multifractal scaling behavior according to the commonly applied statistical tests. The proposed MAP structure is constructed similarly as the unnormalized Haar wavelet transform of finite sequences.

A heuristic fitting method is also proposed to approximate data sets with multifractal scaling behavior by a MAP with the considered structure. Our numerical experiences show a good fitting according to the majority of the

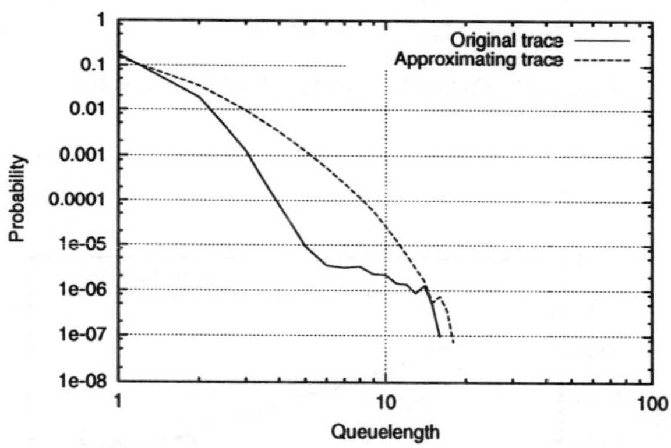

Figure 19. Queue-length distribution at $\rho = 0.2$

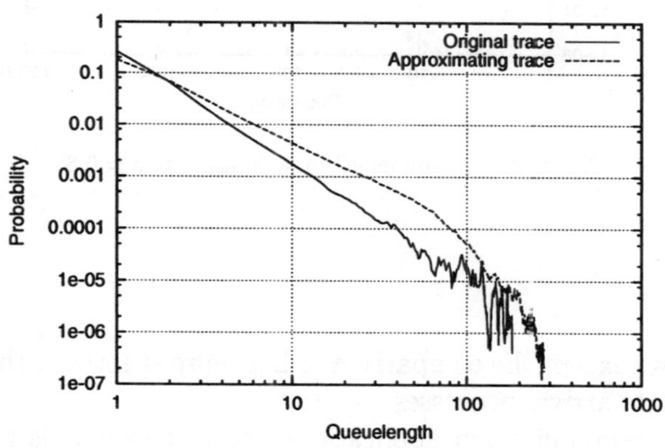

Figure 20. Queue-length distribution at $\rho = 0.4$

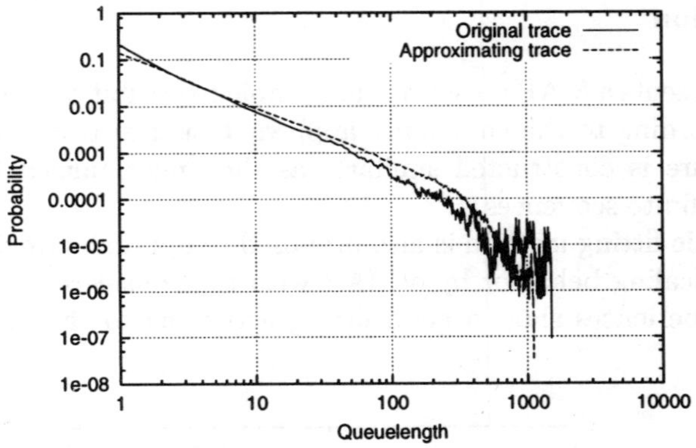

Figure 21. Queue-length distribution at $\rho = 0.6$

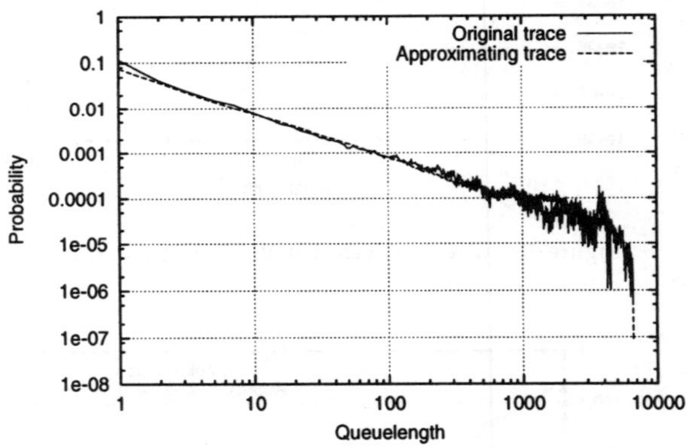

Figure 22. Queue-length distribution at $\rho = 0.8$

performed tests except the comparison of Legendre spectra of the original and the approximate arrival processes.

From telecommunication applications point of view it is promising that the queue length distribution of the original arrival process fits in with the one of the approximate arrival process.

Acknowledgments

This work was partially supported by OTKA grant n. T-30685. András Horváth acknowledges the support of the Italian Ministry for University and Scientific Research, through the Planet-IP project

References

1. W. Willinger, M. S. Taqqu, and A. Erramilli. A bibliographical guide to self-similar traffic and performance modeling for high speed networks. In *Stochastic Networks: Theory and Applications*, pages 339–366. Oxford University Press, 1996.
2. I. Norros. On the use of fractional brownian motion in the theorem of connectionless networks. *IEEE Journal on Selected Areas in Communications*, 13:953–962, 1995.
3. B. B. Mandelbrot and J. W. Van Ness. Fractional Brownian motions, fractional noises and applications. *SIAM Review*, 10:422–437, 1969.
4. G. E. P. Box, G. M Jenkins, and C. Reinsel. *Time Series Analysis: Forecasting and Control*. Prentice Hall, Englewood Cliff, N.J., third edition, 1994.
5. C. W. J. Granger and R. Joyeux. An introduction to long-memory time series and fractional differencing. *Journal of Time Series Analysis*, 1:15–30, 1980.
6. B. Ryu and S. B. Lowen. Point process models for self-similar network traffic, with applications. *Stochastic models*, 14, 1998.
7. A. T. Andersen and B. F. Nielsen. A markovian approach for modeling packet traffic with long-range dependence. *IEEE Journal on Selected Areas in Communications*, 16(5):719–732, 1998.
8. S. Robert and J.-Y. Le Boudec. New models for pseudo self-similar traffic. *Performance Evaluation*, 30:1997, 57-68.
9. A. Horváth, G. I. Rózsa, and M. Telek. A map fitting method to approximate real traffic behaviour. In *8th IFIP Workshop on Performance Modelling and Evaluation of ATM & IP Networks*, pages 32/1–12, Ilkley, England, July 2000.
10. M.F. Neuts. *Structured stochastic matrices of M/G/1 type and their applications*. Marcel Dekker, 1989.
11. G. Latouche and V. Ramaswami. *Introduction to matrix analytic methods in stochastic modeling*. SIAM, 1999.
12. A. Feldman, A. C. Gilbert, and W. Willinger. Data networks as cascades: Investigating the multifractal nature of internet wan traffic. *Computer*

communication review, 28(4):42–55, 1998.
13. R. H. Riedi, M. S. Crouse, V. J. Ribeiro, and R. G. Baraniuk. A multifractal wavelet model with application to network traffic. *IEEE Transactions on Information Theory*, 45:992–1018, April 1999.
14. R. H. Riedi and J. Lévy Véhel. Multifractal properties of TCP traffic: a numerical study. Technical Report 3129, INRIA, February 1997.
15. M. Taqqu, Vadim Teverovsky, and Walter Willinger. Is network traffic self-similar or multifractal? *Fractals*, 5:63–73, 1997.
16. W. E. Leland, M. Taqqu, W. Willinger, and D. V. Wilson. On the self-similar nature of ethernet traffic (extended version). *IEEE/ACM Transactions in Networking*, 2:1–15, 1994.
17. I. Norros. A storage model with self-similar imput. *Queueing Systems*, 16:387–396, 1994.
18. G. Samorodnitsky and M. Taqqu. *Stable Non-Gaussian Random Processes: Stochastic Models with Infinite Variance.* Chapman and Hall, New York, 1994.
19. R. H. Riedi. An introduction to multifractals. Technical report, Rice University, 1997. Available at `http://www.ece.rice.edu/~riedi`.
20. The internet traffic archive. http://ita.ee.lbl.gov/index.html.
21. J. Lévy Véhel and R. H. Riedi. Fractional brownian motion and data traffic modeling: The other end of the spectrum. In C. Tricot J. Lévy Véhel, E. Lutton, editor, *Fractals in Engineering*, pages 185–202. Springer, 1997.

CONVERGENCE OF THE RATIO "VARIANCE OVER MEAN" IN THE IPHP^3

G. LATOUCHE

Université Libre de Bruxelles, Département d'Informatique–CP 212, Blvd du Triomphe, B-1050 Bruxelles, Belgium
E-mail: Guy.Latouche@ulb.ac.be

M.-A. REMICHE

LFG Stochastik, Aachen University of Technology, Wuellnerstr. 3, D-52056 Aachen, Germany
E-mail: mremiche@ulb.ac.be

We consider the counting measure associated with the Isotropic Phase Planar Point Process. We establish the convergence of the ratio of the variance of the count over its mean when the size of the considered set increases to ∞. The proof is based on a spectral decomposition of the involved matrices and is a good example that both matrix analytic and spectral analysis methods are useful tools to analyze MAP-like processes.

Keywords: Planar point processes, Markovian Arrival Process, IPhP^3, spectral decomposition.

1 Description of the context

Isotropic Phase-type Planar Point processes (IPhP^3) with representation (D_0, D_1) are random sets of points in the Euclidean space. Their polar coordinates $(\rho_i, \theta_i), i \in \mathbb{N}_0$, are such that

- $\{\pi\rho_i^2; i \in \mathbb{N}_0\}$ is the counting process of a stationary MAP with representation (D_0, D_1),

- $\{\theta_i; i \in \mathbb{N}_0\}$ is a sequence of independent and uniform random variables (r.v.'s) over $[0, 2\pi)$,

the two processes above are independent of each other. A simulation of a particular IPhP^3 is shown in Figure 1, its representation is given in appendix. We refer to Latouche and Ramaswami, [1,2] Narayana and Neuts, [3] Remiche [4] and Remiche and Latouche [5] for details on MAPs and IPhP^{3}s.

We recall that the characteristics D_0 and D_1 of a MAP are two matrices of order m, with $(D_0)_{ij}$, $i \neq j$ and $(D_1)_{ij}$ nonnegative, $(D_0)_{ii}$ strictly negative, and such that $D\underline{1} = \underline{0}$, where $D = D_0 + D_1$. We assume that D is irreducible and we denote by $\underline{\delta}$ its stationary distribution, that is, the row-vector which

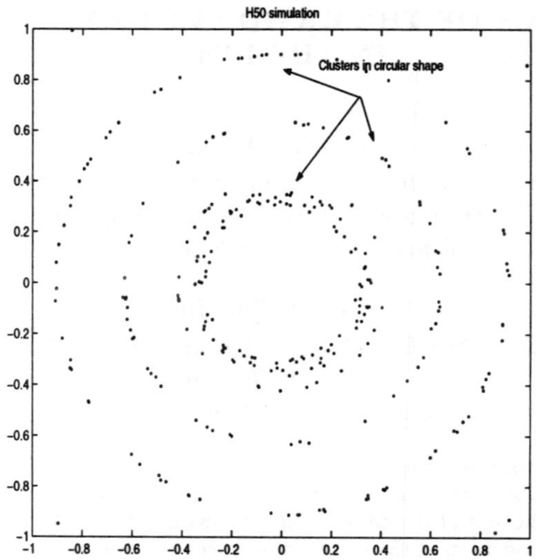

Figure 1. Simulation of Process H50

satisfies

$$\underline{\delta} D = \underline{0}, \qquad \underline{\delta} \underline{1} = 1.$$

These planar point processes are isotropic and mean-stationary only: the origin of the process, i.e., the reference point for locating the points in the plane, plays a central role in their analysis.

The purpose of the present paper is to provide a clear proof of Theorem 1.1 below, which states a property asserted in Remiche [6], and to open the way to a generalization to marked MAP processes.

Theorem 1.1 *Let N be an IPhP*[3] *with representation* (D_0, D_1) *and* $\underline{\delta}$ *be the corresponding stationary probability vector.*

We consider the family $\{\mathcal{C}_r; r \in \mathbb{R}^+\}$ *of circles* $\mathcal{C}_r$ *with radius r and whose center is fixed and located at a distance c from the origin of the process. Let* $N(\mathcal{C}_r) = N \cap \mathcal{C}_r$*, that is the number of points belonging to N located in the circle* $\mathcal{C}_r$*. We have the following asymptotic result:*

$$\lim_{r\to\infty} \frac{\mathrm{Var}[N(\mathcal{C}_r)]}{\mathrm{E}[N(\mathcal{C}_r)]} = \lambda^{(2)}, \tag{1}$$

independently of c, where

$$\lambda^{(2)} = 1 - 2\frac{\lambda^2 + \underline{\delta} D_1 (D - \Delta)^{-1} D_1 \underline{1}}{\lambda}, \tag{2}$$

with $D = D_0 + D_1$, $\Delta = \underline{1}.\underline{\delta}$ *and* $\lambda = \underline{\delta} D_1 \underline{1}$.
The fact that $D - \Delta$ is nonsingular is proved in Lemma 3.1.3 of Latouche and Ramaswami; [2] the quantity λ is the intensity of the planar process, that is, it is the expected number of points per unit area.

This property was first observed through numerical experimentation. In Figure 2 the ratio variance over mean is depicted versus the radius r of the circle for two different processes, named H50 and H100, defined in the appendix. Clearly, there seems to exist an horizontal asymptote to those curves, as stated in Theorem 1.1.

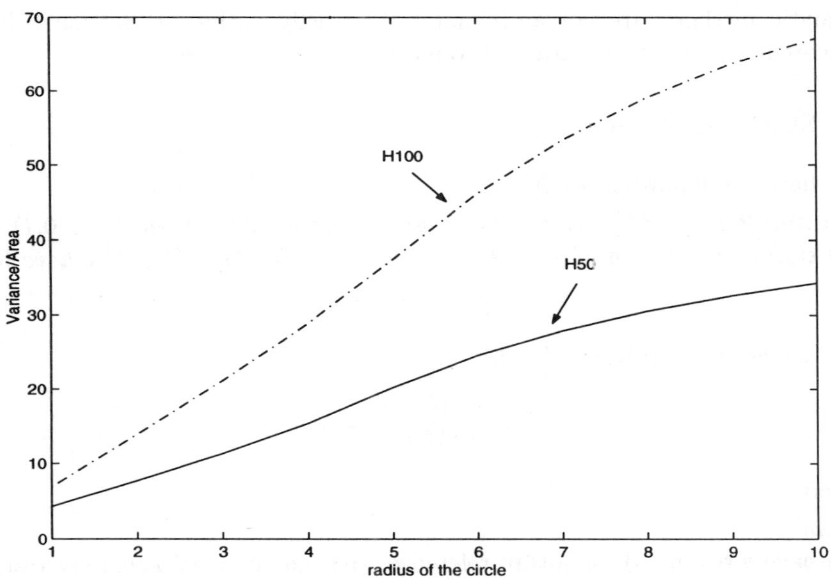

Figure 2. The ratio variance over mean in function of the radius of the circle, the center is located at a distance 4 from the origin

Remiche and Latouche [5] and Remiche [4] have established that far away enough from the origin, any IPhP[3] exhibits Poissonian features, like e.g
Theorem 1.2 *Let N be an IPhP[3] and let* $\{C_r(c); c \in \mathbb{R}^+\}$ *be a family of circles whose radius is fixed and equal to r and whose center is located at a*

distance c from the origin. The following holds

$$\lim_{c\to\infty} \frac{Var[N(\mathcal{C}_r(c))]}{E[N(\mathcal{C}_r(c))]} = 1,$$

independently of r. ■

Theorems 1.1 and 1.2 are complementary results that explain the global and the local behavior of an IPhP[3].

As explained in the next section, Theorem 1.1 is easily proved when the circle is centered at the origin and it is easy to formulate a heuristic argument as to why this property should hold in general. The formal proof is not straightforward, however. We describe in Sections 2 and 3 the difficulties which one encounters when using the probabilistic intuition only; this explains why we provide in Section 4 an analysis using the spectral decomposition of D, showing in the process that spectral decomposition techniques and matrix analytic methods are not antinomic. We briefly outline in Section 5 how Theorem 1.1 may be generalized to marked MAP processes.

2 Stochastic Order

We have the following result.

Lemma 2.1 *Let $M(t)$ be the r.v. counting the number of points in $[0, t)$ for the stationary version of a MAP with representation (D_0, D_1). We have that*

$$\mathrm{E}[M(t)] = \lambda t \tag{3}$$

and we have the following limit:

$$\lim_{t\to\infty} \frac{\mathrm{Var}[M(t)]}{\mathrm{E}[M(t)]} = \lambda^{(2)}, \tag{4}$$

where $\lambda^{(2)}$ is defined in (2).

Proof

We have written (3) for future reference; that equation immediately results from the fact that the MAP is stationary.

We rewrite Equation (10) of Narayana and Neuts [3] with our own notations and obtain that

$$\begin{aligned}\mathrm{Var}[M(t)] = {} & [\lambda - 2\lambda^2 - 2\underline{\delta} D_1 (D - \Delta)^{-1} D_1 \underline{1}]t \\ & -2\underline{\delta} D_1 (D - \Delta)^{-1} (I - \exp\{Dt\})(D - \Delta)^{-1} D_1 \underline{1}.\end{aligned} \tag{5}$$

Since $\exp\{Dt\}$ is a matrix of transition probabilities, the second term in (5) is bounded. Together with (3), this proves the lemma. ■

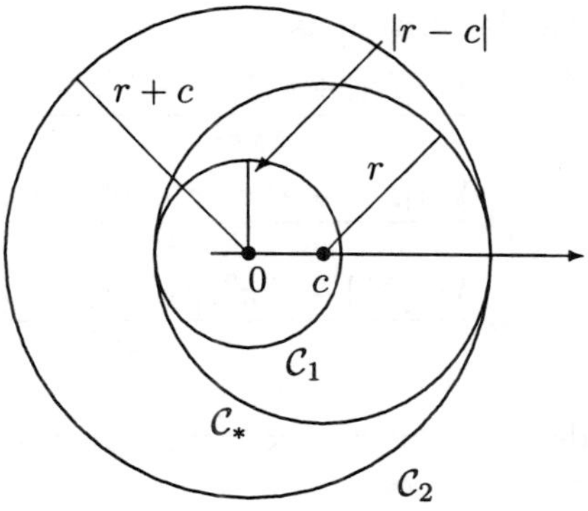

Figure 3. The circle $\mathcal{C}_r$ where the center c is no longer located at the origin of the process

Due to the construction itself of an IPhP[3], one has that

$$N(\mathcal{C}_r(0)) = M(\pi r^2), \tag{6}$$

and Theorem 1.1 immediately results from Lemma 2.1 for those circles which are centered at the origin.

We now consider a circle not centered at the origin and we define the following circles and corresponding counting variables (see Figure 3).

Circle	Count	Center	Radius
$\mathcal{C}_1$	N_1	$(0,0)$	$\lvert r-c\rvert$
$\mathcal{C}_*$	N_*	$(c,0)$	r
$\mathcal{C}_2$	N_2	$(0,0)$	$r+c$

Because of the structure of the sets $\mathcal{C}_1$, $\mathcal{C}_*$ and $\mathcal{C}_2$, we have that

$$N_1 \le N_* \le N_2. \tag{7}$$

Furthermore, $\mathrm{E}[N_*]$ is equal to $\lambda\pi r^2$ and both the means $\mathrm{E}[N_1]$ and $\mathrm{E}[N_2]$ are equal to $\lambda\pi r^2 + O(r)$. Finally, we already know from Lemma 2.1 that

$$\lim_{r\to\infty} \frac{\mathrm{Var}[N_1]}{\mathrm{E}[N_1]} = \lambda^{(2)}, \qquad \text{and} \qquad \lim_{r\to\infty} \frac{\mathrm{Var}[N_2]}{\mathrm{E}[N_2]} = \lambda^{(2)}. \tag{8}$$

so that $\mathrm{Var}[N_1]$ and $\mathrm{Var}[N_2]$ both grow like $\lambda^{(2)}\lambda\pi r^2$. Thus, one expects that the region $\mathcal{C}_2 \setminus \mathcal{C}_1$ becomes negligible as $r \to \infty$, and that $\mathrm{Var}[N_*]$ also grows like $\lambda^{(2)}\lambda\pi r^2$, which would prove Theorem 1.1.

From (3, 7), it results that

$$\frac{\mathrm{E}[N_1^2] - (\lambda\pi r^2)^2}{\mathrm{E}[N_*]} \le \frac{\mathrm{Var}[N_*]}{\mathrm{E}[N_*]} \le \frac{\mathrm{E}[N_2^2] - (\lambda\pi r^2)^2}{\mathrm{E}[N_*]}. \tag{9}$$

The left-most term may be written as

$$\frac{\mathrm{E}[N_1^2] - (\lambda\pi r^2)^2}{\mathrm{E}[N_*]} = \frac{\mathrm{Var}[N_1] + \mathrm{E}[N_1]^2 - \lambda^2\pi^2 r^4}{\lambda\pi r^2}$$
$$= \frac{\mathrm{Var}[N_1] + \lambda^2\pi^2(r-c)^4 - \lambda^2\pi^2 r^4}{\lambda\pi r^2}$$
$$= \frac{\mathrm{Var}[N_1] - \lambda^2\pi^2 c[(r-c)+r][(r-c)^2+r^2]}{\lambda\pi r^2},$$

which obviously tends to $-\infty$ as $r \to \infty$. Similarly, one shows that the right-most term tends to ∞ as $r \to \infty$.

Thus, the inequalities (9) are not tight enough and we need to use the exact expression of $\mathrm{Var}[N_*]$ that we determine in the next section.

3 Preliminary Results

The variance function is explicitly given in the next lemma.

Lemma 3.1 *Let N be an IPhP[3] with representation (D_0, D_1). Let $\mathcal{C}_r$ be the circle of radius r and center c, with $r > c$. Then*

$$\mathrm{Var}[N(\mathcal{C}_r)] = \lambda\pi r^2 + 2\int_0^T \theta(x)dx \int_0^x \underline{\delta} D_1[\exp\{D(y)\} - \Delta]D_1\underline{1}\theta(y)dy \quad (10)$$

where $T = \pi(r+c)^2$ and

$$\theta(x) = \begin{cases} 1 & \textit{for } 0 \le x < \pi(r-c)^2 \\ \frac{1}{\pi}\arccos\frac{x/\pi + c^2 - r^2}{2c\sqrt{x/\pi}} & \textit{for } \pi(r-c)^2 \le x < \pi(r+c)^2 \end{cases} \quad (11)$$

Remark

The assumption that $r > c$ is not restrictive since we are interested in the limit case where r tends to ∞.

Proof

It results from Corollary 3.3.2 in Remiche [6] that

$$\mathrm{E}[N(\mathcal{C}_r)] = \int_0^T \theta(x)\underline{\delta} D_1^\star \underline{1} dx$$

and that

$$\mathrm{E}[N(\mathcal{C}_r)(N(\mathcal{C}_r)-1)] = 2\int_0^T \theta(x)dx\underline{\delta} D_1^\star \int_0^x \theta(y)\exp\{D^\star(x-y)\}D_1^\star\underline{1}dy,$$

where $D^\star = D_0^\star + D_1^\star$, with

$$D_0^\star = \mathrm{diag}(\underline{\delta})^{-1} D_0^T \mathrm{diag}(\underline{\delta}), \qquad D_1^\star = \mathrm{diag}(\underline{\delta})^{-1} D_1^T \mathrm{diag}(\underline{\delta})$$

and diag($\underline{\delta}$) is a diagonal matrix with $\text{diag}(\underline{\delta})_{ii} = \underline{\delta}_i$. The function $\theta(x)$ is defined as in (11).

By transposing the scalar expressions $\underline{\delta}D_1^\star\underline{1}$ and $\underline{\delta}D_1^\star \exp\{D^\star(x-y)\}D_1^\star\underline{1}$, one sees that $\mathrm{E}[N(\mathcal{C}_r)]$ and $\mathrm{E}[N(\mathcal{C}_r)(N(\mathcal{C}_r)-1)]$ are given by the same equations as above, with $D_1^\star$ and $D^\star$ respectively replaced by D_1 and D. Furthermore,

$$\begin{aligned}
2\int_0^T \theta(x)dx \int_0^x \theta(y)dy &= \int_0^T \theta(x)dx(\int_0^x \theta(y)dy + \int_x^T \theta(y)dy) \\
&= \int_0^T \theta(x)dx \int_0^T \theta(y)dy \\
&= \left[\int_0^T \theta(x)dx\right]^2,
\end{aligned}$$

and

$$\lambda^2 = \underline{\delta}D_1\underline{1}\underline{\delta}D_1\underline{1} = \underline{\delta}D_1\Delta D_1\underline{1},$$

so that $\mathrm{E}[N(\mathcal{C}_r)]^2$ may be written as

$$\mathrm{E}[N(\mathcal{C}_r)]^2 = 2\int_0^T \theta(x)dx\underline{\delta}D_1 \int_0^x \Delta D_1\underline{1}\theta(y)dy$$

and it is a simple matter to prove (10) since $\mathrm{Var}[N(\mathcal{C}_r)] = \mathrm{E}[N(\mathcal{C}_r)(N(\mathcal{C}_r) - 1)] - \mathrm{E}[N(\mathcal{C}_r)]^2 + \mathrm{E}[N(\mathcal{C}_r)]$. ■

When the circle is centered at the origin, we have that $c = 0$, $T = \pi r^2$, $\theta(x) = 2\pi$ for all $0 \le x < T$ and the integral in (10) reduces to the right-hand side of (5) where $t = T$.

This observation, in conjunction with the fact that, by (11), $\theta(x)$ becomes equal to 1 over most of its range, as $r \to \infty$, suggests that we might obtain from (10) tighter bounds for $\mathrm{Var}[N(\mathcal{C}_r)]$ than we did in Section 2. Unfortunately, the matrix $\exp\{D(x-y)\} - \Delta$ is of mixed sign and this makes it difficult to obtain sufficiently tight bounds without analyzing it.

In the next section, we use the spectral decomposition of $\exp\{D(x-y)\}$ and directly compute the limit of the right-hand side of (10) as $r \to \infty$.

4 Proof of Theorem 1.1

Define

$$\varphi(r, c) = \mathrm{Var}[N(\mathcal{C}_r)]/\mathrm{E}[N(\mathcal{C}_r)] - 1$$

and

$$g(u) = 2\underline{\delta} D_1[\exp\{D(u)\} - \Delta] D_1 \underline{1}.$$

By (10), we have that

$$\varphi(r,c) = \frac{1}{\lambda \pi r^2} \int_0^T \theta(x) dx \int_0^x g(x-y)\theta(y) dy$$

and we need to prove that $\lim_{r\to\infty} \varphi(r,c) = \lambda^{(2)} - 1$, knowing, as we remarked after the proof of Lemma 2.1 that $\lim_{r\to\infty} \varphi(r,0) = \lambda^{(2)} - 1$.

We may write $\varphi(r,c)$ as

$$\begin{aligned}
\varphi(r,c) &= \frac{1}{\lambda \pi r^2} \int_0^{\pi(r-c)^2} \theta(x) dx \int_0^x g(x-y)\theta(y) dy \\
&\quad + \frac{1}{\lambda \pi r^2} \int_{\pi(r-c)^2}^T \theta(x) dx \int_0^x g(x-y)\theta(y) dy \\
&= \frac{(r-c)^2}{r^2} \varphi(r-c,0) + \frac{1}{\lambda \pi r^2} \int_{\pi(r-c)^2}^T \theta(x) dx \int_0^x g(x-y)\theta(y) dy
\end{aligned}$$

since $\theta \equiv 1$ on $[0, \pi(r-c)^2]$. Thus, the theorem will be proved once we establish that

$$\lim_{r\to\infty} \frac{1}{r^2} \int_{\pi(r-c)^2}^T \theta(x) dx \int_0^x g(x-y)\theta(y) dy = 0. \tag{12}$$

The matrix D being irreducible, conservative and stable, one of its eigenvalues is equal to 0, with algebraic multiplicity 1 and the other eigenvalues have a strictly negative real part. Let β_1, β_2, ..., β_m be the eigenvalues of D, with $\beta_1 = 0 > Re(\beta_2) \geq \cdots \geq Re(\beta_m)$.

The eigenvalues of $\exp\{Dt\}$ are $e^{\beta_1 t} = 1$, $e^{\beta_2 t}$, ..., $e^{\beta_m t}$ and, except for the eigenvalue 1 with multiplicity 1, all decrease exponentially to zero as $t \to \infty$.

The matrix $\exp\{Dt\} - \Delta$ differs little from $\exp\{Dt\}$: one removes from $\exp\{Dt\}$ the eigenvalue 1 and replaces it by 0, without modifying the eigen-characteristics of the matrix in any other way. Thus, $e^{\beta_2 t}$ becomes the dominant eigenvalue as $t \to \infty$. Formally, one has the following property: there exists an integer k with $0 \leq k \leq m-1$, and a nonnegative matrix A such that

$$|\exp\{Dt\} - \Delta| \leq e^{-\mu t} \frac{t^k}{k!} A \qquad \text{for all } t \geq 0,$$

where $-\mu < 0$ is the real part of β_2. The proof is purely technical and is omitted.

We may then write that

$$|g(t)| \leq a e^{-\mu t} \frac{t^k}{k!}, \tag{13}$$

where $a = 2\underline{\delta} D_1 A D_1 \underline{1}$ and that

$$\begin{aligned}
&\left| \frac{1}{r^2} \int_{\pi(r-c)^2}^{T} \theta(x) dx \int_0^x g(x-y)\theta(y) dy \right| \\
&\leq \frac{1}{r^2} \int_{\pi(r-c)^2}^{T} dx \int_0^x |g(y)| dy, \qquad \text{since } 0 \leq \theta(x) \leq 1, \\
&\leq \frac{a}{r^2} \int_{\pi(r-c)^2}^{T} dx \int_0^x e^{-\mu y} \frac{y^k}{k!} dy, \qquad \text{by (13)}, \\
&= \frac{a}{r^2} \int_{\pi(r-c)^2}^{T} [\frac{1}{\mu^{k+1}} - e^{-\mu x} \sum_{0 \leq i \leq k} \frac{x^{k-i}}{(k-i)! \mu^{i+1}}] dx, \\
&\leq \frac{a}{r^2} \int_{\pi(r-c)^2}^{T} \frac{1}{\mu^{k+1}} dx \\
&= \frac{4a\pi c}{\mu^{k+1} r}
\end{aligned}$$

The right-hand side clearly converges to 0 as $r \to \infty$, which proves (12) and completes the proof of Theorem 1.1. ■

5 Conclusion

One interesting aspect of our result is that it may be extended to marked MAP processes, which are as follows. Let $\{T_n : n \in \mathbb{N}\}$ be the set of epochs of arrivals of a MAP process and assume that at time T_n one associates a quantity Y_n called a mark. Define $Y(t) = \sum_{T_n : T_n \leq t} Y_n$, for $t \geq 0$, to be the total marks collected by time t. One shows under suitable conditions that $\mathrm{Var}[Y(t)]/\mathrm{E}[Y(t)]$ converges to a limit as t tends to infinity. This is readily seen if the marks are iid random variables. In other cases, for instance, if the mark at time T_n is a function of T_n, then the proof is along the lines developed in Section 4. This, however, is beyond the scope of the present paper.

Acknowledgment

Part of this research has been supported by a Marie Curie Fellowship of the European Community Program Human Potential under contract number HPMF-CT-1999-00024

Appendix

The map processes H50 and H100 are constructed on the basis of stationary renewal processes where the inter-point distance is hyper-exponential with a coefficient of variation equal to 50 and 100 respectively. Their representation is as follows.

Process	D_0	D_1
H50	$\begin{bmatrix} -17.4387 & 0 \\ 0 & -0.0371 \end{bmatrix}$	$\begin{bmatrix} 16.8283 & 0.6104 \\ 0.0358 & 0.0013 \end{bmatrix}$
H100	$\begin{bmatrix} -7.5884 & 0 \\ 0 & -0.0172 \end{bmatrix}$	$\begin{bmatrix} 7.4746 & 0.1138 \\ 0.0169 & 0.0003 \end{bmatrix}$

References

1. G. Latouche and V. Ramaswami. Spatial point processes of phase type. In V. Ramaswami and P.E. Wirth, editors, *Teletraffic Contributions for the Information Age. Proceedings of the 15th International Teletraffic Congress - ITC 15*, pages 381 – 390. Elsevier, North–Holland, Amsterdam, 1997.
2. G. Latouche and V. Ramaswami. *Introduction to Matrix Geometric Methods in Stochastic Modeling.* ASA-SIAM Series on Statistics and Applied Probability. SIAM, Philadelphia, PA, 1999.
3. S. Narayana and M. F. Neuts. The first two moment matrices of the counts for the Markovian arrival process. *Comm. Statist. Stochastic Models*, 8:459–477, 1992.
4. M.-A. Remiche. Asymptotic Independence of counts in Isotropic Planar Point Processes of phase-type. *Advances in Applied Probability*, 32, 2000.
5. M.-A. Remiche and G. Latouche. Asymptotic Poisson Distribution in Isotropic Ph Planar Point Processes. *Commun. Statist. – Stochastic Models*, 16:259–272, 2000.
6. M.-A. Remiche. On the tractability of the measure associated to the phase-type planar point process. *Methodology and Computing in Applied Probability*, to appear.

APPLICATION OF THE FACTORIZATION PROPERTY TO THE ANALYSIS OF PRODUCTION SYSTEMS WITH A NON-RENEWAL INPUT, BILEVEL THRESHOLD CONTROL, SETUP TIME AND MAINTENANCE

HO WOO LEE

Department of Systems Management Engineering, Sung Kyun Kwan University, Su Won, KOREA 440-746
E-mail: hwlee@yurim.skku.ac.kr

NO IK PARK

Switching & Transmission Tech. Labs, ETRI, Taejon, KOREA
E-mail: nipark@etri.re.kr

JONGWOO JEON

Department of Statistics, Seoul National University, Seoul, KOREA
E-mail: jwjeon@plaza.snu.ac.kr

We consider a production system with a non-renewal batch input, maintenance period, setup period and bilevel threshold control. We model the system by the BMAP/G/1 queue with double thresholds and single vacation. We show how one can apply the factorization property to directly derive the vector generating functions of the queue length (the level of the work-in-process inventory) without going through all the standard procedure. We also present an example that shows the differences in mean queue length between BMAP/G/1 and $M^X/G/1$ queues under the same parameter settings.

1 Introduction

Careful controls of system setup and work-in-process (WIP) inventory are among the most critical factors in the cost-effective operation of a production system. Industrial engineers have long been interested in analyzing the trade-offs between these two factors and providing the conditions under which the system operates most economically in the long run. In their analyses of the various production systems, they usually incorporate the maintenance cost, setup cost and WIP holding cost in their cost model. Queueing models have played important roles in their analytical efforts along this line.

In many production systems, a setup operation takes several days and is very costly. One way to reduce the setup cost per unit time is to delay the production until some raw materials accumulate and this is the well-known N-policy in queueing context. By applying the N-policy, the cycle length becomes larger (which means fewer cycles per unit time) and at the same time, the average WIP level becomes larger. Thus, in real production settings, the N-policy is used to reduce the overall average cost per unit time when the setup cost is excessively high compared to the WIP holding cost.

N-policy queue was first studied by Yadin and Naor [19]. For other works on N-policy queues, see Hersh and Brosh [2], Hofri [3], Kella [4], Lee and Srinivasan [6], Takagi [17], Lee *et al.* [7][8] and Lee and Park [9], to list a few.

In most studies concerning production systems, it has been assumed that the feed process into the production system follows the Poisson process, mainly due to the analytical tractability. But in real production settings, iid exponential interarrival times are hardly found. In this paper, we consider a very general single-machine production system with a BMAP input, bilevel threshold control, setup time and maintenance period.

1.1 The system and objectives

We consider a production system with the following specifications (Figure 1).

1. The raw materials arrive according to the BMAP with m phases of underlying Markov chain (UMC) and parameter matrices $\{\boldsymbol{D}_k,\ k \geq 0\}$.
2. If there are no raw materials to process (time point 1 in Figure 1), the machine undergoes a maintenance period which takes a random length V with distribution function (DF) $V(x)$. After the maintenance,
   a) if the number of units of waiting raw materials is less than α, the operator waits (build-up period) until the number of units reaches or exceeds α before it starts a setup. Or,
   b) if α or more units are waiting, the operator starts a setup (setup period).

The setup time takes a random length H with DF $H(x)$. After the setup,

   a) if the number of units is less than N, the operator waits until the umber of units reaches or exceeds N (stand-by period). Or,
   b) if the number of units is greater than or equal to N, the operator begins to process the units until the system empties (busy period).

3. The processing times are iid with distribution function $S(x)$.
4. The processing times, the maintenance period, the setup period and the arrival process are independent of each other.

If $\alpha = N$, our system becomes the usual N-policy system with a setup. From now on, we will use the terms 'server', 'customer', 'service time' and 'queue length' in exchange with 'operator', 'raw material', 'processing time', and 'WIP level'.

The objective of this study is two-fold.

1. We analyze a production system with BMAP input, the analysis of which can hardly be found in existing literature.
2. In accomplishing objective-(1), we apply the factorization property of the BMAP/G/1 queue with generalized vacations to directly derive the vector generating functions (GF) of the WIP level by saving the effort of going

through the standard procedure. We also derive the average WIP level which is the very first step to deriving the mean operating cost per unit time.

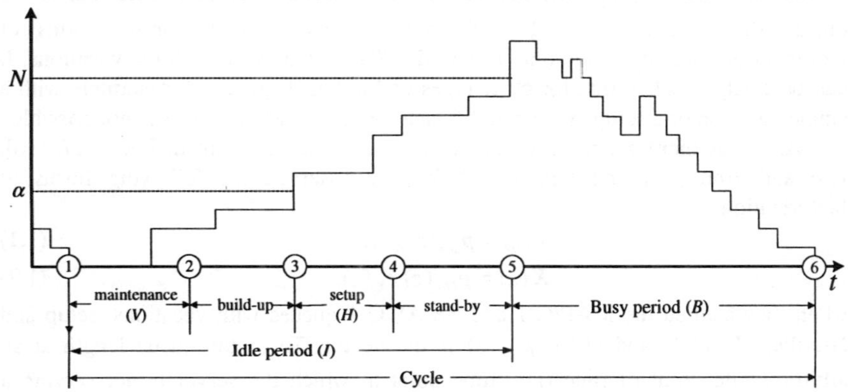

< Figure 1 > The system

1.2 *The factorization property of BMAP/G/1 queues*

The first factorization of queue length GF was observed by Lucantoni *et al.* [13] when they studied the MAP/G/1 queue with multiple vacations. They followed the standard procedure by starting with the analysis of the imbedded Markov renewal process at departure epochs with the following transition probability matrix,

$$\tilde{P}(x) = \begin{pmatrix} \tilde{B}_0(x) & \tilde{B}_1(x) & \tilde{B}_2(x) & \cdots & \cdots \\ \tilde{A}_0(x) & \tilde{A}_1(x) & \tilde{A}_2(x) & \cdots & \cdots \\ 0 & \tilde{A}_0(x) & \tilde{A}_1(x) & \cdots & \cdots \\ 0 & 0 & \tilde{A}_0(x) & \cdots & \cdots \\ \vdots & \vdots & \vdots & \vdots & \vdots \end{pmatrix}$$

in which $\tilde{B}_n(x)$ took a very complicated form due to all the possibilities that may occur during the vacation period (see the proof of their <theorem 1>). Then, they obtained $B(z,\theta) = \sum_{n=0}^{\infty}\left[\int_0^{\infty} e^{-\theta x} d\tilde{B}_n(x)\right] z^n$ and then the vector GF $X(z)$ at departures, and then the vector GF $Y(z)$ at an arbitrary epoch. They finally observed that for the MAP/G/1 queue with multiple vacations, the following factorizations hold,

$$Y(z) = Y_{NV}(z)V^+(z),$$
$$X(z) = X_{NV}(z)V^+(z),$$

where $Y_{NV}(z)$ and $X_{NV}(z)$ are the vector GFs of the MAP/G/1 queue with 'No Vacations', and $V^+(z)$ is the matrix GF of the number of customers that arrive during the remaining (or elapsed) vacation time. But the above forms of factorizations are applicable only to the MAP/G/1 queue with multiple vacations. It can be easily checked that for other types of MAP/G/1 queues, for example with a single vacation or a setup, a meaningful factorization of above forms is not possible.

As far as more general forms of factorizations are concerned, Lee *et al.* [10], Lee and Ahn [12] and Lee and Park [11] observed the following forms of factorizations

$$Y(z) = p_{idle}(z)\chi_Y(z), \tag{1.1}$$

$$X(z) = p_{idle}(z)\chi_X(z), \tag{1.2}$$

when they studied the MAP/G/1 and BMAP/G/1 queues with vacations, setup and N-policy. In (1.1) and (1.2), $p_{idle}(z)$ is the vector GF of the queue length at an arbitrary idle point of time (i.e., time point at which the server is not serving a customer),

$$\chi_Y(z) = (1-\rho)(z-1)A(z)[zI - A(z)]^{-1}, \tag{1.3}$$

and

$$\chi_X(z) = \frac{1}{\lambda}(1-\rho)D(z)A(z)[zI - A(z)]^{-1}. \tag{1.4}$$

In (1.3) and (1.4), $\lambda = \pi\sum_{n=1}^{\infty} nD_n \mathrm{e}$ is the mean arrival rate in which π is the stationary vector of the UMC and $\mathbf{e}$ is the column vector of 1's, $\rho = \lambda E(S)$, $D(z) = \sum_{n=0}^{\infty} D_n z^n$ and $A(z) = \int_0^{\infty} e^{D(z)x} dS(x)$ is the number of customers that arrive during the service time (see Lucantoni [14]).

Later, Chang *et al.* [1] proved that factorizations (1.1) and (1.2) hold for a very broad class of BMAP/G/1 queues with generalized vacations.

Readers can confirm from (1.1) and (1.2) that the following well-known relationship holds (Takine and Takahashi [18]),

$$Y(z)D(z) = \lambda(z-1)X(z). \tag{1.5}$$

The beauty of the factorizations (1.1) and (1.2) is that one does not need to go through all the standard procedure to obtain the queue length GFs of BMAP/G/1-related queues. All one needs to do is to obtain $p_{idle}(z)$.

In this paper, we use the factorization (1.1) to analyze the BMAP/G/1 system depicted in Figure 1.

2 The queue length GF

To obtain the vector GF of the queue length at an arbitrary point of time, we only need to obtain $\boldsymbol{p}_{idle}(z)$, which is the vector GF of the queue length at an arbitrary point of idle period in Figure 1. To this end, we first derive P_{maint}, P_{bu}, P_{su} and P_{sb} which are time-average probabilities that the system is in a maintenance period, in a build-up period, in a setup period and in a stand-by period respectively, under the condition that the system is idle. For these probabilities, let us first derive the mean length $E(I)$ of an arbitrary idle period. In the sequel, we will denote $(\boldsymbol{F})_{ij}$ as the (i, j)-element of a matrix $\boldsymbol{F}$.

Let us define $(\boldsymbol{\Phi}_k^{bu})_{ij}$ as

$(\boldsymbol{\Phi}_k^{bu})_{ij}$ = Pr(the build-up process ever visits level k and the phase of UMC is j at the visiting epoch | UMC phase is i at 1 of Figure 1).

Let $\boldsymbol{\kappa} = (\kappa_1, \kappa_2, \ldots, \kappa_m)$ be the probability vector of the UMC phases at 1. Noting that (i, j)-element of the matrix $(-\boldsymbol{D}_0)^{-1}$ is the mean time the UMC stays in phase j until the next arrival given that the current phase is in i (see for example, Latouche and Ramaswami [5]),

$$\boldsymbol{\kappa}\sum_{k=0}^{\alpha-1}\boldsymbol{\Phi}_k^{bu}(-\boldsymbol{D}_0)^{-1}\mathbf{e}$$

is the mean length of the build-up period.

In the same way, let us define $(\boldsymbol{\Phi}_k^{sb})_{ij}$ as

$(\boldsymbol{\Phi}_k^{sb})_{ij}$ = Pr(the stand-by process ever visits level n of the stand-by period and the phase of UMC is j at the visiting epoch | UMC phase is i at 1).

Then,

$$\boldsymbol{\kappa}\sum_{n=\alpha}^{N-1}\boldsymbol{\Phi}_n^{sb}(-\boldsymbol{D}_0)^{-1}\mathbf{e}$$

is the mean length of an arbitrary stand-by period. Then, we have the mean length of an idle period as

$$E(I) = \boldsymbol{\kappa}\left[E(V)\boldsymbol{I} + \sum_{k=0}^{\alpha-1}\boldsymbol{\Phi}_k^{bu}(-\boldsymbol{D}_0)^{-1} + E(H)\boldsymbol{I} + \sum_{k=\alpha}^{N-1}\boldsymbol{\Phi}_k^{sb}(-\boldsymbol{D}_0)^{-1}\right]\mathbf{e} \tag{2.1}$$

which leads to

$$P_{maint} = \frac{E(V)}{E(I)}, \tag{2.2a}$$

$$P_{bu}=\frac{\kappa\sum_{k=0}^{\alpha-1}\boldsymbol{\Phi}_k^{bu}(-\boldsymbol{D}_0)^{-1}\mathbf{e}}{E(I)},\tag{2.2b}$$

$$P_{su}=\frac{E(H)}{E(I)},\tag{2.2c}$$

and

$$P_{sb}=\frac{\kappa\sum_{k=\alpha}^{N-1}\boldsymbol{\Phi}_k^{sb}(-\boldsymbol{D}_0)^{-1}\mathbf{e}}{E(I)}.\tag{2.2d}$$

Obtaining $\{\boldsymbol{\Phi}_k^{bu},(0\le k\le\alpha-1)\}$ and $\{\boldsymbol{\Phi}_k^{sb},(\alpha\le k\le N-1)\}$ in (2.2b) and (2.2d) will be discussed later.

Let $\boldsymbol{p}_{maint}(z)$, $\boldsymbol{p}_{bu}(z)$, $\boldsymbol{p}_{su}(z)$ and $\boldsymbol{p}_{sb}(z)$ be the vector GFs of the queue length at an arbitrary epoch in each period under the condition that the system is idle. Then we have

$$\boldsymbol{p}_{maint}(z)=P_{maint}\cdot\kappa\boldsymbol{V}^{+}(z),\tag{2.3a}$$

$$\boldsymbol{p}_{bu}(z)=P_{bu}\cdot\frac{\kappa\sum_{k=0}^{\alpha-1}\boldsymbol{\Phi}_k^{bu}(-\boldsymbol{D}_0)^{-1}z^k}{\kappa\sum_{n=0}^{\alpha-1}\boldsymbol{\Phi}_n^{bu}(-\boldsymbol{D}_0)^{-1}\mathbf{e}},\tag{2.3b}$$

$$\boldsymbol{p}_{su}(z)=P_{su}\cdot\kappa\boldsymbol{H}_{\alpha}(z)\boldsymbol{H}^{+}(z),\tag{2.3c}$$

and

$$\boldsymbol{p}_{sb}(z)=P_{sb}\cdot\frac{\kappa\sum_{k=\alpha}^{N-1}\boldsymbol{\Phi}_k^{sb}(-\boldsymbol{D}_0)^{-1}z^k}{\kappa\sum_{n=\alpha}^{N-1}\boldsymbol{\Phi}_n^{sb}(-\boldsymbol{D}_0)^{-1}\mathbf{e}},\tag{2.3d}$$

where

$$\boldsymbol{V}^{+}(z)=\int_{x=0}^{\infty}\int_{t=0}^{x}e^{\boldsymbol{D}(z)t}\frac{1}{x}dt\frac{x\cdot d\boldsymbol{V}(x)}{E(V)}=\frac{[\boldsymbol{V}(z)-\boldsymbol{I}]}{E(V)}\boldsymbol{D}(z)^{-1}\tag{2.4a}$$

and

$$\boldsymbol{H}^{+}(z)=\int_{x=0}^{\infty}\int_{t=0}^{x}e^{\boldsymbol{D}(z)t}\frac{1}{x}dt\frac{x\cdot d\boldsymbol{H}(x)}{E(H)}=\frac{[\boldsymbol{H}(z)-\boldsymbol{I}]}{E(H)}\boldsymbol{D}(z)^{-1}\tag{2.4b}$$

are the matrix GFs of the number of customers that arrive during the elapsed maintenance period and the elapsed setup period respectively in which $\boldsymbol{V}(z)$ and $\boldsymbol{H}(z)$ are the matrix GFs of the number of customers that arrive during the maintenance period and the setup period, and $H_\alpha(z)$ in (2.3c) is the matrix queue length GF at 3 given the UMC phase at 1.

Then, $p_{idle}(z)$ can be obtained from

$$p_{idle}(z) = p_{maint}(z) + p_{bu}(z) + p_{su}(z) + p_{sb}(z). \tag{2.5}$$

Now, to obtain $H_\alpha(z)$ in (2.3c), we need to obtain Φ_n^{bu} first, which is the matrix probability that the idle period process ever visit level n during the build-up period. To this end, we note that the behavior of the queueing process during the build-up period is exactly the same as that of the usual BMAP/G/1 system with N-policy only with starting level dependent on the number of customers that arrive during the maintenance period.

<Lemma 1> (BMAP/G/1/N-policy)

Consider a cycle of a BMAP/G/1 queue with N-policy (Figure 2). Let $(D_k^*)_{ij}$ is the probability that the idle period process ever visits level k and the UMC phase just after the visit is j given that the UMC phase is i at a. Then we have

$$D_0^* = I, \quad D_k^* = \sum_{l=0}^{k-1} D_l^*(-D_0)^{-1}D_{k-l}, \; (1 \le k \le N-1). \tag{2.6}$$

<Proof> $D_0^* = I$ is obvious. Noting that $(-D_0)^{-1}D_k$ is the phase transition probability of the UMC by the arrival of a group of size k, conditioning on the level visited prior to k finishes the proof.

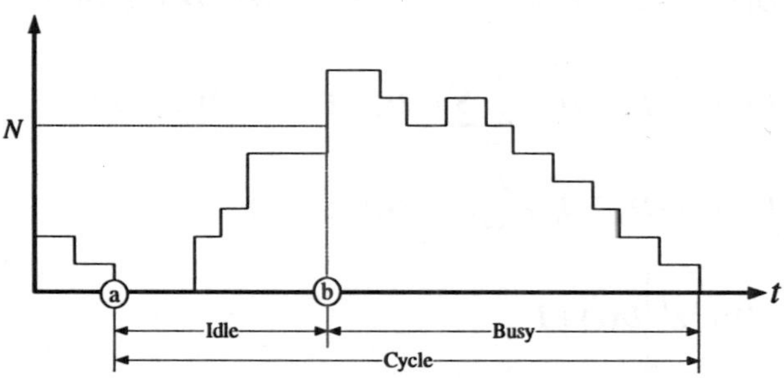

<Figure 2> BMAP/G/1 queue with N-policy

<Lemma 2> (BMAP/G/1/N-policy)

Let $(Q_n^N)_{ij}$, $(n \ge N)$ be the probability that at b of Figure 2 the queue length is n and the UMC phase is j given that UMC phase is in i at a. Then, we have

$$Q_n^N = (-D_0)^{-1}D_n + \sum_{j=1}^{N-1}(-D_0)^{-1}D_j Q_{n-j}^{N-j} \tag{2.7a}$$

and the matrix GF $Q_N(z)$ of Q_n^N becomes

$$Q_N(z) = \sum_{n=N}^{\infty} Q_n^N z^n = \left[\sum_{n=0}^{N-1} D_n^*(-D_0)^{-1} z^n\right] D(z) + I \, . \tag{2.7b}$$

<Proof> (2.7a) can be obtained by conditioning on the size of the first arrival group. To prove (2.7b), we use mathematical induction. For $N=1$, we get

$$Q_1(z) = \sum_{n=1}^{\infty} Q_n^1 z^n = (-D_0)^{-1} \sum_{n=1}^{\infty} D_n z^n = (-D_0)^{-1} D(z) - (-D_0)^{-1} D_0$$
$$= (-D_0)^{-1} D(z) + I$$

which satisfies (2.7b). Assuming (2.7b) holds for $N = 2,3,\ldots,k$, we have

$$\begin{aligned}
&Q_{k+1}(z)\\
&= \sum_{n=k+1}^{\infty} Q_n^{k+1} z^n\\
&= (-D_0)^{-1} \sum_{n=k+1}^{\infty} D_n z^n + (-D_0)^{-1} \sum_{n=k+1}^{\infty} \sum_{j=1}^{k-1} D_j Q_{n-j}^{N-j} z^n\\
&= (-D_0)^{-1} D(z) - (-D_0)^{-1} \sum_{n=0}^{k} D_n z^n + (-D_0)^{-1} \sum_{n=k+1}^{\infty} \sum_{j=1}^{k} D_j Q_{n-j}^{N-j} z^n\\
&= (-D_0)^{-1} D(z) + \sum_{j=1}^{k} (-D_0)^{-1} D_j z^j \left[\sum_{n=k+1}^{\infty} Q_{n-j}^{N-j} z^{n-j} - I\right] - (-D_0)^{-1} D_0\\
&= (-D_0)^{-1} D(z) - (-D_0)^{-1} D_0 + \sum_{j=1}^{k} (-D_0)^{-1} D_j z^j \sum_{k=0}^{k-j} D_k^* (-D_0)^{-1} z^k D(z)\\
&= (-D_0)^{-1} D(z) - (-D_0)^{-1} D_0 + \sum_{j=1}^{k} \sum_{i=1}^{j} (-D_0)^{-1} D_i D_{j-i}^* (-D_0)^{-1} z^j D(z)\\
&= (-D_0)^{-1} D(z) - (-D_0)^{-1} D_0 + \sum_{j=1}^{k} D_j^* (-D_0)^{-1} z^j D(z)\\
&= \left[\sum_{j=0}^{k} D_j^* (-D_0)^{-1} z^j\right] D(z) + I
\end{aligned}$$

which completes the proof.

Now, let us go back to Figure 1. Let $(V_n)_{ij}$ be the probability that n customers arrive during the maintenance period and the UMC phase is j at the end of the maintenance period given that the UMC phase is i at 1. Then, conditioning on the number of arrivals during the maintenance period, we have, without proof, the following theorem.

<Theorem 1> $\Phi_0^{bu} = V_0, \quad \Phi_n^{bu} = \sum_{i=0}^{n} V_i D_{n-i}^*, \ (n \ge 1)$. (2.8)

Now, we are ready to obtain $H_\alpha(z)$.

<Theorem 2> *Let $(H_k^\alpha)_{ij}$, $(k \ge \alpha)$ be the joint probability that at 3 of Figure 1 the queue length is k and the UMC phase is j given that UMC phase is i at* 1. *Then, we have*

$$H_\alpha(z) = \sum_{k=\alpha}^{\infty} H_k^\alpha z^k = V(z) + \sum_{n=0}^{\alpha-1} \Phi_n^{bu} (-D_0)^{-1} z^n D(z). \tag{2.9a}$$

<proof> Conditioning on the number of customers that arrive during the maintenance period and applying <Lemma 2>, we get

$$H_k^\alpha = V_k + \sum_{j=0}^{\alpha-1} V_j Q_{k-j}^{\alpha-j}, \quad (k \ge \alpha). \tag{2.9b}$$

Thus, we get

$$\begin{aligned} H_\alpha(z) &= \sum_{k=\alpha}^{\infty} V_k z^k + \sum_{k=\alpha}^{\infty} z^k \sum_{j=0}^{\alpha-1} V_j Q_{k-j}^{\alpha-j} \\ &= V(z) - \sum_{k=0}^{\alpha-1} V_k z^k + \sum_{j=0}^{\alpha-1} V_j z^j \sum_{k=\alpha-j}^{\infty} Q_k^{\alpha-j} z^k \\ &= V(z) - \sum_{k=0}^{\alpha-1} V_k z^k + \sum_{j=0}^{\alpha-1} V_j z^j \left[\sum_{k=0}^{\alpha-j-1} D_k^* (-D_0)^{-1} z^k D(z) + I \right] \\ &= V(z) + \sum_{j=0}^{\alpha-1} V_j z^j \sum_{k=0}^{\alpha-j-1} D_k^* z^k (-D_0)^{-1} D(z), \end{aligned} \tag{2.9c}$$

where we used (2.7b) in the third equality. Now, by changing the order of terms of $\sum_{j=0}^{\alpha-1} V_j z^j \sum_{k=0}^{\alpha-j-1} D_k^* z^k$ in the last equality, we get

$$\sum_{j=0}^{\alpha-1} V_j z^j \sum_{k=0}^{\alpha-j-1} D_k^* z^k = \sum_{j=0}^{\alpha-1} \sum_{i=0}^{j} V_i D_{j-i}^* z^j. \tag{2.9d}$$

Using (2.9d) in (2.9c), we get

$$H_\alpha(z) = V(z) + \sum_{j=0}^{\alpha-1} \sum_{i=0}^{j} V_i D_{j-i}^* (-D_0)^{-1} z^j D(z). \tag{2.9e}$$

Using (2.8) in (2.9e) completes the proof.

Now, we obtain Φ_n^{sb} contained in (2.1), (2.2d) and (2.3d). To do this, we need to know the queue length at 4 of Figure 1. This can be obtained simply by applying the result of <Theorem 2>. Let $(\boldsymbol{H}_k^+)_{ij}$ be the probability that queue length is k and UMC phase is j at 4 given the UMC phase is i at 1. Let $\boldsymbol{H}_n$ be the probability that n customers arrive during the setup time. Then we get

$$\boldsymbol{H}_i^+ = \sum_{n=\alpha}^{i} \boldsymbol{H}_n^\alpha \boldsymbol{H}_{i-n} \tag{2.10a}$$

where $\boldsymbol{H}_n^\alpha$ was defined in <Theorem 2>. Then, using <Lemma 1>, we get, without proof,

<Theorem 3> $\boldsymbol{\Phi}_k^{sb} = \sum_{i=\alpha}^{k} \boldsymbol{H}_i^+ \boldsymbol{D}_{k-i}^*, \quad (k = \alpha, \ldots, N-1)$. (2.10b)

So far, we have obtained all quantities comprising (2.2a)-(2.3d) except $\boldsymbol{\kappa}$, which is the vector phase probability at 1. Let $\boldsymbol{K}$ be the phase transition probability matrix between 1 and 6. Then, $\boldsymbol{\kappa}$ can be computed from

$$\boldsymbol{\kappa} = \boldsymbol{\kappa K}, \qquad \boldsymbol{\kappa}\mathbf{e} = 1. \tag{2.11}$$

Let $\boldsymbol{K}(z)$ be the matrix GF of the number of customers that are served during a cycle. Then $\boldsymbol{K}$ can be obtained from

$$\boldsymbol{K} = \boldsymbol{K}(z)\,|_{z=1}. \tag{2.12}$$

To obtain $\boldsymbol{K}(z)$, we need to know the matrix GF

$$\boldsymbol{Q}^{(\alpha,N)}(z) = \sum_{n=N}^{\infty} \boldsymbol{Q}_n^{(\alpha,N)} z^n$$

where $(\boldsymbol{Q}_n^{(\alpha,N)})_{ij}$ be the probability that the queue length is $n(\geq N)$ and the UMC phase is j at 5 given that the phase at 1 is i in Figure 1.

<Theorem 4> *We have*

$$\boldsymbol{Q}^{(\alpha,N)}(z) = \boldsymbol{H}_{\alpha,N}^+(z) + \boldsymbol{\Phi}^{sb}(z)(-\boldsymbol{D}_0)^{-1}\boldsymbol{D}(z) \tag{2.13a}$$

where $\boldsymbol{H}_{\alpha,N}^+(z) = \sum_{n=N}^{\infty} \boldsymbol{H}_n^+ z^n$ *and* $\boldsymbol{\Phi}^{sb}(z) = \sum_{n=\alpha}^{N-1} \boldsymbol{\Phi}_n^{sb} z^n$.

<Proof> Conditioning on the queue length at the end of the setup period and using $\boldsymbol{Q}_n^N$ defined in <Lemma 2>, we have

$$\boldsymbol{Q}_n^{(\alpha,N)} = \boldsymbol{H}_n^+ + \sum_{k=\alpha}^{N-1} \boldsymbol{H}_k^+ \boldsymbol{Q}_{n-k}^{N-k}, \quad (n \geq N). \tag{2.13b}$$

Then, we have

$$
\begin{aligned}
Q^{(\alpha,N)}(z) &= \sum_{n=N}^{\infty} H_n^+ z^n + \sum_{n=N}^{\infty}\left[\sum_{k=\alpha}^{N-1} H_k^+ Q_{n-k}^{N-k}\right] z^n \\
&= \sum_{n=N}^{\infty} H_n^+ z^n + \sum_{k=\alpha}^{N-1} H_k^+ z^k \sum_{n=N-k}^{\infty} Q_n^{N-k} z^n \\
&= \sum_{n=N}^{\infty} H_n^+ z^n + \sum_{k=\alpha}^{N-1} H_k^+ z^k \left[\sum_{n=0}^{N-k-1} D_n^* z^n (-D_0)^{-1} D(z) + I\right] \\
&= H_{\alpha,N}^+(z) + \sum_{k=\alpha}^{N-1} H_k^+ z^k \sum_{n=0}^{N-k-1} D_n^* z^n (-D_0)^{-1} D(z)
\end{aligned} \tag{2.13c}
$$

where we used (2.7b) in the third equality. Now, for the last equality of (2.13c), we have

$$
\sum_{k=\alpha}^{N-1} H_k^+ z^k \sum_{n=0}^{N-k-1} D_n^* z^n = \sum_{k=\alpha}^{N-1}\sum_{i=0}^{k} H_k^+ D_{k-i}^* z^k . \tag{2.13d}
$$

Using (2.13d) in (2.13c) after using (2.10b) to (2.13d) finishes the proof.

Now, as a simple consequence of <Theorem 4>, we have, without proof,

<Theorem 5> *We have*

$$
\begin{aligned}
K(z) = Q^{(\alpha,N)}(z)\,|_{z=G(z)} = {} & V(G(z))H(G(z)) \\
& + \sum_{n=0}^{\alpha-1} \Phi_n^{bu} (-D_0)^{-1} [G(z)]^n D(G(z))H(G(z)) \\
& + \sum_{n=\alpha}^{N-1} \Phi_n^{sb} (-D_0)^{-1} [G(z)]^n D(G(z))
\end{aligned} \tag{2.14a}
$$

and

$$
\begin{aligned}
K = K(z)\,|_{z=1} = {} & V(G)H(G) + \sum_{n=0}^{\alpha-1} \Phi_n^{bu} (-D_0)^{-1} G^n D(G) H(G) \\
& + \sum_{n=\alpha}^{N-1} \Phi_n^{sb} (-D_0)^{-1} G^n D(G) .
\end{aligned} \tag{2.14b}
$$

Now, at last, we can obtain $p_{idle}(z)$ from (2.5), and $Y(z)$ and $X(z)$ from (1.1) and (1.2).

3 Queue length probabilities

To calculate the queue length probabilities, we need to obtain $x_0 = (x_{01}, \cdots, x_{0m})$, where x_{0i} is the probability that at the end of a service completion, the queue length is zero and the UMC phase is i. Once $\boldsymbol{x}_0$ is known, we can compute $\{\boldsymbol{x}_n, (n \ge 1)\}$ from the well-known algorithms of Ramaswami [16] (see Lucantoni [14: eq. (49)]). Also the queue length probabilities $\{\boldsymbol{y}_n, (n \ge 0)\}$ at an arbitrary time can be obtained from the well-known relationship (see Lucantoni [14: eq. (34) and (36)])

$$\boldsymbol{y}_0 = \lambda \boldsymbol{x}_0(-\boldsymbol{D}_0)^{-1}, \tag{3.1a}$$

and

$$\boldsymbol{y}_{n+1} = \left[\sum_{j=0}^{n} \boldsymbol{y}_j \boldsymbol{D}_{n+1-j} - \lambda(\boldsymbol{x}_n - \boldsymbol{x}_{n+1})\right](-\boldsymbol{D}_0)^{-1}, \quad (n \ge 0). \tag{3.1b}$$

$\boldsymbol{x}_0$ can be obtained from the following relationship which works for all work-conserving, exhaustive service BMAP/G/1 queues,

$$\boldsymbol{x}_0 = \frac{(1-\rho)\boldsymbol{\kappa}}{\lambda\, E(I)}. \tag{3.2a}$$

Or, we can use the well-known formula (Neuts [15], Lucantoni [14]),

$$\boldsymbol{x}_0 = \frac{\boldsymbol{\kappa}}{\boldsymbol{\kappa}\,\boldsymbol{\kappa}^*} \tag{3.2b}$$

where $\boldsymbol{\kappa}^*$ is the mean number of customers that are served during a cycle.

<Theorem 6> *We have*

$$\begin{aligned}\boldsymbol{\kappa}^* &= \frac{d}{dz}\boldsymbol{K}(z)\mathbf{e} = \frac{\lambda \mathrm{E(V)}}{1-\rho}\mathbf{e} + \sum_{n=0}^{\alpha-1} \boldsymbol{\Phi}_n^{bu}(-\boldsymbol{D}_0)^{-1}\mathbf{e}\frac{\lambda}{1-\rho} + \frac{\lambda E(H)}{1-\rho} \\ &+ \sum_{n=\alpha}^{N-1} \boldsymbol{\Phi}_n^{sb}(-\boldsymbol{D}_0)^{-1}\mathbf{e}\frac{\lambda}{1-\rho} + (\boldsymbol{K}-\boldsymbol{I})[\mathbf{eg}+\boldsymbol{D}(\boldsymbol{G})]^{-1}\sum_{k=1}^{\infty}\boldsymbol{D}_k\sum_{i=0}^{k-1}\boldsymbol{G}^i\boldsymbol{\mu}.\end{aligned} \tag{3.3a}$$

In (3.3a), $\mathbf{g}$ *is the stationary vector of* $\mathbf{G}$ *and* $\boldsymbol{\mu} = \frac{d}{dz}\mathbf{G}(z)\mathbf{e}$, *where* $\mathbf{G}(z)$ *is the number of customers that are served during a fundamental period and* $\mathbf{G} = \mathbf{G}(1)$ *(Neuts [15] and Lucantoni [14]).*

<Proof> Differentiating (2.14a), using $z = 1$, $\boldsymbol{H}(\boldsymbol{G})\mathbf{e} = \mathbf{e}$, $\boldsymbol{V}(\boldsymbol{G})\mathbf{e} = \mathbf{e}$ and $[\mathbf{eg} + \boldsymbol{D}(\boldsymbol{G})]^{-1}\mathbf{e} = \mathbf{e}$, we get

$$\begin{aligned}\boldsymbol{\kappa}^* &= \frac{\lambda E(V)}{1-\rho}\mathbf{e} + (\mathbf{eg}+\boldsymbol{D}(\boldsymbol{G}))^{-1}[\boldsymbol{V}(\boldsymbol{G})-\boldsymbol{I}]\sum_{k=1}^{\infty}\boldsymbol{D}_k\sum_{i=0}^{k-1}\boldsymbol{G}^i\boldsymbol{\mu} \\ &+ \frac{\lambda E(V)}{1-\rho}\mathbf{e} + \boldsymbol{V}(\boldsymbol{G})(\mathbf{eg}+\boldsymbol{D}(\boldsymbol{G}))^{-1}[\boldsymbol{H}(\boldsymbol{G})-\boldsymbol{I}]\sum_{k=1}^{\infty}\boldsymbol{D}_k\sum_{i=0}^{k-1}\boldsymbol{G}^i\boldsymbol{\mu} \\ &+ \sum_{n=0}^{\alpha-1}\boldsymbol{\Phi}_n^{bu}(-\boldsymbol{D}_0)^{-1}\boldsymbol{G}^n\boldsymbol{H}(\boldsymbol{G})\sum_{k=1}^{\infty}\boldsymbol{D}_k\sum_{i=0}^{k-1}\boldsymbol{G}^i\boldsymbol{\mu} + \sum_{n=\alpha}^{N-1}\boldsymbol{\Phi}_n^{sb}(-\boldsymbol{D}_0)^{-1}\boldsymbol{G}^n\sum_{k=1}^{\infty}\boldsymbol{D}_k\sum_{i=0}^{k-1}\boldsymbol{G}^i\boldsymbol{\mu}.\end{aligned} \tag{3.3b}$$

In the above equation, $(\mathbf{eg}+\boldsymbol{D}(\boldsymbol{G}))^{-1}$, $[\boldsymbol{H}(\boldsymbol{G})-\boldsymbol{I}]$, and $[\boldsymbol{V}(\boldsymbol{G})-\boldsymbol{I}]$ commute. Thus, we get

$$\kappa^* = \left[V(G) - I + V(G)H(G) - V(G) + \sum_{n=0}^{\alpha-1} \Phi_n^{bu}(-D_0)^{-1} G^n H(G)[eg + D(G)] \right.$$

$$\left. + \sum_{n=\alpha}^{N-1} \Phi_n^{sb}(-D_0)^{-1} G^n [eg + D(G)] \right] [eg + D(G)]^{-1} \sum_{k=1}^{\infty} D_k \sum_{i=0}^{k-1} G^i \mu \qquad (3.3c)$$

$$+ \frac{\lambda E(V)}{1-\rho} e + \frac{\lambda E(H)}{1-\rho} e \ .$$

Using $g(eg + D(G))^{-1} = g$ completes the proof.

<Theorem 7> *Premultiplying (3.3a) by* κ *and using (3.2b), we get*

$$x_0 = \frac{(1-\rho)\kappa}{\lambda\kappa \left[E(V)I + \sum_{n=0}^{\alpha-1} \Phi_n^{sb}(-D_0)^{-1} + E(H)I + \sum_{n=\alpha}^{N-1} \Phi_n^{sb}(-D_0)^{-1} \right] e} . \qquad (3.4)$$

It is easy to see that (3.4) confirms (3.2a).

4 The mean queue length

From (1.1)-(1.4), we have

$$Y(z)D(z) = (z-1)U(z)$$

where

$$U(z) = \lambda X(z) .$$

Then, following Lucantoni [14], the mean queue length becomes

$$L = Y^{(1)} e = \frac{1}{\lambda} U^{(1)} e - \frac{1}{2\lambda} \pi D^{(2)} e + \frac{1}{\lambda} (\pi D^{(1)} - U)(D + e\pi)^{-1} D^{(1)} e \qquad (4.1a)$$

where, $F = F(z)|_{z=1}$, $F^{(1)} = (d/dz)F(z)|_{z=1}$ and $F^{(2)} = (d^2/dz^2)F(z)|_{z=1}$ for a matrix GF $F(z)$. In (4.1a), π is the stationary probability vector of the UMC that can be obtained from

$$\pi D = 0, \qquad \pi e = 1 , \qquad (4.1b)$$

and λ is the mean arrival rate that is given by

$$\lambda = \pi \sum_{k=1}^{\infty} k D_k e . \qquad (4.1c)$$

Other quantities in (4.1a) are as follows:

$$U = \lambda\pi + \delta \left[\sum_{n=0}^{\alpha-1} \Phi_n^{bu}(-D_0)^{-1} DH + VH + \sum_{n=\alpha}^{N-1} \Phi_n^{sb}(-D_0)^{-1} D - I \right] A(I - A + e\pi)^{-1}$$

where

$$\delta = \frac{(1-\rho)\boldsymbol{\kappa}}{\boldsymbol{\kappa}\left[E(V)\boldsymbol{I} + \sum_{n=0}^{\alpha-1}\boldsymbol{\Phi}_n^{bu}(-\boldsymbol{D}_0)^{-1} + E(H)\boldsymbol{I} + \sum_{n=\alpha}^{N-1}\boldsymbol{\Phi}_n^{sb}(-\boldsymbol{D}_0)^{-1}\right]\mathbf{e}} = \frac{(1-\rho)\boldsymbol{\kappa}}{E(I)}.$$

$$\boldsymbol{U}^{(1)}\mathbf{e} = \frac{1}{(1-\rho)}(E_1 + E_2 + E_3 + E_4 + E_5),$$

where

$$E_1 = \frac{1}{2}\delta\Bigg[2\sum_{n=0}^{\alpha-1} n\boldsymbol{\Phi}_n^{bu}(-\boldsymbol{D}_0)^{-1}\boldsymbol{D}^{(1)}\boldsymbol{H}$$

$$+\sum_{n=0}^{\alpha-1}\boldsymbol{\Phi}_n^{bu}(-\boldsymbol{D}_0)^{-1}\boldsymbol{D}^{(2)}\boldsymbol{H} + \sum_{n=0}^{\alpha-1}\boldsymbol{\Phi}_n^{bu}(-\boldsymbol{D}_0)^{-1}\boldsymbol{D}\boldsymbol{H}^{(2)} + \boldsymbol{V}^{(2)}\boldsymbol{H} + 2\boldsymbol{V}^{(1)}\boldsymbol{H}^{(1)} + \boldsymbol{V}\boldsymbol{H}^{(2)}$$

$$+2\sum_{n=\alpha}^{N-1} n\boldsymbol{\Phi}_n^{sb}(-\boldsymbol{D}_0)^{-1}\boldsymbol{D}^{(1)} + \sum_{n=\alpha}^{N-1}\boldsymbol{\Phi}_n^{sb}(-\boldsymbol{D}_0)^{-1}\boldsymbol{D}^{(2)}\Bigg]\mathbf{e},$$

$$E_2 = \delta\Bigg[\sum_{n=0}^{\alpha-1} n\boldsymbol{\Phi}_n^{bu}(-\boldsymbol{D}_0)^{-1}\boldsymbol{D}\boldsymbol{H} + \sum_{n=0}^{\alpha-1}\boldsymbol{\Phi}_n^{bu}(-\boldsymbol{D}_0)^{-1}\boldsymbol{D}^{(1)}\boldsymbol{H} + \sum_{n=0}^{\alpha-1}\boldsymbol{\Phi}_n^{bu}(-\boldsymbol{D}_0)^{-1}\boldsymbol{D}\boldsymbol{H}^{(1)}$$

$$+\boldsymbol{V}^{(1)}\boldsymbol{H} + \boldsymbol{V}\boldsymbol{H}^{(1)} + \sum_{n=\alpha}^{N-1} n\boldsymbol{\Phi}_n^{sb}(-\boldsymbol{D}_0)^{-1}\boldsymbol{D} + \sum_{n=\alpha}^{N-1}\boldsymbol{\Phi}_n^{sb}(-\boldsymbol{D}_0)^{-1}\boldsymbol{D}^{(1)}\Bigg]\boldsymbol{A}^{(1)}\mathbf{e},$$

$$E_3 = \frac{1}{2}\delta\left[\sum_{n=0}^{\alpha-1}\boldsymbol{\Phi}_n^{bu}(-\boldsymbol{D}_0)^{-1}\boldsymbol{D}\boldsymbol{H} + \boldsymbol{V}\boldsymbol{H} + \sum_{n=\alpha}^{N-1}\boldsymbol{\Phi}_n^{sb}(-\boldsymbol{D}_0)^{-1}\boldsymbol{D} - \boldsymbol{I}\right]\boldsymbol{A}^{(2)}\mathbf{e} + \frac{1}{2}\boldsymbol{U}\boldsymbol{A}^{(2)}\mathbf{e},$$

$$E_4 = -\delta\Bigg[\sum_{n=0}^{\alpha-1} n\boldsymbol{\Phi}_n^{bu}(-\boldsymbol{D}_0)^{-1}\boldsymbol{D}\boldsymbol{H} + \sum_{n=0}^{\alpha-1}\boldsymbol{\Phi}_n^{bu}(-\boldsymbol{D}_0)^{-1}\boldsymbol{D}^{(1)}\boldsymbol{H} + \sum_{n=0}^{\alpha-1}\boldsymbol{\Phi}_n^{bu}(-\boldsymbol{D}_0)^{-1}\boldsymbol{D}\boldsymbol{H}^{(1)}$$

$$+\boldsymbol{V}^{(1)}\boldsymbol{H} + \boldsymbol{V}\boldsymbol{H}^{(1)} + \sum_{n=\alpha}^{N-1} n\boldsymbol{\Phi}_n^{sb}(-\boldsymbol{D}_0)^{-1}\boldsymbol{D} + \sum_{n=\alpha}^{N-1}\boldsymbol{\Phi}_n^{sb}(-\boldsymbol{D}_0)^{-1}\boldsymbol{D}^{(1)}\Bigg]$$

$$\times\boldsymbol{A}(\boldsymbol{I} - \boldsymbol{A} + \mathbf{e}\boldsymbol{\pi})^{-1}(\boldsymbol{I} - \boldsymbol{A}^{(1)})\mathbf{e},$$

and

$$E_5 = -\delta\left[\sum_{n=0}^{\alpha-1}\boldsymbol{\Phi}_n^{bu}(-\boldsymbol{D}_0)^{-1}\boldsymbol{D}\boldsymbol{H} + \boldsymbol{V}\boldsymbol{H} + \sum_{n=\alpha}^{N-1}\boldsymbol{\Phi}_n^{sb}(-\boldsymbol{D}_0)^{-1}\boldsymbol{D} - \boldsymbol{I}\right]$$

$$\times\boldsymbol{A}^{(1)}(\boldsymbol{I} - \boldsymbol{A} + \mathbf{e}\boldsymbol{\pi})^{-1}(\boldsymbol{I} - \boldsymbol{A}^{(1)})\mathbf{e} + \boldsymbol{U}(\boldsymbol{I} - \boldsymbol{A}^{(1)})(\boldsymbol{I} - \boldsymbol{A} + \mathbf{e}\boldsymbol{\pi})^{-1}(\boldsymbol{I} - \boldsymbol{A}^{(1)})\mathbf{e}.$$

5 A numerical example: comparison with $M^X/G/1$ queue

In this section, we present a numerical example that compares the mean queue lengths of BMAP/G/1 and $M^X/G/1$ queues under the same parameter setting. We use, as BMAP parameter matrices,

$$D_0 = \begin{pmatrix} -10.0 & 1.0 \\ 0.4 & -0.8 \end{pmatrix}, \quad D_1 = \begin{pmatrix} 8.0 & 0 \\ 0 & 0.2 \end{pmatrix}, \quad D_2 = \begin{pmatrix} 1.0 & 0 \\ 0 & 0.2 \end{pmatrix}.$$

Then, we have

$$D = D_0 + D_1 + D_2 = \begin{pmatrix} -1.0 & 1.0 \\ 0.4 & -0.4 \end{pmatrix}.$$

From $\pi D = \mathbf{0}$, we get

$$\pi = (\pi_1, \pi_2) = \left(\frac{2}{7}, \frac{5}{7} \right)$$

Thus, the group arrival rate and the total arrival rate become

$$\lambda_g = \pi \sum_{n=1}^{\infty} D_n \mathbf{e} = \frac{20}{7}, \qquad \lambda = \pi \sum_{n=1}^{\infty} D_n \mathbf{e} = \frac{23}{7}.$$

Then, we get

$$g_1 = \frac{\pi D_1 \mathbf{e}}{\lambda_g} = \frac{17}{20}, \quad g_2 = \frac{\pi D_2 \mathbf{e}}{\lambda_g} = \frac{3}{20},$$

where g_i is the probability that an arbitrary group is of size *i*. We will use above λ_g, g_1 and g_2 as the parameters of the $M^X/G/1$ queue. For both BMAP/G/1 and $M^X/G/1$ queues, we use the followings:

(α, N)	:	(3,5),
Maintenance time	:	Exponential with mean 1.0,
Setup time	:	Exponential with mean 1.0,
Service time	:	Erlang (n, μ) with pdf $\frac{\mu^n x^{n-1} e^{-\mu x}}{(n-1)!}$ and mean service time

$$E(S) = \frac{n}{\mu}.$$

We consider seven cases of service times: $\mu = 20$, 15, 12, 10, 9, 8, 7 with order n fixed at 2 for all cases. This arrangement leads to seven cases of traffic intensities, ranging from $\rho = \lambda E(S) = 0.3286$ to $\rho = 0.9388$.

Table 1 shows the mean queue lengths L(BMAP) and L(Poisson) for different values of ρ. The last column shows the ratio $\frac{\text{L(BMAP)}}{\text{L(Poisson)}}$. It can be observed that as ρ is getting closer to 1, the relative difference between two queue lengths is getting larger. Table 1 is graphed on Figure 3.

This simple numerical example tells us that in many real world-world queueing systems including many production systems, a naive Poisson assumption is likely to lead to a severe underestimation of the mean queue length.

	Mean queue length		
ρ	L(BMAP)	L(Poisson)	L(BMAP)/L(Poisson)
0.3286	5.7352	5.2999	1.08
0.4381	6.9645	5.5829	1.25
0.5476	8.9901	5.9896	1.50
0.6571	12.3769	6.6387	1.86
0.7302	16.1710	7.3518	2.20
0.8214	25.2750	9.0423	2.80
0.9388	76.8145	18.5216	4.15

<Table 1> Comparison of mean queue lengths as ρ varies

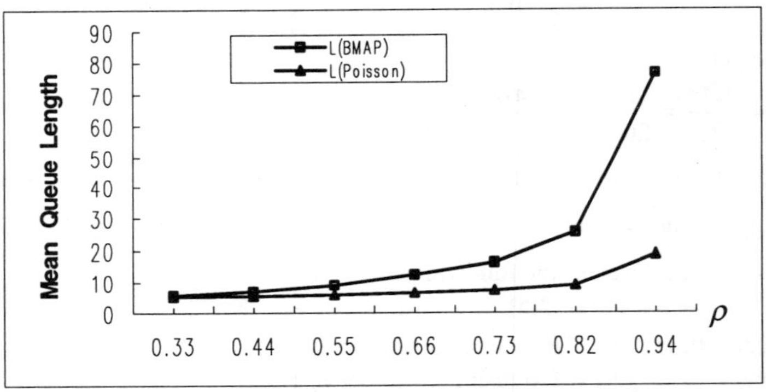

<Figure 3> Comparison of mean queue lengths as ρ varies

6 Summary

In this paper, we applied the factorization property to analyze a production system with BMAP input, maintenance period, bilevel threshold control and setup time. We showed that the factorization property can be applied efficiently and effectively to directly derive the queue length distributions of some complicated BMAP/G/1 queueing systems with generalized vacations by avoiding all the classical standard steps. The procedure shown in this paper can be applied to the analysis of many other BMAP/G/1 queueing systems with more behavioral complexities.

Also, the comparison of the mean queue length between BMAP/G/1 and $M^X/G/1$ queues tells us that a native Poisson assumption may lead to an underestimation of the mean queue length.

7 Acknowledgements

The authors are grateful to the anonymous referee who suggested the inclusion of a numerical example. The example greatly enhanced the quality of this paper. Also, the authors thank professor David Green of the University of Adelaide, Australia, for providing us with the parameter matrices of an MMPP. We used his matrices as the basis to form our BMAP matrices in the numerical example.

This work was supported (in part) by KOSEF through Statistical Research Center for Complex Systems at Seoul National University.

References

1. Chang S. H., Takine T., Chae, K. C. and Lee, H. W., A unified queue length formula for BMAP/G/1 queue with generalized vacations, To appear in *Stochastic Models* (2001).
2. Hersh M. and Brosh I., The Optimal Strategy Structure of an intermittently Operated Service Channel, *Europ. J. Opns. Res.* **5** (1980) pp. 133-141.
3. Hofri M., Queueing Systems with a Procrastinating Server, Performance '86 and ACMSIGMETRICS 1980, *Performance Evaluation Review* **14(1)** (1986) pp. 245–253.
4. Kella O., The Threshold Policy in the M/G/1 Queue with Server Vacations, *Naval Res. Logist.* **36** (1989) pp. 111–123.
5. Latouche G. and Ramaswami V., *Introduction to Matrix Analytic Methods in Stochastic Modeling*, ASA-SIAM series on Statistics and Applied Probability (1999).
6. Lee H. S. and Srinivasan M. M., Control Policies for the $M^X/G/1$ Queueing System, *Mgmt. Sci.* **35(6)** (1989) pp. 708-721.
7. Lee H. W., Lee S. S. and Chae K. C., Operating Characteristics of $M^X/G/1$ Queue with N-Policy, *Queueing Systems* **15** (1994) pp. 387-399.
8. Lee H. W., Lee S. S., Park J. O. and Chae K. C., Analysis of $M^X/G/1$ Queue with N-Policy and Multiple Vacations, *Journal of Applied Probability* **31** (1994) June pp. 467-496.
9. Lee H. W. and Park J. O., Optimal strategy in N-policy system with early setup, *J of Oper Res Soc* **48** (1997) pp. 306-313.
10. Lee H. W., Ahn B. Y. and Park N. I., Decompositions of the queue length distributions in the MAP/G/1 queue under multiple and single vacations with N-policy, *Stochastic Models* **17(2)** (2001) pp. 157-190.
11. Lee H. W. and Park N. I., Analysis of the operational behavior of the BMAP/G/1 queue under N-policy with vacations, submitted for publication (2001).

12. Lee H. W. and Ahn, B. Y., The MAP/G/1 queue under N-policy with single vacation and setup, Accepted for publication in *Appl. Math. & Stoch. Analysis* (2001).
13. Lucatoni D. M., Meier-Hellstern K. and Neuts M. F., A single server queue with server vacations and a class of non-renewal arrival process, *Adv. Appl. Prob.* **22** (1990) pp. 676-705.
14. Lucatoni D. M., New results on the single server queue with a batch Markovian Arrival Process, *Stochastic Models* **7(1)** (1991) pp. 1-46.
15. Neuts M. F., *Structured stochastic matrices of M/G/1 type and their applications*, New York, Marcel Dekker (1989).
16. Ramaswami V., Stable recursion for the steady state vector for Markov chains of M/G/1 type, *Stochastic Models* **4** (1988) pp. 183-188.
17. Takagi H., *Queueing Analysis: A Foundation of Performance Evaluation, Vol I, Vacation and Priority Systems, Part I*, North-Holland (1991).
18. Takine T. and Takahashi Y., On the relationship between queue length at a random time and at a departure in the stationary queue with BMAP arrivals, *Stochastic Models* **14(3)** (1998) pp. 601-610.
19. Yadin M. and Naor P., Queueing System with Removable Service Station, *Opns. Res. Quarterly* **14** (1963) pp. 393-405.

A CONSTRUCTIVE METHOD FOR FINDING β-INVARIANT MEASURES FOR TRANSITION MATRICES OF $M/G/1$ TYPE

QUAN-LIN LI
National Laboratory of Pattern Recognition, Institute of Automation
Chinese Academy of Sciences, Beijing 100080, P.R. China
and
School of Mathematics and Statistics, Carleton University
Ottawa, Ontario, Canada K1S 5B6
E-mail: qli@math.carleton.ca

YIQIANG ZHAO
School of Mathematics and Statistics, Carleton University
Ottawa, Ontario, Canada K1S 5B6
E-mail: zhao@math.carleton.ca

In this paper, we study the transition matrix of $M/G/1$ type. The radius of convergence is discussed, conditions on the α-classification of the states are obtained, and expressions of the β-invariant measure are constructed. The censoring technique is generalized to deal with nonnegative matrices, which may be neither stochastic nor substochastic. This allows us to prove a factorization result for the discounted transition matrix. This factorization provides a unified algorithmic approach for expressing the β-invariant measure for transition matrices with a block-structure, including the matrix of $M/G/1$ type.

Keywords: β-invariant measures, duality, factorizations, $M/G/1$ type, quasi-stationary distributions, radius of convergence.

1 Introduction

We consider an irreducible aperiodic Markov chain $\{X_n; n = 1, 2, \dots\}$ of $M/G/1$ type, whose transition matrix P is partitioned into block-form:

$$P = \begin{bmatrix} D_1 & D_2 & D_3 & D_4 & \cdots \\ D_0 & C_1 & C_2 & C_3 & \cdots \\ & C_0 & C_1 & C_2 & \cdots \\ & & C_0 & C_1 & \cdots \\ & & & \ddots & \ddots \end{bmatrix}, \qquad (1)$$

where D_1 is a matrix of size $m_0 \times m_0$, all C_i are square matrices of finite size m, the sizes of the other block-entries are determined accordingly and all empty entries are zero. P is assumed to be stochastic or strictly substochastic.

By strictly substochastic, we mean that $P \geq 0$, $Pe \leq e$ and $Pe \neq e$, where e is a column vector of ones.

The state space of the above block-partitioned Markov chain can be expressed as $S = \cup_{i=0}^{\infty} L_i$, where $L_0 = \{(0,j); j = 0,1,2,\ldots,m_0\}$ and $L_i = \{(i,j); j = 0,1,2,\ldots,m\}$ for $i \geq 1$. In state (i,j), variable i is called the *level* and variable j, the *phase*. Therefore, L_i is the set of all states at level i. For convenience, we write $L_{\leq i} = \cup_{k=0}^{i} L_k$ and $L_{\geq i}$ for the complement of $L_{\leq(i-1)}$.

Let α be the radius of convergence for the transition matrix $P = (p_{(i,r),(j,s)})$. We know that $\alpha = \sup\{z > 0; \sum_{n=0}^{\infty} z^n p_{(i,r),(j,s)}^{(n)} < \infty\} \geq 1$, where $p_{(i,r),(j,s)}^{(n)}$ is the n-step transition probability and α is independent of states (i,r) and (j,s).

A nonnegative nonzero row vector π is said to be an invariant measure of P if $\pi = \pi P$. For $0 < \beta \leq \alpha$, a nonnegative nonzero row vector π is said to be a β-invariant measure of P if $\pi = \pi \beta P$. Call βP the discounted transition matrix at rate β. Then, a β-invariant measure is simply an invariant measure of the discounted matrix. It follows from the definition that a 1-invariant measure is simply an invariant measure.

For the transition matrix P of $M/G/1$ type, we are interested in

a) the radius of convergence α;

b) the α-classification of the process if $\alpha > 1$; and

c) β-invariant measures for $0 < \beta \leq \alpha$.

There are a number of reasons why the above items are of interest.

1) It is well-known that $\pi = (\pi_i)$ is a quasistationary distribution if and only if for some $\beta > 1$ π is a β-invariant measure satisfying $\sum_i \pi_i < \infty$. The study of quasistationary behavior of a Markov chain is not only theoretically important, but also finds interesting and important applications in many areas, including biology (Scheffer 1951, Holling 1973, Pakes 1987 and Pollett 1987), chemistry (Oppenheim, Shuler and Weiss 1977, Parsons and Pollett 1987 and Pollett 1988), and telecommunications (Schrijner 1995), among others.

2) When the entries π_i in π cannot be summed, the concept of the β-invariant measure is a generalization of invariant measures for a nonergodic chain (Derman 1955, Harris 1957, Latouche, Pearce and Taylor 1998, Gail, Hantler and Taylor 1998, Zhao, Li and Braun 1998). In this case, π can still be interpreted probabilistically in terms of the movement of particles

whose initial states are governed by Poisson distributions (Derman 1955 and Kelly 1983). Also, π can be used to define a time-reversed matrix or dual matrix, which has important applications (Kelly 1979, Ramaswami 1990, Asmussen and Ramaswami 1990, Bright 1996 and Zhao, Li and Alfa 1999).

3) It is well known how important the Perron-Frobenius Theorem is in the theory of finite nonnegative matrices. The decay parameter $\frac{1}{\alpha}$ of P can be considered the Perron-Frobenius eigenvalue of the nonnegative matrix P and an α-invariant measure of P a Perron-Frobenius eigenvector of P.

It is believed that the study of quasistationary behavior was originated by Yaglom (1947). Since then, significant advances in the theory of quasistationarity have been made through the efforts of many researchers. A detailed review on the literature can be found in the Ph.D. dissertation of Schrijner (1995). This study has also been successfully advanced to consider transition matrices with a block-structure since Kijima (1993) made a break through on the determination of the radius of convergence for Markov chains of $GI/M/1$ type and $M/G/1$ type without boundaries. For transition matrices with a block-structure, studies have been centered on obtaining probabilistic measures to express the radius of convergence and quasistationary distributions, including classifications of the states in terms of these measures. People are searching for expressions which are numerically preferable. Results on quasi-birth-and-death (QBD) processes can be found in Kijima (1993), Makimoto (1993), Bean *et al.* (1997), and Bean, Pollett and Taylor (1998, 2000). Some preliminary results on the expressions for the matrices of $GI/M/1$ type and $M/G/1$ type were obtained in Li (1997). A survey on quasistationary distributions of Markov chains arising from queueing processes was given by Kijima and Makimoto (1999).

In this paper, we will study the matrix of $M/G/1$ type with boundary blocks as defined in (1). The issue on the radius of convergence will be addressed by combining a result (see Lemma 5) obtained by Kijima (1993) and the boundary treatment based on censoring. For the case without boundaries, the matrix is always α-transient. With the presence of the boundary, the matrix can be either α-transient or α-recurrent. Conditions on classifications of the transient states will also be discussed in this paper. For the matrix of $M/G/1$ type, we have not noticed the existence of an expression for the β-invariant measure in the literature. We will provide a constructive way of expressing such a measure.

The technique used in this paper to study the radius of convergence and conditions on classifications of the transient states is based on censoring. A censored process is also referred as the imbedded process. For any subset, as the censoring set, of the state space of a process, the censored process is the process obtained by watching the original process only when it travels in the censoring set. This technique has been successfully used in studying many other aspects of block-structured stochastic or strictly substochastic matrices (for example, see Grassmann and Heyman 1990, Latouche 1993, Zhao, Li and Braun 1998, 2001, Zhao, Li and Alfa 1999, Latouche and Ramaswami 1999, and Zhao 2000). In order to use the censoring technique to deal with the issue on the β-invariant measure, we need to generalize results on stochastic or strictly substochastic matrices to that on nonnegative matrices.

What we will use to obtain expressions for the β-invariant measure is the method of factorization, where $I - \beta P$ is factorized into the product of an upper triangular matrix and a lower triangular matrix. We shall call it the *RG*-factorization, since the factors in the factorization involve the *R*- and *G*-measures, two key probabilistic measures in our study, which will be defined later. This factorization may be viewed as an *LU*-factorization for the infinite matrix $I - \beta P$. The procedure of obtaining a solution for the β-invariant measure can be considered a generalization of using an *LU*-factorization to solve a finite system of linear equations. Expressions for the β-invariant measure are different according to the classification of the states and the value of β. When we use the factorization technique, it is a key how to associate the middle factor or the diagonal matrix with either the upper triangular or the lower triangular matrix. Our study will give a way to successively identify two different sets of solutions for the β-invariant measure. When $\beta = 1$, an equivalent form of this factorization was obtained and studied by Heyman (1995), Zhao, Li and Braun (1997, 2000) and Zhao (2000). In Li (1997), the matrix $I - \beta P$ was factored into an equivalent form of the *RG*-factorization without using the *R*-measure. There are three possible difficulties when using the *RG*-factorization on infinite matrices. Firstly, the associativity of matrix multiplications cannot be taken for granted, secondly, the existence of a nontrivial solution to a linear system of infinitely many equations cannot be taken for granted, and thirdly, the method of dealing with a recurrent matrix and a transient matrix should be distinguished. When the Markov chain is positive recurrent, these issues have been successfully addressed in the literature, for example, see Heyman (1995). Ramaswami (1988) presented a stable recursion, equivalent to the factorization of Heyman, for the steady state vector for Markov chains of $M/G/1$ type. Also, Meini (1997) studied the matrix of $M/G/1$ type in terms of a method of factorization. For quasistationary distributions, the method employed by Bean, Pollett and Taylor (2000) to the quasi-birth-and-death process is essentially equivalent to the factorization method used in this paper. However, they did not indicate how the expressions for the β-invariant measure are constructed.

It is our belief that the idea presented here can also be used to study other types of block-structured matrices, for example, matrices of $GI/M/1$ type and, more generally, $GI/G/1$ type.

The rest of the paper is organized as follows.

In Section 2, some basic properties about the matrix βP are provided, including properties on the existence of an inverse of $I - \beta P$, the minimal nonnegative inverse and the fundamental matrix. These properties are needed in later sections.

When P is transient, the states of P can be further classified as α-recurrent or α-transient according to $\widehat{\alpha P} = \sum_{k=0}^{\infty} \alpha^k P^k = \infty$ or $< \infty$, respectively. The matrix $\widehat{\alpha P}$ is referred as the fundamental matrix of αP. If P is α-recurrent, either $\lim_{n\to\infty} \alpha^n p_{(i,r),(j,s)}^{(n)} > 0$ for all states (i,r) and (j,s), or $\lim_{n\to\infty} \alpha^n p_{(i,r),(j,s)}^{(n)} = 0$ for all states (i,r) and (j,s). In the former case, P is called α-positive and in the latter case, α-null. In Section 3, we determine the radius α of convergence and the α-classification of the transient states, based on the combination of the result on determining the radius $\bar{\alpha}$ of convergence for the matrix of $M/G/1$ type without boundaries and a new treatment for the boundary.

In Section 4, the RG-factorization for matrix $I - \beta P$ is proved. We show that

$$I - \beta P = [I - R_U(\beta)][I - U_D(\beta)][I - G_L(\beta)],$$

where $R_U(\beta)$ is a block-form upper triangular matrix involving only the R-measure, $G_L(\beta)$ is a block-form lower triangular matrix involving only the G-measure, and $U_D(\beta)$ is a block-form diagonal matrix. The R-measure is a sequence of matrices defined by (15) and (16) and the G-measure for the matrix $M/G/1$ type consists of two matrices defined by (7) and (17). Probabilistic interpretations for both R- and G-measuers are provided after the definition formulas. In this section, we also show that the RG-factorization exists for the matrix of level-dependent $M/G/1$ type.

In Section 5, based on the RG-factorization, expressions for the β-invariant measure are obtained. There are two different sets of expressions: One is for the α-invariant measure when P is α-recurrent. In this case, the α-invariant measure is unique up to a multiple of a positive constant. For all other cases, we provide a common expression for the β-invariant measure. When the β-invariant measure cannot be summed, this uniqueness is no longer guaranteed.

The Final section, Section 6, consists of concluding remarks.

2 Preliminaries

In this section, we provide some properties of the discounted matrix βP, which will be used in later sections. Most of these results can be viewed as generalizations of the counterparts for a stochastic or strictly substochastic matrix. Proofs of these properties may not be obvious. However, since they can be proved either in the same way as that for a stochastic or strictly substochastic matrix or in a similar fashion, we omit most of the proofs. Relevant references are Seneta (1981), Kemeny, Snell and Knapp (1976), Çinlar (1975) among possible others.

A general statement on the existence and uniqueness of an α-invariant measure can be found in the literature, for example Seneta (1981) which is stated in the following lemma. In order to do so, we need the concept of subinvariant measure (or superregular measure). A row vector x is called a subinvariant measure of P if $x \geq xP$. A row vector x is called a β-subinvariant measure of P if $x \geq \beta xP$. A 1-subinvariant measure is simply subinvariant.

Lemma 1 *For irreducible aperiodic matrix P, there always exists a positive α-subinvariant measure x. If P is α-recurrent, then the unique α-subinvariant measure x, up to a multiple of a positive constant, of P is α-invariant and positive.*

The following are some basic properties about the existence of an inverse, minimal nonnegative inverse and the fundamental matrix.

Lemma 2 *(i) For $0 < \beta < \alpha$ if P is α-recurrent, or for $0 < \beta \leq \alpha$ if P is α-transient, $(I - \beta P)$ is invertible. (ii) If $(I - \beta P)$ is invertible, then*

$$\widehat{\beta P} = \sum_{k=0}^{\infty} \beta^k P^k \tag{2}$$

is the minimal nonnegative inverse of $(I - \beta P)$, which is often referred to as the fundamental matrix of βP. (iii) Let P be partitioned into

$$P = \begin{bmatrix} T & H \\ L & Q \end{bmatrix}. \tag{3}$$

Then, both $(I - \beta T)$ and $(I - \beta Q)$ are invertible for $0 < \beta \leq \alpha$.

The following lemma plays an important role in later sections, which will be used to establish a relationship between block-entries of the fundamental matrix $\widehat{\beta P}$.

Lemma 3 *Let P be partitioned as in (3) and let βP be partitioned accordingly* *as*

$$\beta P = \begin{bmatrix} \beta T & \beta H \\ \beta L & \beta Q \end{bmatrix}, \quad 0 < \beta \le \alpha. \tag{4}$$

Assume that $I - \beta P$ is invertible. Then, the minimal nonnegative inverse $\widehat{\beta P}$ of $(I - \beta P)$ is given by

$$\widehat{\beta P} = \begin{bmatrix} (I - \beta T - \beta H \widehat{\beta Q} \beta L)^{-1}_{\min} & (I - \beta T - \beta H \widehat{\beta Q} \beta L)^{-1}_{\min} \beta H \widehat{\beta Q} \\ \widehat{\beta Q} \beta L (I - \beta T - \beta H \widehat{\beta Q} \beta L)^{-1}_{\min} & \widehat{\beta Q} + \widehat{\beta Q} \beta L (I - \beta T - \beta H \widehat{\beta Q} \beta L)^{-1}_{\min} \beta H \widehat{\beta Q} \end{bmatrix} \tag{5}$$

or equivalently,

$$\widehat{\beta P} = \begin{bmatrix} \widehat{\beta T} + \widehat{\beta T} \beta H (I - \beta Q - \beta L \widehat{\beta T} \beta H)^{-1}_{\min} \beta L \widehat{\beta T} & \widehat{\beta T} \beta H (I - \beta Q - \beta L \widehat{\beta T} \beta H)^{-1}_{\min} \\ (I - \beta Q - \beta L \widehat{\beta T} \beta H)^{-1}_{\min} \beta L \widehat{\beta T} & (I - \beta Q - \beta L \widehat{\beta T} \beta H)^{-1}_{\min} \end{bmatrix}, \tag{6}$$

where $(I - X)^{-1}_{\min} = \sum_{i=0}^{\infty} X^i$ is the minimal nonnegative inverse of $I - X$.

Remark 1 *By a sample path argument or the above lemma, we can show that the fundamental matrix is invariant under censoring. Let E be a subset of the state space. Let βP be partitioned according to E and its complement E^c as in (4). And let the fundamental matrix $\widehat{\beta P}$ of βP be expressed as in (3). Then, the fundamental matrix of the censored matrix $(\beta P)^E$ is equal to the block-entry corresponding to the states in E in the fundamental matrix $\widehat{\beta P}$.*

3 Radius of convergence and classification of states

Let α be the radius of convergence for P. If $\alpha = 1$, the classification of states is conventional. So, we are only interested in the classification of states when $\alpha > 1$. This corresponds to a further classification of the transient states. The main purpose of this section is to determine the radius of convergence α and to provide conditions on classification of the states. To pursue that, we first define the matrix $G(\beta)$ which, together with matrix $G_{1,0}(\beta)$ defined in Section 4, is referred to as the G-measure for the transition matrix P of $M/G/1$ type. The main results in this section will be expressed in terms of the G-measure through the analysis of the fundamental matrix and censored matrices $N(\beta)$ and $N_0(\beta)$. By introducing the G-measure, not only can the theoretical analysis be carried out, but it is also computable.

Partition the discounted transition matrix βP of $M/G/1$ type as in (4) with $\beta T = \beta D_1$, and βH, βL and βQ being determined accordingly. Notice that, in the partition, Q is the transition matrix of $M/G/1$ type without boundaries. Let $\widehat{\beta Q} = (\widehat{Q}_{i,j}(\beta))_{i,j=1,2,\ldots}$ be the fundamental matrix for βQ partitioned in blocks, where $\widehat{Q}_{i,j}(\beta)$ is the (i,j)th block. Write $N(\beta) = \widehat{Q}_{1,1}(\beta)$.

The matrix $G(\beta)$ is defined by

$$G(\beta) = N(\beta)\beta C_0. \tag{7}$$

$G(\beta)$ is a matrix of size m. The (r,s)th entry of $G(\beta)$ can be interpreted as the total expected discounted reward with rate β induced by hitting state (i,s) upon the process entering $L_{\leq i}$ for the first time, given that the process starts in state $(i+1,r)$.

Remark 2 *Though the matrix $G(\beta)$ is defined as the product of $N(\beta)$ and βC_0, we usually first compute $G(\beta)$ and then determine $N(\beta)$ in terms of $G(\beta)$. To do so, we need the following lemma, that says that all the other block-entries in the first block-column in $\widehat{\beta Q}$ can be explicitly expressed in terms of $N(\beta)$, the $(1,1)$st block-entry in $\widehat{\beta Q}$.*

Lemma 4 *For the fundamental matrix $\widehat{\beta Q} = (\widehat{Q}_{i,j}(\beta))_{i,j=1,2,\ldots}$,*

$$\widehat{Q}_{j,1}(\beta) = G(\beta)^{j-1}N(\beta), \quad j \geq 1. \tag{8}$$

Proof: It follows from (3) in Lemma 3 that

$$(\widehat{Q}_{2,1}(\beta)^T, \widehat{Q}_{3,1}(\beta)^T, \ldots)^T = \widehat{\beta Q}\beta L N(\beta).$$

The repeating structure and the property of skip-free-to-left of the transition matrix βQ leads to

$$(\widehat{Q}_{2,1}(\beta)^T, \widehat{Q}_{3,1}(\beta)^T, \ldots)^T = (N(\beta)^T, \widehat{Q}_{2,1}(\beta)^T, \ldots)^T \beta C_0 N(\beta).$$

The proof is completed by the above recursive expression and repeatedly using $N(\beta)\beta C_0 = G(\beta)$. ■

For the discounted transition matrix βP of $M/G/1$ type, we partition the fundamental matrix $\widehat{\beta P}$ of βP according to levels. The block-entries of $\widehat{\beta P}$ are denoted by $\widehat{P}_{i,j}(\beta)$. It is clear that to study the radius of convergence and to classify the states, it is sufficient to only consider an arbitrary block-entry in $\widehat{\beta P}$. For the block-structured transition matrix P in (1), partition P according to (3) with $T = D_1$. It suffices to consider the $(1,1)$st block-entry, denoted by $N_0(\beta)$, in $\widehat{\beta P}$. We express $N(\beta)$ in terms of $G(\beta)$ and $N_0(\beta)$ in terms of $N(\beta)$. This will enable us to determine the radius of convergence α and provide conditions for classification of the states.

Theorem 1 *For the transition matrix of $M/G/1$ type, the $(1,1)$st block-entry $N(\beta)$ in $\widehat{\beta Q}$ can be expressed as*

$$N(\beta) = [I - \sum_{k=1}^{\infty} \beta C_k G(\beta)^{k-1}]^{-1}, \tag{9}$$

or $N(\beta)$ is the fundamental matrix for $U(\beta) = \beta \sum_{k=1}^{\infty} C_k G(\beta)^{k-1}$. The $(1,1)$st block-entry $N_0(\beta)$ in $\widehat{\beta P}$ can be expressed as

$$N_0(\beta) = [I - U_0(\beta)]^{-1}, \tag{10}$$

where

$$U_0(\beta) = \beta D_1 + \sum_{k=1}^{\infty} \beta D_{k+1} G(\beta)^{k-1} N(\beta) \beta D_0, \tag{11}$$

or $N_0(\beta)$ is the fundamental matrix for $U_0(\beta)$.

Proof: Apply Lemma 3 to the discounted transition matrix βQ. It follows from (3) that $N(\beta) = \widehat{Q}_{1,1}(\beta)$ is the fundamental matrix for $\beta T + \beta H \widehat{\beta Q} \beta L$. Then,

$$\begin{aligned} \beta T + \beta H \widehat{\beta Q} \beta L &= \beta C_1 + \beta H (\widehat{Q}_{1,1}(\beta), \widehat{Q}_{2,1}(\beta), \ldots)^T \beta C_0 \\ &= \beta C_1 + \sum_{k=2}^{\infty} \beta C_k \widehat{Q}_{k-1,1}(\beta) \beta C_0. \end{aligned}$$

Noticing that $N(\beta)\beta C_0 = G(\beta)$ and using Lemma 4 will complete the proof to the first assertion.

To prove the second, apply Lemma 3 to the discounted transition matrix βP. Then, $N_0(\beta)$ is the fundamental matrix for $\beta T + \beta H \widehat{\beta Q} \beta L$, where $\beta T = \beta D_1$, $\beta H = \beta(D_2, D_3, \ldots)$, $\widehat{\beta Q}$ is the fundamental matrix of βQ and $\beta L = \beta(D_0, 0, \ldots)$. Therefore,

$$U_0(\beta) = \beta D_1 + \sum_{k=2}^{\infty} \beta D_k \widehat{Q}_{k-1,1}(\beta) \beta D_0.$$

The proof is complete by using Lemma 4. ∎

Remark 3 *It follows from the definition equation (7) and equation (9) that $G(\beta)$ satisfies the following equation:*

$$G(\beta) = \sum_{k=0}^{\infty} \beta C_k G(\beta)^k. \tag{12}$$

We can further prove that $G(\beta)$ is the minimal nonnegative solution to equation (12).

The determination of the radius of convergence α and the conditions on classification of the states given below are based on the combination of the classification result for the matrix without boundaries given by Kijima (1993) and the treatment of the boundary. For convenience, we state two results by Kijima here.

For the transition matrix P of $M/G/1$ type in (1) without boundaries, or all $D_k = C_k$ for $k = 0, 1, \ldots$, Kijima (1993) provided a method for determining the radius of convergence $\bar{\alpha}$ and showed that P is always α-transient.

Lemma 5 (Kijima) *Let $C^*(z)$ be defined by*

$$C^*(z) = \sum_{k=0}^{\infty} C_k z^k, \quad 0 \leq z < z_0. \tag{13}$$

Let $\chi(z)$ be the Perron-Frobenius eigenvalue of $C^*(z)$. If $z_0 > 1$, then there always exists a unique γ such that $\chi(z) \geq \gamma z$ for all $0 < z < z_0$, and there exists some θ with $0 < \theta \leq z_0$ such that $\chi(\theta) = \theta\gamma$. If $\theta = z_0$, then $\gamma = \chi(z_0)/z_0$. Otherwise, γ and θ can be determined by solving the simultaneous equations

$$\chi(\theta) = \gamma\theta \quad \text{and} \quad \chi'(\theta) = \gamma. \tag{14}$$

By using this lemma, Kijima was able to show the following result.

Theorem 2 (Kijima) *For the transition matrix P of $M/G/1$ type without boundaries ($D_k = C_k$ for all $k \geq 0$), if γ is the quantity determined in Lemma 5, then the radius of convergence $\bar{\alpha}$ of P satisfies $\bar{\alpha} = 1/\gamma$ and P is $\bar{\alpha}$-transient.*

Remark 4 *In fact, θ given in the above lemma is the maximal eigenvalue of the $G(\bar{\alpha})$. Makimoto (1993) obtained two types of expressions for the quasistationary distributions of the $PH/PH/c$ queue in terms of θ and γ, Li (1997) and Kijima and Makimoto (1999) generalized those results to the matrix of $GI/M/1$ type without boundaries.*

Remark 5 *Kijima (1993) also related θ and γ to the mean drift. The fact that the matrix of $M/G/1$ type without boundaries is always $\bar{\alpha}$-transient is independent of the mean drift. However, the matrix with boundaries can be α-transient, α-positive recurrent or α-null recurrent.*

For P of $M/G/1$ type in (1) with boundaries, we can perform the spectral analysis on the censored matrix to level 0 to obtain conditions on classifications of the transient states and a determination of the radius of convergence. The censored matrix can be calculated according to Lemma 3 and Remark 1 as $U_0(\beta)$. However, it seems more convenient to reach this goal by considering the relationship between the censored matrix $U_0(\beta)$ and its fundamental matrix $N_0(\beta)$.

Let $u_0(\beta)$ and $n_0(\beta)$ be the maximal eigenvalues of the censored matrix $U_0(\beta)$ and its fundamental matrix $N_0(\beta)$, respectively. Then $n_0(\beta) = \frac{1}{1-u_0(\beta)}$. It follows from results of linear algebra that the first two statements of the following lemma are true, for example, Seneta (1981), and the other two follow from the definitions of the radius of convergence and $N_0(\beta)$.

Lemma 6 *Let $\bar{\alpha}$ and α be the radii of convergence of Q and P respectively. In i) and ii), assume $0 < \beta \leq \bar{\alpha}$.*

i) *Both $u_0(\beta)$ and $n_0(\beta)$ are strictly increasing in β, and*

ii) *$u_0(\beta) < 1$ if and only if $N_0(\beta) < \infty$.*

iii) *$N_0(\beta) < \infty$ if $\beta < \alpha$ and $N_0(\beta) = \infty$ if $\beta > \alpha$.*

iv) *$\alpha \leq \bar{\alpha}$.*

The classification of the states is characterized by the following conditions.

Theorem 3

i) *If for all $0 < \beta \leq \bar{\alpha}$, $u_0(\beta) < 1$, then $N_0(\bar{\alpha}) < \infty$ and $\alpha = \bar{\alpha}$. Therefore, P is α-transient;*

ii) *If there exists a β^* with $0 < \beta^* \leq \bar{\alpha}$ such that $u_0(\beta^*) = 1$, then $\alpha = \beta^*$ and $N_0(\alpha) = \infty$. Therefore, P is α-recurrent.*

Proof: Based on the facts: $n_0(\beta) = \frac{1}{1-u_0(\beta)}$ and $n_0(\beta) < \infty$ if and only if $N_0(\beta) < \infty$, we discuss the following two cases: i) there exists no solution to $1 - u_0(\beta) = 0$ for $0 < \beta \leq \bar{\alpha}$, and ii) there exists a solution β^* to $1 - u_0(\beta) = 0$ for $0 < \beta^* \leq \bar{\alpha}$. In the first case, $n_0(\bar{\alpha}) < \infty$. Hence $N_0(\bar{\alpha}) < \infty$. Therefore, $\alpha \geq \bar{\alpha}$. This, together with iv) of Lemma 6, implies $\alpha = \bar{\alpha}$. Hence, P is α-transient. In the second case, $n_0(\beta^*) = \infty$, hence there exists at least one infinite entry of $N_0(\beta)$. This leads to $\alpha = \beta^* \leq \bar{\alpha}$. Therefore, P is α-recurrent. This completes the proof. ■

Remark 6 *Theorem 3 is also a generalization of classifying an irreducible stochastic matrix into either a recurrent or transient matrix based on censoring. For example, P is recurrent if and only if every censored matrix of P is stochastic. Therefore, the maximal eigenvalue of the censored matrix is one, or $u_0(1) = 1$. P is transient if and only if every censored finite matrix of P is strictly substochastic. Therefore, $u_0(1) < 1$ and $\alpha \geq 1$. If we replace $u_0(1)$ mentioned above by $u_0(\alpha)$, we then have the conditions for α-recurrence and α-transience.*

The above result provides a way to classify the states into either α-transient or α-recurrent and to determine the radius of convergence of P. For an α-recurrent P, the following theorem further provides conditions to determine when it is α-positive or α-null.

Theorem 4 *If $\sum_{k=1}^{\infty} kD_kG(\alpha)^{k-1} < \infty$, $\sum_{k=1}^{\infty} kC_kG(\alpha)^{k-1} < \infty$ and $\alpha < \overline{\alpha}$, then the α-recurrent Markov chain is α-positive; otherwise, it is α-null.*

Proof: This proof is long and needs results in Section 5. Therefore, it is given as an Appendix. ■

Remark 7 *If $\alpha = 1$ and $\overline{\alpha} > 1$, then, the three conditions in Theorem 4 are the same conditions as that in Remark b of Neuts (1989) (pp. 140-141). This is because in this situation, $G(1)$ is stochastic. Therefore, i) $\sum_{k=1}^{\infty} kD_kG(1)^{k-1} < \infty$ if and only if $\sum_{k=1}^{\infty} kD_k < \infty$; ii) $\sum_{k=1}^{\infty} kC_kG(1)^{k-1} < \infty$ if and only if $\sum_{k=1}^{\infty} kC_k < \infty$; and iii) for the recurrent matrix, $\alpha < \overline{\alpha}$ if and only if $I - R^*(1)$ is invertible, which is equivalent to $\rho < 1$.*

Remark 8 *If $D_k = 0$ and $C_k = 0$, $k \geq 3$, then the transition matrix in (1) is a level-independent QBD process with boundary. In this case, Theorem 4 illustrates that the α-recurrent QBD process is α-positive if and only if $\alpha < \overline{\alpha}$. A similar analysis as in Appendix will show that an α-recurrent level-dependent QBD process is α-positive if and only if $\alpha < \overline{\alpha}$. Now, we compare this result with Lemma 15 in Bean, Pollett and Taylor (2000). If $\alpha < \overline{\alpha}$, then since $\alpha < \overline{\alpha} \leq \alpha_2$, where $\overline{\alpha} = \alpha_1$, $N^{(2)}(z)$ is analytic at $z = \alpha$.Thus, $\frac{d}{d\alpha}N^{(2)}(\alpha) = \frac{d}{dz}N^{(2)}(z)_{|z=\alpha} < \infty$. On the contrary, if $\frac{d}{d\alpha}N^{(2)}(\alpha) < \infty$, then $\alpha < \overline{\alpha}$, since $N^{(2)}(\overline{\alpha}) = \infty$ and $N^{(2)}(z)$ is increasing for $1 \leq z < \alpha_2$.*

4 *RG*-factorization

The RG-factorization of $(I - P)$, where P is stochastic or strictly substochastic, is a version of LU-factorization having probabilistic interpretations. This factorization was discussed by Heyman (1995), Zhao, Li and Braun (1997, 2000), and Zhao (2000). Heyman showed how to use this factorization to determine the stationary probability vector of a positive recurrent Markov chain. When studying the quasistationary behavior of transition matrix P of $M/G/1$ type without boundaries, Li (1997) obtained an LU-factorization for $(I - \beta P)$ without using the R-measure defined in this paper.

The RG-factorization of $(I-\beta P)$ can be proved for an arbitrary transition matrix P, with or without a block-structure. However, in this paper, we only concentrate on the transition matrix of $M/G/1$ type defined in (1). We first need to define the R-measure and the matrix $G_{1,0}(\beta)$.

Consider the fundamental matrix $\widehat{\beta Q}$ of βQ. Let the first block-column of $\widehat{\beta Q}$ be $(\widehat{Q}_{1,1}(\beta)^T, \widehat{Q}_{2,1}(\beta)^T, \dots)^T$. The R-measure for the matrix βP in (1) consists of two sequences of matrices $R_{0,k}(\beta)$ and $R_k(\beta)$, $k = 1, 2, \dots$, defined by

$$R_{0,k}(\beta) = \sum_{l=1}^{\infty} \beta D_{k+l} \widehat{Q}_{l,1}(\beta) \tag{15}$$

and

$$R_k(\beta) = \sum_{l=1}^{\infty} \beta C_{k+l} \widehat{Q}_{l,1}(\beta). \tag{16}$$

The (r, s)th entry of $R_{0,k}(\beta)$ can be interpreted as the total expected discounted reward with rate β induced by all visits to state (k, s) before hitting any state in $L_{\leq k-1}$, given that the process starts in state $(0, r)$. Similarly, the (r, s)th entry of $R_k(\beta)$ can be interpreted as the total expected discounted reward with rate β induced by all visits to state $(i + k, s)$ before hitting any state in $L_{\leq i+k-1}$, given that the process starts in state (i, r), where $i \geq 1$.

The G-measure for βP of $M/G/1$ type consists of two matrices, $G(\beta)$ as defined in (7) and $G_{1,0}(\beta)$ defined by

$$G_{1,0}(\beta) = \widehat{Q}_{1,1}(\beta) \beta D_0 = N(\beta) \beta D_0. \tag{17}$$

The (r, s)th entry of $G_{1,0}(\beta)$ can be interpreted as the total expected discounted reward with rate β induced by hitting state $(0, s)$ upon the process entering level 0 for the first time, given that the process starts in state $(1, r)$.

Applying Lemma 4 to (15) and (16), the R-measure can then be expressed as

$$R_{0,k}(\beta) = \sum_{i=1}^{\infty} \beta D_{k+i} G(\beta)^{i-1} N(\beta) \tag{18}$$

and

$$R_k(\beta) = \sum_{i=1}^{\infty} \beta C_{k+i} G(\beta)^{i-1} N(\beta) \tag{19}$$

for $k = 1, 2, \ldots$.

Remark 9 *Up to now, we have obtained all components needed in the factorization equation and expressed then in terms of $G(\beta)$ only.*

For the matrix P of $M/G/1$ type with boundaries, the RG-factorization can be stated in the following theorem.

Theorem 5 *For the matrix P of $M/G/1$ type in (1), $I - \beta P$ can be factorized as*

$$I - \beta P = [I - R_U(\beta)][I - U_D(\beta)][I - G_L(\beta)], \tag{20}$$

where

$$[I - R_U(\beta)] = \begin{bmatrix} I & -R_{0,1}(\beta) & -R_{0,2}(\beta) & -R_{0,3}(\beta) & \cdots \\ & I & -R_1(\beta) & -R_2(\beta) & \cdots \\ & & I & -R_1(\beta) & \cdots \\ & & & I & \cdots \\ & & & & \ddots \end{bmatrix}, \tag{21}$$

$U_D(\beta)$ is the diagonal matrix in block-form with the first block-entry on the diagonal equal to $U_0(\beta)$ and all the other diagonal block-entries equal to $U(\beta)$, or $U_D(\beta) =$ diag $(U_0(\beta, U(\beta), U(\beta), \ldots)$, and

$$[I - G_L(\beta)] = \begin{bmatrix} I & & & & \\ -G_{1,0}(\beta) & I & & & \\ & -G(\beta) & I & & \\ & & -G(\beta) & I & \\ & & & \ddots & \ddots \end{bmatrix}. \tag{22}$$

Proof: We only prove the factorization equation for the first block row and first block-column entries. The remaining can be similarly proved.

The entry $(1, 0)$ on the right-hand side is $-[I - U(\beta)]G_{1,0}(\beta)$, which is equal to $-\beta D_0$ from the definition of $G_{1,0}(\beta)$.

The entry $(0, k)$ with $k \geq 1$ on the right-hand side is

$$\begin{aligned}
& - R_{0,k}(\beta)[I - U(\beta)] + R_{0,k+1}(\beta)[I - U(\beta)]G(\beta) \\
=& - \sum_{i=1}^{\infty} \beta D_{i+k} G(\beta)^{i-1} + \sum_{i=1}^{\infty} \beta D_{i+k+1} G(\beta)^{i-1} G(\beta) \\
=& - \beta D_{k+1},
\end{aligned}$$

where the first equality is due to Lemma 4.

Finally, to see that the entry $(0, 0)$ on the right-hand side is equal to the corresponding entry on the left-hand side, we have

$$\begin{aligned}
& [I - U_0(\beta)] + R_{0,1}(\beta)[I - U(\beta)]G_{1,0}(\beta) \\
=& [I - U_0(\beta)] + \sum_{i=1}^{\infty} \beta D_{i+1} G(\beta)^{i-1} N(\beta) \beta D_0 \\
=& [I - \beta D_1 - \sum_{k=1}^{\infty} \beta D_{k+1} G(\beta)^{k-1} N(\beta) \beta D_0] + \sum_{i=1}^{\infty} \beta D_{i+1} G(\beta)^{i-1} N(\beta) \beta D_0 \\
=& I - \beta D_1.
\end{aligned}$$

where the first equality is due to Lemma 4 and the second one due to (11). ■

Remark 10 *As we mentioned earlier, the RG-factorization can be obtained for an arbitrary transition matrix P. Therefore, the approach of this paper is still valid for using the RG-factorization to obtain expressions for the β-invariant measure of a level-dependent transition matrix of M/G/1 type.*

5 β-invariant measures

In this section, we use the RG-factorization to obtain β-invariant measures for the transition matrix P of $M/G/1$ type with boundaries, where $0 < \beta \leq \alpha$. Since the RG-factorization is a version of the LU-factorization for a matrix of infinite size, the procedure of obtaining an expression for the β-invariant measure is similar to the Gaussian elimination for solving a finite linear system. We present two sets of expressions, one for an α-recurrent matrix with $\beta = \alpha$ and the other for all the other cases. Since for an α-recurrent matrix, its α-invariant measure is unique up to a multiple of a positive constant, the solution given here is a unique solution up to a multiple of a positive constant. When P is α-transient, the β-invariant measure may not be unique. Examples and remarks will be given.

In the RG-factorization in (20), the three matrices, $[I - R_U(\beta)]$, $[I - U_D(\beta)]$ and $[I - G_L(\beta)]$, are associative. We can also prove that they are associative with any nonnegative vector π, which will lead to solutions for the β-invariant measure.

Lemma 7 *Let P be the transition matrix of $M/G/1$ type and let π be any nonnegative row vector. Then,*

$$\begin{aligned}\pi[I-\beta P] &= \{\pi[I-R_U(\beta)]\}\{[I-U_D(\beta)][I-G_L(\beta)]\} \\ &= \{\pi[I-R_U(\beta)][I-U_D(\beta)]\}[I-G_L(\beta)].\end{aligned}$$

Proof: This is clear, for example, from the sufficient conditions provided in Corollary 1-9 of Kemeny *et al.*. ■

5.1 α-recurrent with $\beta=\alpha$

In this case, we solve $\pi(I-\alpha P)=0$ by two steps. In the first step, we let

$$x=\pi[I-R_U(\alpha)]. \tag{23}$$

If $x=(x_0,x_1,\ldots)$ and $\pi=(\pi_0,\pi_1,\ldots)$ are partitioned according to levels, then (23) is equivalent to

$$\begin{aligned}x_0&=\pi_0,\\ x_k&=-\pi_0R_{0,k}(\alpha)-\sum_{i=1}^{k-1}\pi_iR_{k-i}(\alpha)+\pi_k,\quad k\geq 1.\end{aligned}$$

Expressing π_k in terms of x_k, we have

$$\pi_0=x_0, \tag{24}$$

$$\pi_k=\pi_0R_{0,k}(\alpha)+\sum_{i=1}^{k-1}\pi_iR_{k-i}(\alpha)+x_k,\quad k\geq 1. \tag{25}$$

In the second step, we solve

$$x[I-U_D(\alpha)][I-G_L(\alpha)]=0 \tag{26}$$

for a nontrivial nonnegative x. If such a solution exists, then π given in (24) and (25) will be nonnegative and nonzero. According to Lemma 7, the above π is an α-invariant measure of P and it is unique up to a multiple of a positive constant.

Equation (26) is equivalent to

$$\begin{aligned}&x_0[I-U_0(\alpha)]-x_1[I-U(\alpha)]G_{1,0}(\alpha)=0,\\ &x_k[I-U(\alpha)]-x_{k+1}[I-U(\alpha)]G(\alpha)=0,\quad k\geq 1.\end{aligned}$$

Since P is α-recurrent, it follows from Theorem 3 that the maximal eigenvalue of $U_0(\alpha)$ is $u_0(\alpha) = 1$. Therefore, for nonnegative and irreducible $U_0(\alpha)$, there exists a positive x_0 such that

$$x_0[I - U_0(\alpha)] = 0.$$

Hence, $(x_0, 0, 0, \dots)$ is a solution to (26).

Theorem 6 *If P is α-recurrent, then the unique, up to multiplication by a positive constant, α-invariant measure is given by*

$$\pi_0 = x_0, \tag{27}$$

$$\pi_k = \pi_0 R_{0,k}(\alpha) + \sum_{i=1}^{k-1} \pi_i R_{k-i}(\alpha), \tag{28}$$

where x_0 is the unique, up to a multiple of positive constant, solution to $x_0[I - U_0(\alpha)] = 0$.

We may notice that this form of solution is the same as that of the invariant measure for a recurrent Markov chain as obtained using the same procedure in Heyman (1995) or an equivalent method in Ramaswami (1988).

5.2 α-recurrent with $\beta < \alpha$ or α-transient with $\beta \leq \alpha$

In this case, we also proceed in two steps, but the matrices are associated differently. In the first step, let

$$y = \pi[I - R_U(\beta)][I - U_D(\beta)]. \tag{29}$$

This is equivalent to

$$y_0 = \pi_0[I - U_0(\beta)],$$
$$y_1 = [-\pi_0 R_{0,1}(\beta) + \pi_1][I - U(\beta)],$$
$$y_k = \left[-\pi_0 R_{0,k}(\beta) - \sum_{i=1}^{k-1} \pi_i R_{k-i}(\beta) + \pi_k\right][I - U(\beta)], \quad k \geq 2.$$

Since both $[I - U_0(\beta)]$ and $[I - U(\beta)]$ are invertible in this case, we can express π_k in terms of y_k:

$$\pi_0 = y_0[I - U_0(\beta)]^{-1}, \tag{30}$$

$$\pi_1 = \pi_0 R_{0,1}(\beta) + y_1[I - U(\beta)]^{-1}, \tag{31}$$

$$\pi_k = \pi_0 R_{0,k}(\beta) + \sum_{i=1}^{k-1} \pi_i R_{k-i}(\beta) + \pi_k[I - U(\beta)]^{-1}, \quad k \geq 2. \tag{32}$$

In the second step, solve

$$y[I - G_L(\beta)] = 0 \tag{33}$$

for nonnegative nonzero y. If such a solution exists, then π calculated by (30), (31) and (32) is nonnegative and nonzero. According to Lemma 7, the above π is a β-invariant measure of P. Though in many cases such a β-invariant measure is unique up to a multiple of a positive constant, in some other cases, it is simply not unique.

Equation (33) is equivalent to

$$y_0 - y_1 G_{1,0}(\beta) = 0,$$
$$y_k - y_{k+1} G(\beta) = 0, \quad k \geq 1.$$

In the following, we construct a nonnegative nonzero solution y to (33). First, we need the following lemma.

Lemma 8 *For every $0 < \beta \leq \alpha$, there exist a $\theta_\beta > 0$ and a nonnegative nonzero vector z such that*

$$\theta_\beta z = zG(\beta). \tag{34}$$

Proof: Since $G(\beta) \geq 0$, the maximal eigenvalue θ_β of $G(\beta)$ is nonnegative. If $\theta_\beta > 0$, then the lemma is proved by choosing z to be the left eigenvector of $G(\beta)$ associated with θ_β.

It follows from Neuts (1989), by using irreducibility of P, that $\theta_1 > 0$. Therefore, $\theta_\beta > 0$ for all $\beta \geq 1$ since $G(\beta)$ is increasing in β.

For $0 < \beta < 1$, the proof also relies on the irreducibility of P. Suppose that there were an s with $0 < s < 1$ such that $\theta_s = 0$. Then, $\theta_\beta = 0$ for all $0 < \beta \leq s$. Therefore, all the eigenvalues of $G(\beta)$, when $0 < \beta \leq s$, would be zero according to the Perron-Frobenius theorem for nonnegative matrices. It follows from the Cayley-Hamilton theorem that

$$G^m(\beta) = 0, \quad \text{for all } 0 < \beta \leq s, \tag{35}$$

where m is the size of matrix $G(\beta)$. On the other hand, according to the probabilistic interpretation of $G^m(\beta)$ and the assumption of irreducibility on P, $G^m(\beta) \neq 0$, which contradicts (35). ∎

By using Lemma 8 and letting $y_0 = zG_{1,0}(\beta)$, we can easily check that $y = (y_0, z, z/\theta_\beta, z/\theta_\beta^2, \ldots)$ is a nonnegative nonzero solution to (33). Substituting y into (30), (31) and (32), a β-invariant measure is found.

Theorem 7 *For $\beta < \alpha$ if P is α-recurrent, or for $\beta \leq \alpha$ if P is α-transient, a β-invariant measure of P is given by*

$$\pi_0 = y_0 N_0(\beta), \tag{36}$$

$$\pi_1 = z[N(\beta) + G_{1,0}(\beta)N_0(\beta)R_{0,1}(\beta)], \tag{37}$$

$$\pi_2 = \frac{z}{\theta_\beta}\{N(\beta) + G(\beta)N(\beta)R_1(\beta) \tag{38}$$

$$+G(\beta)G_{1,0}(\beta)N_0(\beta)[R_{0,1}(\beta)R_1(\beta) + R_{0,2}(\beta)]\}, \tag{39}$$

$$\pi_3 = \frac{z}{\theta_\beta^2}\{N(\beta) + G(\beta)N(\beta)R_1(\beta) + G(\beta)^2N(\beta)[R_1(\beta)^2 + R_2(\beta)] \tag{40}$$

$$+G(\beta)^2G_{1,0}(\beta)N_0(\beta)[R_{0,1}(\beta)R_2(\beta) + R_{0,1}(\beta)R_1(\beta)^2 \tag{41}$$

$$+R_{0,2}(\beta)R_1(\beta) + R_{0,3}(\beta)]\}$$

$$\cdots\cdots$$

or it can be written as one common expression for $k \geq 1$:

$$\pi_k = \frac{z}{\theta_\beta^{k-1}}\left\{ N(\beta) + \sum_{i=1}^{k-1} G(\beta)^i N(\beta) \sum_{\substack{0\leq j_1\leq\cdots\leq j_i\leq i \\ \sum_t j_t = i}} R_{j_1}(\beta)R_{j_2}(\beta)\cdots R_{j_i}(\beta) \right.$$

$$\left. +G(\beta)^{k-1}G_{1,0}(\beta)N_0(\beta)\sum_{i=1}^{k} R_{0,i}(\beta) \sum_{\substack{0\leq j_1\leq\cdots\leq j_{k-i}\leq k-i \\ \sum_t j_t = k-i}} R_{j_1}(\beta)R_{j_2}(\beta)\cdots R_{j_{k-i}}(\beta) \right\}, \tag{42}$$

where $R_0(\beta) = I$.

Remark 11 *For a QBD process with $D_i = 0$ and $C_i = 0$ for $i \geq 3$ in (1), $R_i = R_{0,i} = 0$ for $i \geq 2$ and*

$$U_0(\beta) = \beta D_1 + R_{0,1}(\beta)\beta D_0 = \beta D_1 + \beta D_2 G_{1,0}(\beta),$$

where

$$R_{0,1}(\beta) = \beta D_2 N(\beta), \; G_{1,0}(\beta) = N(\beta)\beta D_0.$$

The β-invariant measure is then given as

$$\pi_0 = y_0 N_0,$$
$$\pi_k = \frac{z}{\theta_\beta^{k-1}}[N(\beta) + \sum_{i=1}^{k-1} G(\beta)^i N(\beta) R_1(\beta)^i + G(\beta)^{k-1} G_{1,0}(\beta) N_0(\beta) R_{0,1}(\beta) R_1(\beta)^{k-1}]. \tag{43}$$

The expression (43) is the same as the one provided in Theorem 8 of Bean, Pollett and Taylor (2000). To see this, noting that in Theorem 8 of Bean, Pollett and Taylor (2000) we have the relations: $x_k = z$ for $k \geq 1$, $\rho_0 = 1$, $\rho_n = \theta_\beta$ for $n \geq 1$, $G^{(1)}(\beta) = G_{1,0}(\beta)$, $G^{(l)}(\beta) = G(\beta)$ for $l \geq 2$, and $R^{(1)}(\beta) = R_{0,1}(\beta)$, $R^{(l)}(\beta) = R(\beta)$ for $l \geq 2$. Then, by taking $l = 0$,

$$\begin{aligned} m_k &= x_k \left(\prod_{n=0}^{k-1} \rho_n^{-1}\right) \sum_{v=0}^{k} \left(\left[\prod_{u=0}^{k-v-1} G^{(k-u)}(\beta)\right] N^{(v)}(\beta) \left[\prod_{u=0}^{k-v-1} R^{(v+1+u)}(\beta)\right]\right) \\ &= \frac{1}{\theta_\beta^{k-1}} z \sum_{v=1}^{k} G(\beta)^{k-v} N(\beta) R(\beta)^{k-v} \\ &+ \frac{1}{\theta_\beta^{k-1}} z \left[G(\beta)^{k-1} G_{1,0}(\beta)\right] N_0(\beta) \left[R_{0,1}(\beta) R(\beta)^{k-1}\right] \\ &= \frac{z}{\theta_\beta^{k-1}}[N(\beta) + \sum_{i=1}^{k-1} G(\beta)^i N(\beta) R_1(\beta)^i \\ &+ G(\beta)^{k-1} G_{1,0}(\beta) N_0(\beta) R_{0,1}(\beta) R_1(\beta)^{k-1}]. \end{aligned}$$

We also remark that a QBD process can be treated as a matrix of GI/M/1 type, the same approach used in this paper for a matrix of GI/M/1 type will lead to a different expression of the β-measure from (43), e.g., see Makimoto (1993), Li (1997) and Bean, Pollett and Taylor (1998). Furthermore, this expression should be equivalent to the expression obtained in (43).

Remark 12 *For a fixed value of β, $G(\beta)$ can be effectively computed by a similar computational scheme for the case of $\beta = 1$, for example, Ramaswami (1988), Latouche (1994) and Meini (1997). When $G(\beta)$ becomes available, other matrices, including $U(\beta)$, $N(\beta)$, $U_0(\beta)$, $N_0(\beta)$, and the R-measure can be computed. Finally, the β-invariant measure π_k can be computed up to a desired index value. A detailed analysis of the computational scheme has been carried out and computational complexity has been analyzed. We omit all the details here. People should notice that significant efforts should be made for the calculation of $G(\beta)$ when $\beta \to \alpha$. This is because α*

is exactly the point where the underlying series diverges. In other words, it is exactly where the very long sample paths of very low probability get such a large reward that they start to contribute a significant amount. This means that many steps are required and this can involve terms with many exponents multiplied by each other to get terms of reasonable order.

Remark 13 *To see why we need two different sets of expressions for the β-invariant measure, let us consider the scalar case. If P is α-transient, one is not an eigenvalue of $U_0(\alpha)$. Therefore, $x_0[I - U_0(\alpha)] = 0$ only provides the trivial solution. This means that the method used for the case in 5.1 is not valid. If P is α-recurrent, y given in Section 5.2 is zero. In fact, this y cannot satisfy (29) unless $y_0 = 0$. For example, $I - U_0(\alpha) = 1 - 1 = 0$ for the scalar case, which gives $y_0 = 0$.*

While in many cases there exists a unique β-invariant measure up to a multiple of a positive constant, in some other cases, the β-invariant measure is simply not unique. One such example was provided by Gail, Hantler and Taylor (1998).

6 Concluding remarks

In this paper, we considered the matrix of $M/G/1$ type with boundaries. We generalized the censoring technique such that it can be used to deal with the nonnegative matrix βP. Based on the generalized censoring technique, we proposed a method for determining the radius of convergence, we obtained conditions for classifying transient states, and proved a factorization theorem for the matrix $I - \beta P$. This factorization was then used to obtain expressions for the β-invariant measure.

The method developed here can also be used to study the radius of convergence and β-invariant measures for transition matrices with other types of block-structure, such as, for the matrix of $GI/M/1$ type and even for the matrix of $GI/G/1$ type.

Acknowledgements

The authors thank the referees for the valuable suggestions and comments and acknowledge that this work was supported by a research grant from the Natural Sciences and Engineering Research Council of Canada (NSERC). Dr. Li also acknowledges the support from Carleton University.

Appendix

In this appendix, we provide a proof to Theorem 4, which follows from the three lemmas provided here. For simplicity, we assume that the matrix $C^*(1)$ is irreducible and stochastic.

This proof is based on a result in Theorem 6.4 in Seneta (1981), which is restated in the following lemma in the block-partitioned form.

Lemma 9 *Suppose* $\pi = (\pi_0, \pi_1, \pi_2, \cdots)$ *is a* β*-invariant measure and* $v = (v_0^T, v_1^T, v_2^T, \cdots)^T$ *is a* β*-invariant vector of the transition matrix* P*, partitioned according to levels. Then,* P *is* α*-positive if* $\pi v = \sum_{k=0}^{\infty} \pi_i v_i < +\infty$*, in which case* $\beta = \alpha$*,* π *is (a multiple of) the unique* α*-invariant measure of* P *and* v *is (a multiple of) the unique* α*-invariant vector of* P*. Conversely, if* P *is* α*-positive, and* π *and* v *are respectively an invariant measure and vector, then* $\pi v < +\infty$*.*

Based on this lemma, besides the α-invariant measure π provided in Theorem 6, we also need to similarly express the α-invariant vector v according to Subsection 5.1. This is given as

$$v_0 = w_0, \quad v_k = G(\alpha)^{k-1} G_{1,0}(\alpha)\, w_0, \; k \geq 1, \tag{44}$$

where w_0 is the unique, up to a multiplication of a positive constant, solution of $[I - U_0(\alpha)]\, w_0 = 0$.

For convenience, we express the α-invariant measure explicitly in terms of the R-measure, instead of an iterative expression given in Theorem 6. To do this, Let $\Pi^*(z) = \sum_{k=1}^{\infty} z^k \pi_k$, $R^*(z) = \sum_{k=1}^{\infty} z^k R_k(\alpha)$ and $R_0^*(z) = \sum_{k=1}^{\infty} z^k R_{0,k}(\alpha)$. We denote by $t_k * s_k$ the convolution of two sequences $\{t_k\}$ and $\{s_k\}$, and $t_k^{*n} = t_k * t_k^{*(n-1)}$, $n \geq 2$. It follows from Theorem 6 that

$$\Pi^*(z) = x_0 R_0^*(z) [I - R^*(z)]^{-1} = x_0 R_0^*(z) \sum_{n=0}^{\infty} [R^*(z)]^n,$$

which gives

$$\pi_k = x_0 R_{0,k}(\alpha) * \sum_{n=0}^{\infty} R_k(\alpha)^{*n}, \quad k \geq 1. \tag{45}$$

It follows from (44) and (45) that

$$\sum_{k=0}^{\infty} \pi_i v_i = x_0 v_0 + x_0 \sum_{k=1}^{\infty} \left[R_{0,k}(\alpha) * \sum_{n=0}^{\infty} R_k(\alpha)^{*n} \right] G(\alpha)^{k-1} G_{1,0}(\alpha) v_0. \quad (46)$$

Clearly, $\sum_{k=0}^{\infty} \pi_i v_i < \infty$ if and only if

$$\sum_{k=1}^{\infty} \left[R_{0,k}(\alpha) * \sum_{n=0}^{\infty} R_k(\alpha)^{*n} \right] G(\alpha)^{k-1} < \infty. \quad (47)$$

Let g_α and $H(\alpha)$ be the maximal eigenvalue and the associated right eigenvector of $G(\alpha)$, respectively. Since $C^*(1)$ is irreducible, we have $H(\alpha) > 0$. It follows from (47) that

$$\sum_{k=1}^{\infty} \left[R_{0,k}(\alpha) * \sum_{n=0}^{\infty} R_k(\alpha)^{*n} \right] G(\alpha)^{k-1} H(\alpha) = \frac{1}{g_\alpha} R_0^*(g_\alpha) \sum_{n=0}^{\infty} \left[R^*(g_\alpha) \right]^n H(\alpha).$$

Then, (47) is true if and only if, i) $R_0^*(g_\alpha) < \infty$, ii) $R^*(g_\alpha) < \infty$, and iii) the matrix $I - R^*(g_\alpha)$ is invertible.

The following lemma provides the conditions under which, i) $R_0^*(g_\alpha) < \infty$ and ii) $R^*(g_\alpha) < \infty$.

Lemma 10 *i)* $R_0^*(g_\alpha) < \infty$ *if and only if* $\sum_{k=1}^{\infty} k D_k G(\alpha)^{k-1} < \infty$. *ii)* $R^*(g_\alpha) < \infty$ *if and only if* $\sum_{k=1}^{\infty} k C_k G(\alpha)^{k-1} < \infty$.

Proof: We only prove i) and ii) can be similarly proved.

It follows from (20) that

$$R_0^*(g_\alpha) = \sum_{k=1}^{\infty} g_\alpha^k R_{0,k}(\alpha) = \sum_{k=1}^{\infty} g_\alpha^k \sum_{i=1}^{\infty} \alpha D_{k+i} G(\alpha)^{i-1} N(\alpha).$$

Hence we obtain

$$R_0^*(g_\alpha) N(\alpha)^{-1} H(\alpha) = \sum_{k=1}^{\infty} \sum_{i=1}^{\infty} \alpha g_\alpha^{k+i-1} D_{k+i} H(\alpha) = \alpha \sum_{k=1}^{\infty} k g(\alpha)^{k-1} D_k H(\alpha),$$

which illustrates that $R_0^*(g_\alpha) < \infty$ if and only if $\sum_{k=1}^{\infty} k g(\alpha)^{k-1} D_k < \infty$, and if and only if $\sum_{k=1}^{\infty} k D_k G(\alpha)^{k-1} < \infty$. ■

In what follows we provide a condition under which, iii) the matrix $I - R^*(g_\alpha)$ is invertible.

For the discounted matrix αP, a similar analysis to that used by Zhao, Li and Braun (2001) to obtain the RG-factorization in Theorem 14 (also see Theorem 11 in Zhao (2000)) leads to

$$zI - \alpha C^*(z) = [I - R^*(z)][I - U(\alpha)][zI - G(\alpha)]. \tag{48}$$

Let $\chi(z)$ be the maximal eigenvalue of the matrix $C^*(z)$ for $z > 0$. It is clear that property 7 about $\chi(z)$ in Bean, Pollett and Taylor (1998) (pp. 393-394) also holds for the transition matrix of $M/G/1$type. Noting that the matrix $C^*(1)$ is irreducible and stochastic, then the equation $z = \alpha\chi(z)$ has two different roots in $(0, z_0)$ if $1 \leq \alpha < \overline{\alpha}$, and one root repeated twice in $(0, z_0)$ if $\alpha = \overline{\alpha}$, where z_0 is the radius of convergence of $C^*(z)$. Furthermore, the equation $\det(zI - \alpha C^*(z)) = 0$ has two different roots in $(0, z_0)$ if $1 \leq \alpha < \overline{\alpha}$, and one root repeated twice in $(0, z_0)$ if $\alpha = \overline{\alpha}$.

Lemma 11 *i) If $\alpha < \overline{\alpha}$, then the matrix $I - R^*(g_\alpha)$ is invertible. ii) If $\alpha = \overline{\alpha}$, then the matrix $I - R^*(g_\alpha)$ is singular.*

Proof: i) If $\alpha < \overline{\alpha}$, then the equation $\det(zI - \alpha C^*(z)) = 0$ has two different roots in $(0, z_0)$. Since

$$\{0 < z < z_0 : \det(zI - \alpha C^*(z)) = 0\} = \{0 < z < z_0 : \det(I - R^*(z)) = 0\} \cup \{0 < z < z_0 : \det(zI - G(\alpha)) = 0\}$$

and $z = g_\alpha$ is a positive root to the equation $\det(zI - G(\alpha)) = 0$, it is not a positive root to the equation $\det(I - R^*(z)) = 0$. Thus, $I - R^*(g_\alpha)$ is invertible.

ii) If $\alpha = \overline{\alpha}$, then the equation $\det(zI - \alpha C^*(z)) = 0$ has one root repeated twice in $(0, z_0)$. Since $z = g_\alpha$ is a positive and simple root to the equation $\det(zI - G(\alpha)) = 0$, it must be a positive and simple root to the equation $\det(I - R^*(z)) = 0$. Thus, $I - R^*(g_\alpha)$ is singular. ∎

References

1. S. Asmussen and V. Ramaswami, Probabilistic interpretations of some duality results for the matrix paradigms in queueing theory. *Stochastic Models*, **6**, 715–733, 1990.
2. N.G. Bean, L. Bright, G. Latouche, C.E.M. Pearce, P.K. Pollett and P.G. Taylor, The quasi-stationary behavior of quasi-birth-and-death processes. *Ann. of Appl. Prob.*, **7**, 134–155, 1997
3. N.G. Bean, P.K. Pollett and P.G. Taylor, The quasistationary distributions of level-independent quasi-birth-and-death processes. *Stochastic Models*, **14**, 389–406, 1998.

4. N.G. Bean, P.K. Pollett and P.G. Taylor, Quasistationary distributions for level-dependent quasi-birth-and-death processes. *Stochastic Models*, **16**, 511–541, 2000.
5. L. Bright, *Matrix-Analytic Methods in Applied Probability*. Ph.D. Thesis, University of Adelaide, Australia, 1996.
6. E. Çinclar, *Introduction to Stochastic Processes*. Prentice-Hall, 1975.
7. C. Derman, Some contributions to the theory of denumerable Markov chains. *Trans. Amer. Math. Soc.*, **79**, 541–555, 1955.
8. H.R. Gail, S.L. Hantler and B.A. Taylor, Matrix-geometric invariant measures for $G/M/1$ type Markov chains. *Stochastic Models*, **14**, 537–569, 1998.
9. W.K. Grassmann and D.P. Heyman, Equilibrium distribution of block-structured Markov chains with repeating rows. *J. Appl. Prob.*, **27**, 557–576, 1990.
10. T.E. Harris, Transient Markov chains with stationary measures. *Proc. Amer. Math. Soc.*, **8**,937–942, 1957.
11. D.P. Heyman, A decomposition theorem for infinite stochastic matrices. *J. Appl. Prob.*, **32**, 893–901, 1995.
12. C.S. Holling, Resilience and stability of ecological systems. *Ann. Rev. Ecol. Systematics*, **4**, 1–23, 1973.
13. F.P. Kelly, *Reversibility and Stochastic Networks*. Wiley, London, 1979.
14. F.P. Kelly, Invariant measures and the Q-matrix, probability, statistics and analysis. *London Math. Soc. Lecture Notes*, Kingman, J.F.C. and Reuter, G.E.H. (eds), **79**, 143–160, 1983.
15. J.G. Kemeny, J.L. Snell and A.W. Knapp, *Denumerable Markov Chains,* 2nd edn, Springer-Verlag, New York, 1976.
16. M. Kijima, Quasi-stationary distributions of single-server phase-type queues. *Math. of Oper. Res.*, **18**, 423–437, 1993.
17. M. Kijima, *Markov Processes for Stochastic Modeling*. Chapman & Hall, London, 1997.
18. M. Kijima and N. Makimoto, Quasi-stationary distributions of Markov chains arising from queueing processes. *Applied probability and stochastic processes*, J. G. Shanthikumar and U. Sumita (eds), 277-311, Kluwer Academic Publishers, 1999.
19. G. Latouche, Algorithms for infinite Markov chains with repeating columns. *Linear Algebra, Queueing Models and Markov Chains*, Meyer, C.D. and Plemmons, R.J. (eds), 231–265, Springer-Verlag, New York, 1993.
20. G. Latouche, C.E.M. Pearce and P.G. Taylor, Invariant measures for quasi-birth-and-death processes. *Stochastic Models*, ***14***, 443–460, 1998.

21. G. Latouche and V. Ramaswami, *Introduction to Matrix Analytic Methods in Stochastic Modeling.* SIAM, Philadelphia, 1999.
22. Q.L. Li, *Stochastic Integral Functionals and Quasi-Stationary Distributions in Stochastic Models.* Ph.D. Thesis, Inst. of Appl. Math, Chinese Academy of Sciences, China, 1997.
23. N. Makimoto, Quasi-stationary distributions in a $PH/PH/c$ queue. *Stochastic Models*, **9**, 195–212, 1993.
24. B. Meini, An improved FFT-based version of Ramaswami' formula. *Stochastic Models*, **13**, 223–238, 1997.
25. M.F. Neuts, *Structured Stochastic Matrices of $M/G/1$ Type and Their Applications.* Marcel Decker Inc., New York, 1989.
26. I. Oppenheim, K.E. Shuler and G.H. Weiss, Stochastic theory of nonlinear rate processes with multiples stationary states. *Phys.*, A *88*, 191–214, 1977.
27. A.G. Pakes, Limit theorems for the population size of a birth and death process allowing catastrophes. *J. Math. Biol.*, **25**, 307–325, 1987.
28. R.W. Parsons and P.K. Pollett, Quasistationary distributions for autocatalytic reactions. *J. Statist. Phys.*, **46**, 249–254, 1987.
29. P.K Pollett, On the long-term behaviour of a population that is subject to large-scale mortality or emigration. *Proceedings of the 8th National Conference of the Australian Society for Operations Research*, Kumar, S. (ed), 196–207, 1987.
30. P.K. Pollett, Reversibility, invariance and μ-invariance. *Adv. in Appl. Prob.*, **20**, 600-621, 1988.
31. V. Ramaswami, A duality theorem for the matrix paradigm in queueing theory. *Stochastic Models*, **6**, 151–161, 1990.
32. V.B. Scheffer, The rise and fall of a reindeer herd. *Sci. Monthly*, **73**, 356–362, 1951.
33. P. Schrijner, *Quasi-Stationarity of Discrete-Time Markov Chains.* Ph.D. Thesis, University of Twente, The Netherlands, 1995.
34. E. Seneta, *Non-Negative Matrices and Markov Chains.* Springer-Verlag, New York, 1981.
35. A.M. Yaglom, Certain limit theorems of the theory of branching stochastic processes (in Russian). *Dokl. Akad. Nauk SSSR*, **56**, 795–798, 1947.
36. Y.Q. Zhao, Censoring technique in studying block-structured Markov chains. To appear in *Proceedings of The Third International Conference on Matrix-Analytic Methods*, 2000.
37. Y.Q. Zhao, W. Li and A.S. Alfa, Duality results for block-structured transition matrices. *J. Appl. Prob.*, **36**, 1045–1057, 1999.
38. Y.Q. Zhao, W. Li and W.J. Braun, On a decomposition for infinite tran-

sition matrices. *Queueing Systems*, **27**, 127–130, 1997.

39. Y.Q. Zhao, W. Li and W.J. Braun, Infinite block-structured transition matrices and their properties. *Adv. Appl. Prob.*, **30**, 365–384, 1998.
40. Y.Q. Zhao, W. Li and W.J. Braun, Correction to: On a decomposition for infinite transition matrices. *Queueing Systems*, **35**, 399, 2000.
41. Y.Q. Zhao, W. Li and W.J. Braun, Censoring, factorization, and spectral analysis for transition matrices with block-repeating entries. Technical report (No. 355), Laboratory for Research in Statistics and Probability, Carleton University and University of Ottawa, 2001.

A PARADIGM OF MARKOV ADDITIVE PROCESSES FOR QUEUES AND THEIR NETWORKS

MASAKIYO MIYAZAWA
Department of Information Sciences, Tokyo University of Science, Noda, Chiba 278-8510, Japan
E-mail: miyazawa@is.noda.tus.ac.jp

From the title of this paper, one may wonder why we need to go back to the Markov additive process, which had been actively studied long time ago. In the theory of stochastic processes, this may be a right question. However, the situation is different in the queueing theory. As long as the author knows, a modeling power of this additive process has not yet been fully utilized. Recent developments of the queueing theory are motivated by algorithmic solutions and rare events. For example, the matrix geometric approach belongs to the former class, while the large deviation technique belongs to the latter. We are also compelled by those two motivations. It is apparent that certain additive structure is always a clue to study them. For example, phase transitions have a nice additive structure in the $M/G/1$ type queues. So far, we shall attempt to discover new things by studying the older MAP, i.e., the Markov additive process, through scrutiny of the old. The reader will see nice structure of the Markov additive process in queues and their networks.

1 MAP is not MAP in stochastic processes

In the queueing theory, MAP is one of most popular technical terms today. This MAP stands for Markovian Arrival Process, originally proposed by Neuts[26] and developed by his colleagues (see, e.g., Ramaswami[27] and Lucantoni[20]). It is an additive process governed by a continuous-time Markov chain with a finite state space, possibly together with auxiliary jump distributions. However, MAP has been used in a different sense. In the theoretical study of stochastic processes, it stands for Markov additive process, which is a much more general process than the MAP, currently used in the queueing theory. Namely, the MAP of queueing people is a special case of the MAP of stochastic people. In the recent book of Asmussen[6], the former MAP is rephrased as *MArP*. We here refer to it as it is or *MAP in queues*, while the MAP of stochastic people is full spelled as *Markov additive process* to prevent possible confusions.

The Markov additive process has been studied mainly from theoretical interests for considering its subordinated processes such as ladder processes and excursions, while the MAP in queues has been used for modeling analysis. This MAP in queues is often combined with matrix computations, whose algorithms are of major interest (see, e.g. Takine and Hasegawa[33]). It is true that the MAP in queues covers a large class of additive processes in the sense of approximations in distribution. However, this does not mean that the additive processes that are obtained as limits of the MAP in queues are well studied through the MAP themselves. One can easily find counter examples. Furthermore, the framework of the MAP in queues is too restrictive to describe a wide class of queueing models including queueing networks.

The purpose of this paper is to revisit structural properties of the Markov additive process and to combine them with algorithmic approach of the MAP in queues in order to get stationary distributions in queues and their networks. We also study asymptotic behavior of the tails of those distributions, using this approach.

Roughly speaking, the Markov additive process is composed of two components, a background Markov process and an additive process that is a functional of the background Markov process and an independent incremental process. We assume that the additive process is real valued, and study the background processes indexed by the ascending and descending ladder heights of the additive process. Assuming that the additive process is skip free to the downward direction, we observe that the background process indexed by the ladder height is a Markov process. We aim to identify the generator of this Markov process, which can be used to compute the hitting probabilities. These probabilities are known to be useful to get the stationary distribution of interest in queueing models.

We are not concerned with full generality of the Markov additive process, but limited to the case that the Markov additive process is driven from the MAP in queues. This MAP is basically used to describe the inputs of queues and their networks as well as service speeds in them. There are two reasons for this. First, it may be only feasible to describe the input by a finite number of parameters and a finite number of distributions. Secondly, it simplify analyses, although the background process may be still complicated.

The reader may wonder why the Markov additive process with one dimensional additive process is useful for multi-dimensional queues such as queueing networks. A key issue here is that the background process can have a countable or more general state space. Hence, all the complexities arising in multi-dimensional queueing phenomena can be pushed into the background states. An important observation here is that there are a variety of choices for the additive component and the corresponding background process. Unlike the conventional approach for multi-dimensional queues, we need to consider a set of one dimensional distributions to see properties of the multidimensional distribution. For instance, this may be an only way to rightly consider decay rates in a multidimensional distribution. Of course, there would be hard problems to pursue this approach since the background states become complicated. Indeed, we are under way in several places, but hope to overcome them by further studies on the present line.

This paper is organized by seven sections. In Section 2, we introduce the Markov additive process that is driven by the MAP in queues. Four examples of the Markov additive process in queueing models are discussed in Section 3. Section 4 is devoted for identifying the generator of the background process that is indexed by the descending ladder height. An identity related to the Wiener-Hope factorization is obtained in Section 5. Section 6 review Markov renewal approach on the asymptotic behaviors of the hitting probabilities. Finally, some concluding remarks are given in Section 7.

2 Markov additive process driven by MAP in queues

We first introduce the MAP in queues. We also introduce the Markov additive process, but we specialize it in such a way that it is driven by the MAP. In queueing applications, this means that the MAP describes an input process while the Markov additive process describes a whole system behavior that has this input.

Let $M(t)$ be a continuous-time Markov chain with a finite state space S_0. Let C_{ij} for $i \neq j$ and D_{ij} for all $i, j \in S_0$ be nonnegative numbers and let

$$C_{ii} = -\left(\sum_{j\neq i} C_{ij} + \sum_{j\in S_0} D_{ij}\right), \qquad i \in S_0.$$

For convenience, it is assumed that $c(i) \equiv -C_{ii} > 0$ for all $i \in S_0$. Let C and D be the matrices whose ij entry is C_{ij} and D_{ij}, respectively. We choose $S_0 \times S_0$ matrix $C + D$ for the rate matrix of Markov chain $M(t)$, and assume that $C + D$ is irreducible. Since S_0 is finite, this irreducibility implies that the Markov chain has the unique stationary distribution, which is denoted by $\{\pi_i\}$. This distribution is also denoted by row vector $\boldsymbol{\pi}$ whose i-th entry is π_i.

Note that the Markov chain $M(t)$ may include transitions that do not change their states, since D_{ii} may be positive. Of course, these transitions are irrelevant to the sample path of $M(t)$, but we do need them to describe jumps associated with $M(t)$. Let N be the point process generated by all the transition instants due to C and D. Let S_J be a finite dimensional vector space. We assume that the transition due to D_{ij} generates a positive jump which takes a value in S_J subject to distribution function F^{D}_{ij}, where $F^{D}_{ij}(\mathbf{0}) = 0$, where $\mathbf{0}$ is null vector in S_J. This jump is independent of everything else.

Define $D_{ij}(\boldsymbol{x})$ as

$$D_{ij}(\boldsymbol{x}) = D_{ij}F^{\mathrm{D}}_{ij}(\boldsymbol{x}), \qquad i,j \in S_0, \boldsymbol{x} \in S_{\mathrm{J}}.$$

Thus $M(t)$ together with those jumps constitutes a marked point process whose marks are states of $M(t)$ and jump sizes. Denote this marked point process by (M, J), where $J(t)$ denotes the latest jump size at or before time t. Throughout the paper, we assume that $M(t)$ and therefore (M, J) are stationary processes on the whole line $\mathbb{R} = (-\infty, +\infty)$. As usual, $M(t)$ is assumed to be right continuous. Thus, we can choose a probability space $(\Omega, \mathcal{F}, P)$ on which there is a shift operator group $\{\theta_t\}$ such that

$$\begin{aligned} &P(\{\theta_t(\omega) \in A\}) = P(A), \qquad A \in \mathcal{F}, \\ &M(t)(\theta_s(\omega)) = M(t+s)(\omega), \\ &J(t)(\theta_s(\omega)) = J(t+s)(\omega), \end{aligned}$$

for $\omega \in \Omega, s, t \in \mathbb{R}$. For the operations of θ_s, we shall use an abbreviate notation such that $M(t) \circ \theta_s$ for $M(t)(\theta_s(\omega))$.

We next define the Markov additive process driven by (M, J). Let $X(t)$ be a Markov process with a state space S such that

(2.a) for a given $X(s)$ with $s \geq 0$, $\{X(u+s); 0 \leq u \leq t\}$ is a functional of $\{(M(u+s), J(u+s)); 0 \leq u \leq t\}$, and this functional is independent of s.

Condition (2.a) implies that $X(t)$ has a time homogenous transition probabilities and $X(s) \circ \theta_t = X(s+t)$, if we define $X(0) \circ \theta_t = X(t)$. Usually, $X(t)$ includes $M(t)$ as a component. Let $Y(t)$ be a real valued right-continuous process such that

(2.b) $Y(0) = 0$,

(2.c) $Y(s+t) - Y(s)$ is real valued, and has independent increments for given $\{(X(u), J(t)); s < u \leq s+t\}$ for each $s, t \geq 0$, and does not depend on s for given $X(s)$.

Condition (2.c) implies that $Y(t)$ is an additive process, i.e., under the convention that $X(0) \circ \theta_t = X(t)$, we have

$$Y(s) \circ \theta_t = Y(s+t) - Y(t), \qquad s, t \geq 0. \tag{1}$$

Then, $\{(X(t), Y(t)); t \geq 0\}$ together with (M, J) is referred to as a *Markov additive process* driven by the MAP (M, J). Furthermore, $\{X(t)\}$ is said to be a background process. If condition (2.a) is removed and if $J(t)$ in

(2.c) is dropped, then we have the standard Markov additive process with the real valued additive process in the literature (see Çinler[9,10] for a formal definition). In general, $Y(t)$ may be vector valued, but we do not need this generality. Roughly speaking, $Y(t)$ is an independently incremental process, given sample paths of $X(t)$. An important thing here is the simple structure of $Y(t)$, given $X(t)$ and $J(t)$, where $X(t)$ accommodates all the remaining complexity of characteristics of interest.

Markov additive process $\{(X(t), Y(t)); t \geq 0\}$ driven by (M, J) is determined by transition probabilities:

$$p_x^{(t)}(A, B) \equiv P(X(t) \in A, Y(t) \in B | X(0) = x),$$
$$t \geq 0, x \in S, A \in \mathcal{S}, B \in \mathcal{B}(\mathbb{R}),$$

where $\mathcal{B}(\mathbb{R})$ is the Borel field on $\mathbb{R}$. In this paper, we shall not formally use a filtration, which is a sequence of sigma-fields on the sample space Ω, but it is helpful to make arguments transparent. So we briefly introduce it. Let $\{\mathcal{F}_t; t \geq 0\}$ be a filtration such that $\mathcal{F}_t$ includes the sigma field $\sigma((X(u), M(u)); 0 \leq u \leq t)$, i.e., includes all events generated by $X(u)$ and $M(u)$ for $0 \leq u \leq t$. This $\mathcal{F}_t$ is chosen in such a way that all events on $Y(u)$ for $u \leq t$ belongs to it. Since $Y(t)$ may have independently incremental components due to the jump process J, $\mathcal{F}_t$ is larger than $\sigma((X(u), M(u)); 0 \leq u \leq t)$ in general.

Another convenient notion is a stopping time. A nonnegative random variable T is said to be a stopping time with respect to $\{\mathcal{F}_t\}$, or simply a stopping time, if $\{T \leq t\} \in \mathcal{F}_t$ for all $t \geq 0$. In our discussions, we do not need to use a stopping time manifestly, but this notion enables us to see what are essential features in arguments (see Section 5). In our applications, the state space S is finite, countable, or finite dimensional Euclidean vector space, so it is equipped with natural topologies. As usual, it is assumed without loss of generality that $X(t)$ and $Y(t)$ are right-continuous and have right-hand limits in t with respect these topologies.

Yet another important notion is a dual process for the Markov additive process. To define the dual process, we need to extend Markov process $X(t)$ to the negative time direction. For $t > 0$, we define $\tilde{X}(t)$ in such a way that $\tilde{X}(t) = X(0) \circ \theta_{-t}$ when $\tilde{X}(0) = X(t)$. In other words, $\{\tilde{X}(t-u); 0 \leq u \leq t\}$ is the same functional as for $\{X(u); 0 \leq u \leq t\}$ but with variable $\{M(-u); 0 \leq u \leq t\}$ instead of $\{M(u); 0 \leq u \leq t\}$. By condition (2.a), $\tilde{X}(t)$ is well defined for $t \geq 0$, and a Markov process. Similarly but changing a sign, we define

$$\tilde{Y}(t) = -(Y(0) - Y(-t)).$$

Since $\tilde{Y}(t) = -Y(t) \circ \theta_{-t}$ by (1), we have

$$\begin{aligned}\tilde{Y}(s) \circ \theta_{-t} &= -Y(s) \circ \theta_{-s-t} \\ &= -(Y(s) + Y(t) \circ \theta_s) \circ \theta_{-s-t} + Y(t) \circ \theta_{-t} \\ &= -Y(s+t) \circ \theta_{-s-t} + Y(t) \circ \theta_{-t} \\ &= \tilde{Y}(s+t) - \tilde{Y}(t) \qquad\qquad s, t \geq 0,\end{aligned}$$

where $X(0) \circ \theta_t = X(t)$ is assumed. Thus, $\tilde{Y}(t)$ is an additive process in reversed time. We refer to $(\tilde{X}(t), \tilde{Y}(t))$ as a dual process of $(X(t), Y(t))$. Throughout the paper, tilde ˜ indicates a corresponding characteristics of the dual process.

Clearly, the dual process is again a Markov additive process driven by $(\tilde{M}, \tilde{J})$, where $(\tilde{M}(t), \tilde{J}(t)) \equiv (M(-t), J(-t))$. Since $M(t)$ is stationary with the stationary distribution $\boldsymbol{\pi}$, Markov process $\tilde{M}(t)$ is also stationary with $\boldsymbol{\pi}$, and has the following transition rates $\tilde{C}$ and $\tilde{D}(x)$ corresponding with C and $D(x)$.

$$\tilde{C} = \Delta_{\boldsymbol{\pi}}^{-1} C^{\mathrm{T}} \Delta_{\boldsymbol{\pi}}, \qquad \tilde{D}(\boldsymbol{x}) = \Delta_{\boldsymbol{\pi}}^{-1} D(\boldsymbol{x})^{\mathrm{T}} \Delta_{\boldsymbol{\pi}},$$

where $\Delta_{\boldsymbol{a}}$ is $S_0 \times S_0$ diagonal matrix whose i-th diagonal entry is given by the i-th entry of vector $\boldsymbol{a}$, and T in C^{T} stands for the transpose of matrix C.

It is obvious to see that the MAP in queues is a special case of the Markov additive process. In this paper, we are mainly concerned with continuous time processes. However, it is immediate to define a discrete time Markov additive process by replacing continuous time t by discrete time n. To prevent possible confusions, we write discrete time processes like X_n and Y_n.

3 Examples

Although the Markov additive processes may not have been well recognized in queueing models, they have indeed appeared there. For example, the workload process in the $MAP/G/1$ queue and the buffer content process in a Markov modulated fluid queue are simple functions of them (see Example 3.1 and Example 3.3 below). In those Markov additive processes, there are three key assumptions: the background state space is finite, and the additive process is real valued and skip free to either downward or upward. For the discrete time case, similar properties hold for the $M/G/1$ type queues, in which the additive component is integer valued and represents phases of the whole state space (see, e.g., Neuts[26] and Takine[32]).

In this section, we show that the Markov additive processes can be found not only single node queues but also networks of queues. Here, we can retain the two properties that the additive process is real valued and skip free at least to one direction. However, the background state space is no longer finite, which is a cost to pay for accommodating a larger class of queueing models, particularly, queueing networks. In the following examples, the jumps are positive numbers, i.e., $S_{\mathrm{J}} = [0, \infty)$.

Example 3.1 ($MAP/G/1$ queue) Let $W(t)$ be the workload process of a single server queue with the MAP input process (M, J), which is called $MAP/G/1$ queue. Let $Y(t)$ be defined as

$$Y(t) = \sum_{n=1}^{N(t)} J_n - t,$$

where J_n is the jump size at the n-th jump instant after time t. Then, $W(t)$ satisfies

$$W(t) = W(0) + Y(t) + \int_0^t 1(W(u-) = 0)du,$$

which has the solution

$$W(t) = Y(t) + \max\{W(0), -\inf_{0 \le u \le t} Y(u)\}.$$

In particular, $W \equiv \sup\{\tilde{Y}(u); u \geq 0\}$ is subject to the stationary distribution of $W(t)$. So, the stationary distribution is obtained by

$$P(W > x) = P(\tilde{\tau}_x^{0-} < \infty), \qquad x \geq 0. \tag{2}$$

Let $X(t) = M(t)$, then $(X(t), Y(t))$ is a Markov additive process driven by (M, J). Obviously, $(W(t), M(t))$ is generated by this Markov additive process.

Example 3.2 ($MAP/M/1 \to /G/1$ queue) This is a tandem queue with single servers which has MAP arrivals with exponential service times at the first queue and independently and identically distributed service times at the second queue. Let $L(t)$ be the number of customers at the first queue. Let $Y(t)$ be defined as

$$Y(t) = \sum_{n=1}^{N_d(t)} J_n^{(2)} - t,$$

where $N_d(t)$ is the number of departures from the first queue up to time t, and $J_n^{(2)}$ is the n-th service time of the second queue, which is independent of everything else. Let $X(t) = (M(t), L(t))$, then $(X(t), Y(t))$ is a Markov additive process driven by M. In this case, $S = S_0 \times \mathbb{Z}_+$, where $\mathbb{Z}_+$ is the set of nonnegative integers. Similar to the previous example, $(X(t), Y(t))$ generates the workload process $W(t)$ at the second queue. Joint process $(L(t), W(t))$ may be interesting in applications.

Example 3.3 (Markov modulated fluid queue with batch fluids) This is a fluid queue with a leaky bucket and continuous as well as batch fluid inputs. These inputs are controlled by the Markov process $M(t)$. Let $v(i)$ be the difference of the input and output rates when the background state is i and the bucket is not empty. When $M(t)$ has a transition with an associated jump, the queue receives a batch fluid whose quantity is the jump size. Thus, we define $Y(t)$ as

$$Y(t) = \int_0^t v(M(u))du + \sum_{n=1}^{N(t)} J_n,$$

where J_n is the jump size of the n-th arrival batch. Then $(X(t), Y(t))$ with $X(t) = M(t)$ is a Markov additive process driven by (M, J). As in Example 3.1, $Y(t)$ generates a buffer content process $W(t)$. This model includes the model of Example 3.1 as a special case in which $v(i) \equiv -1$ and the conventional Markov modulated fluid model (see, e.g., Elwalid and Stern[13]).

Example 3.4 (Markov modulated fluid network) Consider a network of m fluid queues. These queues are called nodes and numbered as $1, 2, \ldots, m$. All the exogenous input rates and the output rates for non-empty buffers are specified by $M(t)$. Namely, those at node k are $p_k(i)$ and $q_k(i)$ for $k = 1, 2, \ldots, m$ when the background state is i. Let $r_{k\ell}$ be a routing ratio from node k to ℓ for $k, \ell = 0, 1, \ldots, m$, where node 0 represents the outside. It is assumed that $\sum_{\ell=0}^{m} r_{k\ell} = 1$. Namely, $r_{k\ell}$ is the portion of the fluid that goes from node k to node ℓ. For simplicity, it is assumed that $r_{kk} = 0$ for all k. Let $d_k(i)$ be the total arrival rate at node k when the background state is i. Clearly,

$$d_k(i) = p_k(i) + \sum_{k=1}^{m} q_j(i) r_{jk}.$$

Let $W_k(t)$ be the buffer content at node k, and let $\boldsymbol{W}(t) = W_1(t), \ldots, W_m(t))$. It is easy to see that $(M(t), \boldsymbol{W}(t))$ is a Markov process. Let $(M(t), \boldsymbol{V}(t))$ be another Markov process such that its k-th entry is:

$$V_k(t) = \int_0^t (d_k(M_k(u)) - q_k(M_k(u))du, \qquad k = 1, 2, \ldots, m,$$

Let $\boldsymbol{c} = (c_1, \ldots, c_m)$ be a arbitrarily given nonnegative vector with unit length, and let $Y(t) = \max\{u \geq 0; V_k(t) - uc_k \geq 0, k = 1, \ldots, m\}$, and define $X(t)$ as

$$X(t) = (M(t), \boldsymbol{V}(t) - Y(t)\boldsymbol{c}).$$

Then, $(X(t), Y(t))$ is a Markov additive process driven by M. Note that the vector $Z(t) \equiv \boldsymbol{V}(t) - Y(t)\boldsymbol{c}$ in $X(t)$ is always in the set $\mathcal{L} \equiv \cup_{\emptyset \neq U \subset \{1,2,\ldots,m\}} \mathcal{L}_U$, where $\mathcal{L}_U = \{\boldsymbol{x} \geq \boldsymbol{0}; x_k = 0, k \in U\}$. $\mathcal{L}$ is called a phase space. Thus, $S = S_0 \times \mathcal{L}$. It is not hard to see that joint process $(M(t), \boldsymbol{W}(t))$ is generated by $(X(t), Y(t))$ with boundary conditions on $\mathcal{L}$.

4 Ladder height indexed processes

In this section, we consider the hitting probabilities of the additive process at given levels, i.e., the probabilities that the additive process $Y(t)$ is firstly across the levels. These hitting probabilities may also include the distribution of the background state at the hitting instant. Those probabilities can be computed from the joint distributions of the ladder epoch and ladder height together with the background state. In particular, those distributions for the dual process are useful to get the stationary distributions of interests in many applications (see, e.g., Examples 3.1 and 3.3). Since the dual process is also a Markov additive process, it is easy to transform results on the original Markov additive process to its dual. So, we are mainly concerned with the Markov additive process $(X(t), Y(t))$ itself.

As we mentioned in Section 3, we are particularly interested in the case that the additive process $Y(t)$ is skip free at least for one direction. So, we shall assume it later, but we start without this assumption. For each $x \geq 0$, define hitting times τ_x^+ and τ_x^{0-} as

$$\tau_x^+ = \inf\{u > 0; Y(u) > x\}, \qquad \tau_x^{0-} = \inf\{u > 0; Y(u) \leq -x\},$$

where these hitting times take ∞ if there are no time u to satisfy the conditions. We refer to τ_x^+ and τ_x^{0-} as the hitting times of the additive process $Y(t)$ at levels $x \geq 0$ and $-x \leq 0$, respectively.

Assume that the Markov additive process $(X(t), Y(t))$ is strong Markov with respect to $\mathcal{F}_t$, which is automatically satisfied in many cases in our applications.

Then, we define the following processes indexed by $x \geq 0$.

$$\xi^+(x) \equiv (X(\tau_x^+), Y(\tau_x^+)), \qquad \xi^{0-}(x) \equiv (X(\tau_x^{0-}), Y(\tau_x^{0-}))$$

These process are Markov by the strong Markov property, and well defined for all $x \geq 0$ such that $\tau_x^+ < \infty$ and $\tau_x^{0-} < \infty$, respectively. Namely, they may be terminated within finite times. Our main problem is to compute the probabilities that $\xi^+(x)$ and $\xi^{0-}(x)$ take values in appropriate bounded sets.

Since the level indexed processes $\xi^+(x)$ and $\xi^{0-}(x)$ are Markov, the problem is reduced to compute generators of these Markov processes. In general, this is not an easy task, but the skip free property of the additive component makes it to be worked out in many cases, as we aforementioned.

So, we assume the following assumption from now on.

(4.a) $Y(t)$ is skip free to the downward.

Observe that $X(\tau_x^{0-})$ itself is a Markov process under this assumption. If the background state space is finite or countable, then the generator of the Markov process $X(\tau_x^{0-})$ is given by a rate or subrate matrix. Here, a matrix is said to be subrate if its row sums are all not greater than zero and if all off diagonal entries are nonnegative, while a rate matrix is the subrate matrix whose row sums are all zero. Thus, the problem becomes easier in this case.

For convenience, we write $X(t)$ as $(M(t), Z(t))$, where $Z(t)$ may or may not be discrete valued. Let S_1 be the state space of $Z(t)$. We separately consider the two cases: S_1 is discrete and continuous. For the latter, S_1 is assumed to be a subset of $\mathbb{R}^m$.

4.1 Discrete state case

Suppose S_1 is countable. Without loss of generality, we can assume that there are real valued function v on $S = S_0 \times S_1$, a function s from $S \times S_0$ to S_1, and distribution functions $F_{iki'}$ for $(i, k, i') \in S_0 \times S_1 \times S_0$ such that

(4.1.a) During $X(t) = (i, k)$, $Y(t)$ changes with rate $v(i, k)$,

(4.1.b) When $M(t) = i$ with $Z(t) = k$ changes to i', $Z(t)$ changes to $k' = s(i, k, i')$,

(4.1.c) When $M(t) = i$ with $Z(t) = k$ changes to i' due to the rate $D_{ii'}$, $Y(t)$ has a nonnegative jump whose size is subject to $F_{iki'}$.

Note that (4.1.c) holds due to the skip free condition (4.a).

Although $X(t)$ has the detailed transition structure, its transition kernel is basically the same as the MAP in queues. However, there are two essential differences: $X(t)$ may have countable states, i.e., C and D may have infinite dimensions, and $X(t)$ may not have the stationary distribution. We here assume that $X(t)$ has the unique stationary distribution, although we believe this condition can be removed. Then, we can put $X(t) = M(t)$ but $S = S_0$ may be countable. This case was recently studied by Miyazawa and Takada[25]. In what follows we briefly review their results and some from Miyazawa[24].

We first introduce some convenient notation for vectors and matrices. Let

$$S^+ = \{i \in S; v(i) > 0\} \quad \text{and} \quad S^- = \{i \in S; v(i) < 0\}.$$

We assume

(I.a) $v(i) \neq 0$ for all $i \in S$, so $S = S^+ \cup S^-$,

(I.b) $|v(i)C_{ii}|$ is uniformly bounded and uniformly away from 0 for all $i \in S$.

Assumption (I.a) can be removed, while assumption (I.b) is needed to overcome a technical difficulty. We use superscript $^-$ and $^+$ to indicate the sizes of vectors and matrices. For instance, C^{+-} is the $S^+ \times S^-$ matrix whose ij entry is C_{ij} for $i \in S^+$ and $j \in S^-$. Let

$$R_{ij}(x) = P\left(X(\tau_x^{0-}) = j | X(0) = i\right), \qquad x \geq 0,\ i, j \in S, \tag{3}$$

where the event $X(\tau_x^{0-}) = j$ includes $\tau_x^{0-} < \infty$. By $R(x)$, we denote the $S \times S$ matrix whose ij entry is given by (3). Since this distribution may be defective, we call it hitting probabilities. Since the background state must be in S^- at time τ_x^{0-}, we have

$$R^{++}(x) = 0^{++}, \qquad R^{-+}(x) = 0^{-+}.$$

Hence we only need to consider $R^{+-}(x)$ and $R^{--}(x)$. Obviously, $R^{--}(0) = I^{--}$, where I is the identity matrix. Since we shall frequently use $R^{+-}(0)$, we simply denote it by R^{+-}. That is,

$$R_{ij}^{+-} == P\left(X(\tau_0^{0-}) = j | X(0) = i\right), \qquad i \in S^+, j \in S^-.$$

Since $Y(t)$ is skip free for the downward direction, we have

$$R^{--}(x+y) = R^{--}(x)R^{--}(y), \qquad x, y \geq 0, \tag{4}$$
$$R^{+-}(x+y) = R^{+-}(x)R^{--}(y), \qquad x, y \geq 0. \tag{5}$$

The first equation implies that $R^{--}(x) = e^{xQ^{(-)}}$ for some subrate matrix $Q^{(-)}$.

Observe the sample path of $Y(t)$ during the time interval $[0, \Delta t]$ for a small $\Delta t > 0$. We then can see that $X(\tau_x^{0-})$ is obtained from $X(\tau_{x+v(i)\Delta t}^{0-})$ starting at time Δt, if $X(t) = i$ for $[0, \Delta t]$ with sufficiently small Δt. Hence,

$$\begin{aligned}[R(x)]_{ij} &= (1 + C_{ii}\Delta t)[R(x + v(i)\Delta t)]_{ij} \\ &\quad + \sum_{k \neq i, k \in S} C_{ik}\Delta t[R(x + v(i)\Delta t)]_{kj} \\ &\quad + \sum_{k \in S} \int_0^\infty D_{ik}(du)\Delta t[R(x + v(i)\Delta t + u)]_{kj} + o(\Delta t).\end{aligned}$$

Thus, we intuitively get

$$-\Delta_{\boldsymbol{v}} \frac{d}{dx} R(x) = CR(x) + \int_0^\infty D(du)R(x+u). \tag{6}$$

Combining (4), (5) and (6), we get the following results, which is formally proved in Miyazawa and Takada[25].

Theorem 4.1 *There exist minimal subrate matrices* $Q^{(-)}$ *and* R^{+-} *satisfying the following matrix equations.*

$$Q^{(-)} = (-\Delta_v^{--})^{-1}\left(C^{--} + C^{-+}R^{+-} + \int_0^\infty D^{--}(du)e^{uQ^{(-)}} + \int_0^\infty D^{-+}(du)R^{+-}e^{uQ^{(-)}}\right), \tag{7}$$

$$-R^{+-}Q^{(-)} = (\Delta_v^{++})^{-1}\left(C^{+-} + C^{++}R^{+-} + \int_0^\infty D^{+-}(du)e^{uQ^{(-)}} + \int_0^\infty D^{++}(du)R^{+-}e^{uQ^{(-)}}\right), \tag{8}$$

Then, we have, for $x > 0$,

$$\begin{pmatrix} R^{--}(x) \\ R^{+-}(x) \end{pmatrix} = \begin{pmatrix} I^{--} \\ R^{+-} \end{pmatrix} e^{xQ^{(-)}}. \tag{9}$$

Remark 4.1 As is remarked in Takada[30], equations (7) and (8) extend the Wiener-Hopf factorization for a Markov chain due to Rogers[28] (see also Asmussen[4]).

It remains how to get $Q^{(-)}$ and R^{+-}. To this end, we iteratively solve equations (7) and (8). Following Takada[30], for $\eta \geq \max_{i\in S^+} |C_{ii}|$, we define $Q_n^{(-)}$ and R_n^{+-} inductively by

$$Q_{n+1}^{(-)} = (-\Delta_v^{--})^{-1}\left(C^{--} + C^{-+}R_n^{+-} + \int_0^\infty D^{--}(du)e^{uQ_n^{(-)}} + \int_0^\infty D^{-+}(du)R_n^{+-}e^{uQ_n^{(-)}}\right), \tag{10}$$

$$R_{n+1}^{+-} = (\Delta_v^{++})^{-1}\left(C^{+-} + (\eta I^{++} + C^{++})R_n^{+-} + \int_0^\infty D^{+-}(du)e^{uQ_n^{(-)}} + \int_0^\infty D^{++}(du)R_n^{+-}e^{uQ_n^{(-)}}\right)\left(\eta I^{--} - Q_n^{(-)}\right)^{-1}, \tag{11}$$

where $Q_0^{(-)} = C^{--}$ and $R_0^{+-} = 0^{+-}$, provided $(\eta I^{--} - Q_n^{(-)})^{-1}$ exists. The following theorem can be proved in the exactly same way as in Takada[30].

Theorem 4.2 *$Q_n^{(-)}$ and R_n^{+-} are well defined and non-decreasing in component-wise, and we have*

$$\begin{pmatrix} Q^{(-)} \\ R^{+-} \end{pmatrix} = \lim_{n\to\infty} \begin{pmatrix} Q_n^{(-)} \\ R_n^{+-} \end{pmatrix}. \tag{12}$$

Theorem 4.2 has been used to prove Theorem 4.1. If $S = S_0$ is finite, the iterations (10) and (11) are feasible. However, this is not the case otherwise. Nevertheless, there should be some ways to overcome this, since the deriving source $M(t)$ is of finite states. This will be discussed somewhere else. The following facts are easy consequences of Theorem 4.1.

Corollary 4.1 *$\{X(\tau_x^{0-}); x \geq 0\}$ on S^- is a continuous-time Markov chain with transition subrate matrix $Q^{(-)}$, which is obtained as a minimal solution of (7) and (8). Furthermore, $Q^{(-)}$ is a rate matrix if and only if*

$$E(Y(1)) = \boldsymbol{\pi}\Delta_{\boldsymbol{v}}\mathbf{e} + \boldsymbol{\pi}\int_0^\infty xD(dx)\mathbf{e} < 0, \tag{13}$$

where $\mathbf{e}$ is S-vector all of whose entries are unit

We next consider the hitting time τ_x^+ for the upward direction. In this case, $X(\tau_x^+)$ itself is not Markov, because of upward jumps of $Y(t)$. However, there is a nice duality relation. Let

$$\begin{aligned}
\sigma^-(t) &= \sup\{u < t; N([u,t)) > 0\}, \\
\tilde{R}_{ij}(x) &= P\left(\tilde{X}(\tilde{\tau}_x^{0-}) = j \,\middle|\, \tilde{X}(0+) = i\right), \\
U_{ij}(x) &= C_{ij}\delta_0(x) + D_{ij}(x), \\
\tilde{U}_{ij}(x) &= \tilde{C}_{ij}\delta_0(x) + \tilde{D}_{ij}(x),
\end{aligned}$$

where $\delta_0(x)$ is the distribution function with a unit jump at the origin. Then, Miyazawa[24] obtained the following result (see Lemma 3.1 of that paper).

Lemma 4.1 *For $i \in S^-$, $j,k \in S$ and $x \geq 0$,*

$$\begin{aligned}
&P\left(X(\tau_0^+-) = j, X(\tau_0^+) = k, 0 < Y(\tau_0^+) \leq x \,\middle|\, X(0) = i\right) \\
&\qquad = \frac{\pi_k}{\pi_i}\int_0^x dw \int_w^\infty \tilde{D}_{kj}(dy)\tilde{R}_{ji}(y-w)\frac{-1}{v(i)},
\end{aligned} \tag{14}$$

and, for $i \in S^-$, $j \in S$ and $k \in S^+$,

$$\begin{aligned}
&P\left(X(\sigma^-(\tau_0^+)-) = j, X(\tau_0^+) = k, Y(\tau_0^+) = 0 \,\middle|\, X(0) = i\right) \\
&\qquad = \frac{\pi_k}{\pi_i}\int_0^\infty e^{-c(k)w/v(k)}dw \int_0^\infty \tilde{U}_{kj}(dy)\tilde{R}_{ij}(w+y)\frac{-1}{v(i)},
\end{aligned} \tag{15}$$

where $c(i) = -C_{ii}$.

The following result is a dual version of Theorem 4.1.

Lemma 4.2 *There are $S^- \times S^-$ ML matrix $K^{(-)}$ and $S^- \times S^+$ nonnegative matrix L^{-+} such that they are minimal solutions of the following equation.*

$$-K^{(-)}\left(I^{--}, L^{-+}\right) = \int_0^\infty e^{uK^{(-)}}\left(I^{--}, L^{-+}\right) U(du)\Delta_{\boldsymbol{v}}^{-1}, \tag{16}$$

where a square matrix is said to be ML if its off diagonal entries are nonnegative

These lemmas with some auxiliary results enable us to get the hitting probabilities. Let

$$\Psi_{ij}(x) = P\left(X(\tau_x^+) = j \middle| X(0) = i\right), \tag{17}$$

and denote the component-wise convolution of two matrix functions $A(x)$ and $B(x)$ by $(A * B)(x)$. Then, the following result is obtained in Miyazawa[24].

Theorem 4.3 *$\Psi(x)$ is obtained as*

$$\Psi(x) = \sum_{n=0}^{\infty}\left(H^{(n*)} * \bar{G}\right)(x), \qquad x \geq 0, \tag{18}$$

where H is and $\bar{G}$ are given by the following equations. For $x \geq 0$,

$$H_{ij}(x) = \begin{cases} \dfrac{1}{-v(i)}\left(\displaystyle\int_0^x dw\left[\int_w^\infty e^{(y-w)K^{(-)}}\left(I^{--}, L^{-+}\right) D(dy)\right]_{ij}\right. \\ \left. +\displaystyle\sum_{k\in S^+}\int_0^x L_{ik}^{-+}\left(1 - e^{-c(k)(x-y)/v(k)}\right)\frac{v(k)}{c(k)}U_{kj}(dy)\right), & i \in S^-, \\ \displaystyle\int_0^x (1 - e^{-c(i)(x-y)/v(i)})\frac{1}{c(i)}U_{ij}(dy), & i \in S^+. \end{cases} \tag{19}$$

For $i \in S^+, j \in S$,

$$\bar{G}_{ij}(x) = 1(i=j)e^{-xc(i)/v(i)} + \int_0^x \frac{c(i)}{v(i)}e^{-c(i)y/v(i)}dy\int_{x-y}^\infty \frac{1}{c(i)}U_{ij}(dw), \tag{20}$$

and, for $i \in S^-, j \in S$,

$$\begin{aligned}\bar{G}_{ij}(x) = \frac{-1}{v(i)}\Bigg[&\int_x^\infty dw\left[\int_w^\infty e^{(y-w)K^{(-)}}\left(I^{--}, L^{-+}\right) D(dy)\right]_{ij} \\ &+ L_{ij}^{-+}v(j)e^{xC_{jj}/v(j)} \\ &+ \sum_{k\in S^+} L_{ik}^{-+}\frac{v(k)}{c(k)}\int_0^x\int_{x-y}^\infty U_{kj}(dw)\frac{c(k)}{v(k)}e^{-yc(k)/v(k)}dy\Bigg].\end{aligned} \tag{21}$$

Remark 4.2 The expression (18) typically appears in the $M/G/1$ type queue, e.g., in the $MAP/G/1$ queue (see. e.g., Asmussen[3] and Takine[31]). Those formulas are sometimes referred to as *Pollaczek-Khinchine formulas.* The corresponding results can be obtained for the ruin probabilities in the Markov modulated risk processes (see, e.g., Asmussen[5]).

Remark 4.3 Function $H_{ij}(x)$ represents a certain ascending ladder height distribution stating with background state i and ending up by j. If there is no continuous movement to the upward, then it is the exactly ascending ladder height distribution.

4.2 Continuous state case

We consider the case that S_1 is a subset of $\mathbb{R}^m$. A typical setting is

$$S_1 = \{\boldsymbol{z} \in \mathbb{R}^m; \boldsymbol{z} \geq \boldsymbol{0}, z_\ell = 0 \text{ for some } \ell \in \{1, 2, \dots m\}\}.$$

We assume the following dynamic between transition instants of $M(t)$. Suppose that such an interval include $[t, t + \Delta t]$ for sufficiently small Δt, then there are real and vector valued functions a and $\boldsymbol{b}$ such that, for $M(t) = i$,

$$\text{(II.a)} \quad Y(t + \Delta t) = Y(t) + a(i, Z(t))\Delta t + o(\Delta t),$$

$$\text{(II.b)} \quad Z(t + \Delta t) = Z(t) + \boldsymbol{b}(i, Z(t))\Delta t + o(\Delta t),$$

where $o(\Delta t)$ is the small order of Δt, i.e., $o(\Delta t)/\Delta t$ goes to zero as Δt goes to zero. For example, a and $\boldsymbol{b}$ of Example 3.4 for $m = 2$ are obtained for $\boldsymbol{z} \neq \boldsymbol{0}$ as

$$a(i, \boldsymbol{z}) = \begin{cases} v_2(i)/c_2, & \boldsymbol{z} \in \mathcal{L}_1, \\ v_1(i)/c_1, & \boldsymbol{z} \in \mathcal{L}_2, \end{cases}$$

$$\boldsymbol{b}(i, \boldsymbol{z}) = \begin{cases} (1 - c_1)(v_1(i), 0), & \boldsymbol{z} \in \mathcal{L}_1, \\ (1 - c_2)(0, v_2(i)), & \boldsymbol{z} \in \mathcal{L}_2, \end{cases}$$

where $v_k(i) = d_k(i) - q_k(i)$ for $k = 1, 2$. For $\boldsymbol{z} = \boldsymbol{0}$, they are similarly obtained depending on $(v_1(i), v_2(i)) \geq \boldsymbol{0}$ or not.

For simplicity, we here also assume that matrix D is null, i.e, there is no jump associated with $M(t)$. Hence, $Y(t)$ is skip free to both of the upward and downward directions. Unfortunately, even this simple case has some technical problems that have not yet been resolved. So we only outline an idea to identify the generator of Markov chain $X(\tau_x^{0-})$.

Let, for $\boldsymbol{s}, \boldsymbol{z} \in S_1$ and $x \geq 0$,

$$\hat{R}_{ij}(\boldsymbol{s}, x|\boldsymbol{z}) = E\left(e^{\boldsymbol{s} \cdot Z(\tau_x^{0-})} 1(M(\tilde{\tau}_x^{0-}) = j) \middle| M(0) = i, Z(0) = \boldsymbol{z}\right),$$

where $\boldsymbol{s} \cdot \boldsymbol{z}$ is the inner product of vectors $\boldsymbol{s}$ and $\boldsymbol{z}$, and $1(\cdot)$ is the indicator function of the statement "$\cdot$". Similar to (6), observing the sample path of $(X(t), Y(t))$ starting at Δt, we get

$$-a(i,\boldsymbol{z})\frac{\partial}{\partial x}\hat{R}_{ij}(\boldsymbol{s},x|\boldsymbol{z}) = \left[C\hat{R}(\boldsymbol{s},x|\boldsymbol{z})\right]_{ij} + \left(\boldsymbol{b}(i,\boldsymbol{z}) \cdot \frac{\partial}{\partial \boldsymbol{z}}\right)\hat{R}_{ij}(\boldsymbol{s},x|\boldsymbol{z}),$$

where $\boldsymbol{b}(i,\boldsymbol{z}) \cdot \frac{\partial}{\partial \boldsymbol{z}} = \sum_{n=1}^{d} \boldsymbol{b}_n(i,\boldsymbol{z})\frac{\partial}{\partial z_n}$. Hence, denoting the generator of $X(\tau_x^{0-})$ by $\hat{Q}^{(-)}(\boldsymbol{s}|\boldsymbol{z})$ and letting $\hat{R}_{ij}(\boldsymbol{s}|\boldsymbol{z}) = \hat{R}_{ij}(\boldsymbol{s},0|\boldsymbol{z})$, we get

$$-a(i,\boldsymbol{z})\hat{Q}_{ij}^{(-)}(\boldsymbol{s}|\boldsymbol{z}) = \left[C^{--} + C^{-+}\hat{R}(\boldsymbol{s}|\boldsymbol{z})^{+-}\right]_{ij} + \delta_{ij}\boldsymbol{b}(i,\boldsymbol{z}) \cdot \frac{\partial}{\partial \boldsymbol{z}}, \tag{22}$$

$$-a(i,\boldsymbol{z})\left[\hat{R}(\boldsymbol{s}|\boldsymbol{z})^{+-}\hat{Q}^{(-)}(\boldsymbol{s}|\boldsymbol{z})\right]_{ij} = \left[C^{--} + C^{-+}\hat{R}(\boldsymbol{s}|\boldsymbol{z})^{+-}\right]_{ij}. \tag{23}$$

Here, the domain of $\hat{Q}^{(-)}(\boldsymbol{s}|\boldsymbol{z})$ is d-dimensional vector valued function of $\boldsymbol{z}$ whose components are partially differentiable with respect to z_n for $n = 1, 2, \ldots, d$. Denote this domain by $\mathcal{D}^{(1)}$.

Thus, we first determine $\hat{Q}_{ij}^{(-)}(\boldsymbol{s}|\boldsymbol{z})$ and $\hat{R}(\boldsymbol{s}|\boldsymbol{z})^{+-}$ restricting to the domain of constant functions, then $\hat{Q}_{ij}^{(-)}(\boldsymbol{s}|\boldsymbol{z})$ is obtained for domain $\mathcal{D}^{(1)}$. Once the generator $\hat{Q}_{ij}^{(-)}(\boldsymbol{s}|\boldsymbol{z})$ is identified, it is conceptually possible to compute the hitting probability $\hat{R}_{ij}(\boldsymbol{0},x|\boldsymbol{z})$. However, a feasible algorithm have not yet been obtained for this computation.

5 An identity related to Wiener-Hopf factorization

One of fascinating results on the ladder height problems in the Markov additive process is the Wiener-Hopf factorization. In the discrete time case, this identity factorizes the transition probability kernel into the ascending ladder height distribution and the dual of the descending ladder height distribution, and uniquely determine those distributions on the ladder heights. These results are originally obtained for a random walk on the line (see, e.g., Feller[14]), then extended to the Markov additive process (see, e.g., Dinges[12] and Arjas and Speed[1]).

Unfortunately, a little is known about the continuous time case except simple cases such that the background Markov chain is of finite states (see Greenwood[15] and Kaspi[16]). Since the Wiener-Hopf factorization aims to uniquely identify the ladder height and ladder epoch distributions, Theorem 4.1 and equations (7) and (8) are referred to as the Wiener-Hopf factorization as well (see Remark 4.1). The difficulty that arises in the continuous time case is that we can not deal with one step transition kernel, but have to work with the generator of a Markov process.

In this section, we derive a certain kind of decomposition formulas along the line of the symmetric Wiener-Hopf factorization of Arjas and Speed[1], and use it to get an identity concerning the ladder height and background state. The results of Arjas and Speed[1] are of discrete time, so we need to change it to be of continuous time.

Suppose that the Markov additive process $(X(t), Y(t))$ with $X(t) = (M(t), Z(t))$ satisfies assumptions (II.a) and (II.b) of Subsection 4.2, where $Z(t)$ takes values in S_1 there. However, this time we allow jumps at the transition instants of $M(t)$. We specify the dynamics of $Z(t)$ and $Y(t)$ at the jump instants in the following way. Suppose that a transition of $M(t)$ due to D occurs at time t and its state changes from i to j by this transition, then, for some real valued function φ and S_1-valued function $\boldsymbol{\psi}$,

(III.a) $\Delta Y(t) = Y(t-) + \varphi(i, Z(t-), \boldsymbol{J}(i,j))$,

(III.b) $\Delta Z(t) = Z(t-) + \boldsymbol{\psi}(i, Z(t-), \boldsymbol{J}(i,j))$,

where Δ stands for the difference such that $\Delta Y(t) = Y(t) - Y(t-)$, and $\boldsymbol{J}(i,j) \in S_J$ is the jump size vector subject to the distribution function $F^{D}_{ij}(\boldsymbol{x})$. Let T be any stopping time and let $f(i, \boldsymbol{z})$ be a real valued partially differentiable function with respect to $\boldsymbol{z} \in S_1$, then we have, for $\theta \in \mathbb{R}$, where θ is not the shift operator here,

$$\begin{aligned} e^{\theta Y(T)} f(X(T)) - f(X(0)) &= \int_0^T \theta a(X(u)) e^{\theta Y(u)} f(X(u))du \\ &+ \int_0^T e^{\theta Y(u)} \boldsymbol{b}(X(u)) \cdot \frac{\partial}{\partial \boldsymbol{z}} f(M(u), \boldsymbol{z})|_{\boldsymbol{z}=Z(u)} du \\ &+ \int_0^T (e^{\theta Y(u)} f(X(u)) - e^{\theta Y(u-)} f(X(u-)))N(du), \end{aligned}$$

where N is the point process generated by all the transition epochs of $M(t)$. Since $M(t)$ is a finite state Markov process, taking the conditional expectation of the above formula given $X(0-) = \boldsymbol{x}$, which is denoted by $E_{\boldsymbol{x}}$ yields

Lemma 5.1

$$\begin{aligned} E_{\boldsymbol{x}}\left(e^{\theta Y(T)} f(X(T))\right) - f(\boldsymbol{x}) &= E_{\boldsymbol{x}}\Big(\int_0^T \theta a(X(u)) e^{\theta Y(u)} f(X(u))du \Big) \\ &+ E_{\boldsymbol{x}}\Big(\int_0^T e^{\theta Y(u)} \boldsymbol{b}(X(u)) \cdot \frac{\partial}{\partial \boldsymbol{z}} f(M(u), \boldsymbol{z})|_{\boldsymbol{z}=Z(u)} du \Big) \qquad (24) \\ &+ E_{\boldsymbol{x}}\Big(\int_0^T c(X(u)) e^{\theta Y(u)} \big(E(e^{\theta \Delta Y(u)} f(X(u)) \big| X(u-)) - f(X(u-))\big)du \Big). \end{aligned}$$

This lemma is a continuous time version of Proposition (2.8) of Arjas and Speed[1]. It immediately yields

Proposition 5.1 *If there are f and θ such that, for any $i \in S_0$ and $z \in S_1$,*

$$\theta a(i,z)f(i,z) + \boldsymbol{b}(i,z) \cdot \frac{\partial}{\partial z} f(i,z)$$
$$+ \sum_{j \in S_0} C_{ij} f(j,z) + \sum_{j \in S_0} \int_{S_J} D_{ij}(dv) e^{\theta \varphi(i,z,v)} f(j, z + \psi(i,z,v)) = 0. \quad (25)$$

then we have, for $i \in S_0$ and $z \in S_1$,

$$E_{(i,z)}\left(e^{\theta Y(T)} f(M(T), Z(T))\right) = f(i,z). \quad (26)$$

Let $T = \tau_x^+$ in (26). Since $Y(\tau_x^+) \geq x$, (26) with nonnegative f implies

$$E_{(i,z)}\left(f(M(\tau_x^+), Z(\tau_x^+))\right) \leq e^{-\theta x} f(i,z). \quad (27)$$

Hence, if $\theta > 0$, (27) gives an exponential bound for the hitting probability as the hitting level goes to infinity. If the component $Z(t)$ is dropped, similar formulas are obtained in risk processes, in which $a(i,z) \equiv -1$ (see, e.g., Asmussen[5]). Note that a typical solution of (25) is of product form:

$$f(i,z) = f(i)\exp(\eta \cdot z), \qquad \eta \in \text{ a subset of } \mathbb{R}^m. \quad (28)$$

A computation procedure to determine f for each given θ will be discussed somewhere else.

6 Markov renewal approach for the asymptotic behaviors

In this section, we study asymptotic behavior of the hitting probability when the hitting level is going to infinity. From a technical difficulty, we are here limited to the case that the background state space S is finite, although most parts of our arguments are applicable to the countable case. In what follows, we cite results from Miyazawa[24].

We first introduce Markov renewal theorem. This theorem is originally obtained for the countable background state case (see, e.g., Çinler[11]). We also only require this level of the theorem below. However, it may be worthwhile mentioning about more general cases, because such results may require to consider the continuous background state space case. Kesten[17] studied Markov renewal theorem allowing the background state space to be very general, provided certain strong recurrence conditions on the background process. This result is revisited by Athreya, McDonald and Ney[8], using a nice coupling argument (see also Athreya and Ney[7]). So far, the following arguments may be extended for these general cases.

Suppose S be finite. Let $S \times S$ matrix $P(x) = \{P_{ij}(x)\}$ be a Markov renewal kernel that may be defective, i.e., $P(x)\mathbf{e} \le \mathbf{e}$. Denote the moment generating function of $P(x)$ by

$$\hat{P}(\theta) \equiv \int_0^\infty e^{\theta x} P(dx).$$

Since $\hat{P}(\theta)$ is nonnegative and substochastic matrix, it has a positive eigen value $\gamma(\theta)$ such that the absolute values of all other eigen values are less than $\gamma(\theta)$ and the associated right and left eigen vectors are nonnegative (see, e.g., Seneta[29]). Denote these associated eigen vectors by $\boldsymbol{\nu}^{(\theta)}$ and $\boldsymbol{h}^{(\theta)}$, respectively. Suppose that a $S \times S$ matrix functions $\Psi(x)$ and $B(x)$ for $x \ge 0$ satisfy the Markov renewal equation:

$$\Psi(x) = B(x) + P * \Psi(x), \qquad x \ge 0, \tag{29}$$

where $\Phi(x)$ is assumed to be nondecreasing. Since P may not be proper, we can not apply the Markov renewal theorem directly. So, using nonnegative θ, we change it to

$$P^{(\theta)}(x) \equiv \int_0^x e^{\theta u} P(du), \qquad x \ge 0,$$

and find a θ and a invariant vector $\boldsymbol{h}^{(\theta)}$ such that

$$P^{(\theta)}(\infty)\boldsymbol{h}^{(\theta)} = \boldsymbol{h}^{(\theta)}. \tag{30}$$

This is a basic idea to derive the following result, which is a slightly extended version of the one obtained in Asmussen[2] as Theorem 2.6 of Chapter X (see Miyazawa[23] for the present context).

Lemma 6.1 *Suppose the following conditions.*

(6.a) $P(x)$ has a single irreducible recurrent class that can be reached from any state in S with probability one, and the return time to each state in the irreducible recurrent class has a non-arithmetic distribution.

(6.b) There exists a positive α such that $\gamma(\alpha) = 1$.

(6.c) Each entry of $e^{\alpha x}B(x)$ is directly Riemann integrable.

(6.d) $\boldsymbol{\nu}^{(\alpha)}\hat{P}_{(1)}(\alpha)\boldsymbol{h}^{(\alpha)}$ is finite.

Then, we have

$$\lim_{x\to\infty} e^{\alpha x}\Psi(x) = \frac{1}{\boldsymbol{\nu}^{(\alpha)}\hat{P}_{(1)}(\alpha)\boldsymbol{h}^{(\alpha)}} \boldsymbol{h}^{(\alpha)}\boldsymbol{\nu}^{(\alpha)} \int_0^\infty e^{\alpha u} B(u)du \tag{31}$$

where $\hat{P}_{(1)}(\alpha) = \frac{d}{d\theta}\hat{P}(\theta)\Big|_{\theta=\alpha}$.

Remark 6.1 For this lemma, we do not need to compute $\gamma(\theta)$. What all we need is to find $\alpha > 0$ such that $\hat{P}(\alpha)$ has eigen value 1 and the associated left and right eigen vectors are positive. This is because such eigen vectors are uniquely determined up to multiplicative constants, from the irreducibility condition (6.a). For the existence of the solution α, we need the light tail condition that $\hat{P}(\theta)$ exists for some $\theta > 0$, which will be assumed in this section. See Takine[32] for the discrete time version of this theorem.

Consider the Markov additive process $(X(t), Y(t))$ of Section 4.1 with finite background space S. Let us set $P(x) = H(x)$. Then, from Theorem 4.3, $\Psi(x)$ of (17) satisfies the renewal equation (29) with $B(x) = \bar{G}(x)$. Clearly, condition (6.a) is satisfied by the irreducibility of $C + D$. Note that this $P(x)$ may not be irreducible, since the states at the ascending ladder heights may be limited. This is the reason why we use the slightly weaker condition (6.a) than the irreducibility of $P(x)$.

We next compute moment generating function $\hat{H}(\theta)$ of $H(x)$ from Theorem 4.3.

Lemma 6.2 *For* $i \in S^-$,

$$\hat{H}_{ij}(\theta) = \delta_{ij} + \frac{-1}{v(i)}\left[\left(\theta I^{--} - K^{(-)}\right)^{-1}(I^{--}, L^{-+})\left(C + \hat{D}(\theta) + \theta\Delta_{\boldsymbol{v}}\right)\right]_{ij}$$
$$+ \frac{-1}{v(i)}\left[(0^{--}, L^{-+})\,\Delta_{\boldsymbol{v}}\Delta^{-1}_{\boldsymbol{c}-\theta\boldsymbol{v}}\left(C + \hat{D}(\theta) + \theta\Delta_{\boldsymbol{v}}\right)\right]_{ij}, \tag{32}$$

and, for $i \in S^+$,

$$\hat{H}_{ij}(\theta) = \left[I + \Delta^{-1}_{\boldsymbol{c}-\theta\boldsymbol{v}}\left(C + \hat{D}(\theta) + \theta\Delta_{\boldsymbol{v}}\right)\right]_{ij}, \tag{33}$$

where $\Delta_{\boldsymbol{c}-\theta\boldsymbol{v}}$ *is the diagonal matrix whose i-th diagonal entry is* $c(i) - \theta v(i)$. *That is, in matrix form,*

We now assume the following light tail assumption on $D(x)$ and related regularity assumption.

(6.e) Stability condition (13) holds, i.e., $E(Y(1)) < 0$.

(6.f) All entries of matrix $\hat{D}(\theta) \equiv \int_0^\infty e^{\theta u} D(du)$ are finite for some $\theta > 0$.

(6.g) $\sup\{\hat{D}'_{ij}(\theta) < \infty; i, j \in S,\ \theta > 0\} = \infty$

Condition (ii) can be relaxed, but it is sufficient for most of conventional distributions with light tails. Note that condition (6.e) implies that $\tilde{Q}^{(-)}$ is a rate matrix.

Since $C + \hat{D}(\theta) + \theta\Delta_{\boldsymbol{v}}$ is an ML matrix for each $\theta > 0$, it has a real eigen value $\kappa(\theta)$ such that it dominates the real parts of all other eigen values, and the associated left and right eigen vectors are positive. Denote these eigen vectors by $\boldsymbol{\mu}^{(\theta)}$ and $\boldsymbol{h}^{(\theta)}$, respectively. Let $\boldsymbol{k}^-$ be the right eigen vector of $K^{(-)}$ for eigen value 0, where $\boldsymbol{k}^-$ is unique and positive (see Lemma 4.2). We normalize $\boldsymbol{\mu}^{(\theta)}$ and $\boldsymbol{h}^{(\theta)}$ so that

$$(\boldsymbol{\mu}^{(\theta)})^- \boldsymbol{k}^- = 1, \qquad \boldsymbol{\mu}^{(\theta)} \boldsymbol{h}^{(\theta)} = 1.$$

Since $\kappa(\theta) = \boldsymbol{\mu}^{(\theta)} \left(\theta\Delta_{\boldsymbol{v}} + C + \hat{D}(\theta)\right) \boldsymbol{h}^{(\theta)}$, we have

$$\begin{aligned}\kappa'(\theta) &= \kappa(\theta) \left((\boldsymbol{\mu}^{(\theta)})' \boldsymbol{h}^{(\theta)} + \boldsymbol{\mu}^{(\theta)} (\boldsymbol{h}^{(\theta)})'\right) + \boldsymbol{\mu}^{(\theta)} \left(\Delta_{\boldsymbol{v}} + \hat{D}'(\theta)\right) \boldsymbol{h}^{(\theta)} \\ &= \boldsymbol{\mu}^{(\theta)} \left(\Delta_{\boldsymbol{v}} + \hat{D}'(\theta)\right) \boldsymbol{h}^{(\theta)},\end{aligned}$$

where we have used the fact that $\boldsymbol{\mu}^{(\theta)} \boldsymbol{h}^{(\theta)} = 1$. This implies that $\kappa'(0) = E(Y(1)) < 0$ because $\boldsymbol{\mu}^{(0)} = \boldsymbol{\pi}$. Hence, from the fact that $\kappa(0) = 0$ and $\kappa(\theta)$ is a convex function (see Kingman[19]), condition (ii) guarantees that $\kappa(\theta) = 0$ has the unique positive solution. Thus, condition (6.b) is satisfied. Denote this solution by α.

Define $\boldsymbol{\nu}^{(\alpha)}$ as

$$\boldsymbol{\nu}^{(\alpha)} = -\boldsymbol{\mu}^{(\alpha)} \begin{bmatrix} \alpha I^{--} - K^{(-)} & L^{-+}\Delta^{++}_{\boldsymbol{C}-\alpha\boldsymbol{v}} (\Delta^{++}_{\boldsymbol{v}})^{-1} + (\alpha I^{--} - K^{(-)}) L^{-+} \\ 0^{+-} & -\Delta^{++}_{\boldsymbol{C}-\alpha\boldsymbol{v}} (\Delta^{++}_{\boldsymbol{v}})^{-1} \end{bmatrix} \Delta_{\boldsymbol{v}}.$$

Since

$$\begin{aligned}&\begin{bmatrix} (\alpha I^{--} - K^{(-)})^{-1} & (\alpha I^{--} - K^{(-)})^{-1} L^{-+} + L^{-+}\Delta^{++}_{\boldsymbol{v}} (\Delta^{++}_{\boldsymbol{C}-\alpha\boldsymbol{v}})^{-1} \\ 0^{+-} & -\Delta^{++}_{\boldsymbol{v}} (\Delta^{++}_{\boldsymbol{C}-\alpha\boldsymbol{v}})^{-1} \end{bmatrix}^{-1} \\ &\qquad = \begin{bmatrix} \alpha I^{--} - K^{(-)} & L^{-+}\Delta^{++}_{\boldsymbol{C}-\alpha\boldsymbol{v}} (\Delta^{++}_{\boldsymbol{v}})^{-1} + (\alpha I^{--} - K^{(-)}) L^{-+} \\ 0^{+-} & -\Delta^{++}_{\boldsymbol{C}-\alpha\boldsymbol{v}} (\Delta^{++}_{\boldsymbol{v}})^{-1} \end{bmatrix},\end{aligned}$$

it can be seen that $\boldsymbol{\nu}^{(\alpha)}$ is the left invariant vector of $\hat{H}(\alpha)$. It can be also proved that $\boldsymbol{\nu}^{(\alpha)}$ is positive.

It is easy to see that $\boldsymbol{h}^{(\alpha)}$ is the right positive eigen vector of $\hat{H}(\alpha)$ for eigen value 1. Thus, in the view of Remark 6.1 we are now in a position to apply Lemma 6.1. Take H and $\bar{G}$ for P and B, respectively, in Lemma 6.1. We also requires the following integrations. For $i \in S^+$, (20) yields

$$\int_0^\infty e^{\alpha x} \bar{G}_{ij}(x)dx = \frac{1}{\alpha} \left[\Delta^{-1}_{\boldsymbol{C}-\alpha\boldsymbol{v}} \left(C + \hat{D}(\alpha) + \alpha\Delta_{\boldsymbol{v}} - (C + D)\right)\right]_{ij}. \quad (34)$$

Similarly, for $i \in S^-$, we have

$$\int_0^\infty e^{\alpha x}\bar{G}_{ij}(x)dx = \frac{-1}{\alpha v(i)}\Bigg[(\alpha I^{--} - K^{(-)})^{-1}(I^{--}, L^{-+})\left(C + \hat{D}(\alpha) + \alpha\Delta_{\boldsymbol{v}}\right)$$
$$+(0^{--}, L^{-+})\Delta_{\boldsymbol{v}}\Delta_{\boldsymbol{c}-\alpha\boldsymbol{v}}^{-1}\left(C + \hat{D}(\alpha) + \alpha\Delta_{\boldsymbol{v}} - (C + D)\right)$$
$$-\boldsymbol{k}^-\boldsymbol{\pi}^-(I^{--}, L^{-+})\left(\Delta_{\boldsymbol{v}} + \int_0^\infty yD(dy)\right)$$
$$-(\boldsymbol{k}^-\boldsymbol{\pi}^- - K^{(-)})^{-1}(I^{--}, L^{-+})(C + D)\Bigg]_{ij}. \tag{35}$$

Then, applying Lemma 6.1, we have the following asymptotics.

Theorem 6.1 *Suppose the stability condition (6.e) and the light tail conditions (6.f) and (6.g). Then, the maximal eigen value $\kappa(\theta)$ of $C + \hat{D}(\theta) + \theta\Delta_{\boldsymbol{v}}$ attains 0 for a unique $\theta = \alpha > 0$, and we have, for $i, j \in S$,*

$$\lim_{x\to\infty} e^{\alpha x}P(X(\tau_x^+) = j|X(0) = i) = \frac{1}{\eta^{(\alpha)}}\left[\boldsymbol{h}^{(\alpha)}\boldsymbol{\mu}^{(\alpha)}\left(\Gamma(\alpha) - \frac{1}{\alpha}(C + D)\right)\right]_{ij} \tag{36}$$

where $\eta^{(\alpha)} = \boldsymbol{\mu}^{(\alpha)}\left(\hat{D}_{(1)}(\alpha) + \Delta_{\boldsymbol{v}}\right)\boldsymbol{h}^{(\alpha)}$, where $\hat{D}_{(1)}(\alpha)$ is the first derivative of $\hat{D}(\theta)$ at $\theta = \alpha$, and $\Gamma(\alpha)$ is a $S \times S$-matrix such that its $S^+ \times S$ sub-matrix and $S^- \times S$ sub-matrices are defined as

$$\left(\Gamma(\alpha)^{+-}, \Gamma(\alpha)^{++}\right) = (0^{-+}, 0^{++}),$$
$$\left(\Gamma(\alpha)^{--}, \Gamma(\alpha)^{-+}\right) = -\boldsymbol{k}^-\boldsymbol{\pi}\left(\Delta_{\boldsymbol{v}} + \hat{D}'(0)\right)$$
$$-\frac{1}{\alpha}(\alpha I^{--} - \boldsymbol{k}^-\boldsymbol{\pi}^-)\left(\boldsymbol{k}^-\boldsymbol{\pi}^- - K^{(-)}\right)^{-1}(I^{--}, L^{-+})(C + D).$$

In particular, for $i \in S$,

$$\lim_{x\to\infty} e^{\alpha x}P(\tau_x^+ < \infty|X(0) = i) = \frac{-\eta^{(0)}h_i^{(\alpha)}}{\eta^{(\alpha)}}(\boldsymbol{\mu}^{(\alpha)})^-\boldsymbol{k}^-, \tag{37}$$

where $\eta^{(0)} = E(Y(1))$.

We apply these results to the buffer content process $W(t)$ in Example 3.3. Let W be a random variable subject to its stationary distribution. Then, (2) leads to the following results.

Theorem 6.2 *Under the same conditions of Theorem 6.1, we have, for $i \in S$,*

$$\lim_{x\to\infty} e^{\alpha x}P(W > x, X = i) = \frac{-\eta^{(0)}\mu_i^{(\alpha)}}{\eta^{(\alpha)}}\boldsymbol{\beta}^-(\boldsymbol{h}^{(\alpha)})^-, \tag{38}$$

where $\boldsymbol{\beta}^-$ is the stationary probability vector of the rate matrix $Q^{(-)}$ that is obtained as the minimal rate matrix of the following matrix equation together with R^{+-}.

$$-\begin{pmatrix} I^{--} \\ R^{+-} \end{pmatrix} Q^{(-)} = \Delta_{\boldsymbol{v}}^{-1} \int_0^{\infty} (\delta_0(du)C + D(du)) \begin{pmatrix} I^{--} \\ R^{+-} \end{pmatrix} e^{uQ^{(-)}}. \quad (39)$$

7 Epilog

The story of this paper has not yet ended. We have not fully discussed how the hitting probabilities can be used to compute the stationary distributions in queueing models. This issue has already been answered for single node queues (see (2) and Neuts[26]). But, a little has been studied for queueing networks in the literature. We believe that a similar argument to the one dimensional case can go through for a wide class of queueing networks using the dual Markov additive process. This seems true at least by intuition, but needs a formal proof.

Another issue that remains not answered is the decay rate of a multi-dimensional distribution in queueing networks. There are also few studies on this subject (see, e.g., McDonald[21]). Note that (26) with (28) and (30) have essentially the same form. Since (26) can be obtained under a multidimensional setting, it would be useful to find the decay rate in queueing networks. Here is one problem that θ of (26) may not be uniquely determined. So we need to find a maximum θ among feasible set of them, which may be restricted by the network dynamics at the boundaries. There are some suggestive results for this (see Miyazawa[22] and Kella and Miyazawa[18]). A research for these issues are now under way.

Acknowledgments

This research is supported in part by JSPS under grant No. 13680532.

References

1. E. Arjas and T. P. Speed, Symmetric Wiener-Hopf factorizations in Markov additive processes. *Z. Wahrscheinlich. verw Geb.* **26**, 105-118 (1973).
2. S. Asmussen, *Applied Probability and Queues* (John Wiley & Sons, Chichester, 1987).
3. S. Asmussen, Ladder heights and the Markov-modulated $M/G/1$ queue. *Stochastic Processes and their Applications* **37**, 313-326 (1991).

4. S. Asmussen, Stationary distributions for fluid flow models with or without brownian noise. *Stochastic Models* **11**, 21-49 (1995).
5. S. Asmussen, *Ruin Probabilities*, (World Scientific, Singapore, 2000).
6. S. Asmussen, *Applied Probability and Queues*, 2nd edition, to be published by Springer-Verlag 2002.
7. K. B. Athreya and P. Ney, A renewal approach to the Perron-Frobenius Theory of Non-negative kernels on general state spaces. *Math. Zeitschrift* **179**, 507-529 (1982).
8. K. B. Athreya, D. McDonald and P. Limit theorems for semi-Markov processes and renewal theory for Markov chains. *Ann. Prob.* **6**, 788-797 (1978).
9. E. Çinler, Markov additive processes I. *Z. Wahrscheinlich. verw. Geb.* **24**, 85-93 (1972).
10. E, Çinler, Markov additive processes II. *Z. Wahrscheinlich. verw. Geb.* **24**, 95-121 (1972).
11. E. Çinler, *Introduction to Stochastic Processes*, (Prentice-Hall, Englewood Cliffs, New Jersey, 1975).
12. H. Dinges, Wiener-Hopf-Faktorisierung für substochastische Übergangsfunktionen in angeordneten Räumen. *Z. Wahrscheinlich. verw. Geb.* **11**, 152-164 (1969).
13. A. I. Elwalid and T. E. Stern, Analysis of separable Markov-modulated rate models for information-handling systems. *Adv. in Appl. Prob.* **23**, 105-139 (1991).
14. W. Feller, *An introduction to Probability Theory and its Applications vol. II* (Wiley, New York 1971)
15. P. Greenwood, Wiener-Hopf Decomposition of Random walks and Lévy processes. *Z. Wahrscheinlich. verw. Geb.* **34**, 193-198 (1976).
16. H. Kaspi, On the symmetric Wiener-Hopf factorization for Markov additive processes. *Z. Wahrscheinlich. verw. Geb.* **59**, 179-196 (1982).
17. H. Kesten, Renewal theory for functional of a Markov chain with general state space. *Ann. Prob.* **2**, 355-386 (1974).
18. O. Kella and M. Miyazawa, Parallel fluid queues with constant inflows and simultaneous random reductions, *J. Appl Prob.* **38**, 609-620 (2001).
19. J.F.C. Kingman, A convexity property of positive matrices. *The Quarterly Journals of Mathematics, Oxford , Second Series* **12**, 283-284 (1961).
20. D. M. Lucantoni, New results on the single server queue with a batch Markovian arrival process *Stoch. Models* **7**, 1-46 (1991).
21. D. R. McDonald, Asymptotics of first passage times for random walk in an orthant. *Ann. Appl. Prob.* **9**, 110-145 (1999).
22. M. Miyazawa, Conjectures on decay rates of tail probabilities in general-

ized Jackson and batch movement networks. Submitted for publication (2001).

23. M. Miyazawa, A Markov renewal approach to the asymptotic decay of the tail probabilities in risk and queueing processes. To appear in *Probability in the Engineering and Informational Sciences* 2002/1.
24. M. Miyazawa, The upward hitting probabilities in a Markov modulated additive process with linear and upward jump components. Preprint (2002).
25. M. Miyazawa and H. Takada, A matrix exponential form for hitting probabilities and its application to a Markov modulated fluid queue with downward jumps. Submitted for publication (2000).
26. M. F. Neuts, *Structured Stochastic Matrices of M/G/1 Type and Their Applications* (Mercel Dekker, New York 1989).
27. V. Ramaswami, From the matrix-geometric to the matrix-exponential. *Queueing Systems* **6**, 229-260 (1990).
28. L. C. G. Rogers, Fluid models in queueing theory and Wiener-Hopf factorization of Markov chains. *Ann. Appl. Prob.* **4**, 390-413 (1994).
29. E. Seneta, *Non-negative Matrices and Markov chains*, (Second edition, Springer-Verlag, New York 1994).
30. H. Takada, Markov Modulated Fluid Queues with Batch Fluid Arrivals. *J. of Operations Society of Japan* **44**, 344-365 (2001).
31. T. Takine, A recent progress in algorithmic analysis of FIFO queues with Markovian arrival streams, to appear in *J. of the Korean Mathematical Society* (2001).
32. T. Takine An alternative formula for the steady-state solution of Markov chains of M/G/1 type and its geometric and subexponential asymptotics. Preprint (2001).
33. T. Takine and T. Hasegawa, The workload in the $MAP/G/1$ queue with state-dependent services: Its application to a queue with preemptive resume priority. *Stochastic Models* **10**, 183-204 (1994).

SPECTRAL METHODS FOR A TREE STRUCTURE MAP

SHOICHI NISHIMURA
Department of Mathematics and Information Science, Tokyo University of Science, Japan 162-8601.
E-mail: nishimur@rs.kagu.tus.ac.jp

To model the quasi second order self-similarity, a tree structure Markovian arrival process ($TSMAP$) is introduced and effective calculations of spectral methods based on eigenvalues and eigenvectors for a $TSMAP/G/1$ are considered. Let $D(z)$ be the z-transform of arrival rate matrices. For a $TSMAP$ it is shown that for a real z all the eigenvalues of $D(z)$ are real. Then the eigenvalues of $D(1)$, the zero points of $\det(zI - A(z))$ on the unit disk and the minimal zero point of $\det(zI - A(z))$ greater than 1 are calculated by a bisection method, where $A(z)$ is the probability generating matrix function of arrival numbers during the service time. Eigenvectors of $D(1)$ and null vectors of $zI - A(z)$ are inductively calculated by null vectors of submatrices. And for a $TSMAP/G/1$ queue, the stationary distribution at service completion epochs is effectively calculated by the Fourier inversion formula. Finally, we consider a numerical example of a $TSMAP$ which has a second-order self-similar property. The stationary distribution and the asymptotic decreasing stationary distribution are numerically calculated.

1 Introduction

To model traffic with burst in telecommunication systems, a self-similar property has been widely studied. However, it is difficult to obtain analytical characteristics of a queueing system whose arrival process is self-similar, e.g. the mean or the stationary distribution of the queue length. Moreover, real traffic is not self-similar in the *strict* sense because arrival rate is to be bounded by mechanical constraints and random environment is to be changed over a long time period.

By using a Markovian arrival process (MAP) introduced by Neuts [8] a self-similar process is unable to be modeled in the strict sense since a state space of an underlying process of a MAP is finite. However, MAPs are greatly flexible and performance measures of a $MAP/G/1$ queue are analytically obtained. So it has been studied that traffic with the self-similarity over several time scales is modeled by a MAP [1 3 9 11 15]. Nishimura [11] considers a birth and death Markovian arrival process ($BDMAP$), where the underlying phase process is formulated as a birth and death process. As an extension of a $BDMAP$ we introduce a MAP with a tree structure ($TSMAP$), whose underlying phase process has a tree structure.

We consider a $MAP/G/1$ queue. Let $D(z)$ and $A(z)$ be the z-transform of arrival rate matrices of a MAP and the probability generating matrix function of arrival numbers during the service time, respectively. Spectral methods based on eigenvalues and eigenvectors of $D(z)$ and $A(z)$ have been studied [4 5 10 11 12 16]. For a $TSMAP$ queue, by virtue of the tree structure, effective algorithms for calculation of eigenvalues and eigenvectors are given. For a numerical example its second order similarity is shown and its stationary distribution is effectively calculated.

In Section 2, applying the Fourier inversion formula to the probability generating vector function of the queue size at service completion epochs, we calculate the stationary probability distribution of the queue size. In Section 3 we introduce a $TSMAP$ and prove that for a real z all the eigenvalues of $D(z)$ are real. In Section 4 for any fixed t we derive inductive calculations for $\det(tI - D(z))$ and the numbers of the eigenvalues greater than t, where I is the identity matrix. In Section 5 we consider three problems where characteristics are calculated by spectral methods. And we show that all the eigenvalues are calculated by a bisection method. In Section 6 we show that an eigenvector corresponding to each eigenvalue is inductively calculated by null vectors of submatrices. In Section 7 we consider an algorithm for calculations of eigenvalues on the unit circle and corresponding eigenvectors for the Fourier inversion formula. Particularly, eigenvalues are calculated by Newton's method. Finally, we consider a numerical example of a $TSMAP$ which has a second-order self-similar property. The stationary distribution and the asymptotic decreasing stationary distribution are numerically calculated.

2 A $MAP/G/1$ queue

Let the symbol $\boldsymbol{T}$ be the transposition of a matrix and I be the identity matrix. We define vectors as $\mathbf{0} = (0, \ldots, 0)$, $\mathbf{1} = (1, 0, \ldots, 0)^{\boldsymbol{T}}$ and $\boldsymbol{e} = (1, \ldots, 1)^{\boldsymbol{T}}$. Sizes of I, $\mathbf{0}$, $\mathbf{1}$ and $\boldsymbol{e}$ are appropriately determined in the later discussions. We consider a Markovian arrival process whose underlying state space is $\{i : i = 1, \ldots, M\}$. Let D_k $(k = 0, 1, 2, \ldots)$ be the $M \times M$ arrival rate matrix of batch size k. As usual, assume that for $|z| \leq 1$ $D(z) = \sum_{k=0}^{\infty} D_k z^k$ is analytic and $D = D(1) = \sum_{k=0}^{\infty} D_k$ is the irreducible infinitesimal generator of the underlying process. And let $\boldsymbol{\pi} = (\pi_1, \pi_2, \ldots, \pi_M)$ as the stationary probability vector of D $(\boldsymbol{\pi} D = \mathbf{0}$ and $\boldsymbol{\pi e} = 1)$. The mean arrival rate of customers is given by $\lambda = \boldsymbol{\pi} D(1)'\boldsymbol{e} = \boldsymbol{\pi} \sum_{k=1}^{\infty} k D_k \boldsymbol{e}$. Let $H(t)$ be the distribution function of the service time with the mean service time $1/\mu$. Let $h(\alpha) = \int_0^{\infty} e^{\alpha t} dH(t)$ and $h^{-1}(z)$ be the moment generating function of the service time and its inverse function, respectively. Then $A(z) = \int_0^{\infty} e^{D(z)t} dH(t)$ is the probability generating matrix function of arrival numbers during the service time. Assume that the traffic intensity $\rho = \frac{\lambda}{\mu} < 1$.

Let $\alpha_i(z)$ $(i = 1, \ldots, M)$ be the eigenvalues of $D(z)$ and $\boldsymbol{u}_i(z)$ and $\boldsymbol{v}_i(z)$ be corresponding left and right eigenvectors ($\boldsymbol{u}_i(z)D(z) = \alpha_i(z)\boldsymbol{u}_i(z)$ and $D(z)\boldsymbol{v}_i(z) = \alpha_i(z)\boldsymbol{v}_i(z)$). We assume the following.

Assumption 1
For any fixed $|z| \leq 1$ all the eigenvalues $\alpha_i(z)$ $(i = 1, \ldots, M)$ are simple.

So, we may assume eigenvectors are normalized as $U(z)V(z) = I$, where

$$U(z) = \begin{pmatrix} \boldsymbol{u}_1(z) \\ \vdots \\ \boldsymbol{u}_M(z) \end{pmatrix} \quad \text{and} \quad V(z) = (\boldsymbol{v}_1(z), \ldots, \boldsymbol{v}_M(z)).$$

Particularly, $\alpha_1(1) = 0$, $\boldsymbol{u}_1(1) = \boldsymbol{\pi}$ and $\boldsymbol{v}_1(1) = \boldsymbol{e}$.

Let $\boldsymbol{p}_k = (p_{k,1}, \ldots, p_{k,M})$ $(k \geq 0)$ be the M-dimensional row vector whose ith entry is the stationary joint probability that just after service completion epochs, the arrival phase is i and the number of customers in the system is k. Then $\boldsymbol{p} = (\boldsymbol{p}_0, \boldsymbol{p}_1, \ldots)$ is the stationary probability vector just after service completion epochs. Define its probability generating vector function as $\boldsymbol{p}(z) = \sum_{k=0}^{\infty} \boldsymbol{p}_k z^k$. As in Neuts [8] let G and $\boldsymbol{g}$ be the phase transition stochastic matrix of the first passage time from the level $k+1$ to the level k and its invariant probability vector ($\boldsymbol{g}G = \boldsymbol{g}$, $\boldsymbol{g}\boldsymbol{e} = 1$), respectively. In Lucantoni [7], $\boldsymbol{p}(z)$ is given by

$$\boldsymbol{p}(z) = \lambda^{-1}(1-\rho)\boldsymbol{g}D(z)A(z)(zI - A(z))^{-1}. \tag{1}$$

For a $M/G/1$ type Markov chain, in order to calculate the stationary distribution at service completion epochs, the matrix analytic method introduced in Ramaswami [14] is effective. We consider a spectral method of the Fourier inversion formula for calculation of the stationary distribution discussed in Tijms [18] page 29 and Nishimura [11]. Let N be an integer such that the tail probability $\sum_{l=N}^{\infty} \boldsymbol{p}_l \boldsymbol{e}$ is negligible and define

$$\tilde{\boldsymbol{p}}_n = \sum_{l=0}^{\infty} \boldsymbol{p}_{lN+n} \quad (n = 0, \ldots, N-1)$$

$$\tilde{\boldsymbol{p}}(z) = \sum_{k=0}^{N-1} \tilde{\boldsymbol{p}}_k z^k.$$

Let $\omega = e^{2\pi \boldsymbol{i}/N}$ be the Nth root of the unity ($\boldsymbol{i}$ is the imaginary unit). Since for $n = lN + k$ $\omega^n = \omega^k$ then $\tilde{\boldsymbol{p}}(\omega^k) = \boldsymbol{p}(\omega^k)$. From the Fourier inversion formula, we have for $n = 0, \ldots, N-1$

$$p_n \approx \tilde{p}_n = \frac{1}{N}\sum_{k=0}^{N-1} \tilde{p}(\omega^k)\omega^{-kn} = \frac{1}{N}\sum_{k=0}^{N-1} p(\omega^k)\omega^{-kn}$$

Finally, from the spectral representation of (1) we have

$$p_n \approx \frac{\lambda^{-1}(1-\rho)g}{N} \cdot \sum_{k=0}^{N-1} V(\omega^k) \begin{bmatrix} \frac{\alpha_1(\omega^k)h(\alpha_1(\omega^k))}{\omega^k - h(\alpha_1(\omega^k))} & & \\ & \ddots & \\ & & \frac{\alpha_M(\omega^k)h(\alpha_M(\omega^k))}{\omega^k - h(\alpha_M(\omega^k))} \end{bmatrix} U(\omega^k)\omega^{-kn}. \quad (2)$$

In (2), for $k = 0$ and $i = 1$, we define as

$$\frac{\alpha_i(\omega^k)h(\alpha_i(\omega^k))}{\omega^k - h(\alpha_i(\omega^k))} = \frac{\lambda}{1-\rho}.$$

For a $TSMAP/G/1$ queue an effective method for calculation of eigenvalues and eigenvectors will be discussed in Section 7.

3 A Markovian arrival process with a tree structure

Consider a tree T with the root node 1 that a node represented as a string $J = j_1j_2, \ldots, j_n$ is connected with one ancestor node $J' = j_1j_2, \ldots, j_{n-1}$ and $N(J)$ children nodes $j_1j_2, \ldots, j_nj_{n+1}$ ($j_{n+1} = 1, \ldots, N(J)$). Let M be the number of nodes in T. For a node J, let $T(J)$ be a subtree of T with the node J and its descendant nodes. Let $M(J)$ be the number of nodes in $T(J)$. Then $M(J) = \sum_{j_{n+1}=1}^{N(J)} M(Jj_{n+1}) + 1$.

For a $TSMAP$ with a tree structure T, direct transitions from a node (or *phase*) $J = j_1j_2, \ldots, j_n$ are restricted to its ancestor node $J' = j_1j_2, \ldots, j_{n-1}$ and to children nodes $j_1j_2, \ldots, j_nj_{n+1}$ ($j_{n+1} = 1, \ldots, N(J)$). Transition rates are defined as:

$a_{J,k}$= phase transition rate from J' to J with k ($k \geq 1$) batch arrivals,

$b_{J,k}$= phase transition rate from J to J' with k ($k \geq 1$) batch arrivals,

$c_{J,k}$= arrival rate with k ($k \geq 1$) batch arrivals and without phase transition at J,

$a_{J,0}$= phase transition rate from J' to J without arrival,

$b_{J,0}$= phase transition rate from J to J' without arrival,

$c_{J,0}$=$-\{\sum_{k=0}^{\infty}\sum_{j=1}^{N(Jj)} a_{Jj,k} + \sum_{k=0}^{\infty}\sum_{j=1}^{N(Jj)} b_{Jj,k} + \sum_{k=1}^{\infty} c_{J,k}\}$,

$a_J(z) = \sum_{k=0}^{\infty} a_{J,k} z^k$, $b_J(z) = \sum_{k=0}^{\infty} b_{J,k} z^k$ and $c_J(z) = \sum_{k=0}^{\infty} c_{J,k} z^k$.

Assume phases of the rate matrix D_k $(k = 0, 1, 2, \ldots)$ are arranged as the lexicographic order

$$1, 11, 111, 1111, \cdots, 112, \cdots, 11N(11), \cdots, 1N(1), \cdots, \tag{3}$$

where (J', J), (J, J') and (J, J) elements are $a_{J,k}$, $b_{J,k}$ and $c_{J,k}$, respectively. Then $D(z) = \sum_{k=0}^{\infty} D_k z^k$ is the matrix function whose elements are $a_J(z)$, $b_J(z)$ and $c_J(z)$ arranged as a tree structure T of the lexicographic order in (3).

For $J = j_1 j_2, \ldots, j_n$ define

$$\tau_J(z) = \prod_{l=2}^{n} \frac{a_{j_1 j_2, \ldots, j_l}(z)}{b_{j_1 j_2, \ldots, j_l}(z)},$$

where $\tau_1(z) = 1$. Let $\boldsymbol{\tau}(z)$ be the row vector of size M whose element corresponding to node J is $\tau_J(z)$ arranged as (3). Let

$$X(z) = \Delta(\boldsymbol{\tau}(z))$$

be the diagonal matrix whose (J, J) diagonal element is $\tau_J(z)$. Set $d_J(z) = \sqrt{a_J(z) b_J(z)}$.

Proposition 1 *Let R_D be the convergence radius of $D(z)$. For $|z| < R_D$ define*

$$F(z) = X(z)^{\frac{1}{2}} D(z) X(z)^{-\frac{1}{2}}.$$

Then $F(z)$ is a symmetric matrix with the tree structure T such that for $J = j_1 j_2, \ldots, j_n$ and $J' = j_1 j_2, \ldots, j_{n-1}$, (J', J) and (J, J') elements are $d_J(z)$, (J, J) element is $c_J(z)$ and otherwise is 0.

For any real $-R_D < z < R_D$, $D(z)$ and $F(z)$ have the same eigenvalues and all the eigenvalues are real.

Let $D_J(z), F_J(z)$ and $X_J(z)$ be $M(J) \times M(J)$ submatrices of $D(z), F(z)$ and $X(z)$ corresponding the subtree $T(J)$, respectively. In these notations, we have $D(z) = D_1(z)$, $F(z) = F_1(z)$, $X(z) = X_1(z)$ and $F_J(z) = X_J(z)^{\frac{1}{2}} D_J(z) X_J(z)^{-\frac{1}{2}}$. Therefore, for submatrices $D_J(z)$, $F_J(z)$ and $X_J(z)$, the same results in Proposition 1 are also satisfied.

4 Eigenvalues of $D(z)$

For a subtree $T(J)$ let the determinant of the submatrix $tI - F_J(z)$ defined as

$$f_J(t, z) \equiv \det(tI - F_J(z)),$$

where $f_1(t,z) = f(t,z) \equiv \det(tI - F(z))$. From the direct calculation of the determinant of a matrix with a tree structure we have the following lemma.

Lemma 1 *From a tree structure, $f_J(t,z)$ is inductively obtained by $f_{Jj}(t,z)$ and $f_{Jjl}(t,z)$ as*

$$f_J(t,z) = (t - c_J(z)) \prod_{j=1}^{N(J)} f_{Jj}(t,z)$$
$$- \sum_{j=1}^{N(J)} d_{Jj}(z)^2 \prod_{l=1}^{N(Jj)} f_{Jjl}(t,z) \prod_{j'\neq j, j'=1}^{N(J)} f_{Jj'}(t,z), \tag{4}$$

where $\prod_{j=1}^{N(J)} f_{Jj}(t,z) = 1$ if $N(Jj) = 0$.

Remark 1. In Section 7 for calculation of the eigenvalues $\alpha_i(z)$ for $|z| = 1$ we consider Newton's method. In this algorithm, effective calculations of $f(t,z)$ and $\frac{\partial}{\partial t} f(t,z)$ are necessary. (4) is the inductive calculations for $f(t,z)$ using determinants $f_J(t,z)$ of submatrices. Using the same discussion in Lemma 1, we also have the inductive calculations of $\frac{\partial}{\partial t} f(t,z)$ using both $f_J(t,z)$ and $\frac{\partial}{\partial t} f_J(t,z)$ by the partial differentiation of (4) with respect to t.

Let z be any fixed number in $(0,1]$. Let us define as
$n(\alpha, z)$ = the number of the zero points t of $f(t,z)$ which are greater than α.
$n_J(\alpha, z)$ = the number of the zero points t of $f_J(t,z)$ which are greater than α.
$n(\alpha, z) \equiv n_1(\alpha, z)$.
Equivalently, $n_J(\alpha, z)$ is the number of the eigenvalues of $F_J(z)$ which are greater than α since a zero point of $f_J(t,z)$ is an eigenvalue of $F_J(z)$. It is proved in Nishimura [11] that for a $BDMAP$ all the zero points of $f_J(t,z)$ are simple. However, in general there is a $TSMAP$ whose zero points of $f_J(t,z)$ have multiplicities. Let us assume the following.

Assumption 2
For any fixed z and J, all the zero points t of both $f_J(t,z)$ and $f_{Jj}(t,z)$ $(j = 1, \ldots, N(J))$ are distinct.

Theorem 1 *Under Assumption 2*

$$n_J(\alpha, z) = \begin{cases} \sum_{j=1}^{N(J)} n_{Jj}(\alpha, z) & \text{if } f_J(\alpha, z) \prod_{j=1}^{N(J)} f_{Jj}(\alpha, z) > 0 \\ \sum_{j=1}^{N(J)} n_{Jj}(\alpha, z) + 1 & \text{if } f_J(\alpha, z) \prod_{j=1}^{N(J)} f_{Jj}(\alpha, z) < 0 \\ \sum_{j=1}^{N(J)} n_{Jj}(\alpha, z) & \text{if } f_J(\alpha, z) = 0 \\ \sum_{j=1}^{N(J)} n_{Jj}(\alpha, z) + 1 & \text{if } \prod_{j=1}^{N(J)} f_{Jj}(\alpha, z) = 0. \end{cases}$$

This theorem is a natural extension of the Sturm sequence property of a symmetric tridiagonal matrix discussed in Wilkinson [19] (p.300-302) to a symmetric matrix with a tree structure. Under Assumption 2, zero points of $f_J(\alpha, z)$ are simple. From counting zero points of $f_{Jj}(\alpha, z)$ for subtrees we may derive the above equation. From Theorem 1, for a fixed z $n(\alpha, z)$ is inductively calculated by determinants of submatrices.

5 A bisection method

In this section we consider the following three problems.

Problem (A) **The vector g:**

The first problem for calculation of the stationary probability p_k in (1) is to obtain a numerically precise vector g. In Gail et al. [5] it is proved that if $\rho < 1$,

$$\det(zI - A(z)) = 0 \tag{5}$$

has the M zero points on the unit disk ($|z| \leq 1$) such that 1 is a simple zero point and there are the $M-1$ zero points counting multiplicities in the open unit disk ($|z| < 1$).

From Assumption 1 all the zero points z_i ($i = 1, \ldots, M$) of (5) on the unit disk are distinct. Let η_i be a right null vector of $zI - A(z)$ corresponding to z_i. Particularly, $z_1 = 1$ and $\eta_1 = e$. Then the matrix G is represented as

$$G = Y \begin{bmatrix} z_1 & & \\ & \ddots & \\ & & z_M \end{bmatrix} Y^{-1},$$

where $Y = [\eta_1, \ldots, \eta_M]$. From $gG = g$ and $ge = 1$, the vector g is calculated.

Problem (B) **Variance of an aggregated process:**

For a MAP let $X(0,t)$ ($t \in (0, \infty)$) and $V[X(0,t)]$ be the number of arrivals of a stationary during $(0,t)$ and its variance, respectively. In Neuts [8], it is proved that

$$\mathrm{V}[X(0,t)] = \boldsymbol{\pi} M_2(t)\boldsymbol{e} + \boldsymbol{\pi} M_1(t)\boldsymbol{e} - (\boldsymbol{\pi} M_1(t)\boldsymbol{e})^2, \tag{6}$$

where

$$M_1(t) = \int_0^t \exp(Du)D'\exp(D(t-u))du,$$

$$M_2(t) = 2\int_0^t \int_0^u \exp(D\nu)D'\exp(D(u-\nu))d\nu D'\exp(D(t-u))du + \int_0^t \exp(Du)D''\exp(D(t-u))du.$$

It is proved in Sato et al. [16] that substituting the spectral decomposition

$$\exp(Dt) = \sum_{i=1}^{M} \exp\{\alpha_i(1)t\}\boldsymbol{v}_i(1)\boldsymbol{u}_i(1)$$

into (6), we have the spectral representation of $\mathrm{V}[X(0,t)]$ as

$$\mathrm{V}[X(0,t)] = \lambda t + 2\sum_{i=2}^{M} \frac{\exp(\alpha_i(1)t) - 1 - \alpha_i(1)t}{\alpha_i(1)^2}\xi_{1i}\xi_{i1} + \bar{\xi}t,$$

where $\xi_{ii'} = \boldsymbol{u}_i(1)D'\boldsymbol{v}_{i'}(1)$ and $\bar{\xi} = \boldsymbol{u}_1(1)D''\boldsymbol{v}_1(1)$.

Problem (C) **The asymptotic behavior:**

We consider the asymptotic behavior of the stationary distribution $\boldsymbol{p}_n$. Let τ be the minimal zero point of $\det(\tau I - A(\tau))$ which is greater than 1. There are two eigenvalues 1 and τ satisfying that for a real $\tau > 0$ the Perron-Frobenius eigenvalue of $A(\tau)$ is equal to τ. Therefore, τ is the Perron-Frobenius eigenvalue $A(\tau)$ with $\tau > 1$. Using Corollary 2 in Seneta [17], the Taylor expansion of $(zI - A(z))^{-1}$ in $\boldsymbol{p}(z)$ of (1) at τ is given. Under the regularity condition it is proved in Falkenberg [4] and Tijms [18] that for a sufficient large n

$$\boldsymbol{p}_n \approx \boldsymbol{q}_n \equiv \frac{(1-\rho)\alpha_1(\tau)}{\lambda(h'(\alpha_1(\tau))\boldsymbol{u}_1(\tau)D'(\tau)\boldsymbol{v}_1(\tau) - 1)}\boldsymbol{g}\boldsymbol{v}_1(\tau)\boldsymbol{u}_1(\tau)\tau^{-n}.$$

Next, for a *TSMAP* we consider algorithms for calculations of z_i in the problem (A), $\alpha_i(1)$ in the problem (B) and τ in the problem (C).

Calculation for all the zero points of $\det(zI - A(z))$ on the unit disk:

Theorem 2 *On the unit disk, the zero points of* $\det(zI - A(z))$ *and* $\det(h^{-1}(z)I - D(z))$ *are the same, simple and real in* $(0,1]$. *Let the zero points be numbered as* $0 < z_M < z_{M-1} < \cdots < z_1 = 1$. *Then* z_i *is the unique solution of the equation*

$$h(z)^{-1} = \alpha_i(z)$$

in $z \in (0,1]$.

For a fixed $z' = h(\alpha')$, $n(\alpha', h(\alpha'))$ *is the number of the zero points of* $\det(zI - A(z))$ *greater than* z', *where* $n(\alpha, h(\alpha)) = n_1(\alpha, h(\alpha))$ *is inductively obtained in Theorem 1.*

Proof. For a real z, all the eigenvalues of $D(z)$ are real and simple. Particularly, eigenvalues of D are nonpositive. Since $h^{-1}(1) = 0$ and $\lim_{z\to 0+} h^{-1}(z) = -\infty$, for each i there exists the unique solution z_i of $\alpha_i(z) = h^{-1}(z)$, which is a zero point of $\det(zI - A(z))$. Therefore, all the zero points z_i $(i = 1, \ldots, M)$ on the unit disk are real and simple in $(0,1]$, see Figure 1. This completes the proof of the first part of this theorem.

The second part is that for a fixed $z' \in (0,1]$ we have :

the number of the zero points $z_1, \ldots, z_i$ *of* $\det(zI - A(z))$ *greater than* z'
$=$ *the number of the zero points* $z_1, \ldots, z_i$ *of* $\det(h(z)^{-1}I - D(z))$ *greater than* z'
$=$ *the number of the eigenvalues* $\alpha_1(z'), \ldots, \alpha_i(z')$ *of* $D(z')$ *greater than* $h(z')^{-1}$
$= n(h(z')^{-1}, z')$
$= n(\alpha', h(\alpha'))$. □

From Lemma 1 and Theorem 1, $f_J(t,z)$ and $n_J(\alpha,z)$ are inductively obtained by $f_{Jj}(t,z)$ and $n_{Jj}(\alpha,z)$, respectively. The existence region of eigenvalues is $(-2\delta, 0]$, where $\delta \equiv \max_J c_J(1)$. The number $n(w, h(w))$ is calculated by induction and by using a bisection method all the zero points in (0,1] are effectively calculated as follows.

Algorithm (A): Calculation of all the zero points of $\det(zI - A(z))$ on the unit disk.

1. Fix a positive error bound ϵ.

2. Put $i = M$, $n = 0$, $x_0 = -2\delta$ and $y_0 = 0$.

3. Put $w_n = (x_n + y_n)/2$.

   (i) If $n(w_n, h(w_n)) < i$ then $x_{n+1} = x_n$ and $y_{n+1} = w_n$.

   (ii) If $n(w_n, h(w_n)) \geq i$ then $x_{n+1} = w_n$ and $y_{n+1} = y_n$.

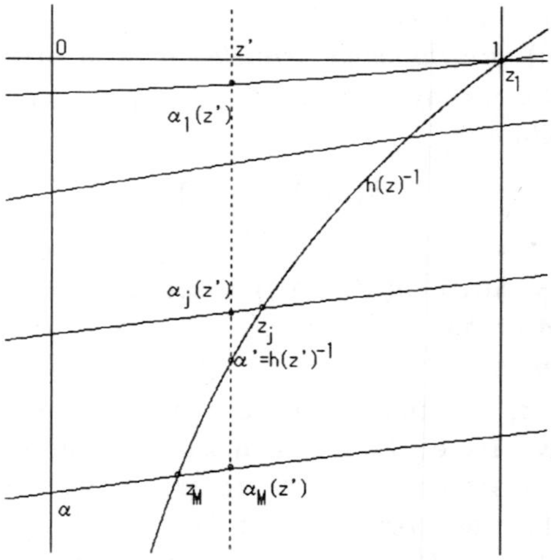

Figure 1. Graph of $\alpha_j(z)$ and $h^{-1}(z)$

4. (i) If $x_{n+1} - y_{n+1} \geq \epsilon$, set n as $n+1$ and return to 3.

   (ii) If $i \geq 3$ and $x_{n+1} - y_{n+1} < \epsilon$, put $\alpha_i = (x_{n+1} + y_{n+1})/2$ and $z_i = h(\alpha_i)$. Put $x_0 = \alpha_i$, $y_0 = 0$ and $n = 0$. Set i as $i-1$ and return to 3.

   (iii) If $i = 2$ and $x_{n+1} - y_{n+1} < \epsilon$, put $\alpha_2 = (x_{n+1} + y_{n+1})/2$ and $z_2 = h(\alpha_2)$. And put $\alpha_1 = 0$ and $z_1 = 1$.

5. End.

By a little modification of the algorithm for Problem (A), algorithms for Problem (B) and (C) are given below.

Algorithm (B): Calculation of all the eigenvalues of D:

Replace $n(w_n, h(w_n))$ to $n(w_n, 1)$ in (A).

Algorithm (C): Calculation of the Perron-Frobenius eigenvalue τ:

1. Fix a positive error bound ϵ.

2. Put $n = 0$ and $x_0 = 0$. Select a sufficiently large positive number such that $y_0 > \tau$.

3. Put $w_n = (x_n + y_n)/2$.

   (i) If $n(w_n, h(w_n)) > 0$ then $x_{n+1} = x_n$ and $y_{n+1} = w_n$.

   (ii) If $n(w_n, h(w_n)) = 0$ then $x_{n+1} = w_n$ and $y_{n+1} = y_n$.

4. (i) If $x_{n+1} - y_{n+1} \geq \epsilon$, set n as $n+1$ and return to 3.

   (ii) If $x_{n+1} - y_{n+1} < \epsilon$, then put $\tau = (x_{n+1} + y_{n+1})/2$.

5. End.

6 Eigenvector of $F(z)$

For spectral methods it is very important to effectively calculate eigenvectors as well as eigenvalues. Particularly, in calculation of Fourier inversion formula a lot of eigenvectors should be calculated. We derive an algorithm for calculation of an eigenvector inductively obtained by null vectors of submatrices. From Assumption 1 for a fixed $|z| \leq 1$ all the eigenvalues of $D(z)$ are simple. To simplify notations, abbreviating z, i and α we define $R = \alpha_i(z)I - F(z)$ as the $M \times M$ symmetric matrix with rank$(R) = M - 1$ whose (J, J) element is $\gamma_J \equiv \alpha_i(z) - c_J(z)$ and both (J, Jj) and (Jj, J) elements are $\delta_{Jj} \equiv -d_{Jj}(z)$, respectively. We assume that for all Jj $\delta_{Jj} \neq 0$. Our problem is to derive a nonzero left null vector $\boldsymbol{\psi}$ of R satisfying

$$\boldsymbol{\psi} R = \mathbf{0}. \tag{7}$$

Let $R_J \equiv (\boldsymbol{f}_J, F_J)$ be the $M(J) \times M(J)$ submatrix of R corresponding to a subtree T_J. Suppose that for each subtree T_{Jj} $(j = 1, \ldots, N(J))$ a nonzero row vector $\boldsymbol{x}_{Jj}$ of size $M(Jj)$ satisfying

$$\boldsymbol{x}_{Jj} F_{Jj} = \mathbf{0}, \tag{8}$$

is given. Let x_{Jj} be the first element of $\boldsymbol{x}_{Jj}$ and define

$$w_{Jj} = \boldsymbol{x}_{Jj} \boldsymbol{f}_{Jj}.$$

From the given nonzero vectors $\boldsymbol{x}_{Jj}$ $(j = 1, \ldots, N(J))$ we will derive a nonzero vector $\boldsymbol{y}_J = (y_J, \boldsymbol{y}_{J1}, \ldots, \boldsymbol{y}_{JN(J)})$ satisfying

$$\boldsymbol{y}_J F_J = \mathbf{0}. \tag{9}$$

From the tree structure and rank$(R) = M - 1$, rank$(R_{Jj}) = M(Jj)$ or $M(Jj) - 1$. If and only if rank$(R_{Jj}) = M(Jj) - 1$, R_{Jj} is singular. Let us define the number of singular matrices R_{Jj} as

$$L_J = \sharp\{R_{Jj} : \text{rank}(R_{Jj}) = M(Jj) - 1 \ \ (j = 1, \ldots, N(J))\}.$$

Proposition 2 *Suppose that for each* $j = 1, \ldots, N(J)$ $\mathrm{rank}(F_{Jj}) = M(Jj) - 1$ *is satisfied and a nonzero vector* $\boldsymbol{x}_{Jj}$ *satisfying (8) is given. Then we have* $L_J = 0, 1, 2$. *For each case* $\boldsymbol{y}_J$ *satisfying (9) is given as follows.*

1. *If* $L_J = 0$, $\mathrm{rank}(F_J) = M(J) - 1$ *and a nonzero vector* $\boldsymbol{y}_J$ *is given by*

$$\boldsymbol{y}_J = t(1, \bar{\boldsymbol{x}}_{J1}, \ldots, \bar{\boldsymbol{x}}_{JN(J)}),$$

where $\bar{\boldsymbol{x}}_{Jj} = (-\delta_{Jj}/w_{Jj})\boldsymbol{x}_{Jj}$.

2. *If* $L_J = 1$, *that is,* $\mathrm{rank}(R_{Jl}) = M(Jl) - 1$, *then* $\mathrm{rank}(F_J) = M(J) - 1$ *and a nonzero vector* $\boldsymbol{y}_J$ *is given*

$$\boldsymbol{y}_J = t(0, \mathbf{0}, \ldots, \mathbf{0}, \boldsymbol{x}_{Jl}, \mathbf{0}, \ldots, \mathbf{0}).$$

3. *If* $L_J = 2$, *that is,* $\mathrm{rank}(R_{Jj}) = M(Jj) - 1$ (if $j = l_1$ or l_2) *then a nonzero vector* $\boldsymbol{y}$ *is given by*

$$\boldsymbol{y}_J = (0, \mathbf{0}, \ldots, \mathbf{0}, s_1\boldsymbol{x}_{(Jl_1)}, \mathbf{0}, \ldots, \mathbf{0}, \mathbf{0}, s_2\boldsymbol{x}_{(Jl_2)}, \mathbf{0}, \ldots, \mathbf{0}).$$

Moreover, $\boldsymbol{\psi}$ *in (7) is given as*

$$\boldsymbol{\psi} = t(\mathbf{0}, \ldots, \mathbf{0}, \boldsymbol{y}_J, \mathbf{0}, \ldots, \mathbf{0}),$$

where $t \neq 0$, $s_1 = 1$ *and* $s_2 = -\frac{x_{Jl_1}\delta_{Jl_1}}{x_{Jl_2}\delta_{Jl_2}}$.

Remark 2. A nonzero left null vector $\boldsymbol{\psi}$ is inductively derived by left null vectors $\boldsymbol{y}_J$ of submatrices F_J. Suppose that for a submatrix F_J in the inductive procedure, there does not occur the case 3, that is, for all Jj ($j = 1, \ldots, N(J)$) and their descendant nodes, null vectors are obtained by the cases 1 and 2. Then we have that for j ($j = 1, \ldots, N(J)$) $\mathrm{rank}(F_{Jj}) = M(Jj) - 1$. Therefore, $L_J = \sharp\{j : w_{Jj} = 0 \ \ (j = 1, \ldots, N(J))\}$ because $w_{Jj} = 0$ if and only if $\mathrm{rank}(R_{Jj}) = \mathrm{rank}(\boldsymbol{f}_{Jj}, F_{Jj}) = \mathrm{rank}(F_{Jj}) = M(Jj) - 1$. So if for some J, $\sharp\{j : w_{Jj} = 0 \ \ (j = 1, \ldots, N(J))\} = 2$, then from the case 3 we immediately obtain $\boldsymbol{\psi}$.

7 Calculation of eigenvalues and eigenvectors for the Fourier inversion formula

The discrete Fourier inversion formula for calculations of the stationary probability $\boldsymbol{p}_n$ is discussed in Section 2. Both eigenvalues of $D(z)$ ($|z| = 1$) and corresponding eigenvectors should be effectively calculated.

Calculation of eigenvalues

From Assumption 1 for each fixed $|z| = 1$ $\alpha_i(z)$ $(i = 1, \ldots, M)$ are simple. Since $\alpha_i(z)$ is continuous in z and N in (2) is large. $\alpha_i(\omega^k)$ is very close to $\alpha_i(\omega^{k+1})$, where $\omega = e^{2\pi \boldsymbol{i}/N}$. Therefore, we consider Newton's method [11]. Suppose that for a fixed k $\alpha_i(\omega^k)$ $(i = 1, \ldots, M)$ are obtained. To calculate $\alpha_i(\omega^{k+1})$, set $\alpha_i^{(0)}(\omega^{k+1}) = \alpha_i(\omega^k)$ as the initial condition. Define

$$\alpha_i^{(\nu+1)}(\omega^{k+1}) = \alpha_i^{(\nu)}(\omega^{k+1}) - \frac{f(\alpha_i^{(\nu)}(\omega^{k+1}), \omega^{k+1})}{\frac{\partial}{\partial \alpha} f(\alpha, \omega^{k+1}) \mid_{\alpha = \alpha_i^{(\nu)}(\omega^{k+1})}} \quad \nu = 0, 1, 2, \ldots \tag{10}$$

as Newton's iterations until $\alpha_i^{(\nu)}(\omega^{k+1})$ converges. The inductive calculations of $f(\alpha_i^{(\nu)}(\omega^{k+1}), \omega^{k+1})$ and $\frac{\partial}{\partial \alpha} f(\alpha, \omega^{k+1}) \mid_{\alpha = \alpha_i^{(\nu)}(\omega^{k+1})}$ are given in Lemma 1 and Remark 1. As the initial condition for $k = 1$ we set $\alpha_i^{(0)}(\omega) = \alpha_i(0)$ $(i = 1, \ldots, M)$.

Convergence of $\{\alpha_i^{(\nu)}(\omega^{k+1})\}$ in (10): Since $D(z)$ is uniformly bounded on the unit disk and the characteristic function $f(\alpha, z)$ is a polynomial in α with degree M, both $f(\alpha, z)$ and $\frac{\partial^2}{\partial \alpha^2} f(\alpha, z)$ are uniformly bounded. And since $\alpha_i(z)$ is simple, there exists a positve number L such that

$$|\frac{\partial}{\partial \alpha} f(\alpha_i(z), z)| \geq L > 0 \tag{11}$$

on the compact set $|z| \leq 1$. Then for any $|z| \leq 1$ there exists a uniform radius r_α such that for any $\alpha \in \{\alpha; |\alpha - \alpha_i(z)| < r_\alpha\}$

$$|\frac{f(\alpha, z) \frac{\partial^2}{\partial \alpha^2} f(\alpha, z)}{\frac{\partial}{\partial \alpha} f(\alpha, z)}| < 1.$$

That is, for any initial value $\alpha \in \{\alpha; |\alpha - \alpha_i(z)| < r_\alpha\}$ the sequence $\{\alpha_i^{(\nu)}(z)\}$ quadratically converges to $\alpha_i(z)$.

On the other hand, $D(z)'$ is a uniformly bounded matrix on the unit disk because $D'(1)$ is a finite matrix. Then there exists a positive number U such that

$$|\frac{\partial}{\partial z} f(\alpha, z)| \leq U. \tag{12}$$

It follows from the theory of implicit functions that for any $|z| \leq 1$ $\alpha_i(z)$ is analytic and

$$\frac{d}{dz} \alpha_i(z) = -\frac{\frac{\partial}{\partial \alpha} f(\alpha_i(z), z)}{\frac{\partial}{\partial z} f(\alpha_i(z), z)}.$$

From (11) and (12) the derivative is bounded as

$$|\frac{d}{dz}\alpha_i(z)| \leq \frac{U}{L}.$$

There exists a positive number r_z such that for $|z'-z| < r_z$, $|\alpha_i(z')-\alpha_i(z)| < r_\alpha$. Therefore, for a sufficient large N, $|\alpha_i(\omega^k) - \alpha_i(\omega^{k+1})| < r_\alpha$ and the sequence $\{\alpha_i^{(\nu)}(\omega^{k+1})\}$ converges to $\alpha_i(\omega^{k+1})$ as $\nu \to \infty$.

Calculation of eigenvectors

Suppose that the eigenvalues $\alpha_i(\omega^k)$ $(i = 1, \ldots, M)$ of the symmetric matrix $F(\omega^k)$ are obtained. From Proposition 2 the left null vector $\boldsymbol{\psi}_i(\omega^k)$ of $F(\omega^k)$ is inductively obtained with the normalized condition $\boldsymbol{\psi}_i(\omega^k)\boldsymbol{\psi}_i(\omega^k)^{\boldsymbol{T}} = 1$. Then from the decomposition $F(z) = X(z)^{\frac{1}{2}}\ D(z)\ X(z)^{-\frac{1}{2}}$ in Proposition 1 we have normalized left and right eigenvectors of $D(\omega^k)$ as

$$\boldsymbol{u}_i(\omega^k) = \boldsymbol{\psi}_i(\omega^k)X^{\frac{1}{2}}(\omega^k)$$
$$\boldsymbol{v}_i(\omega^k) = X^{-\frac{1}{2}}(\omega^k)\boldsymbol{\psi}_i(\omega^k)^{\boldsymbol{T}}$$

with $U(\omega^k)V(\omega^k) = I$, respectively.

8 A numerical example

As a numerical model, we consider the following $TSMAP$ with two dimensional phases (k, l) $(k = 1, \ldots, K, l = 1, \ldots, L)$ arranged as

$$\begin{array}{ccccccc}
(1,1) & - (1,2) & - \cdots - & (1,l) & - \cdots - & (1,L) \\
| & & & & & \\
\vdots & & & & & \\
| & & & & & \\
(k,1) & - (k,2) & - \cdots - & (k,l) & - \cdots - & (k,L) \\
| & & & & & \\
\vdots & & & & & \\
| & & & & & \\
(K,1) & - (K,2) & - \cdots - & (K,l) & - \cdots - & (K,L).
\end{array}$$

Its transition and arrival rates are defined as
transition rate from (k, l) to $(k, l + 1)$ without arrival $= 1$ if $l < L$
transition rate from (k, l) to $(k, l - 1)$ without arrival $= 1$ if $l > 1$

Table 1. In the case of $(K, L) = (10, 10)$ $\alpha_i(1)$, z_i and g_i.

i	$\alpha_i(1)$	z_i	g_i
1	0	1	1.8574360202888E-2
2	-7.55023962308249E-3	.98357864838016	1.9302459845886E-2
⋮	⋮	⋮	⋮
⋮	⋮	⋮	⋮
99	-4.99999997814935	.15588274527097	1.4128801150203E-3
100	-5.24668954570343	.14766384796854	1.3963135796600E-3

transition rate from (k, l) to $(k+1, l)$ without arrival $= 1$ if $k < K$
transition rate from (k, l) to $(k-1, l)$ without arrival $= 1$ if $k > 1$
arrival rate at (k, l) without transition of phase $= k - 1$.
Note that this index of phases is different from that of nodes in Section 3 but is suitable for this $TSMAP$ because k and l represent the level of arrival rate and the depth of the level, respectively. If the depth l is large, the arrival system tends to keep the same arrival level k. If $L = 1$, this $TSMAP$ is formulated as a $BDMAP$ discussed in Nishimura [11]. For several pairs of parameters (K, L) we numerically analyze a $TSMAP/G/1$ queue by using spectral methods. To make easy comparison we assume a exponential service time with traffic intensity $\rho = 0.5$ but these assumptions are not essential for numerical calculations as was discussed for a $BDMAP$ [11]. In our calculations, programs are written by Decimal BASIC (Double Precision) and executed by a Pentium 3 PC.

8.1 Computational remarks for the bisection method

In the case of $K = 10$ and $L = 10$, numerical results of $\alpha_i(1)$, z_i and g_i $i = 1, \ldots, 100$ are listed in Table 1. In this case the Perron-Frobenius eigenvalue $\tau = 1.12348415714507$. In a short computation time these values are calculated. In author's experiences, for a MAP with 100×100 matrix size it is impossible to precisely calculate the vector $\boldsymbol{g}$ by Lucantoni's method [7] using a personal computer. For calculation of the above characteristics of a $TSMAP$ the bisection method is very efficient.

8.2 Self-similarity of second order

In traffic of telecommunication systems a self-similar property is observed [6 13]. One of the important properties is the second order similarity based

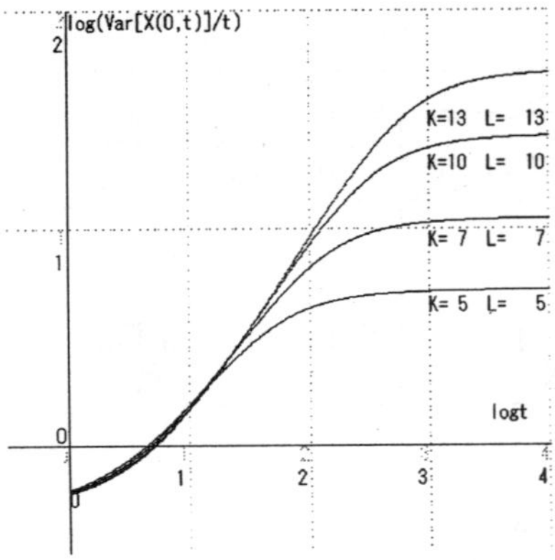

Figure 2.

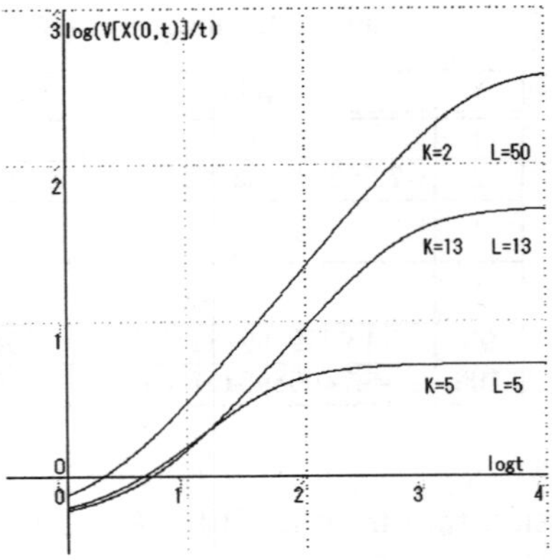

Figure 3.

on the analysis of the variance for arrival numbers of aggregated processes. We consider a self-similar process with the Harst parameter $H = 1 - \beta/2$ $(1 > \beta > 0)$. Let Y_t $(t \in (0, \infty))$ and $\mathrm{V}[Y_t]$ be the number of arrivals of the stationary version during $(0, t)$ and its variance, respectively. Then $\log_{10}(\mathrm{V}[Y_t]/t) = \log_{10}\sigma^2 + (\beta - 1) \times \log_{10} t$.

For a $TSMAP$ by the problem (B) in Section 5, $\mathrm{V}[X(0,t)]$ has the spectral representation. In Figure 2 and 3, several curves of $\log_{10}(\mathrm{V}[X(0,t)]/t)$ are illustrated with the horizontal axis $\log_{10} t$. Roughly speaking, the slope of each curve is nearly equal to 1. The Harst parameter $H \approx 1/2$ $(\beta \approx 1)$ over several time scales. In Figure 2 the self-similar time scale is longer if K is larger. For instance, the longest time scale in these four cases is about 10^3 when $K = L = 13$. In Figure 3 when $K = 2$ and $L = 50$ the longest time scale is about 10^4. If $K = 2$, there are two states *on* and *off*. It seems that in the cases of the constant matrix size $M = 2L$, the longest time scale is accomplished by the above on-off model with the depth L, which is an natural extension of an interrupted Poisson process. This type of a $TSMAP$ is formulated as a $BSMAP$. To show the caudal characteristics of a QBD, the similar input source is considered in Example 1 [2].

8.3 The stationary distribution of a $TSMAP/G/1$ queue

The stationary distribution $\boldsymbol{p}_n$ and the asymptotic approximated behavior $\boldsymbol{q}_n$ are calculated by the Fourier inversion formula (2) and the Perron-Frobenius eigenvalue τ in the problem (C), respectively. In the case of $K = 2$ and $L = 60$, 120 traces of the eigenvalues $\alpha_i(z)$ $(|z| = 1)$ are illustrated in Figure

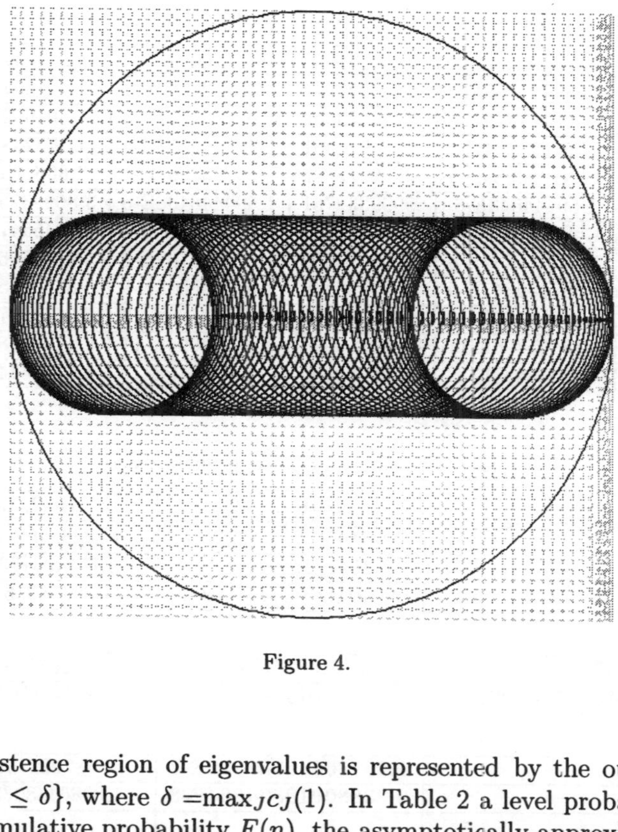

Figure 4.

4. The existence region of eigenvalues is represented by the outside circle $\{\alpha : |\alpha + \delta| \leq \delta\}$, where $\delta = \max_J c_J(1)$. In Table 2 a level probability $p_n = \boldsymbol{p}_n \boldsymbol{e}$, its cumulative probability $F(n)$, the asymptotically approximated level probability $q_n = \boldsymbol{q}_n \boldsymbol{e}$ and the relative error between p_n and q_n as $(q_n - p_n)/p_n$ are listed. From this the accuracy of the stationary probability are checked by $F(n)$ and q_n.

In Table 3, the stationary level probabilities p_n are given in the three cases of a Poisson arrival, (10,10) and (2,50). Let $b(10^{-8})$ be the buffer size such that the loss probability with a finite buffer queue are less than 10^{-8} [3]. For the same traffic intensity $\rho = 0.5$ the buffer sizes are quite different (see, $n = 23, 112$ and 442).

9 Concluding remarks

Under Assumption 1 and 2 we consider the bisection method and Newton's method for calculations of the eigenvalues. We do not know whether our numerical example satisfies these assumptions, theoretically. But all the eigen-

Table 2. The stationary distribution at service completion epochs

n	p_n	$F(n)$	q_n	$(q_n - p_n)/p_n$
0	3.681727460E-2	3.681727E-2	.0201829599	-.45180733
100	1.906231829E-3	.9203279342	1.902791171E-3	-1.804952E-3
200	1.793917604E-4	.9924929401	1.793896557E-4	-1.173238E-5
300	1.691233989E-5	.9992922586	1.691233861E-5	-7.606175E-8
400	1.59444643E-6	.9999332761	1.594446436E-6	3.8099325E-9
500	1.503197972E-7	.9999937094	1.503198047E-7	4.9946547E-8
600	1.417170854E-8	.9999994069	1.417171702E-8	5.9819666E-7
700	1.336059347E-9	.9999999440	1.336068548E-9	6.8866765E-6
800	1.25951404E-10	.9999999947	1.25960683E-10	7.3673867E-5
900	1.18655706E-11	.9999999994	1.18752094E-11	8.1233018E-4
1000	1.13658885E-12	.9999999999	1.14631370E-12	8.5561684E-3

Table 3. The stationary level probability p_n

n	Poisson	K=10, L=10	K=2, L=50
0	.5	.310357270680539	4.23282460408955E-2
1	.25	.164272193796172	3.7814031700177E-2
2	.125	.104430737702228	3.45692495622615E-2
⋮	⋮	⋮	⋮
22	**1.19209289550781E-7**	3.50404388712015E-3	.1405937685638E-2
23	**5.96046447753906E-8**	3.09827019691307E-3	.13599115845769E-2
⋮	⋮	⋮	⋮
111	—	**1.03796539181892E-7**	1.06822217527898E- 3
112	—	**9.23880018894864E-8**	1.03864380337084E-3
⋮	⋮	⋮	
441	—	1.43824229327343E-14	**1.01631507276042E- 7**
442	—	1.44630672455764E-14	**9.88193622934901E-8**
⋮	⋮	⋮	⋮

values are distinct numerically and characteristics are calculated by spectral methods. It seems to the author that these assumptions are not so strong in applications since in his experiences two assumptions are satisfied numerically.

The matrix analytic methods given in Lucantoni [7] are applicable to a general $MAP/G/1$ queueing model. On the other hand, to apply spectral methods to a $MAP/G/1$ queue, effective calculations for all the eigenvalues and corresponding eigenvectors are necessary. Though computational costs of spectral methods discussed in this paper are not estimated, we may show that spectral methods is more convenient than the above matrix analytic methods.

To model burst traffic of telecommunication systems by using a MAP with a large number of phases, we should estimate many parameters. To avoid this difficulty, one of the useful methods is to introduce a structure to a MAP and to find effective algorithms for estimation of its parameters and calculations of its characteristics.

Acknowledgments

The author would like to thank two anonymous referees for their valuable comments.

References

1. Andersen, A. T. and Nielsen, B. F. A Markovian approach for modeling packet traffic with long range dependence. *IEEE JSAC.* **16**, 719-732 (1998).
2. Bean, N. G., Jian-min Li and Taylor, P. G. Caudal characteristics of QBDs with decomposable phase spaces In *Advances in Algorithmic Methods for Stochastic Models*, eds. G. Latouche and P.G. Taylor (Notable Publications, 279–294, 2000).
3. Choudhury, G. L., Lucantoni, D. M. and Whitt, W. Squeezing the most out of ATM. *IEEE Trans. Commun.* **44**, 203- 217 (1996).
4. Falkenberg, E. On the asymptotic behaviour of the stationary distribution of Markov chains of MAP/G/1-type. *Commun. Statist. Stochastic Models* **10**, 75-97 (1994).
5. Gail, H. R., Hantler, S. L. and Taylor, B. A. Solutions of the basic matrix equation for $M/G/1$ and $G/M/1$ type Markov chains. *Commun. Statist. Stochastic Models* **10**, 1-43 (1994).
6. Leland, W. E.; Taqqu, M. S.; Willinger, W.; Wilson, D. V. On the self-similar nature of Ethernet traffic (extended version). *IEEE Trans. Networking* **2**, 1-15 (1994).
7. Lucantoni, D. M. New results on the single server queue with a batch Markovian arrival process. *Commun. Statist. Stochastic Models* **7**, 1-46 (1991).
8. Neuts, M. F. *Structured Stochastic Matrices of $M/G/1$ Type and Their Applications* (Marcel Dekker, New York, 1989).
9. Nielsen, B. F. Modelling long-range dependent and heavy-tailed phenomena by matrix analytic method. In *Advances in Algorithmic Methods for*

Stochastic Models eds. G. Latouche and P.G. Taylor (Notable Publications, 279–294, 2000).

10. Nishimura, S. A spectral analysis for a $MAP/D/N$ queue. In *Advances in Algorithmic Methods for Stochastic Models*, eds. G. Latouche and P.G. Taylor (Notable Publications, 279–294, 2000).
11. Nishimura, S. A $MAP/G/1$ queue with a birth and death underlying process. (submitted)
12. Nishimura, S.; Jiang, Y. Spectra of matrix generating function for a MAP/SM/1 queue. *Commun. Statist. Stochastic Models* **16**, 99-120 (2000).
13. Paxson, V.; Floyd, S. Wide area traffic: the failure of Poisson modeling. *IEEE Trans. Networking* **3**, 226-244 (1995).
14. Ramaswami, V. A stable recursion for the steady state vector in Markov chains of $M/G/1$ type. *Commun. Statist. Stochastic Models* **4**, 183-188 (1988).
15. Robert, S.; Le Boudec, J. New models for pseudo self-similar traffic. *Performance Evaluation* **30**, 57-68 (1997).
16. Sato, H.; Nishimura, S. Time-dependent moments of the counts on a $BMAP$. *Mathematical and Computer Modelling* **31**, 307-312 (2000).
17. Seneta, E. *Non-negative matrices and Markov chains,* 2nd ed. (Springer-Verlag, New York, 1980).
18. Tijms, H. *Stochastic Models,* (Wiley, New York, 1994).
19. Wilkinson, J. *The algebraic eigenvalue problem,* (Clarendon Press, 1965).

SOJOURN AND PASSAGE TIMES IN MARKOV CHAINS

CLÁUDIA NUNES AND ANTÓNIO PACHECO
Department of Mathematics and CEMAT
Instituto Superior Técnico, Av. Rovisco Pais, 1049-001 Lisboa, Portugal
E-mails: {cnunes,apacheco}@math.ist.utl.pt

We consider a finite state Markov chain whose state space is partioned in two sets, E_0 and E_1, and define a *cycle* as being constituted by a sojourn period in E_0 and another one in E_1. We characterize the bivariate sequences of sojourn times in and passage times into E_0 (E_1) and states visited at entrance times into E_0 (E_1), along with the sequence of cycle durations and the states visited at the beginning of cycles. We derive bivariate probability generating matrix functions of sojourn times in E_0 and E_1, and of cycle durations. Using these, we obtain expressions for moments of sojourn times in E_0 and E_1 and of cycle durations, including their expected values, variances and covariances. Stationary versions of the Markov chain are also discussed. The derived results are illustrated in a variant of the *success runs for Bernoulli trials* setting, where changes between two different success probabilities are allowed. The analysis shows that non-negligible correlations between sojourn times in a set of states and between cycle durations may arise, as illustrated through the considered example.

1 Introduction

Sojourn times in sets of states and passage times into sets of states are among the most important functionals of Markov chains and have been the subject of much research; see, e.g., the article of Rubino and Sericola [10] and the books of Latouche and Ramaswami [6], Kijima [4] and Wasan [11], and references therein.

However, traditionally sojourn times and passage times have been characterized probabilistically only through their marginal distributions. This severely limits the usefulness of the results derived, as sequences of sojourn times in sets of states and passage times into sets of states in Markov chains are dependent sequences in a very natural way.

In this paper we characterize the dependence structure of sequences of sojourn times in and passage times into a given set of states, say E_0, of a discrete time Markov chain (DTMC). This characterization is done by considering simultaneously the sequence of soujourn times and passage times into the given set, E_0, and its complement, say E_1.

In order to better motivate and explain the work done, let us consider $J = \{J_n, n \geq 0\}$ to be an homogeneous irreducible DTMC with finite state space $E = E_0 \cup E_1$, with E_0 and E_1 being disjoint sets. We consider, w.l.o.g., the states of E to be numbered such that the first s_0 states belong to E_0 and the remaining $s_1 = s - s_0$ states belong to E_1. This induces the following partition of the transition probability matrix P of J:

$$P = \begin{bmatrix} P_{00} & P_{01} \\ P_{10} & P_{11} \end{bmatrix} \tag{1}$$

where P_{00} is the $s_0 \times s_0$ matrix of transition probabilities between states in E_0.

As an example, consider that $K = \{K_n, n \geq 0\}$ is the phase sequence of a discrete time Markovian arrival process (DMAP) with parameter matrices D_0 and D_1, and having m phases, say $0, 1, 2, \ldots, m-1$. The matrices D_0 and D_1 are substochastic matrices, and are such that $D = D_0 + D_1$ is a stochastic matrix and is the transition probability matrix of K; see, e.g., Johnson and Narayana [3] for further details on the DMAP.

For $n \in \mathbb{N}_0$, let N_n denote the number of arrivals until time n, and define

$$L_n = \begin{cases} 0 & N_n = N_{n-1} \\ 1 & \text{otherwise} \end{cases} \tag{2}$$

with $N_{-1} = 0$. Then $J = (L, K) = \{(L_n, K_n), n \geq 0\}$ is a DTMC on $\{0,1\} \times \{0, 1, \ldots, m-1\}$ with transition probability matrix

$$P = \begin{bmatrix} D_0 & D_1 \\ D_0 & D_1 \end{bmatrix}. \tag{3}$$

Thus, P has the form (1) for $E_0 = \{0\} \times \{0, 1, \ldots, m-1\}$ and $E_1 = \{1\} \times \{0, 1, \ldots, m-1\}$. The sequence of sojourn times in E_1 is the sequence of number of consecutive instants with arrivals. This sequence has been studied by Li and Neuts [7]; more precisely, these authors investigated waiting times for (pre-specified length) runs of consecutive arrivals, or *success runs* in their terminology.

The model described by (2)-(3) corresponds to a generalization of a sequence of independent Bernoulli trials. The results for a sequence of independent Bernoulli trials corresponds to the case where $m = 1$, in which case we may consider $E_0 = \{0\}$ and $E_1 = \{1\}$ and a success run may be identified as a sequence of consecutive ones that is followed by a zero.

We now describe a generalization of a sequence of independent Bernoulli trials where the success probability may change between the values p_0 and p_1. Changes in the success probability are modeled by independent Bernoulli trials. These changes are only allowed when a success occurs after a (nonempty) sequence of failures. Moreover, the probability of a change in the success probability after a success may depend on the previous state (i.e., success probability) of the DTMC that drives the dependent sequence of Bernoulli trials.

In a more precise way, we consider that the success probability changes are modulated by the DTMC $J = \{J_n, n \geq 0\}$, on the state space $E = \{1, 2, 3, 4\}$, with transition probability matrix

$$P = \begin{bmatrix} 1-p_0 & 0 & p_0(1-\alpha_0) & p_0\alpha_0 \\ 0 & 1-p_1 & p_1\alpha_1 & p_1(1-\alpha_1) \\ 1-p_0 & 0 & p_0 & 0 \\ 0 & 1-p_1 & 0 & p_1 \end{bmatrix}. \tag{4}$$

The states 3 and 4 represent successes and the states 1 and 2 represent failures. If a success is obtained when departing from state 1 (2), the probability of a change in the success probability from p_0 to p_1 (p_1 to p_0) is α_0 (α_1). If we let $E_1 = \{3, 4\}$, the sequence of sojourn times in E_1 is the sequence of success run lengths. Note that the transition probability matrix (4) is not of the form (3). This example will be investigated in more depth in Section 6.

The investigation in this paper is also connected to busy-period cycles in Markovian type queues. As an illustration, let N_n be the number of customers in the DMAP/Geometric(p)/1/k system at time n, for $n \in \mathbb{N}_0$, with the arrival DMAP having matrix parameters D_0 and D_1 of size $m \times m$. Let $\{0, 1, \ldots, m-1\}$ denote the phase space of the arrival DMAP and let K_n be the corresponding phase at time n, so that K is an homogeneous DTMC with transition probability matrix P given by (3).

It follows that the process $J = (N, K) = \{(N_n, K_n), n \geq 0\}$ is an homogeneous DTMC on the state space $E = \{0, 1, \ldots, k\} \times \{0, 1, \ldots, m-1\}$ with block-partitioned transition probability matrix

$$\begin{bmatrix} D_0 & D_1 & \mathbf{0} & \mathbf{0} & \ldots & \mathbf{0} & \mathbf{0} & \mathbf{0} & \mathbf{0} \\ D_0 p & C & D_1^\star & \mathbf{0} & \ldots & \mathbf{0} & \mathbf{0} & \mathbf{0} & \mathbf{0} \\ \mathbf{0} & D_0 p & C & D_1^\star & \ldots & \mathbf{0} & \mathbf{0} & \mathbf{0} & \mathbf{0} \\ \ldots & \ldots & \ldots & \ldots & \ldots & \ldots & \ldots & \ldots & \ldots \\ \mathbf{0} & \mathbf{0} & \mathbf{0} & \mathbf{0} & \ldots & D_0 p & C & D_1^\star & \mathbf{0} \\ \mathbf{0} & \mathbf{0} & \mathbf{0} & \mathbf{0} & \ldots & \mathbf{0} & D_0 p & C & D_1^\star \\ \mathbf{0} & \mathbf{0} & \mathbf{0} & \mathbf{0} & \ldots & \mathbf{0} & \mathbf{0} & D_0 p & D_0^\star + D_1 \end{bmatrix} \tag{5}$$

where $D_0^\star = D_0(1-p)$, $D_1^\star = D_1(1-p)$, $C = D_0^\star + D_1 p$ and $\mathbf{0}$ is a matrix with all entries equal to zero.

Note that if we let $E_0 = \{(0, j) : 0 \leq j \leq m-1\}$ and $E_1 = E \setminus E_0$, then the sequence of sojourn times in E_1 (E_0) is the sequence of busy (idle) period durations, and a *cycle* constituted by a visit to E_1 and another one to E_0 is what is commonly called a busy period cycle.

In this work we generalize most of the results presented by Rubino and Sericola [10] for DTMCs. However, these authors only study the marginal distribution of the sojourn times in one of the sets of states of interest, say E_0. By contrast, we characterize the bivariate sequences of sojourn times in and passage times into E_0 (E_1) and states visited at entrance times into E_0 (E_1), along with the sequence of cycle durations and the states visited at the beginning of cycles. Moreover, we also investigate the dependence structure of these sequences.

Our work is also related to the work of Johnson and Narayana [3] on the burstiness of DMAPs, as the transition probability matrix they analyze, which models the phase of the DMAP at the start of *bursts* and *gaps*, has the same structure as (3).

Due to the partition of the set of states (E) of the DTMC of interest, $J = \{J_n, n \geq 0\}$, in two sets (E_0 and E_1), and for ease of notation, we will next define some matrix operations related with block partition and self-extraction that will be used many times along the paper.

Accordingly, given a matrix $A = [A_{ij}]_{i \in K_0, j \in K_1}$ and two sets, $C_0 \subseteq K_0$ and $C_1 \subseteq K_1$, we denote by $A_{C_0 C_1}$ or $[A]_{C_0 C_1}$ the submatrix formed by the entries in rows belonging to C_0 and columns belong to C_1; i.e.,

$$A_{C_0 C_1} = [A]_{C_0 C_1} = [A_{ij}]_{i \in C_0, j \in C_1}. \tag{6}$$

Moreover, if $B = \{B(m) = [B_{ij}(m)]_{i \in K_1, j \in K_2}\}$ is a matrix function, we let $B_{C_0 C_1}(m) = [B(m)]_{C_0 C_1}$, for $C_0 \subseteq K_0$ and $C_1 \subseteq K_1$. Thus, the sets C_0 and C_1 induce the block partitionings

$$A = \begin{bmatrix} A_{C_0 C_1} & A_{C_0 \bar{C}_1} \\ A_{\bar{C}_0 C_1} & A_{\bar{C}_0 \bar{C}_1} \end{bmatrix} \quad \text{and} \quad B(m) = \begin{bmatrix} B_{C_0 C_1}(m) & B_{C_0 \bar{C}_1}(m) \\ B_{\bar{C}_0 C_1}(m) & B_{\bar{C}_0 \bar{C}_1}(m) \end{bmatrix} \tag{7}$$

where $\bar{C}_0 = K_0 \setminus C_0$ and $\bar{C}_1 = K_1 \setminus C_1$. In addition, if $C_0 = \{i\}$ and $C_1 = \{j\}$, we denote $A_{\{i\}\{j\}}$ simply by A_{ij} or $[A]_{ij}$.

Finally, whenever a matrix D may be written as a function of blocks of A as given in (7), so that $D = g(A_{C_0 C_1}, A_{C_0 \bar{C}_1}, A_{\bar{C}_0 C_1}, A_{\bar{C}_0 \bar{C}_1})$ for some function g, we denote by $\boxed{0 \leftrightarrow 1}(D)$ the matrix $g(A_{C_1 C_0}, A_{C_1 \bar{C}_0}, A_{\bar{C}_1 C_0}, A_{\bar{C}_1 \bar{C}_0})$.

When $K_0 = K_1 = E$ and $C_0 = E_0, E_1$ and $C_1 = E_0, E_1$, we replace E_0 by $\mathbf{0}$ and E_1 by $\mathbf{1}$, so that, e.g., $A_{\mathbf{00}} = [A]_{\mathbf{00}} = [A]_{E_0 E_0}$ and $B_{\mathbf{00}}(m) = [B(m)]_{E_0 E_0}$. This leads, in particular, to the representation of the transition probability matrix P given in (1).

We end the introduction with an overview of the paper's plan. In Section 2 we define cumulative cycle times and sojourn times in sets of states of Markov chains and characterize the associated processes. In Section 3 we characterize the associated sequences of individual cycle and sojourn times, and in Section 4 we compute moments of these processes, including expected values, variances and covariances. The results previously obtained are: specialized to the stationary case in Section 5 and illustrated in Section 6 for the sequence of dependent Bernoulli trials with transition probability matrix given by (4), where changes between two different success probabilities are allowed.

2 Cycles on Markov chains

Let $J = \{J_n, n \geq 0\}$ be an homogeneous irreducible discrete time Markov chain with finite state space E and probability transition matrix P of the form (1). In this section we introduce processes which are derived from this original Markov chain J, and we characterized them as Markov processes.

2.1 Reentrance times and states

The sequence of successive (re)entrance times of J in E_0 and E_1 is $T^{[\cdot]} = \{T_n^{[\cdot]}, n \geq 0\}$, where $T_0^{[\cdot]} = 0$ and

$$T_{n+1}^{[\cdot]} = \inf\left\{k > T_n^{[\cdot]} : (J_{k-1} \in E_1 \wedge J_k \in E_0) \vee (J_{k-1} \in E_0 \wedge J_k \in E_1)\right\} \quad (8)$$

for $n \in \mathbb{N}_0$. The sequence of states visited at the successive entrances of J in E_0 and E_1 is $J^{[\cdot]} = \{J_n^{[\cdot]}, n \geq 0\}$, with $J_n^{[\cdot]} = J_{T_n^{[\cdot]}}$.

Using standard techniques from the theory of DTMC, it is easy to prove that the process $(T^{[\cdot]}, J^{[\cdot]}) = \{(T_n^{[\cdot]}, J_n^{[\cdot]}), n \geq 0\}$ is a persistent homogeneous Markov renewal process [a] (MRP), on the state space $\mathbb{N}_0 \times E$, whose transition probability (matrix) function is

$$Q^{[\cdot]}(m) = \left[P(T_1^{[\cdot]} = m, J_1^{[\cdot]} = j | J_0^{[\cdot]} = i)\right]_{i,j \in E} = \begin{bmatrix} \mathbf{0} & P_{\mathbf{00}}^{m-1} P_{\mathbf{01}} \\ P_{\mathbf{11}}^{m-1} P_{\mathbf{10}} & \mathbf{0} \end{bmatrix} \quad (9)$$

for $m \in \mathbb{N}$, and $J^{[\cdot]}$ has transition probability matrix

$$P^{[\cdot]} = \begin{bmatrix} \mathbf{0} & [I - P_{\mathbf{00}}]^{-1} P_{\mathbf{01}} \\ [I - P_{\mathbf{11}}]^{-1} P_{\mathbf{10}} & \mathbf{0} \end{bmatrix}. \quad (10)$$

Remark that $J^{[\cdot]}$ alternates between states in E_0 and states in E_1; it stays in E_0 (E_1) a phase-type time with probability matrix $P_{\mathbf{00}}$ ($P_{\mathbf{11}}$), and consequently having absorption probability vector $P_{\mathbf{01}}\mathbf{1}$ ($P_{\mathbf{10}}\mathbf{1}$), which follows from

[a] For the definition of MRPs and their main properties see, e.g., Çinlar [2], Kulkarni [5] or Prabhu [9].

(9). Thus, it follows that the mean transition time vector from E_0 to E_1 (E_1 to E_0) is $[I - P_{00}]^{-1} P_{01}\mathbf{1}$ ($[I - P_{11}]^{-1} P_{10}\mathbf{1}$), which is a consequence of (10).

Consider now the sequences of (re)entrance times of J into E_0 and into E_1, $T^{[0]} = \{T_n^{[0]}, n \geq 0\}$ and $T^{[1]} = \{T_n^{[1]}, n \geq 0\}$, respectively, where

$$T_0^{[0]} = \inf\{k \geq 0 : J_k \in E_0\} \quad T_{n+1}^{[0]} = \inf\left\{k > T_n^{[0]} : J_{k-1} \in E_1 \wedge J_k \in E_0\right\}$$

$$T_0^{[1]} = \inf\{k \geq 0 : J_k \in E_1\} \quad T_{n+1}^{[1]} = \inf\left\{k > T_n^{[1]} : J_{k-1} \in E_0 \wedge J_k \in E_1\right\}.$$

Remark that

$$T_n^{[0]} = \begin{cases} T_{2n}^{[\cdot]} & T_0^{[\cdot]} \in E_0 \\ T_{2n+1}^{[\cdot]} & T_0^{[\cdot]} \in E_1 \end{cases} \quad T_n^{[1]} = \begin{cases} T_{2n+1}^{[\cdot]} & T_0^{[\cdot]} \in E_0 \\ T_{2n}^{[\cdot]} & T_0^{[\cdot]} \in E_0 \end{cases}.$$

The sequence of states visited at successive entrances into E_0 (E_1) is $J^{[0]} = \{J_n^{[0]}, n \in \mathbb{N}_0\}$ ($J^{[1]} = \{J_n^{[1]}, n \in \mathbb{N}_0\}$), with $J_n^{[0]} = J_{T_n^{[0]}}$ ($J_n^{[1]} = J_{T_n^{[1]}}$), whose transition probability function is

$$P^{[0]} = [I - P_{00}]^{-1} P_{01} [I - P_{11}]^{-1} P_{10} \tag{11}$$

($P^{[1]} = \boxed{0 \leftrightarrow 1}(P^{[0]})$, i.e., $P^{[1]} = [I - P_{11}]^{-1} P_{10} [I - P_{00}]^{-1} P_{01}$). We note that (11) may be identified with Theorem 2.4 of Rubino and Sericola [10] and follows from $P^{[0]}$ being equal to $[(P^{[\cdot]})^2]_{00}$.

2.2 *Sojourn times*

The sequence of accumulated sojourn times of J in E_0 is $\{S_n^{[0]}, n \geq 0\}$, where $S_0^{[0]} = 0$ and $S_n^{[0]} = S_{n-1}^{[0]} + [V_n^{[0]} - U_n^{[0]}]$, for $n \in \mathbb{N}$, with

$$U_n^{[0]} = \inf\{k \geq T_{2(n-1)}^{[\cdot]} : J_k \in E_0\} \tag{12}$$

$$V_n^{[0]} = \inf\{k > U_n^{[0]} : J_{k-1} \in E_0 \wedge J_k \in E_1\}. \tag{13}$$

Thus $Y_n^{[0]} = V_n^{[0]} - U_n^{[0]}$ is the amount of time J stays in E_0 during the nth visit to the set, i.e., the nth sojourn time in E_0, and $S_n^{[0]}$ is the accumulated sojourn time of J in E_0 during the first n visits. In a similar way, define $S_n^{[1]}$ and $Y_n^{[1]}$, $n \in \mathbb{N}$, for E_1.

We consider that a *cycle* is constituted by a visit to E_0 and another one to E_1, with the cycles being counted from time zero irrespective of the initial state, J_0, belonging to E_0 or E_1. For $n \in \mathbb{N}$, $Y_n^{[*]} = Y_n^{[0]} + Y_n^{[1]}$ is the duration of the nth cycle, and $S_n^{[*]} = S_n^{[0]} + S_n^{[1]} = T_{2n}^{[\cdot]}$ is the duration of the first n cycles. Then, $S^{[*]} = \{S_n^{[*]}, n \geq 0\}$ is the sequence of accumulated cycle times and $J^{[*]} = \{J_n^{[*]}, n \geq 0\}$, with $J_n^{[*]} = J_{2n}^{[\cdot]}$, is the sequence of states visited at the beginning of cycles.

It follows that $(S^{[0]}, S^{[1]}, J^{[*]})$ is a bivariate MRP [b], on the state space $\mathbb{N}_0^2 \times E$, having transition probability function $\{R(l,m), (l,m) \in \mathbb{N}^2\}$ with

$$R(l,m) = \begin{bmatrix} P_{\mathbf{00}}^{l-1} P_{\mathbf{01}} P_{\mathbf{11}}^{m-1} P_{\mathbf{10}} & \mathbf{0} \\ \mathbf{0} & P_{\mathbf{11}}^{m-1} P_{\mathbf{10}} P_{\mathbf{00}}^{l-1} P_{\mathbf{01}} \end{bmatrix} \tag{14}$$

and that $J^{[*]}$ has transition probability function

$$P^* = \begin{bmatrix} P^{[0]} & \mathbf{0} \\ \mathbf{0} & P^{[1]} \end{bmatrix} \tag{15}$$

since necessarily $P^* = (P^{[\cdot]})^2$. Remark that if $J_0 \in E_0$, then $J^{[*]}$ takes only values in E_0, whereas if $J_0 \in E_1$, then $J^{[*]}$ takes only values in E_1.

Moreover the marginal processes $(S^{[0]}, J^{[*]})$, $(S^{[1]}, J^{[*]})$ and $(S^{[*]}, J^{[*]})$ are also MRPs on the state space $\mathbb{N}_0 \times E$, whose transition probability functions are $R^{[0]} = \{R^{[0]}(l), l \in \mathbb{N}\}$, $R^{[1]} = \{R^{[1]}(m), m \in \mathbb{N}\}$ and $R^* = \{R^*(m), m \geq 2\}$, respectively, with

$$R^{[0]}(l) = \begin{bmatrix} P_{\mathbf{00}}^{l-1} P_{\mathbf{01}} [I - P_{\mathbf{11}}]^{-1} P_{\mathbf{10}} & \mathbf{0} \\ \mathbf{0} & [I - P_{\mathbf{11}}]^{-1} P_{\mathbf{10}} P_{\mathbf{00}}^{l-1} P_{\mathbf{01}} \end{bmatrix} \tag{16}$$

$$R^{[1]}(m) = \begin{bmatrix} [I - P_{\mathbf{00}}]^{-1} P_{\mathbf{01}} P_{\mathbf{11}}^{l-1} P_{\mathbf{10}} & \mathbf{0} \\ \mathbf{0} & P_{\mathbf{11}}^{l-1} P_{\mathbf{10}} [I - P_{\mathbf{00}}]^{-1} P_{\mathbf{01}} \end{bmatrix} \tag{17}$$

and $R^*(m) = \sum_{l=1}^{m-1} R(l, m-l)$.

Finally, the sequences $(S^{[0]}, J^{[0]})$ and $(S^{[1]}, J^{[1]})$ are MRPs, on the state spaces $\mathbb{N}_0 \times E_0$ and $\mathbb{N}_0 \times E_1$, whose transition probability functions are $R_{\mathbf{00}}^{[0]} = \{R_{\mathbf{00}}^{[0]}(l), l \geq 1\}$ and $R_{\mathbf{11}}^{[1]} = \{R_{\mathbf{11}}^{[1]}(l), l \geq 1\}$, respectively, with $R_{\mathbf{00}}^{[0]}(l) = P_{\mathbf{00}}^{l-1} P_{\mathbf{01}} [I - P_{\mathbf{11}}]^{-1} P_{\mathbf{10}}$ and $R_{\mathbf{11}}^{[1]}(l) = P_{\mathbf{11}}^{l-1} P_{\mathbf{10}} [I - P_{\mathbf{00}}]^{-1} P_{\mathbf{01}}$.

3 Distributions of sojourn and cycle times

This section is divided in two parts. In Subsection 3.1 we derive transition probabilities associated with the sequences of sojourn times in E_0 and cycle durations. In Subsection 3.2 we derive their probability generating functions.

We give explicit expressions only for results concerning sojourn times in E_0. The results for the sequence of sojourn times in E_1 are similar to the ones

---

[b] A sequence $(Z, K) = \{(Z_n, K_n), n \geq 0\}$ whose state space is a subset of $(\mathbb{R}_0^+)^k \times F$ is called a k-variate MRP if it is a Markov additive sequence with Markov component K taking values on F, so that the additive component Z is k-dimensional. For the definition and properties of Markov additive processes see Çinlar [1] and Pacheco and Prabhu [8].

that we will obtain for the sequence of sojourn times in E_0; they just require the replacement of the matrices $P_{\mathbf{00}}$, $P_{\mathbf{01}}$, $P_{\mathbf{10}}$ and $P_{\mathbf{11}}$ by $P_{\mathbf{11}}, P_{\mathbf{10}}, P_{\mathbf{01}}$ and $P_{\mathbf{00}}$, respectively.

3.1 Transition probability functions

Before deriving the results of this subsection, it is convenient to introduce some additional notation. Let $\mathbf{X} = \{\mathbf{X}_n, n \geq 1\}$ be a random sequence taking values in $I\!N_0^r$ and $K = \{K_n, n \geq 0\}$ a DTMC taking values on F. We let

$$P^{(n)}_{(\mathbf{X},K)}(\mathbf{l}) = [P(\mathbf{X}_n = \mathbf{l}, K_n = j | K_0 = i)]_{i,j\in F} \tag{18}$$

$$P^{(n,m)}_{(\mathbf{X},K)}(\mathbf{l}, \mathbf{k}) = [P(\mathbf{X}_n = \mathbf{l}, \mathbf{X}_{n+m} = \mathbf{k}, K_{n+m} = j | K_0 = i)]_{i,j\in F} \tag{19}$$

for $\mathbf{l}, \mathbf{k} \in I\!N_0^r$ and $n, m \in I\!N$, whenever these probability matrices are well defined. We call $P^{(n)}_{(\mathbf{X},K)}(.)$ the n-step transition probability function of $(\mathbf{X}, K)$, and $P^{(n,n)}_{(\mathbf{X},K)}(.,.)$ the joint (n, m)-step transition probability function of $(\mathbf{X}, K)$.

Lemma 1 *Let $(\mathbf{X}, K) = \{(\mathbf{X}_n, K_n), n \geq 0\}$ be a k-variate MRP on $I\!N_0^r \times F$, with transition probability function $B(\mathbf{l})$, for $\mathbf{l} \in I\!N_0^r$, and denote*

$$\mathbf{Z}_n = \mathbf{X}_n - \mathbf{X}_{n-1}, \quad n \in I\!N.$$

Then, the sequence $(\mathbf{Z}, K)$ has transition probability functions

$$P^{(n)}_{(\mathbf{Z},K)}(\mathbf{l}) = \bar{B}^{n-1} B(\mathbf{l}) \tag{20}$$

$$P^{(n,m)}_{(\mathbf{Z},K)}(\mathbf{l}, \mathbf{k}) = P^{(n)}_{(\mathbf{Z},K)}(\mathbf{l})\, P^{(m)}_{(\mathbf{Z},K)}(\mathbf{k}) \tag{21}$$

for $\mathbf{l}, \mathbf{k} \in I\!N_0^r$ and $n, m \in I\!N$, where $\bar{B}$ is the transition probability matrix of K; i.e., $\bar{B} = \sum_{\mathbf{k}\in I\!N_0^r} B(\mathbf{k})$.

Proof: For $i, j \in F$, $n \in I\!N$, and $\mathbf{l} \in I\!N_0^r$,

$$\begin{aligned}
\left[P^{(n)}_{(\mathbf{Z},K)}(\mathbf{l})\right]_{ij} &= \sum_{k\in F} P(\mathbf{Z}_n = \mathbf{l}, K_n = j, K_{n-1} = k | K_0 = i) \\
&= \sum_{k\in F} P(K_{n-1} = k | K_0 = i) P(\mathbf{Z}_n = \mathbf{l}, K_n = j | K_{n-1} = k) \\
&= \sum_{k\in F} [\bar{B}^{n-1}]_{ik} B_{kj}(\mathbf{l}) = [\bar{B}^{n-1} B(\mathbf{l})]_{ij}
\end{aligned}$$

where in the second equality we used the Markov renewal property of $(\mathbf{X}, K)$ and in the third equality the fact that K is a Markov chain with transition probability matrix $\bar{B}$. The proof of (21) is similar. ■

Note that, taking into account (21), it is sufficient to derive n-step transition probability functions, as joint (n, m)-step transition probability functions may be obtained by a simple matrix multiplication from those.

In view of Lemma 1, we have the following result for the sequence of sojourn times in E_0 and E_1 along with the states visited at the beginning of cycles along with a corollary concerning the sequence of cycle durations.

Theorem 1 *The sequence* $(Y^{[0]}, Y^{[1]}, J^{[*]})$ *takes values on* $\mathbb{N}^2 \times E$ *and has* n*-step transition probability function*

$$P^{(n)}_{(Y^{[0]},Y^{[1]},J^{[*]})}(\mathbf{l}) = \begin{bmatrix} P^{(n)}_{\mathbf{00}}(\mathbf{l}) & \mathbf{0} \\ \mathbf{0} & P^{(n)}_{\mathbf{11}}(\mathbf{l}) \end{bmatrix} \tag{22}$$

with $P^{(n)}_{\mathbf{11}}(l_1, l_2) = \boxed{0 \leftrightarrow 1}(P^{(n)}_{\mathbf{00}}(l_2, l_1))$ *and*

$$P^{(n)}_{\mathbf{00}}(\mathbf{l}) = (P^*_{\mathbf{00}})^{n-1} P^{l_1-1}_{\mathbf{00}} P_{\mathbf{01}} P^{l_2-1}_{\mathbf{11}} P_{\mathbf{10}}. \tag{23}$$

Proof: From the definition, $(Y^{[0]}, Y^{[1]}, J^{[*]})$ takes values on $\mathbb{N}^2 \times E$. Moreover, (22) follows from (20) with $\bar{B} = P^*$ and $B(\mathbf{l}) = R(\mathbf{l})$. ■

Corollary 1 *The sequence* $(Y^{[*]}, J^{[*]})$ *takes values on* $\{2, 3, \ldots\} \times E$ *and has* n*-step transition probability function*

$$\begin{aligned} P^{(n)}_{(Y^{[*]},J^{[*]})}(l) &= \begin{bmatrix} P^{*(n)}_{\mathbf{00}}(l) & \mathbf{0} \\ \mathbf{0} & P^{*(n)}_{\mathbf{11}}(l) \end{bmatrix} \\ &= \begin{bmatrix} \sum_{w=1}^{l-1} P^{(n)}_{\mathbf{00}}(w, l-w) & \mathbf{0} \\ \mathbf{0} & \sum_{w=1}^{l-1} P^{(n)}_{\mathbf{11}}(w, l-w) \end{bmatrix}. \end{aligned} \tag{24}$$

Proof: Let $l, k \in \{2, 3, \ldots\}$, $n, m \in \mathbb{N}$ and $i, j \in E$. As $Y^{[*]}_n = Y^{[0]}_n + Y^{[1]}_n$, and in view of Theorem 1, we conclude that $(Y^{[*]}, J^{[*]})$ takes values on $\{2, 3, \ldots\} \times E$ and

$$\begin{aligned} P(Y^{[*]}_n = l, J^{[*]}_n = j | J^{[*]}_0 = i) &= \sum_{w=1}^{l-1} P(Y^{[0]}_n = w, Y^{[1]}_n = l - w, J^{[*]}_n = j | J^{[*]}_0 = i) \\ &= \sum_{w=1}^{l-1} P^{(n)}_{(Y^{[0]},Y^{[1]},J^{[*]})}(w, l-w). \end{aligned}$$

Thus, (24) follows using (22). ■

Note that $P_{\mathbf{11}}^{(n)}(l_1, l_2) = \boxed{0\leftrightarrow 1}(P_{\mathbf{00}}^{(n)}(l_2, l_1))$, but $P_{\mathbf{11}}^{\star(n)}(l) = \boxed{0\leftrightarrow 1}(P_{\mathbf{00}}^{\star(n)}(l))$. We next derive the transition probability function for the sequence of sojourn times in E_0 along with the states visited at the entrance times in E_0.

Theorem 2 *The sequence $(Y^{[0]}, J^{[0]})$ takes values on $\mathbb{N} \times E_0$ and has n-step transition probability function*

$$P_{(Y^{[0]}, J^{[0]})}^{(n)}(l) = (P_{\mathbf{00}}^{*})^{n-1} P_{\mathbf{00}}^{l-1} P_{\mathbf{01}}[I - P_{\mathbf{11}}]^{-1} P_{\mathbf{10}}. \tag{25}$$

Proof: From the definition, $(Y^{[0]}, J^{[0]})$ takes values on $\mathbb{N} \times E_0$, and (25) follows from (20) with $\bar{B} = P_{\mathbf{00}}^{*}$ and $B(l) = R_{\mathbf{00}}^{[0]}(l)$. ■

Since $[I - P_{\mathbf{11}}]^{-1} P_{\mathbf{10}} = P_{\mathbf{10}}^{[\cdot]}$ is a stochastic matrix, a simple deduction shows that (25) implies Theorem 2.6 of Rubino and Sericola [10].

3.2 Probability generating functions

If $\mathbf{X} = \{\mathbf{X}_n, n \geq 1\}$ is a random sequence taking values in $\mathbb{N}_0^r$ and $K = \{K_n, n \geq 0\}$ is a DTMC taking values in F, we define the probability generating function (matrices)

$$G_{(\mathbf{X},K)}^{(n)}(\mathbf{z}) = \left[\mathrm{E}[\mathbf{z}^{\mathbf{X}_n} \mathrm{I}_{\{K_n = j\}} | K_0 = i]\right]_{i,j \in F} \tag{26}$$

$$G_{(\mathbf{X},K)}^{(n,m)}(\mathbf{z}, \mathbf{w}) = \left[\mathrm{E}[\mathbf{z}^{\mathbf{X}_n} \mathbf{w}^{\mathbf{X}_{n+m}} \mathrm{I}_{\{K_{n+m} = j\}} | K_0 = i]\right]_{i,j \in F} \tag{27}$$

for $\mathbf{z}, \mathbf{w} \in \mathbb{R}^r$, $\mathbf{0} \leq \mathbf{z}, \mathbf{w} \leq \mathbf{1}$, and $n, m \in \mathbb{N}$, where $\mathbf{z}^{\mathbf{l}} = \prod_{i=1}^{r} z_i^{l_i}$ and I_A is the indicator function of set A. We call $G_{(\mathbf{X},K)}^{(n)}(\mathbf{z})$ ($G_{(\mathbf{X},K)}^{(n,m)}(\mathbf{z}, \mathbf{w})$) the n-step ((n, m)-step) probability generating function of $(\mathbf{X}, K)$.

Proceeding as in Lemma 1, we conclude that

$$G_{(\mathbf{X},K)}^{(n,m)}(\mathbf{z}, \mathbf{w}) = G_{(\mathbf{X},K)}^{(n)}(\mathbf{z})\, G_{(\mathbf{X},K)}^{(m)}(\mathbf{w}). \tag{28}$$

Thus, it is sufficient to derive n-step probability generating functions.

In the next theorem we characterize the probability generating functions of the sequence of joint sojourn times in E_0 and E_1 along with the states visited at the beginning of cycles.

Theorem 3 *The sequence $(Y^{[0]}, Y^{[1]}, J^{[*]})$ has n-step probability generating function*

$$G_{(Y^{[0]}, Y^{[1]}, J^{[*]})}^{(n)}(\mathbf{z}) = \begin{bmatrix} G_{\mathbf{00}}^{(n)}(\mathbf{z}) & \mathbf{0} \\ \mathbf{0} & G_{\mathbf{11}}^{(n)}(\mathbf{z}) \end{bmatrix} \tag{29}$$

where $G^{(n)}_{\mathbf{11}}(z_1,z_2)=\boxed{0\leftrightarrow 1}(G^{(n)}_{\mathbf{00}}(z_2,z_1))$ *and*

$$G^{(n)}_{\mathbf{00}}(\mathbf{z})=\mathbf{z^1}(P^*_{\mathbf{00}})^{n-1}\left[I-z_1P_{\mathbf{00}}\right]^{-1}P_{\mathbf{01}}\left[I-z_2P_{\mathbf{11}}\right]^{-1}P_{\mathbf{10}}. \tag{30}$$

Proof: As $G^{(n)}_{(Y^{[0]},Y^{[1]},J^{[*]})}(\mathbf{z})=\sum_{\mathbf{l}\in\mathbf{N}^2}\mathbf{z^l}P^{(n)}_{(Y^{[0]},Y^{[1]},J^{[*]})}(\mathbf{l})$, then, by using Theorem 1, it follows that $[G^{(n)}_{(Y^{[0]},Y^{[1]},J^{[*]})}(\mathbf{z})]_{\mathbf{01}}$ and $[G^{(n)}_{(Y^{[0]},Y^{[1]},J^{[*]})}(\mathbf{z})]_{\mathbf{10}}$ are null matrices,

$$\begin{aligned}[G^{(n)}_{(Y^{[0]},Y^{[1]},J^{[*]})}(\mathbf{z})]_{\mathbf{00}} &= \sum_{l_1,l_2\in\mathbb{N}} z_1^{l_1}z_2^{l_2}(P^*_{\mathbf{00}})^{n-1}P_{\mathbf{00}}^{l_1-1}P_{\mathbf{01}}P_{\mathbf{11}}^{l_2-1}P_{\mathbf{10}}\\ &= \mathbf{z^1}(P^*_{\mathbf{00}})^{n-1}\left[I-z_1P_{\mathbf{00}}\right]^{-1}P_{\mathbf{01}}\left[I-z_2P_{\mathbf{11}}\right]^{-1}P_{\mathbf{10}}\end{aligned}$$

and $G^{(n)}_{\mathbf{11}}(z_1,z_2)=\boxed{0\leftrightarrow 1}(G^{(n)}_{\mathbf{00}}(z_2,z_1))$. ■

As a consequence, we have the following result for the sequence of cycle durations along with the states visited at the beginning of cycles.

Corollary 2 *The sequence* $(Y^{[*]},J^{[*]})$ *has n-step probability generating function*

$$G^{(n)}_{(Y^{[*]},J^{[*]})}(z)=\begin{bmatrix}G^{\star(n)}_{\mathbf{00}}(z) & \mathbf{0}\\ \mathbf{0} & G^{\star(n)}_{\mathbf{11}}(z)\end{bmatrix} \tag{31}$$

where $G^{\star(n)}_{\mathbf{11}}(z)=\boxed{0\leftrightarrow 1}(G^{\star(n)}_{\mathbf{00}}(z))$ *and*

$$G^{\star(n)}_{\mathbf{00}}(z)=z^2(P^*_{\mathbf{00}})^{n-1}\left[I-zP_{\mathbf{00}}\right]^{-1}P_{\mathbf{01}}\left[I-zP_{\mathbf{11}}\right]^{-1}P_{\mathbf{10}}. \tag{32}$$

Proof: As $Y^{[*]}_n=Y^{[0]}_n+Y^{[1]}_n$, the result follows from Theorem 3 by considering $\mathbf{z}=(z,z)$. ■

We next characterize the n-step probability generating function of the sojourn times in E_0 along with the states visited at the entrance times in E_0.

Corollary 3 *The sequence* $(Y^{[0]},J^{[0]})$ *has n-step probability generating function*

$$G^{(n)}_{(Y^{[0]},J^{[0]})}(z)=z(P^*_{\mathbf{00}})^{n-1}[I-zP_{\mathbf{00}}]^{-1}P_{\mathbf{01}}[I-P_{\mathbf{11}}]^{-1}P_{\mathbf{10}}. \tag{33}$$

Proof: Let $0\le z\le 1$. For $i\in E_0$, $J^{[0]}_n|J^{[0]}_0=i$ has the same distribution as $J^{[*]}_n|J^{[*]}_0=i$. Thus, from (30), and since $G^{(n)}_{(Y^{[0]},J^{[0]})}(z)=G^{(n)}_{\mathbf{00}}(z,1)$, we get (33). ■

4 Moments of sojourn times and cycle durations

Let $\mathbf{X} = \{\mathbf{X}_n, n \geq 1\}$ be a random sequence taking values on $\mathbb{N}_0^r$ and $K = \{K_n, n \geq 0\}$ a DTMC taking values on F; we define the matrices of expected values

$$E^{(n)}_{(\mathbf{X},K)}(\mathbf{l}) = \left[E(\mathbf{X}_n^{\mathbf{l}} \mathrm{I}_{\{K_n=j\}} | K_0 = i)\right]_{i,j\in F} \tag{34}$$

$$\mathrm{E}^{(n,m)}_{(\mathbf{X},K)}(\mathbf{l},\mathbf{k}) = \left[E(\mathbf{X}_n^{\mathbf{l}} \mathbf{X}_{n+m}^{\mathbf{k}} \mathrm{I}_{\{K_{n+m}=j\}} | K_0 = i)\right]_{i,j\in F} \tag{35}$$

for $\mathbf{l},\mathbf{k} \in \mathbb{N}^r$ and $n, m \in \mathbb{N}$. $E^{(n)}_{(\mathbf{X},K)}(\mathbf{l})$ ($\mathrm{E}^{(n,m)}_{(\mathbf{X},K)}(\mathbf{l},\mathbf{k})$) may be regarded as the n-step (joint (n,m)-step) conditional moment of order $\mathbf{l}$ ($(\mathbf{l},\mathbf{k})$) of $(\mathbf{X}, K)$.

We next compute first and second order conditional moments of cycle durations. For convenience, we define for $n \in \mathbb{N}$

$$K_{\mathbf{00}}(n) = n[I - P_{\mathbf{00}}]^{-1} P_{\mathbf{01}} P_{\mathbf{11}}^n [I - P_{\mathbf{11}}]^{-(n+1)} P_{\mathbf{10}} \tag{36}$$

$$L_{\mathbf{00}}(n) = \{I + [I - P_{\mathbf{00}}]^{-1} + \ldots + n[I - P_{\mathbf{00}}]^{-n}\} \tag{37}$$

$$M_{\mathbf{00}}(n) = K_{\mathbf{00}}(n) + L_{\mathbf{00}}(n) P^*_{\mathbf{00}} \tag{38}$$

and, similarly, $K_{\mathbf{11}}(n) = \boxed{0\leftrightarrow 1}(K_{\mathbf{00}}(n))$, $L_{\mathbf{11}}(n) = \boxed{0\leftrightarrow 1}(L_{\mathbf{00}}(n))$ and $M_{\mathbf{11}}(n) = \boxed{0\leftrightarrow 1}(M_{\mathbf{00}}(n))$. Let $K_{\mathbf{00}}$ ($K_{\mathbf{11}}$), $L_{\mathbf{00}}$ ($L_{\mathbf{00}}$) and $M_{\mathbf{00}}$ ($M_{\mathbf{11}}$) stand for $K_{\mathbf{00}}(1)$ ($K_{\mathbf{11}}(1)$), $L_{\mathbf{00}}(1)$ ($L_{\mathbf{11}}(1)$) and $M_{\mathbf{00}}(1)$ ($M_{\mathbf{11}}(1)$), respectively.

Lemma 2 *For $i = 1, 2$ and $n \in \mathbb{N}$,*

$$E^{(n)}_{(\mathbf{Y}^{[\bullet]},J^{[\bullet]})}(i) = \begin{bmatrix} E^{\star(n)}_{\mathbf{00}}(i) & \mathbf{0} \\ \mathbf{0} & E^{\star(n)}_{\mathbf{11}}(i) \end{bmatrix} \tag{39}$$

where $E^{\star(n)}_{\mathbf{11}}(1) = \boxed{0\leftrightarrow 1}(E^{\star(n)}_{\mathbf{00}}(1))$ *and* $E^{\star(n)}_{\mathbf{11}}(2) = \boxed{0\leftrightarrow 1}(E^{\star(n)}_{\mathbf{00}}(2))$, *with*

$$E^{\star(n)}_{\mathbf{00}}(1) = (P^*_{\mathbf{00}})^{n-1} M_{\mathbf{00}}, \; E^{\star(n)}_{\mathbf{00}}(2) = (P^*_{\mathbf{00}})^{n-1}[M_{\mathbf{00}}(2) + 2M_{\mathbf{00}} + K_{\mathbf{00}}]. \tag{40}$$

Moreover, the matrix of conditional expected values of the product of the nth and $(n+m)$th cycle durations, $\mathrm{E}^{(n,m)}_{(\mathbf{Y}^{[\bullet]},J^{[\bullet]})}(1,1)$, is

$$\begin{bmatrix} (P^*_{\mathbf{00}})^{n-1} M_{\mathbf{00}} (P^*_{\mathbf{00}})^{m-1} M_{\mathbf{00}} & \mathbf{0} \\ \mathbf{0} & (P^*_{\mathbf{11}})^{n-1} M_{\mathbf{11}} (P^*_{\mathbf{11}})^{m-1} M_{\mathbf{11}} \end{bmatrix}. \tag{41}$$

Proof: From (31), it follows that $[E^{(n)}_{(\mathbf{Y}^{[\bullet]},J^{[\bullet]})}(i)]_{\mathbf{01}}$ and $[E^{(n)}_{(\mathbf{Y}^{[\bullet]},J^{[\bullet]})}(i)]_{\mathbf{10}}$ are null matrices, which implies (39). Moreover since

$$E^{\star(n)}_{\mathbf{00}}(1) = \frac{\mathrm{d}}{\mathrm{d}z} G^{\star(n)}_{\mathbf{00}}(z)\bigg|_{z=1}, \; E^{\star(n)}_{\mathbf{00}}(2) = \left[\frac{\mathrm{d}^2}{\mathrm{d}z^2} G^{\star(n)}_{\mathbf{00}}(z) + \frac{\mathrm{d}}{\mathrm{d}z} G^{\star(n)}_{\mathbf{00}}(z)\right]_{z=1}$$

(40) follows, after some matrix manipulations, using the fact that, from (32)

$$\frac{\mathrm{d}}{\mathrm{d}z}G^{\star(n)}_{\mathbf{00}}(z) = P_{\mathbf{01}}[I - P_{\mathbf{11}}]^{-1}P_{\mathbf{10}} + z^2(P^{*}_{\mathbf{00}})^{n-1}[I - zP_{\mathbf{00}}]^{-1} \times P_{\mathbf{01}}P_{\mathbf{11}}[I - P_{\mathbf{11}}]^{-2}P_{\mathbf{10}}$$

$$\frac{\mathrm{d}^2}{\mathrm{d}z^2}G^{\star(n)}_{\mathbf{00}}(z)) = 2z^2(P^{*}_{\mathbf{00}})^{n-1}P^2_{\mathbf{00}}[I - zP_{\mathbf{00}}]^{-3}P_{\mathbf{01}}[I - P_{\mathbf{11}}]^{-1}P_{\mathbf{10}} + 2z^2(P_{\mathbf{00}}[I - zP_{\mathbf{00}}]^{-2}P_{\mathbf{01}}P_{\mathbf{11}}[I - zP_{\mathbf{11}}]^{-2}P_{\mathbf{10}} + [I - zP_{\mathbf{00}}]^{-1}P_{\mathbf{01}}P^2_{\mathbf{11}}[I - zP_{\mathbf{11}}]^{-3}P_{\mathbf{10}}).$$

The proof of the equation for $E^{\star(n)}_{\mathbf{00}}(2)$ is similar. Finally, (41) follows by the fact that, using an argument similar to the one used to derive (28), we conclude that

$$\mathrm{E}^{(n,m)}_{(\mathbf{Y}^{[*]},J^{[*]})}(1,1) = E^{(n)}_{(\mathbf{Y}^{[*]},J^{[*]})}(1)\, E^{(m)}_{(\mathbf{Y}^{[*]},J^{[*]})}(1). \quad \blacksquare$$

The next result characterizes the conditional moments associated to the sequences of sojourn times in E_0 along with the states visited at the entrance times in E_0.

Lemma 3 *The first and the second order conditional moments of the nth sojourn time in E_0 have the form*

$$E^{(n)}_{(\mathbf{Y}^{[0]},J^{[0]})}(1) = (P^{*}_{\mathbf{00}})^{n-1}[I - P_{\mathbf{00}}]^{-1}P^{*}_{\mathbf{00}} \tag{42}$$

$$E^{(n)}_{(\mathbf{Y}^{[0]},J^{[0]})}(2) = (P^{*}_{\mathbf{00}})^{n-1}[I + P_{\mathbf{00}}][I - P_{\mathbf{00}}]^{-2}P^{*}_{\mathbf{00}}. \tag{43}$$

Moreover, the matrix of the expected values of the product of the nth and $(n+m)$th sojourn times in E_0 is

$$\mathrm{E}^{(n,m)}_{(Y^{[0]},J^{[0]})}(1,1) = (P^{*}_{\mathbf{00}})^{n-1}[I - P_{\mathbf{00}}]^{-1}(P^{*}_{\mathbf{00}})^{m}[I - P_{\mathbf{00}}]^{-1}P^{*}_{\mathbf{00}}. \tag{44}$$

Proof: Deriving once and twice (33), and making $z = 1$, we end up with (42)-(43). Moreover, (44) follows from (42) since

$$\mathrm{E}^{(n,m)}_{(Y^{[0]},J^{[0]})}(1,1) = E^{(n)}_{(\mathbf{Y}^{[0]},J^{[0]})}(1)\, E^{(m)}_{(\mathbf{Y}^{[0]},J^{[0]})}(1). \quad \blacksquare$$

5 Stationary versions

In this section we derive results for stationary versions of the processes defined in the previous sections. First we characterize the stationary distributions of the embedded Markov chains that have been studied. Afterwards we derive moments associated to the sequences of sojourn times in E_0 and cycle durations.

5.1 Stationary embedded Markov chains

In this subsection we derive the stationary distributions for the Markov chains of entrance states into E_0 (E_1), $J^{[0]}$ ($J^{[1]}$), and states visited at the beginning of cycles, $J^{[*]}$, and at transitions times between E_0 and E_1, $J^{[\cdot]}$.

Theorem 4 *The Markov chains $J^{[0]}$ and $J^{[1]}$ have unique stationary probability row vectors, $\pi^{[0]}$ and $\pi^{[1]}$, respectively, satisfying*

$$\pi^{[0]} = \pi^{[0]} [I - P_{\mathbf{00}}]^{-1} P_{\mathbf{01}} [I - P_{\mathbf{11}}]^{-1} P_{\mathbf{10}} \tag{45}$$

$$\pi^{[1]} = \pi^{[1]} [I - P_{\mathbf{11}}]^{-1} P_{\mathbf{10}} [I - P_{\mathbf{00}}]^{-1} P_{\mathbf{01}}. \tag{46}$$

Moreover, the reducible Markov chain $J^{[*]}$ has stationary probability vectors $\{\pi^{[*]}(\rho),\ \rho \in [0,1]\}$ with

$$\pi^{[*]}(\rho) = [\rho\pi^{[0]}\ (1-\rho)\pi^{[1]}] \tag{47}$$

and the Markov chain $J^{[\cdot]}$ has unique stationary probability vector

$$\pi^{[\cdot]} = \frac{1}{2}[\pi^{[0]}\ \pi^{[1]}]. \tag{48}$$

Proof: It is easy to conclude that $J^{[0]}$ has exactly one recurrent class. Since it is a finite Markov chain, this implies that it has unique stationary probability row vector $\pi^{[0]}$, solution of (45), in view of (11). The result for $J^{[1]}$ follows in a similar way.

Let $\alpha = [\alpha^{[0]}\ \alpha^{[1]}]$ denote a probability vector, with $\alpha^{[0]}$ having dimension s_0 and $\alpha^{[1]}$ dimension s_1. Using (15), $[\alpha^{[0]}\ \alpha^{[1]}]P^* = [\alpha^{[0]}P^*_{\mathbf{00}}\ \alpha^{[1]}P^*_{\mathbf{11}}]$. Thus, if α is a stationary probability vector of $J^{[*]}$, then $\alpha^{[0]} = \alpha^{[0]}P^*_{\mathbf{00}}$ and $\alpha^{[1]} = \alpha^{[1]}P^*_{\mathbf{11}}$. As a consequence, $\alpha^{[0]} = p_0\pi^{[0]}$ and $\alpha^{[1]} = p_1\pi^{[1]}$, for some $p_0, p_1 \geq 0$. Moreover, as α is a probability vector, $p_0+p_1 = 1$ and so $\alpha = [\rho\pi^{[0]}\ (1-\rho)\pi^{[1]}]$, for some $\rho \in [0,1]$. On the other hand, any vector of the form $[\rho\pi^{[0]}\ (1-\rho)\pi^{[1]}]$ with $\rho \in [0,1]$ is a stationary vector of $J^{[*]}$, in view of (47), since $\pi^{[0]} = \pi^{[0]}P^*_{\mathbf{00}}$ and $\pi^{[1]} = \pi^{[1]}P^*_{\mathbf{11}}$.

¿From (10), (15), and (11), $P^{[\cdot]}$ and $(P^{[\cdot]})^2$ are of the following form

$$P^{[\cdot]} = \begin{bmatrix} \mathbf{0} & P^{[\cdot]}_{\mathbf{01}} \\ P^{[\cdot]}_{\mathbf{10}} & \mathbf{0} \end{bmatrix}, \qquad (P^{[\cdot]})^2 = P^* = \begin{bmatrix} P^*_{\mathbf{00}} & \mathbf{0} \\ \mathbf{0} & P^*_{\mathbf{11}} \end{bmatrix}. \tag{49}$$

If α is a stationary probability vector of $J^{[\cdot]}$, then $\alpha = \alpha(P^{[\cdot]})^2 = \alpha P^*$. Thus, as explained before, $\alpha = [\rho\pi^{[0]}\ (1-\rho)\pi^{[1]}]$, for some $\rho \in [0,1]$. Moreover, if $[\rho\pi^{[0]}\ (1-\rho)\pi^{[1]}] = [\rho\pi^{[0]}\ (1-\rho)\pi^{[1]}]P^{[\cdot]}$, then $\rho\pi^{[0]} = (1-\rho)\pi^{[1]}P^{[\cdot]}_{\mathbf{10}}$. By postmultiplying by $P^{[\cdot]}_{\mathbf{01}}\mathbf{1}$, we get $\rho\pi^{[0]}P^{[\cdot]}_{\mathbf{01}}\mathbf{1} = (1-\rho)\pi^{[1]}P^{[\cdot]}_{\mathbf{10}}P^{[\cdot]}_{\mathbf{01}}\mathbf{1}$, so that

$\rho = 1 - \rho$ and, therefore, $\alpha = (1/2)[\pi^{[0]}\ \pi^{[1]}]$. On the other hand, the vector $(1/2)[\pi^{[0]}\ \pi^{[1]}]$ is a stationary probability vector of $J^{[\cdot]}$. ∎

We note that (45) may be identified with Lemma 2.2 of Rubino and Sericola [10]. Thus, the main results of Theorem 4 are (47) and (48).

5.2 *Moments of sojourn times and cycle durations*

We use the results from Section 4 and the stationary distributions for the involved Markov chains to derive unconditional moments of sojourn times and cycle durations.

Accordingly, for the sequences $(Y^{[*]}, J^{[*]})$ and $(Y^{[0]}, J^{[0]})$, we will denote by $(\bar{Y}^{[*]}, \bar{J}^{[*]})$ and $(\bar{Y}^{[0]}, \bar{J}^{[0]})$ the corresponding stationary versions; i.e, $\bar{J}^{[*]}$, $\bar{J}^{[0]}$ are the stationary versions of $J^{[*]}$, $J^{[0]}$.

We first present some stationary moments for the sequence of cycle durations. Namely we derive the stationary expected value and variance of the duration of a cycle, and the covariance of the duration of two distinct cycles.

Theorem 5 *The stationary sequence of cycle durations, $(\bar{Y}^{[*]}, \bar{J}^{[*]})$, with $\bar{J}^{[*]}$ having initial probability vector $\pi^{[*]}(\rho) = [\rho\pi^{[0]}\ (1-\rho)\pi^{[1]}]$, $0 \le \rho \le 1$, has expected value, variance and covariances given by:*

$$\mathrm{E}[\bar{Y}_n^{[*]}] = \{\pi^{[0]}[I - P_{\mathbf{00}}]^{-1} + \pi^{[1]}[I - P_{\mathbf{11}}]^{-1}\}\mathbf{1} \tag{50}$$

$$\begin{aligned}\mathrm{Var}[\bar{Y}_n^{[*]}] &= \pi^{[0]}\{\rho[L_{\mathbf{00}}(2) + [I + 2L_{\mathbf{00}}]K_{\mathbf{00}}] + (1-\rho)P_{\mathbf{00}}^2[I - P_{\mathbf{00}}]^{-2}\}\mathbf{1} \\ &\quad + \pi^{[1]}\{(1-\rho)[L_{\mathbf{11}}(2) + [I + 2L_{\mathbf{11}}]K_{\mathbf{11}}] + \rho P_{\mathbf{11}}^2[I - P_{\mathbf{11}}]^{-2}\}\mathbf{1} \\ &\quad - (\pi^{[0]}[I - P_{\mathbf{00}}]^{-1}\mathbf{1} + \pi^{[1]}[I - P_{\mathbf{11}}]^{-1}\mathbf{1})^2 \end{aligned} \tag{51}$$

$$\begin{aligned}\mathrm{Cov}(\bar{Y}_n^{[*]}, \bar{Y}_{n+m}^{[*]}) =& \{\rho\pi^{[0]}M_{\mathbf{00}}(P_{\mathbf{00}}^*)^{m-1}M_{\mathbf{00}} + (1-\rho)\pi^{[1]}M_{\mathbf{11}}(P_{\mathbf{11}}^*)^{m-1}M_{\mathbf{11}}\}\mathbf{1} \\ & - \{\pi^{[0]}[I - P_{\mathbf{00}}]^{-1}\mathbf{1} + \pi^{[1]}[I - P_{\mathbf{11}}]^{-1}\mathbf{1}\}^2. \end{aligned} \tag{52}$$

Proof: Since $\pi^{[*]}(\rho) = [\rho\pi^{[0]}\ (1-\rho)\pi^{[1]}]$, $0 \le \rho \le 1$, it follows that $\mathrm{E}[\bar{Y}^{[*]}] = \rho\pi^{[0]}E_{\mathbf{00}}^{*(n)}(1)\mathbf{1} + (1-\rho)\pi^{[1]}E_{\mathbf{11}}^{*(n)}(1)\mathbf{1}$. Thus, in view of (40),

$$\mathrm{E}[\bar{Y}^{[*]}] = \rho\pi^{[0]}(P_{\mathbf{00}}^*)^{n-1}M_{\mathbf{00}}\mathbf{1} + (1-\rho)\pi^{[\cdot]}(P_{\mathbf{11}}^*)^{n-1}M_{\mathbf{11}}\mathbf{1}.$$

Remark that $\pi^{[0]}(P_{\mathbf{00}}^*)^{n-1} = \pi^{[0]}$ and $\pi^{[1]}(P_{\mathbf{11}}^*)^{n-1} = \pi^{[1]}$, $\forall n \in I\!N$. In addition, $\pi^{[0]}[I - P_{\mathbf{00}}]^{-1}P_{\mathbf{01}} = \pi^{[1]}$ and, conversely, $\pi^{[1]}[I - P_{\mathbf{11}}]^{-1}P_{\mathbf{10}} = \pi^{[0]}$. Moreover, $P_{\mathbf{00}}^*\mathbf{1} = \mathbf{1}$, $P_{\mathbf{11}}^*\mathbf{1} = \mathbf{1}$, $[I - P_{\mathbf{11}}]^{-1}P_{\mathbf{10}}\mathbf{1} = P_{\mathbf{10}}^{[\cdot]}\mathbf{1} = \mathbf{1}$, and $[I - P_{\mathbf{00}}]^{-1}P_{\mathbf{01}}\mathbf{1} = P_{\mathbf{01}}^{[\cdot]}\mathbf{1} = \mathbf{1}$. Therefore, using these relations, we conclude that

$$\begin{aligned}
\mathrm{E}[\bar{Y}^{[*]}] &= \rho\{1 + \pi^{[1]} P_{\mathbf{11}}[I - P_{\mathbf{11}}]^{-1}\mathbf{1} + \pi^{[0]}[I - P_{\mathbf{00}}]^{-1}\mathbf{1}\} \\
&\quad + (1-\rho)\{1 + \pi^{[0]} P_{\mathbf{00}}[I - P_{\mathbf{00}}]^{-1}\mathbf{1} + (1-p)\pi^{[1]}[I - P_{\mathbf{11}}]^{-1}\mathbf{1}\} \\
&= \pi^{[0]}[I - P_{\mathbf{00}}]^{-1}\mathbf{1} + \pi^{[1]}[I - P_{\mathbf{11}}]^{-1}\mathbf{1}
\end{aligned}$$

where in the last equality we have used the fact that $P_{\mathbf{00}}[I - P_{\mathbf{00}}]^{-1} = [I - P_{\mathbf{00}}]^{-1} - I$ and $P_{\mathbf{11}}[I - P_{\mathbf{11}}]^{-1} = [I - P_{\mathbf{11}}]^{-1} - I$. Similarly, since

$$\mathrm{Var}[\bar{Y}^{[*]}] = \rho\pi^{[0]} E^{\star(1)}_{\mathbf{00}}(2)\mathbf{1} + (1-\rho)\pi^{[1]} E^{\star(1)}_{\mathbf{11}}(2)\mathbf{1} - E^2[\bar{Y}^{[*]}]$$

(51) follows from (40) and (50). In the same way, (52) comes from (41) and (50), since $\mathrm{Cov}(\bar{Y}^{[*]}_n, \bar{Y}^{[*]}_{n+m})$ is equal to

$$p\pi^{[0]}[\mathrm{E}^{(n,m)}_{(\bar{Y}^{[*]},\bar{J}^{[*]})}(1,1)]_{\mathbf{00}}\mathbf{1} + (1-p)\pi^{[1]}[\mathrm{E}^{(n,m)}_{(\bar{Y}^{[*]},\bar{J}^{[*]})}(1,1)]_{\mathbf{11}}\mathbf{1} - E^2[\bar{Y}^{[*]}]. \quad \blacksquare$$

The next theorem gives the stationary expected value, variance and covariance of the sojourn times in E_0.

Theorem 6 *The stationary sequence of sojourn times in E_0, $(\bar{Y}^{[0]}, \bar{J}^{[0]})$, has the following moments:*

$$\mathrm{E}[\bar{Y}^{[0]}_n] = \pi^{[0]}[I - P_{\mathbf{00}}]^{-1}\mathbf{1} \tag{53}$$

$$\mathrm{Var}[\bar{Y}^{[0]}_n] = \pi^{[0]}[I + P_{\mathbf{00}}][I - P_{\mathbf{00}}]^{-2}\mathbf{1} - (\pi^{[0]}[I - P_{\mathbf{00}}]^{-1}\mathbf{1})^2 \tag{54}$$

$$\begin{aligned}
\mathrm{Cov}(\bar{Y}^{[0]}_n, \bar{Y}^{[0]}_{n+m}) &= \pi^{[0]}[I - P_{\mathbf{00}}]^{-1}(P^*_{\mathbf{00}})^{n-1}[I - P_{\mathbf{00}}]^{-1}\mathbf{1} \\
&\quad - (\pi^{[0]}[I - P_{\mathbf{00}}]^{-1}\mathbf{1})^2.
\end{aligned} \tag{55}$$

Proof: Equation (53) comes from (42). In addition, from (42)-(44), we obtain (54) and (55). ■

6 Example

Let $J = \{J_n, n \geq 0\}$ be the generalization of a sequence of independent Bernoulli trials described in Section 1. Thus, J has state space $E = \{1, 2, 3, 4\}$ and transition probability matrix given by (4). As explained in the introduction, the success probability changes between the values p_0 and p_1. Moreover, the set $E_0 = \{1, 2\}$ ($E_1 = \{3, 4\}$) is the *failure set* (*success set*) and α_0 (α_1) is the probability of a change in the success probability from p_0 to p_1 (p_1 to p_0), when such a change is possible, i.e., when a transition out from E_0 occurs from state 1 (2).

In this section we apply the results derived in the previous section to J assuming that the initial states for $J^{[0]}$, $J^{[1]}$ and $J^{[*]}$ are chosen accordingly to the respective stationary distributions.

The case $p_0 = p_1 = p$, $0 < p < 1$, reduces to the classic independent Bernoulli trials model with success probability p. The sojourn times in E_1 correspond to the length of sequences of successive success trials, and thus have a geometric distribution with success probability $1-p$. Conversely, the sojourn times in E_0 correspond to the length of sequences of successive failure trials and have a geometric distribution with success probability p.

Naturally, for the classic independent Bernoulli trials model the sequences of sojourn times in E_1 and in E_0 are independent sequences of i.i.d. random variables. As a result, the cycle durations consist of the sum of two independent geometrically distributed random variables with success probabilities p and $1-p$, irrespective of the initial state. Thus

$$\mathrm{E}[Y_n^{[0]}] = \frac{1}{p}, \qquad \mathrm{Var}[Y_n^{[0]}] = \frac{1-p}{p^2}, \qquad \mathrm{Corr}(Y_n^{[0]}, Y_{n+1}^{[0]}) = 0,$$

$$\mathrm{E}[Y_n^{[*]}] = \frac{1}{p(1-p)}, \quad \mathrm{Var}[Y_n^{[*]}] = \frac{1-3p(1-p)}{(1-p)^2}, \quad \mathrm{Corr}(Y_n^{[*]}, Y_{n+1}^{[*]}) = 0.$$

6.1 Dependent Bernoulli trials with $p_1 = 1 - p_0$

In this subsection we will consider two cases of dependent Bernoulli trials with $p_1 = 1 - p_0$. In the first case $\alpha_0 = \alpha_1 = \alpha$ and in the second case $\alpha_0 = 1 - \alpha_1 = \alpha$. For both cases $0 < p, \alpha < 1$.

Case $\alpha_0 = \alpha_1 = \alpha$: Since $p_1 = 1 - p_0$ and $\alpha_1 = \alpha_0 = \alpha$, the transition probability matrix of J is

$$P(p,\alpha) = \begin{bmatrix} 1-p & 0 & p(1-\alpha) & p\alpha \\ 0 & p & (1-p)\alpha & (1-p)(1-\alpha) \\ 1-p & 0 & p & 0 \\ 0 & p & 0 & 1-p \end{bmatrix} \tag{56}$$

and has parameters p and α. Using (45), it follows that

$$\pi^{[0]} = \pi^{[1]} = [0.5 \;\; 0.5]. \tag{57}$$

Thus, in view of (53)-(55), $\mathrm{E}[\bar{Y}_n^{[0]}]$, $\mathrm{Var}[\bar{Y}_n^{[0]}]$, and $\mathrm{Corr}(\bar{Y}_n^{[0]}, \bar{Y}_{n+1}^{[0]})$ are respectively equal to

$$\frac{1}{2p(1-p)}, \qquad \frac{3-10p(1-p)}{4(1-p)^2p^2}, \quad \text{and} \quad \frac{(1-2p)^2}{3-10p(1-p)}.$$

Remark that the last three formulas are symmetric functions with respect to $p = 0.5$, as they are functions of $p(1-p)$ and $(1-2p)$. For an explanation of this fact consider the relabelling of states $(1', 2', 3', 4') = (2, 1, 4, 3)$. The corresponding transition probability matrix with indices on $\{1', 2', 3', 4'\}$ is

$$P'(p,\alpha) = \begin{bmatrix} p & 0 & (1-p)(1-\alpha) & (1-p)\alpha \\ 0 & (1-p) & p\alpha & p(1-\alpha) \\ p & 0 & 1-p & 0 \\ 0 & 1-p & 0 & p \end{bmatrix}.$$

As $P'(p,\alpha) = P(1-p,\alpha)$, and, by (57), $\pi_1^{[0]} = \pi_2^{[0]}$ and $\pi_1^{[1]} = \pi_2^{[1]}$,

$$(\bar{Y}^{[0]}, J^{[0]})(p,\alpha) \stackrel{\mathrm{d}}{=} (\bar{Y}^{[0]}, J^{[0]})(1-p,\alpha)$$
$$(\bar{Y}^{[1]}, J^{[1]})(p,\alpha) \stackrel{\mathrm{d}}{=} (\bar{Y}^{[1]}, J^{[1]})(1-p,\alpha)$$

where $(\bar{Y}^{[0]}, J^{[0]})(q,\beta)$ $((\bar{Y}^{[1]}, J^{[1]})(q,\beta))$ is the stationary version of $(Y^{[0]}, J^{[0]})$ $((Y^{[1]}, J^{[1]}))$ with parameters $p = q$ and $\alpha = \beta$, and $\stackrel{\mathrm{d}}{=}$ denotes equality in distribution. Moreover

$$\bar{Y}_n^{[0]} \stackrel{\mathrm{d}}{=} \pi_1^{[0]}\mathrm{Geo}(p) \otimes (1-\pi_1^{[0]})\mathrm{Geo}(1-p)$$
$$\bar{Y}_n^{[1]} \stackrel{\mathrm{d}}{=} \pi_2^{[1]}\mathrm{Geo}(p) \otimes (1-\pi_2^{[1]})\mathrm{Geo}(1-p)$$

where $\mathrm{Geo}(\beta)$ denotes a random variable with a geometric distribution with parameter β and $\otimes$ denotes the mixture of random varibles. As $\pi_1^{[0]} = \pi_2^{[1]}$ (in fact $\pi_1^{[0]} = \pi_2^{[0]} = \pi_1^{[1]} = \pi_2^{[1]} = 1/2$), it follows that $\bar{Y}_n^{[0]}$ and $\bar{Y}_n^{[1]}$ are equally distributed, and consequently their expected values and variances are equal. Moreover we may conclude that the covariances associated to $\bar{Y}^{[0]}$ and to $\bar{Y}^{[1]}$ are also equal.

In view of (47), the stationary distributions of $J^{[*]}$ are of the form $\pi^{[*]}(\rho) = [\rho\pi^{[0]} \; (1-\rho)\pi^{[1]}]$, $0 \le \rho \le 1$. Consider the parameter ρ fixed. Then, from (50)-(52), it follows that

$$\mathrm{E}[\bar{Y}_n^{[*]}] = \frac{1}{p(1-p)}, \qquad \mathrm{Var}[\bar{Y}_n^{[*]}] = \frac{1+\rho\alpha-(3+4\rho\alpha)p(1-p)}{p^2(1-p)^2}$$

$$\mathrm{Corr}(\bar{Y}_n^{[*]}, \bar{Y}_{n+1}^{[*]}) = -2\rho\alpha^2 p(1-p)(1-2p)^2/\{1-\rho\alpha-2p[2+\rho(-3+\alpha)\alpha]$$
$$+ 2p^2[3+\rho\alpha(-7+5\alpha)] - 4p^3[1+4\rho(-1+\alpha)\alpha] + 2p^4[1+4\rho(-1+\alpha)\alpha]\}.$$

Note that the expected duration of a cycle does not depende on α and

$$\bar{Y}_n^{[*]} \stackrel{\mathrm{d}}{=} \pi_1^{[*]}(\rho)\alpha[\mathrm{Geo}(p) \oplus \mathrm{Geo}(p)] \otimes \pi_2^{[*]}(\rho)\alpha[\mathrm{Geo}(1-p) \oplus \mathrm{Geo}(1-p)]$$
$$\otimes \{1 - [\pi_1^{[*]}(\rho) + \pi_2^{[*]}(\rho)]\alpha\}[\mathrm{Geo}(p) \oplus \mathrm{Geo}(1-p)]$$

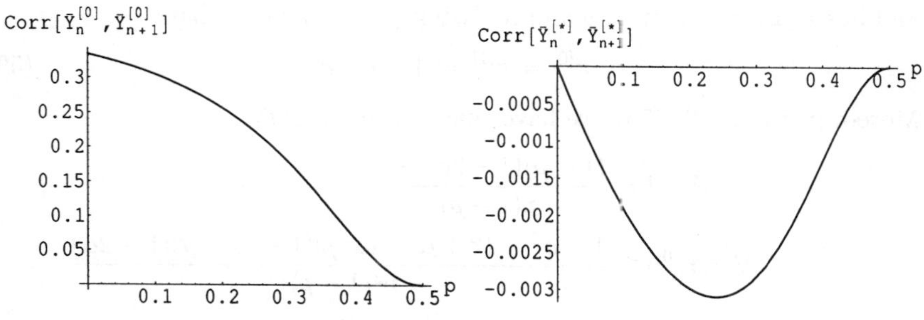

Figure 1. Correlations for the case $p_0 = 1 - p_1 = p$ and $\alpha_0 = \alpha_1 = 0.1$.

where $\oplus$ denotes the convolution of two random variables.

If, in particular, $\rho = 0$, so that the initial state belongs to E_1), the distribution of $\bar{Y}_n^{[*]}$ does not depend on α, as

$$\bar{Y}_n^{[*]} \stackrel{\mathrm{d}}{=} \mathrm{Geo}(p) \oplus \mathrm{Geo}(1-p).$$

In addition, as the distributions of cycles initiated in state 3 and in state 4 are equal, then the sequence $\{\bar{Y}_n^{[*]}, n \in I\!N_0\}$ is a renewal sequence and consequently the correlation of any two cycles is zero. Therefore, when the cycles start in E_1, in addition to the fact that the expected value and the variance of the cycle duration do not depend on α, the correlation of the duration of two consecutive (or any other) cycles is zero and

$$\mathrm{Var}[\bar{Y}_n^{[*]}] = \frac{1-3p(1-p)}{p^2(1-p)^2}, \quad \mathrm{Corr}(\bar{Y}_n^{[*]}, \bar{Y}_{n+1}^{[*]}) = 0.$$

In Figure 1 we plot the correlation of two consecutive sojourn times in E_0 and the durations of two consecutive cycles when $\alpha = 0.1$. Remark that for $\alpha = 0.1$ the correlation of sojourn times in E_0 (or, equivalently, in E_1) is always non-negative, whereas the correlation of cycle durations is non-positive. The correlation the sojourn times in E_0 is always non-negative and the only case where it is zero is when $p = 0.5$, independently of the value of α.

Case $\alpha_0 = 1 - \alpha_1 = \alpha$: Since $p_1 = 1 - p_0$ and $\alpha_0 = 1 - \alpha_1 = \alpha$, the transition probability matrix of J is

$$P(p,\alpha) = \begin{bmatrix} 1-p & 0 & p(1-\alpha) & p\alpha \\ 0 & p & (1-p)(1-\alpha) & (1-p)\alpha \\ 1-p & 0 & p & 0 \\ 0 & p & 0 & 1-p \end{bmatrix} \qquad (58)$$

and has again parameters p and α. Using (45), it follows that

$$\pi^{[0]} = \pi^{[1]} = [1-\alpha \;\; \alpha]. \tag{59}$$

Moreover, from (53)-(55), we have that, for any $n \in \mathbb{N}$,

$$\mathrm{E}[\bar{Y}_n^{[0]}] = \frac{(1-p)(1-2\alpha)+\alpha}{p(1-p)}$$

$$\mathrm{Var}[\bar{Y}_n^{[0]}] = \frac{1-\alpha^2-(3+\alpha-4\alpha^2)p(1-p)-p^3(1-2\alpha)}{p^2(1-p)^2}$$

$$\mathrm{Corr}(\bar{Y}_n^{[0]}, \bar{Y}_{n+1}^{[0]}) = \frac{(1-2p)^2(\alpha-1)\alpha}{1-\alpha^2-(3+\alpha-4\alpha^2)p(1-p)-p^3(1-2\alpha)}.$$

By using (58)-(59), and similarly to the previous subsection, we get

$$(\bar{Y}_n^{[0]}, \bar{J}_n^{[0]})(p,\alpha) \stackrel{\mathrm{d}}{=} (\bar{Y}_n^{[1]}, \bar{J}_n^{[1]})(1-p,\alpha)$$

$$\bar{Y}_n^{[0]}(p,\alpha) \stackrel{\mathrm{d}}{=} (1-\alpha)\mathrm{G}eo(p) \otimes \alpha\mathrm{G}eo(1-p)$$

$$\bar{Y}_n^{[1]}(p,\alpha) \stackrel{\mathrm{d}}{=} \alpha\mathrm{G}eo(p) \otimes (1-\alpha)\mathrm{G}eo(1-p).$$

Moreover, from (50)-(52), it follows that

$$\mathrm{Var}[\bar{Y}_n^{[*]}] = \frac{1+2\rho(1-\alpha)\alpha-[3+8\rho\alpha(1-\alpha)]p(1-p)}{p^2(1-p)^2},$$

$$\mathrm{Corr}(\bar{Y}_n^{[*]}, \bar{Y}_{n+1}^{[*]}) = [-\alpha(1-\alpha)\rho p(1-p)(1-2p)^2]/\{(1-\alpha)(1-\rho)+\rho\alpha^2 + p[-4(1-\alpha)(1-\rho)+\rho\alpha(1-5\alpha)] + 3p^2[2(1-\alpha)(1-\rho)-\alpha(1-3\alpha)] + 4p^3[(1-\rho)(1-\alpha)+\alpha\rho(1-2\alpha)] + p^4[1-4\rho\alpha(1-\alpha)]\}.$$

In particular, if $\rho = 0$, so that the initial state is in E_1, then

$$\mathrm{Var}[\bar{Y}_n^{[*]}] = \frac{1-3p(1-p)}{p^2(1-p)^2}, \quad \mathrm{Corr}(\bar{Y}_n^{[*]}, \bar{Y}_{n+1}^{[*]}) = 0.$$

Figure 2 exhibits the correlations of consecutive sojourn times in E_0 and of the durations of consecutive cycles starting in E_0 (i.e., $\rho = 1$), assuming $\alpha = 0.1$. The main conclusion that may be drawn from Figure 2 is that considerably large positive (negative) correlations may arise between consecutive sojourn times in E_0 (cycle durations) when p is large (small). By contrast, the correlations of sojourn times in E_0 (or E_1) are zero when $p = 0.5$, independently of the value of α, and the correlations of cycle durations are zero when $\rho = 0.5$.

In general, it can be proved that if $p_1 = 1 - p_0$ the correlation of the durations of two consecutive cycles is always non-positive, for every $0 < \alpha_0, \alpha_1 < 1$.

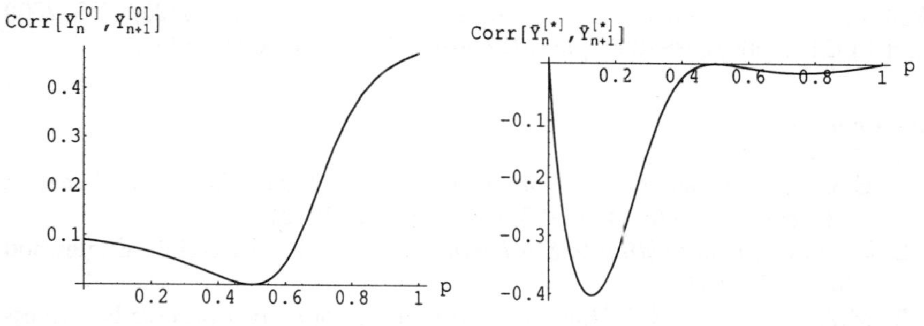

Figure 2. Correlations for the case $p_1 = 1 - p_0 = p$, $\alpha_1 = 1 - \alpha_0 = 0.1$ and $\rho = 1$.

Thus, we get the intuitive result that after a "long" cycle we should expect a "short" cycle. Conversely, the correlation of two consecutive sojourn times in E_0 (or E_1) is always non-negative.

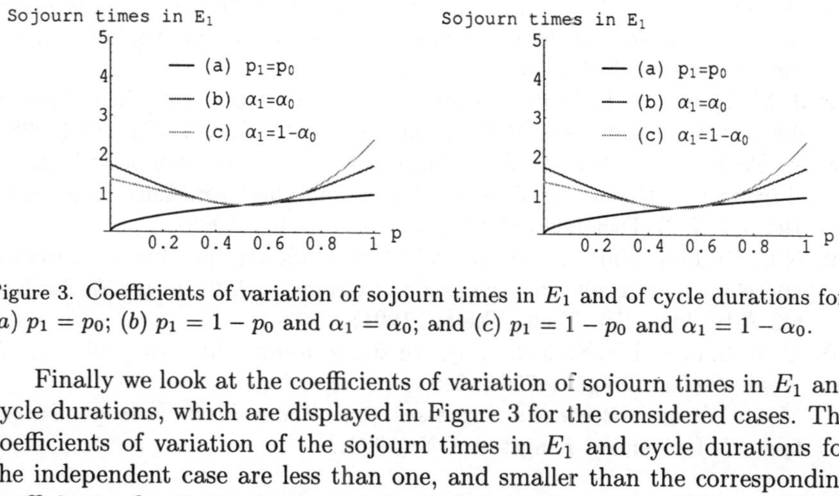

Figure 3. Coefficients of variation of sojourn times in E_1 and of cycle durations for: (a) $p_1 = p_0$; (b) $p_1 = 1 - p_0$ and $\alpha_1 = \alpha_0$; and (c) $p_1 = 1 - p_0$ and $\alpha_1 = 1 - \alpha_0$.

Finally we look at the coefficients of variation of sojourn times in E_1 and cycle durations, which are displayed in Figure 3 for the considered cases. The coefficients of variation of the sojourn times in E_1 and cycle durations for the independent case are less than one, and smaller than the corresponding coefficients of variation in the two dependent cases considered. Note that for small values of p the coefficient of variation of sojourn times in E_1 assumes considerable large values when $\alpha_1 = 1 - \alpha_0$.

Acknowledgments

The authors would like to thank the anonymous referees for their valuables comments and suggestions. This research was supported in part by

Fundação para a Ciência e a Tecnologia, the projects POSI/34826/CPS/2000 and POSI/42069/CPS/2001, and the grant SFRH/BSAB/251/01.

References

1. E. Çinlar, Markov additive processes. I, II, *Z. Wahrscheinlichkeitstheorie und Verw. Gebiete* **24**, 85–93; ibid. 95–121 (1972).
2. E. Çinlar, *Introduction to Stochastic Processes*, Prentice-Hall, Englewood Cliffs, NJ (1975).
3. M.A. Johnson and S. Narayana, Descriptors of arrival-process burstiness with application to the discrete Markovian arrival process, *Queueing Systems* **23**, 107–130 (1996).
4. M. Kijima, *Markov Processes for Stochastic Modeling*, Chapman and Hall, London (1997).
5. V.G. Kulkarni, *Modeling and Analysis of Stochastic Systems*, Chapman and Hall, London (1995).
6. G. Latouche and V. Ramaswami, *Introduction to Matrix Analytic Methods in Stochastic Modeling*, Society for Industrial and Applied Mathematics (SIAM), Philadelphia, PA (1999).
7. J.-M. Li and M.F. Neuts, Waiting times for success runs in a class of discrete point processes, *Bull. Hong Kong Math. Soc.* **2**, 131–142 (1998).
8. A. Pacheco and N.U. Prabhu, Markov-additive processes of arrivals, in *Advances in Queueing: Theory, Methos and Open Problems*, pages 167–194, ed. J. H. Dshalalow (CRC, Boca Raton, FL, 1995).
9. N.U. Prabhu, Markov-renewal and Markov-additive processes – a review and some new results, in *Analysis and Geometry 1991*, pages 57–94 (Korea Adv. Inst. Sci. Tech., Taejŏn, 1991).
10. G. Rubino and B. Sericola, Sojourn times in finite Markov processes, *J. Appl. Probab.* **26**, 744–756 (1989).
11. M.T. Wasan, *Stochastic Processes and Their First Passage Times*, Queen's University, Kingston, ON (1994).

MATRIX-ANALYTIC ANALYSIS OF A MAP/PH/1 QUEUE FITTED TO WEB SERVER DATA

ALMA RISKA
Dept. of Comp. Sci., College of William & Mary, Williamsburg, VA 23187, USA
E-mail: riska@cs.wm.edu

MARK S. SQUILLANTE, SHUN-ZHENG YU, ZHEN LIU, LI ZHANG
IBM Thomas J. Watson Research Center, Yorktown Heights, NY 10598, USA
E-mail: {mss,shuzheng,zhen,zhangl}@watson.ibm.com

We consider a MAP/PH/1 queue whose interarrival time and service time processes are fitted to measurement data from commercial Web sites. A methodology is developed for this fitting of Web data to various instances of Markovian Arrival Processes (MAP). Numerical experiments are used to evaluate our approach and to analyze several performance measures of the MAP/PH/1 queue.

1 Introduction

As the Internet continues to grow at an extremely rapid pace, Web servers are playing an ever increasing and important role in our daily life by providing access to a wide variety of commercial services. Critical issues for continued and successful growth concern the performance of such Web servers, which must provide reliable, scalable and efficient access to Internet applications and services.

A significant body of research has considered the network-packet or client-request patterns for different Web server environments (with greater attention given to the former), and has developed mathematical models to characterize these (packet or request) traffic patterns; e.g., see [1,2,3,4] and the references therein. Conversely, far less research has focused on analytic queueing-theoretic models of Web server performance based on such mathematical traffic models, and even less research has evaluated the quality of these traffic models with respect to various measures of Web server performance (as opposed to statistical tests of fit between the traffic data and the models). In fact, given the complexity of Web traffic patterns, it is commonly believed that Markovian methods cannot be used to model Web server traffic patterns or Web server performance with sufficient accuracy.

In this paper we present an initial study that develops a methodology for fitting measurement data from Web sites to the interarrival time and service time processes of a MAP/PH/1 queue used to model Web server performance.

Our study is based on an analysis of the client-request patterns found at real commercial Web sites from the retail industry, which demonstrates complex traffic patterns. We also observe from the measurement data that there are (peak and off-peak) traffic periods which appear to be stationary for relatively long intervals of time (on the order of five hours), and thus it is reasonable for us to focus on the steady-state performance measures that are of most interest in current Web server performance analysis and capacity planning studies. By modeling Web server performance as a MAP/PH/1 queue, we can further exploit matrix-analytic methods to obtain such stationary performance measures in a computationally efficient and numerically stable manner.

A considerable body of research has focused on developing strategies for fitting data sets to various stochastic models. Much of these efforts have examined the fitting of independent data sets to phase-type distributions (e.g., refer to [5,6,7,8] and the references cited therein) and the fitting of correlated data sets to MMPPs and BMAPs. In each case, the basic approach is based either on matching the moments of the data set with the moments of the stochastic model, or on using maximum likelihood estimators (MLEs). The moment matching approaches tend to be computationally more efficient, but the models tend to be somewhat more restrictive with respect to the number of states and the interactions among states that comprise the underlying Markov chain. Meier-Hellstern[9] uses the approach of MLEs for fitting data sets to 2-state MMPPs. Optimization methods have been used by Ryden[10] as part of an MLE-based approach for MMPP parameter estimation. Horvath et al.[11] develop a heuristic fitting method for MAPs based on the superposition of phase-type and interrupted Poisson processes. An estimation procedure for BMAPs based on the EM algorithm has been proposed by Breuer.[12,13] We refer the interested reader to [14], the above references, and the references therein for additional technical details.

An important distinction between the present study and previous related work is our focus here on client-request patterns (as opposed to network-packet patterns) found at real commercial Web sites. Our methodology must be sufficiently scalable (in time and space) to be able to handle the very large data sets of such environments, whose sizes exceed those of many previous studies by several orders of magnitude. Moreover, as demonstrated in a recent study,[15] an analysis of Web server performance based on the batch arrival process obtained from Web site data can significantly overestimate the response time and queue length performance measures of the corresponding Web server system. We therefore develop a hierarchical approach that first considers the batch arrival process at a relatively coarse time scale and then considers the underlying interarrival process at a finer time scale. In particular, we exploit standard Hidden Markov Model (HMM) methods to identify and characterize the dependence structure of, and some of the variability in, the batch process data set. This includes identifying the set of control states for the MAP and defining the interactions among these control states, at a coarser time scale than the batch process data set. Then we use an additional statistical analysis of the data set at a finer time scale to characterize the sojourn time and interarrival process for each control state. This includes exploiting the EM algorithm for fitting phase-type distributions to subsets of the data set. The result is a MAP that captures the complexities of client-request patterns found at the Web server environments of interest.

The remainder of the paper is organized as follows. After covering some technical preliminaries in Section 2, we then present our methodology for fitting Web server measurement data to instances of MAPs. Section 4 provides a representative set of results from a large number of numerical experiments. Our concluding remarks are given in Section 5.

2 Technical Preliminaries

2.1 Web Server Environments

We consider commercial Web sites from the retail industry whose identities will remain anonymous for obvious confidentiality restrictions. The Web sites generally consist of multiple single-server computing nodes to which incoming requests are routed by a set of front-end routers. Each router attempts to balance the Web site load across the set of single-server computing nodes by sending more traffic to less loaded nodes.

The access logs generated at one of the server nodes from each of these commercial Web sites are used as the basis for our study. Previous research has shown that the set of routers has the effect of smoothing out and equalizing (in a statistical sense) the arrival process observed at each of the server nodes when the client requests are relative small and have relatively low variability.[16,2] We have verified that the content served at the commercial Web sites of interest in our present study are consistent with the content served at the Web site considered in [16,2], at least in this respect, and thus our analysis of the access logs from one of the server nodes is assumed to be representative of the arrival process found at the other server nodes comprising the commercial Web sites of interest.

A large set of detailed measurements were performed on an isolated IBM SP2 uniprocessor node to obtain the resource requirements of each page. These measurement-based experiments reveal that the time to serve static

Web pages fits very well to a linear function of the page size, at least for the Web environments under consideration. We therefore use in our performance model the corresponding linear function fitted against the measurement data, $f_s(\cdot)$, in order to accurately estimate the resource requirements of each static page request based on the size of the request provided in the access log.

More information on the Web server environments considered in our study can be found in [16,2] and the references cited therein.

2.2 Web Server Data

Each access log contains several pieces of useful information about every client request served by the corresponding Web server node. This includes the time epoch of the nth request, which we denote by A_n, and the number of bytes comprising the nth request, which we denote by B_n, $n \in \mathbb{Z}^+$. The unit of time in the access logs available to us is one second, which is quite standard. There are typically tens or even hundreds of requests within a second for the commercial Web sites of interest. Thus, the access log data set directly provides us with the discrete-time batch process for the number of client requests per second, which we denote by $\mathcal{B}(t)$, $t \in \mathbb{Z}^+$.

There is an important problem with the coarse time granularity of this batch arrival process for our purposes. As demonstrated by a recent study,[15] the client-request response time and queue length measures obtained under the batch process $\mathcal{B}(\cdot)$ can significantly overestimate the performance measures in the real system (by more than an order of magnitude). To address this problem, we apply the methodology developed in [15] to construct the corresponding interarrival process at finer time scales from the batch process at the coarser time scale of 1 second. A discussion of this methodology, which basically exploits the statistical properties of the batch arrival process $\mathcal{B}(\cdot)$ to construct the interarrival process, is beyond the scope of the present paper and we refer the interested reader to [15] for the technical details.

The above procedure provides an accurate transformation from the coarse time-scale batch process data set to the interarrival process data set at a finer time scale (which is a millisecond for the data sets used in our study). We shall henceforth focus on the latter (finer time-scale) versions of the Web server data sets. Let $\mathcal{A}_n$ denote the time epoch of the nth request in each of these data sets, $n \in \mathbb{Z}^+$. Let T_n and S_n respectively denote the interarrival time and the service time of the nth request. We then have $T_n = \mathcal{A}_n - \mathcal{A}_{n-1}$ and $S_n = f_s(B_n)$ for $n \in \mathbb{Z}^+$, where $T_1 \equiv 0$.

Given the main objectives of our study, we identify and focus on sufficiently long stationary intervals of (peak and off-peak) traffic periods found in each of the commercial Web site data sets of interest. Thus, the corresponding processes $\{T_n\}$ and $\{S_n\}$ derived from these data sets are stationary sequences in which each of the generic random variables $T \stackrel{\mathrm{d}}{=} T_n$ and $S \stackrel{\mathrm{d}}{=} S_n$ follow common marginal distributions $F_T(\cdot)$ and $F_S(\cdot)$, respectively. We have selected three representative data sets to be used in our study. The first, which we call *Trace A*, represents the peak traffic period for one of the Web sites of interest. This data set is long-range dependent with a Hurst parameter H_A of approximately 0.78. Another data set, called *Trace B*, represents the off-peak traffic period for one of the Web sites, which is long-range dependent with a Hurst parameter $H_B \approx 0.64$. The third data set, *Trace C*, is somewhat artificial but included here to represent a more extreme case where the Hurst parameter H_C is around 0.9. Each of these data sets is comprised of traffic periods whose lengths are on the order of five hours and consist of more than 500,000 data points.

Our analysis of the measurement data further suggests that the service time process has a negligible dependence structure. We shall therefore assume herein that the client-request service times are i.i.d. according to a phase-type distribution. This further explains our use of a MAP/PH/1 queue to model Web server performance.

2.3 MAP/PH/1 Queue

Following standard notation, let $(\mathbf{D}_0, \mathbf{D}_1)$ be the MAP descriptors of order m_A with mean $\lambda = (\mathbf{x}\mathbf{D}_1\mathbf{e})^{-1}$, where $\mathbf{e}$ is the column vector of appropriate dimension containing all ones and $\mathbf{x}$ is the invariant probability vector of the generator $\mathbf{D} = \mathbf{D}_0 + \mathbf{D}_1$, i.e., the (unique) solution of $\mathbf{x}\mathbf{D} = \mathbf{0}$ and $\mathbf{x}\mathbf{e} = 1$. Let $(\boldsymbol{\beta}, \mathbf{S})$ be the phase-type service time distribution parameters of order m_B with mean $\mu^{-1} = -\boldsymbol{\beta}\mathbf{S}^{-1}\mathbf{e}$. The infinitesimal generator matrix $\mathbf{Q}$ for the corresponding MAP/PH/1 queue has a structure given by

$$\mathbf{Q} = \begin{bmatrix} \mathbf{B}_1 & \mathbf{B}_0 & \mathbf{0} & \mathbf{0} & \mathbf{0} & \cdots \\ \mathbf{B}_2 & \mathbf{A}_1 & \mathbf{A}_0 & \mathbf{0} & \mathbf{0} & \cdots \\ \mathbf{0} & \mathbf{A}_2 & \mathbf{A}_1 & \mathbf{A}_0 & \mathbf{0} & \cdots \\ \mathbf{0} & \mathbf{0} & \mathbf{A}_2 & \mathbf{A}_1 & \mathbf{A}_0 & \cdots \\ \mathbf{0} & \mathbf{0} & \mathbf{0} & \mathbf{A}_2 & \mathbf{A}_1 & \cdots \\ \vdots & \vdots & \vdots & \vdots & \vdots & \ddots \end{bmatrix}, \tag{1}$$

where

$$\mathbf{B}_0 = \mathbf{D}_1 \otimes \boldsymbol{\beta}, \qquad \mathbf{B}_1 = \mathbf{D}_0, \qquad \mathbf{B}_2 = \mathbf{I}_{m_A} \otimes \mathbf{S}^0, \tag{2}$$

$$\mathbf{A}_0 = \mathbf{D}_1 \otimes \mathbf{I}_{m_B}, \qquad \mathbf{A}_1 = \mathbf{I}_{m_A} \otimes \mathbf{S} + \mathbf{D}_0 \otimes \mathbf{I}_{m_B}, \qquad \mathbf{A}_2 = \mathbf{I}_{m_A} \otimes \mathbf{S}^0\boldsymbol{\beta}, \tag{3}$$

$\mathbf{I}_k$ is the order k identity matrix, $\mathbf{S}^0 = -\mathbf{Se}$, and $\otimes$ denotes the Kroneker product.

Assuming this QBD process to be irreducible and positive recurrent, then the components of its stationary probability vector $\boldsymbol{\pi}$ are given by

$$(\boldsymbol{\pi}_0, \boldsymbol{\pi}_1)\begin{bmatrix}\mathbf{B}_{00} & \mathbf{B}_{01} \\ \mathbf{B}_{10} & \mathbf{B}_{11} + \mathbf{RA}_2\end{bmatrix} = \mathbf{0}, \tag{4}$$

$$\boldsymbol{\pi}_{k+1} = \boldsymbol{\pi}_1\mathbf{R}^k, \quad k \in \mathbb{Z}_+, \tag{5}$$

$$\boldsymbol{\pi}_0\,\mathbf{e} + \boldsymbol{\pi}_1(\mathbf{I}-\mathbf{R})^{-1}\mathbf{e} = 1, \tag{6}$$

where $\mathbf{R}$ is the minimal non-negative solution of the equation

$$\mathbf{R}^2\mathbf{A}_2 + \mathbf{RA}_1 + \mathbf{A}_0 = \mathbf{0}. \tag{7}$$

Define $\mathbf{A} \equiv \mathbf{A}_0 + \mathbf{A}_1 + \mathbf{A}_2$. We shall henceforth assume that the matrix $\mathbf{A}$ is irreducible, since this is indeed the case for all instances of the MAP/PH/1 queue considered in our study.

Given the components of the invariant vector $\boldsymbol{\pi}$, we can obtain various performance measures of interest. In particular, the tail distribution of the number of Web server requests in the system can be expressed as

$$\mathsf{P}[Q > x] = \sum_{k=x+1}^{\infty} \boldsymbol{\pi}_k\mathbf{e} = \boldsymbol{\pi}_1\mathbf{R}^x(\mathbf{I}-\mathbf{R})^{-1}\mathbf{e}, \quad x \ge 0, \tag{8}$$

with the corresponding expectation given by

$$\mathsf{E}[Q] = \boldsymbol{\pi}_1\sum_{k=0}^{\infty}\mathbf{R}^k\mathbf{e} + \boldsymbol{\pi}_1\sum_{k=0}^{\infty}k\mathbf{R}^k\mathbf{e} = \boldsymbol{\pi}_1(\mathbf{I}-\mathbf{R})^{-1}\mathbf{e} + \boldsymbol{\pi}_1\mathbf{R}(\mathbf{I}-\mathbf{R})^{-2}\mathbf{e}. \tag{9}$$

The expected response time of Web server requests in the system can then be calculated using Little's law[17] and (9), which yields

$$\mathsf{E}[R] = \lambda^{-1}\left(\boldsymbol{\pi}_1(\mathbf{I}-\mathbf{R})^{-1}\mathbf{e} + \boldsymbol{\pi}_1\mathbf{R}(\mathbf{I}-\mathbf{R})^{-2}\mathbf{e}\right). \tag{10}$$

Let η denote the spectral radius of the matrix $\mathbf{R}$, which is often called the *caudal characteristic.*[18] In addition to providing the stability condition for the MAP/PH/1 queue, η is indicative of the tail behavior of the stationary queue length distribution. Let $\mathbf{u}$ and $\mathbf{v}$ be the left and right eigenvectors corresponding to η normalized by $\mathbf{ue} = 1$ and $\mathbf{uv} = 1$. Under the above assumptions, it is known that[19]

$$\mathbf{R}^x = \eta^x\,\mathbf{v}\cdot\mathbf{u} + o(\eta^x), \quad \text{as } x \to \infty,$$

which together with equation (5) yields

$$\boldsymbol{\pi}_x \mathbf{e} \;=\; \boldsymbol{\pi}_1 \mathbf{v}\, \eta^{x-1} \;+\; \mathrm{o}(\eta^{x-1}), \quad \text{as } x \to \infty. \tag{11}$$

It then follows that

$$\mathsf{P}\,[Q > x] \;=\; \frac{\boldsymbol{\pi}_1 \mathbf{v}}{1-\eta}\, \eta^{x} \;+\; \mathrm{o}(\eta^{x}), \quad \text{as } x \to \infty, \tag{12}$$

and thus

$$\lim_{x\to\infty} \frac{\mathsf{P}\,[Q > x]}{\eta^x} \;=\; \frac{\boldsymbol{\pi}_1 \mathbf{v}}{1-\eta}, \tag{13}$$

or equivalently

$$\mathsf{P}\,[Q > x] \;\sim\; \frac{\boldsymbol{\pi}_1 \mathbf{v}}{1-\eta}\, \eta^{x}, \quad \text{as } x \to \infty. \tag{14}$$

The caudal characteristic can be obtained without having to first solve for the matrix $\mathbf{R}$. We define the matrix $\mathbf{A}^*(s) = \mathbf{A}_0 + s\mathbf{A}_1 + s^2\mathbf{A}_2$, for $0 < s \leq 1$. Since the generator matrix $\mathbf{A}$ is irreducible, this matrix $\mathbf{A}^*(s)$ is irreducible with nonnegative off-diagonal elements. Let $\chi(s)$ denote the spectral radius of the matrix $\mathbf{A}^*(s)$. Then, under the above assumptions, η is the unique solution in (0,1) of the equation $\chi(s) = 0$. This solution can be directly computed, which is the approach taken for all of the experiments in Section 4. A more efficient method is developed in [20].

The maximum throughput of the Web server system is given by the maximum value of λ that yields a stable MAP/PH/1 queue. Since the matrix $\mathbf{A}$ is irreducible, the stability condition is given by[21]

$$\mathbf{x}\mathbf{A}_0\mathbf{e} \;<\; \mathbf{x}\mathbf{A}_2\mathbf{e}, \tag{15}$$

where $\mathbf{x}$ is the invariant probability vector of $\mathbf{A}$. This equation can be used to solve for the maximum throughput without having to first solve for the matrix $\mathbf{R}$. In this case we define $\widehat{\rho} = (\mathbf{x}\mathbf{A}_0\mathbf{e})/(\mathbf{x}\mathbf{A}_2\mathbf{e})$, where from equation (15) $\widehat{\rho} < 1$. Alternatively, the caudal characteristic η can be used to determine an effective measure of maximum throughput that is useful in practice.

3 Methodology

A primary objective of this paper is to develop a general parameter estimation methodology for fitting Web server data to MAPs as part of deriving a matrix-analytic analysis of Web server performance models. Our methodology can be viewed to be a hierarchical approach, based on the observation that a MAP essentially consists of a set of *control states* with arbitrary interactions among them and a set of independent interarrival time processes, one for each of the control states. We first employ standard HMM methods to identify the set of control states and define the interactions among these states, at a coarser time scale than the data set, in an attempt to capture the correlations among the data set points as well as some of the variability. Then we either use these results directly to construct an MMPP, or we exploit these results together with an additional statistical analysis of the data set and the EM algorithm for fitting phase-type distributions in order to construct a more general MAP. The interarrival process for each control state is an i.i.d. stochastic sequence, which introduces additional states in the underlying Markov chain of the MAP process. We refer to these additional states as ***phase-type states.***

An overview of the basic steps of our hierarchical fitting methodology is provided in Figure 1. Note that the algorithm input consists of the data set and a few assumptions and initial values (discussed in Section 3.1), but that our methodology does not place any restrictions on the structure of the underlying Markov chain of the MAP. The first step produces the set of outputs described in **1b** under the assumption of either exponential or hyperexponential sojourn time distributions for each control state. In the latter case, we have $\boldsymbol{\mu}_i = [\boldsymbol{\mu}_{i,1}, \ldots, \boldsymbol{\mu}_{i,m_H}]$ for each control state i, $1 \leq i \leq m_C$, where m_C denotes the number of control states and m_H denotes the number of phases in the hyperexponential distributions; otherwise, $\boldsymbol{\mu}_i$ is a scalar. The corresponding MMPP is constructed directly from the output of step **1**, whereas the remaining steps are used together with the output of step **1** to construct the corresponding MAP. In step **3**, the fitting of the data set sequence $\{S_i\}$ for control state i to a Coxian distribution uses a slightly modified version of an implementation[22] of the EM algorithm developed in [5]. The remaining details of the basic steps in Figure 1 are explained in Sections 3.1 and 3.2.

3.1 Hidden Markov Model for Parameter Estimation

An HMM with explicit state duration is a doubly (embedded) stochastic process, whose intensity is controlled by a finite-state discrete-time Markov chain $\{J_n : n \in Z^+\}$ on the state space $\{i : 1 \leq i \leq m_C\}$ representing the set of control states. The amount of time that the process has been in the current state J_n at time n is denoted by τ_n, and the number of arrivals per unit time associated with state J_n is denoted by r_n, which is the observable output associated with state J_n. It is usually assumed that the control states $\{J_n\}$ and the observations $\{r_n\}$ are conditionally independent with the conditional distribution of r_n dependent on J_n only.

Since this semi-Markov process is not directly observable, the state se-

1. Use HMM methods to construct the control states and their interactions
a. Input
- data set with N data entries
- desired number of control states
- assumption for the arrival process per control state (coarse time scale)
  - Poisson
- assumption for the sojourn times per control state
  - Exponential
  - Hyperexponential : desired number of phases

b. Output
- number of control states m_C
- transition probability matrix $\mathbf{P} = [\mathbf{P}_{i,j}]_{1 \leq i,j \leq m_C}$ for the control states
- mean interarrival times $\boldsymbol{\lambda}_i^{-1}$ for $1 \leq i \leq m_C$; $\boldsymbol{\lambda} = [\boldsymbol{\lambda}_1, \ldots, \boldsymbol{\lambda}_{m_C}]$
- (vector of) mean sojourn times $\boldsymbol{\mu}_i^{-1}$ for $1 \leq i \leq m_C$; $\boldsymbol{\mu} = [\boldsymbol{\mu}_1, \ldots, \boldsymbol{\mu}_{m_C}]$
- sojourn time probability vectors $\mathbf{p}_i^{\boldsymbol{\mu}}$ for $1 \leq i \leq m_C$, if using hyperexponential distribution; $\mathbf{p}^{\boldsymbol{\mu}} = [\mathbf{p}_1^{\boldsymbol{\mu}}, \ldots, \mathbf{p}_{m_C}^{\boldsymbol{\mu}}]$; $\mathbf{p}_i^{\boldsymbol{\mu}} = [\mathbf{p}_{i,1}^{\boldsymbol{\mu}}, \ldots, \mathbf{p}_{i,m_H}^{\boldsymbol{\mu}}]$
- mapping $\{J_n\}$ of each data entry to its corresponding control state

2. Construct a data set sequence $\{\mathcal{S}_i\}$ per control state i using mapping $\{J_n\}$ from step **1** and the original data set
3. Feed each sequence $\{\mathcal{S}_i\}$ to EM algorithm to generate a Coxian distribution for the interarrival process of each control state
4. Compute the probability of state change upon arrival using mapping $\{J_n\}$
5. Construct $\mathbf{D}_0$ using:
- transition probability matrix from step **1**
- model for sojourn times from step **1**
- Coxian model for interarrivals from step **3**

6. Construct $\mathbf{D}_1$ using:
- Coxian model for interarrivals from step **3**
- transition probability matrix from step **1**
- model for sojourn times from step **1**
- probability vector computed in step **4**

Figure 1. Overview of our MAP parameter estimation methodology.

quence $\{J_n, \tau_n\}$ and the model parameters (i.e., the transition probability matrix $\mathbf{P}$ for the control states, the mean interarrival times $\boldsymbol{\lambda}_i^{-1}$, the vector of mean sojourn times $\boldsymbol{\mu}_i^{-1}$, the sojourn time probability vector $\mathbf{p}_i^{\boldsymbol{\mu}}$ for each control state i, and the control state sequence $\{J_n\}$ of the data set) are estimated from the observed sequence $\{r_n\}$. The main steps of the standard recursive procedure for HMM with explicit state duration are summarized as follows:

- Given an initial set of assumptions for the HMM model parameters, obtain refined maximum likelihood estimators for the model parameters by applying the *HMM re-estimation algorithms* with explicit state duration to the given observation sequence $\{r_n\}$.

- Apply one of the many *HMM forward-backward algorithms* with explicit state duration to find the maximum a posteriori state estimate, $\{J_n\}$, for the given observation sequences $\{r_n\}$.

We refer the interested reader to [23] for an overview of the details on these standard HMM algorithms. Additional technical details can be found in [24,25,26] and the references cited therein.

Following the above procedure, we can obtain the maximum likelihood model parameters for the given observation sequence $\{r_n\}$ and the state space $\{1, \ldots, m_C\}$. Let $\widehat{H}_i(\tau)$ denote the estimated non-parametric probability mass function for the sojourn time τ of state i, and let $\widehat{O}_i(r)$ denote the estimated non-parametric probability mass function for the observation r of state i.

The total number of model parameters can be reduced if the observation distribution or the state sojourn time distribution is approximated by some parametric distributions such as Gaussian, Poisson or gamma distributions.[27,28,29] In this case, one only needs to estimate a few parameters that specify the selected distribution functions. Ferguson[30] has shown that the parameters for the parametric sojourn time distribution $H_i(\tau)$ and the parametric observation distribution $O_i(r)$ for state i can be found by maximizing $\sum_\tau \widehat{H}_i(\tau)\ln(H_i(\tau))$ and $\sum_r \widehat{O}_i(r)\ln(O_i(r))$ subject to the stochastic constraints $\sum_\tau H_i(\tau) = 1$ and $\sum_r O_i(r) = 1$.

Under the assumptions that the arrival process for each control state (at a coarse time scale) is Poisson and that the per-control-state sojourn times follow a hyperexponential distribution, i.e.,

$$O_i(r) = \frac{(\boldsymbol{\lambda}_i t)^r}{r!} e^{-\boldsymbol{\lambda}_i t}, \tag{16}$$

$$H_i(\tau) = \sum_{j=1}^{m_H} \mathbf{p}_{ij}^{\boldsymbol{\mu}}\, \boldsymbol{\mu}_{ij}\, e^{-\boldsymbol{\mu}_{ij}(\tau-1)}, \tag{17}$$

where $\boldsymbol{\lambda}_i \geq 0$, $\boldsymbol{\mu}_{ij} \geq 0$ and $\sum_j \mathbf{p}_{ij}^{\boldsymbol{\mu}} = 1$, then the arrival rate $\boldsymbol{\lambda}_i$ of the Poisson process for state i can be estimated by $\widehat{\boldsymbol{\lambda}}_i = \sum_r \widehat{O}_i(r) r$.[30,27] The parameters

$\mathbf{p}_{ij}^{\boldsymbol{\mu}}$ and $\boldsymbol{\mu}_{ij}$ of the hyperexponential distribution can be determined numerically via equation (17) and the optimization problem described above.

Finally, as part of the initialization step, we set the total number of control states to a prespecified input parameter (which was analyzed and tuned in our experiments). When this parameter is a sufficiently large integer, then in the re-estimation procedure, the states that are never visited will be deleted from the state space, so that the value of m_C is reduced to a number of control states that match the data set. This led to a maximum of 20 control states for the data sets used in our study. We also need to initialize the elements of the transition probability matrix, the control state sojourn time distributions, and the initial control state probability vector. An often used choice is to assume that these initial values for the model parameters are uniformly distributed, which was the choice made in our study. In addition, we assume the initial values of the control-state arrival rates to be proportional to the state index, i.e., $\boldsymbol{\lambda}_i = r_{max}\frac{i}{m_C}$ where i is the index of the control state, r_{max} is the maximum value of r, and m_C is the total number of control states.

3.2 Generation of MAP from HMM output

The above HMM methods produce the output described in Figure 1, which includes the sequence $\{J_n\}$, $1 \leq n \leq N$, representing the mapping of the entries in the data set to the set of control states (at a coarse time scale). That is, as part of the HMM analysis, each interarrival time in the data set is assigned to one of the m_C control states. We first construct a new data set sequence $\{S_i\}$, $1 \leq i \leq m_C$, that consists of all of the interarrival times from the data set where $J_n = i$, which are then combined in the same relative order as in the original data set. We also compute from the sequence $\{J_n\}$, $1 \leq n \leq N$, the probability vector $\widehat{\mathbf{p}}$ of dimension m_C where the element $\widehat{\mathbf{p}}_i$ denotes the probability that upon an arrival from control state i the process switches to another control state. Specifically, we have

$$\widehat{\mathbf{p}}_i = \frac{\sum_{j=2}^{N} I_{J_j \neq i,\, J_{j-1}=i}}{\sum_{j=2}^{N} I_{J_j = i,\, J_{j-1}=i} + \sum_{j=2}^{N} I_{J_j \neq i,\, J_{j-1}=i}}, \tag{18}$$

where I_A denotes the indicator function for event A having the value 1 if A occurs and the value 0 otherwise. Thus, with probability $1 - \widehat{\mathbf{p}}_i$ the process immediately returns to the same control state i.

To help facilitate the description of the procedure for generating MAPs from the above set of variables, we need the following definitions:

- $\boldsymbol{\Lambda} = \mathbf{diag}(\boldsymbol{\lambda}_1, \ldots, \boldsymbol{\lambda}_{m_C})$ is a diagonal matrix of order m_C whose diagonal elements are the elements of vector $\boldsymbol{\lambda}$;

- $\mathbf{P}_i^{\boldsymbol{\mu}}$ is the $m_H \times m_H$ matrix whose rows are all equal to $\mathbf{p}_i^{\mu}$, $1 \le i \le m_C$;
- $\boldsymbol{\Phi} = \mathbf{diag}(\boldsymbol{\Phi}_1, \ldots, \boldsymbol{\Phi}_{m_C})$ is a (block) diagonal matrix of order $m_C \cdot m_H$, where $\boldsymbol{\Phi}_i = \mathbf{diag}(\boldsymbol{\mu}_{i,1}, \ldots, \boldsymbol{\mu}_{i,m_H})$ is a diagonal matrix of order m_H;
- $\mathbf{col}(M, k, j)$ is a matrix function that partitions the columns of the matrix M into blocks of size k and then extracts the jth such block of columns of size k from matrix M (the resulting matrix has the same number of rows as M and k columns);
- $\mathbf{row}(M, k, j)$ is a matrix function that partitions the rows of the matrix M into blocks of size k and then extracts the jth such block of rows of size k from matrix M (the resulting matrix has k rows and the same number of columns as M);
- $\mathbf{V} = [\mathbf{V}_1\ \mathbf{V}_2\ \ldots\ \mathbf{V}_{m_C}]$, where $\mathbf{V}_i = \mathbf{col}(\mathbf{P}, 1, i) \otimes \mathbf{P}_i^{\boldsymbol{\mu}}$, $1 \le i \le m_C$.

The off-diagonal elements of the matrix $\mathbf{D}_0^{\text{MMPP}}$ and the matrix $\mathbf{D}_1^{\text{MMPP}}$ for an MMPP with exponential interarrival times and hyperexponential sojourn times per control state can be expressed as

$$\mathbf{D}_0^{\text{MMPP}} = \boldsymbol{\Phi}\mathbf{V}, \qquad \mathbf{D}_1^{\text{MMPP}} = \boldsymbol{\Lambda} \otimes \mathbf{I}_{m_H}. \tag{19}$$

Then the diagonal element of each row of $\mathbf{D}_0^{\text{MMPP}}$ is computed as the negative sum of the non-diagonal elements on the same row in the matrix $\mathbf{D}_0^{\text{MMPP}} + \mathbf{D}_1^{\text{MMPP}}$. The number of states in this MMPP is $m_C \cdot m_H$. During the numerical experiments we observed that some of the values of the vectors $\mathbf{p}_i^{\boldsymbol{\mu}}$, $1 \le i \le m_C$, are very small, i.e., less than a desired tolerance of accuracy. The states that are represented by these probabilities are removed from the set of states during the generation of $\mathbf{D}_0$ and $\mathbf{D}_1$ to avoid numerical problems in the solution of the MAP/PH/1 queues. We note that the size of the state space is decreased by up to 30% with this simple state reduction technique.

From the HMM output we can also construct a more general MAP by using the same matrix $\mathbf{D}_0$ from the above MMPP and modifying the matrix $\mathbf{D}_1$ by making use of the probability vector $\widehat{\mathbf{p}}$. Define $\widehat{\mathbf{P}} = \mathbf{diag}(\widehat{\mathbf{p}}_1, \ldots, \widehat{\mathbf{p}}_{m_C})$ to be the order m_C diagonal matrix corresponding to the vector $\widehat{\mathbf{p}}$. We then have

$$\mathbf{D}_0^{\text{MAP}} = \mathbf{D}_0^{\text{MMPP}}, \tag{20}$$

$$\mathbf{D}_1^{\text{MAP}} = \left(\mathbf{I}_{m_C \cdot m_H} - \mathbf{diag}(((\widehat{\mathbf{P}} \otimes \mathbf{I}_{m_H})\mathbf{V}\mathbf{e})^T) + (\widehat{\mathbf{P}} \otimes \mathbf{I}_{m_H})\mathbf{V}\right) \mathbf{D}_1^{\text{MMPP}}. \tag{21}$$

We compute from the sequence $\{J_n\}$, $1 \le n \le N$, only the probability of leaving a control state upon arrival. The probability of reaching any other control state is not computed from this sequence, but rather from the probability transition matrix $\mathbf{V}$ since it is estimated using a similar analysis (which is reflected in the definition of matrix $\mathbf{D}_1^{\text{MAP}}$ in equation (21)).

Another set of MAP processes is obtained by incorporating the results of fitting the interarrival times in each of the data set sequences $\{\mathcal{S}_i\}$, $1 \le i \le m_C$, to a Coxian distribution using the EM algorithm.[5] This computes for each control state i the corresponding vector $\widehat{\boldsymbol{\alpha}}_i$ and matrix $\widehat{\mathbf{T}}_i$, $1 \le i \le m_C$, both of order m_X. We define

$$\mathbf{U}_i = \mathbf{row}(\mathbf{D}_0^{\text{MAP}}, m_H, i) \otimes \hat{\mathbf{T}}^i,$$
$$\mathbf{X}_i = \mathbf{row}(\mathbf{I}_{m_C m_H} - \mathbf{diag}(((\hat{\mathbf{P}} \otimes \mathbf{I}_{m_H})\mathbf{Ve})^T) + (\hat{\mathbf{P}} \otimes \mathbf{I}_{m_H})\mathbf{V}, m_H, i) \otimes \hat{\mathbf{T}}_i^o \hat{\boldsymbol{\alpha}}_i,$$
$$\mathbf{U} = \left[\mathbf{U}_1^T, \cdots, \mathbf{U}_{m_C}^T\right]^T, \qquad \mathbf{X} = \left[\mathbf{X}_1^T, \cdots, \mathbf{X}_{m_C}^T\right]^T.$$

Then the matrices $\mathbf{D}_0^{\text{MAP-C}}$ and $\mathbf{D}_1^{\text{MAP-C}}$, where MAP-C stands for the MAP with Coxian interarrival processes for each control state, can be expressed as

$$\mathbf{D}_0^{\text{MAP-C}} = \mathbf{U}, \qquad \mathbf{D}_1^{\text{MAP-C}} = \mathbf{X}. \tag{22}$$

The total number of states for this MAP is $m_C \cdot m_H \cdot m_X$. In the same manner as described above, the states with probabilities that are very close to 0 are removed from the final version of the MAP.

4 Numerical Results

Our objectives in this paper have been to develop a general methodology for fitting measurement data from Web sites to the interarrival time and service time processes of a MAP/PH/1 queue used to model Web server performance. To this end, a large number of numerical experiments have been performed where we apply the methodology of Section 3 to the commercial Web site data sets of Section 2, and then we solve the resulting MAP/PH/1 queue. These experiments are primarily used to explore two key issues: the accuracy of our methodology for fitting Web data to MAPs; and the performance characteristics of the corresponding MAP/PH/1 queues.

We consider fitting the interarrival times of each data set with a wide variety of models that can be obtained from our methodology as follows:

- an exponential (Exp) or Y-phase Coxian (Cox$Y$) distribution for the interarrival times associated with each control state;
- an exponential or Y-phase hyperexponential (HrY) distribution for the sojourn times of each control state.

To facilitate the presentation of results, we shall use the notation MMPP(s,d) and MAP(s,a,d) where s denotes the number of control states, d denotes the distribution of sojourn times for each control state, and a denotes the distribution that models the interarrival times associated with each control state.

In order to evaluate the accuracy of our approach, we compare various steady-state performance measures of the MAP/PH/1 queue against the corresponding measures obtained by simulating the well-known Lindley equation[31] under the input sequence $\{(T_n, S_n)\}$ from the Web data set. We assume that requests are served in an FCFS manner and that the server depletes work at rate $\mathcal{C}$, where $\mathcal{C}$ is a deterministic constant. By varying the parameter $\mathcal{C}$, different server loads $\rho \equiv \lambda \mathsf{E}[S]/\mathcal{C}$ can be considered, where $\lambda = \mathsf{E}[T]^{-1}$. For stability of the queue, we also assume $\rho < 1$. The value of $\mathcal{C}$ can only be reduced up to points that still maintain a stable regenerative G/G/1 queue (in the sense that the system empties with probability 1 and that there are a sufficiently large number of such regeneration points). Moreover, the spectrum of traffic intensities of interest from a practical perspective is well within the range over which the mean response time remains below $100 \times \mathsf{E}[S]$. We therefore restrict our attention to this spectrum of traffic intensities, which also ensures that the regenerative G/G/1 queue is stable for each of the cases considered.

Before presenting a representative sample of our results, it is important to point out that each Web data set available to us in our study represents a single stationary interval of traffic from a specific Web site, i.e., a single sample path of an underlying stationary stochastic process. We have considerable evidence suggesting that the statistical properties of each of these stationary sequences is representative of the per-weekday (peak and off-peak) traffic intervals found at the corresponding commercial Web site. However, our ongoing research is pursuing the use of multiple independent and statistically identical data sets from each of the specific Web sites as replicas in our simulation of the Lindley equation to compare against our results obtained via matrix-analytic methods, as well as the use of such i.i.d. data sequences in our fitting methodology.

4.1 Peak Traffic Period (Trace A)

We vary the number of control states used by the HMM algorithm as part of our methodology for fitting the interarrival times of Trace A to various MAPs. This makes it possible for us to examine the impact of the size of the underlying Markov chain on the accuracy of the model fitting. We start with 2 control states, and then increase to 5 and 10 states. The mean response times under a small subset of these MAPs as a function of the traffic intensity ρ are plotted in Figure 2(a), together with the corresponding simulation results for Trace A. Our results clearly demonstrate that the accuracy of the fitting improves significantly with increases in the number of control states, as expected. (Note that the MAP(5,Cox2,Exp) results are somewhat more accurate than the MAP(2,Cox2,Exp) results, which are not shown.) This is because more control states in the underlying Markov chain provide greater flexibility which makes it possible to better capture not only the dependence structure but also the variability of the arrival process. The output of the HMM algorithm for larger numbers of control states exhibit small probabilities for entering a few of the control states (which represent extremes of the arrival rate values), together with small transition rates for leaving these control states once entered. This suggests that such control states improve the ability of the MAP to capture the tail of the interarrival process, and that the degree to which this is possible improves with increases in the number of control states.

In order to isolate, to some extent, the impact of the dependence structure on mean response time measures, we have ignored such dependencies and fitted the interarrival times of Trace A to a phase-type distribution. The mean response time measures for this PH/PH/1 queue are also provided in Figure 2(a). It can be clearly observed from these results that the PH/PH/1 fitting is very poor and, with the exception of very light traffic intensities, the PH/PH/1 queue is simply not capable of capturing the performance of the queueing system under Trace A. Conversely, the MAP models that capture the dependence structure do a much better job of matching the queueing system performance, particularly at heavier loads, where the accuracy increases with the complexity of the MAP.

In a similar manner, our methodology is used for fitting Trace A to various MMPPs, and the corresponding mean response times under a small subset of these MMPPs are plotted in Figure 2(b). The results from simulation are also included in the figure for comparative purposes. We continue to observe that the larger the number of control states, the more accurate the fitting. Once again, more control states in the underlying Markov chain provide greater flexibility that makes it possible to better capture both the dependence structure and the variability of the arrival process, for the reasons described above.

We observe that one of the MMPP models provides the most accurate results in comparison with simulation for light loads. Under heavier loads, however, one of the MAPs tends to provide the most accurate results. Specifically, MAP(10,Cox2,Exp) provides the best accuracy with a relative error always less than 20%. We further observe that the models using Coxian dis-

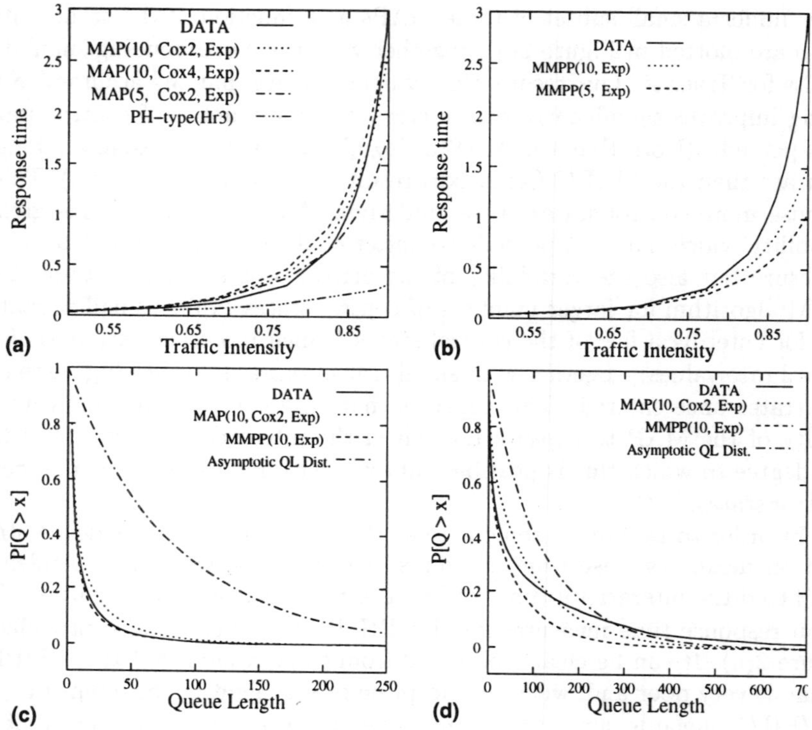

Figure 2. Response time as function of traffic intensity for (a) MAP models, (b) MMPP models; and Queue length tail distribution of fitted models for (c) moderate, (d) high system loads; all for Trace A.

tributions for the interarrival time process of each control state tend to provide better fits. Based on a simple statistical analysis of the sequence $\{J_n\}$ defined in Section 3.1, we further observe that the interarrivals for each control state are not exponential, which is why we use the Coxian distribution to model each of these control-state interarrival processes.

We also study the tail behavior of the queue length distribution, using equation (8), for the best MAP and MMPP models, i.e., MAP(10,Cox2,Exp) and MMPP(10,Exp). For comparison we also consider the asymptotic queue length tail distribution that corresponds to the MAP(10,Cox2,Exp) model, using equation (14) based on the caudal characteristic η, and the tail of the queue length distribution from the simulation of Trace A. Figure 2(c) plots these five queue length tail distributions for the traffic intensity $\rho = 0.77$, which represents a case where the system is moderately loaded. The corresponding set of results for a traffic intensity of $\rho = 0.90$, which represents a heavily loaded system, are presented in Figure 2(d). Note that in each of these figures we only plot the asymptotic queue length tail distribution for the MAP model because the corresponding curve for the MMPP model is quite close to that of the MAP model.

We observe that the queue length tail distributions obtained under the MAP and MMPP models provide a reasonably close match with the corresponding tail distribution obtained from simulation over a relatively wide range. In the case of moderate load, i.e., $\rho = 0.77$, the tail distribution from the MMPP model closely matches the simulation results for small queue length values, which helps to explain the low relative error values at light to moderate loads. Conversely, the tail distribution from the MAP model overestimates the simulation results for all but very large queue length values. This causes the MAP to yield poorer relative errors at light to moderate loads, although still always less than 20%. In the case of heavier load, i.e., $\rho = 0.90$, the tail distribution from the MMPP model continues to provide a close match with the simulation results but only for very small queue length values. The tail distribution from the MAP model continues to overestimate the simulation results over a relatively large range of queue length values, crossing at around a queue length of 140. In fact, the accuracy of the expected response time under the MAP model is achieved in part by pushing this crossover point relatively far to the right (to larger queue lengths). This further explains why the expected response times under the MAP model underestimate those obtained from simulation at heavier loads. More accurate response times under the MAP model at heavier loads can be obtained by increasing the number of (control and/or phase-type) states in the underlying Markov chain.

We also observe that the asymptotic queue length tail distribution based on the caudal characteristic η dominates all other tail distributions, for both $\rho = 0.77$ and $\rho = 0.90$, across a relatively wide range of queue length values. The tail of the queue length distribution from simulation eventually crosses the asymptotic tail distribution, as expected due to the dependence structure and variability in Trace A, but these crossover points occur at queue length values greater than 250 which can be considered to be a relatively high value of queue length.

We note that the maximum response time value plotted for the MAP(10,Cox2,Exp) model represents $\rho = 0.9$, $\widehat{\rho} = 0.94$ and $\eta = 0.99$. This illustrates and quantifies the degree to which the latter two variables provide a better measure of effective system load than the standard traffic intensity ρ for Trace A, at least from a practical perspective.

4.2 *Off-Peak Traffic Period (Trace B)*

The same set of experiments and analysis as those described in Section 4.1 are performed on Trace B. Since the results in Section 4.1 demonstrate that a model with 10 control states fits the interarrival process much more accurately than the corresponding models with fewer control states, here we focus solely on models with 10 control states in the analysis of Trace B.

Based on equations (19), (20), (21), and (22), several different MAP and MMPP models are fitted to the interarrival process of Trace B. The mean response times under a small subset of these MAPs and MMPPs as a function of the traffic intensity ρ are respectively plotted in Figures 3(a) and 3(b). In Figure 3(a) we observe that the gap between the PH/PH/1 model and some of the MAP and MMPP models is smaller than the corresponding results of the previous section for Trace A. This is directly related to the weaker long-range dependence of the interarrival process of Trace B. From among the set of MAP models, MAP(10,Cox2,Exp) performs the best with a relative error always less than 12%. On the other hand, the MMPP(10,Exp) model performs slightly better with a worst case relative error of 10%. This supports the notion that the MMPP is a good model for data sets which have a relatively weak long-range dependence structure, or a short-range dependence structure.

The queue length tail distributions for the models MMPP(10,Exp) and MAP(10,Cox2,Exp), as well as the asymptotic behavior (as characterized by equation (14) in terms of the caudal characteristic η) for the model MAP(10,Cox2,Exp), are compared with the queue length tail distribution obtained from the simulation of Trace B in Figure 3(c) for a traffic intensity of 0.88. The corresponding curves for a traffic intensity of 0.95 are shown in Figure 3(d). These plots illustrate that for both moderate and relatively heavy traffic intensities MMPP(10,Exp) is a slightly better fit than MAP(10,Cox2,Exp). However, for heavier traffic intensities, the MAP(10,Cox2,Exp) curve follows the simulation curve for larger values of queue length than does the MMPP(10,Exp) curve.

Note that the maximum response time value plotted for the MAP(10,Cox2,Exp) model represents $\rho = 0.97$, $\widehat{\rho} = 0.99$ and $\eta = 0.99$.

4.3 *Strong Long-Range Dependence (Trace C)*

We perform the same set of experiments and analysis on the Trace C data set as those considered in Sections 4.1 and 4.2. Figure 4(a) presents the mean response time as a function of the traffic intensity for a small subset of the fitted MAP models in comparison with the corresponding simulation curve for Trace C. We note that the MAPs do a very good job of accurately capturing

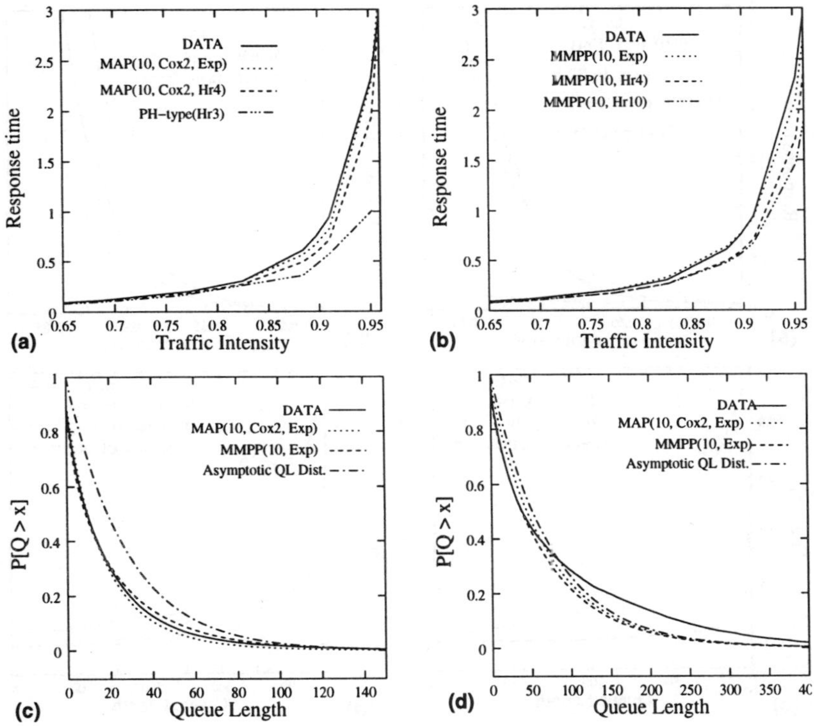

Figure 3. Response time as function of traffic intensity for (a) MAP models, (b) MMPP models; and Queue length tail distribution of fitted models for (c) moderate, (d) high system loads; all for Trace B.

the queueing system performance even though the dependence structure of Trace C is much stronger than that found in Traces A and B. It is this strong long-range dependence in Trace C that causes the MAP models with hyper-exponential sojourn time distributions for the control states to perform much better than the cases where the sojourn times are assumed to be exponentially distributed. Moreover, unlike the 2-phase Coxian distributions that are used to fit the interarrival process for each of the control states for Traces A and B, we choose a 4-phase Coxian distribution for fitting the interarrival process of each control state for Trace C. These 4-phase Coxian distributions are essentially Erlang distributions because we find that the coefficient of variation for each of the constructed control-state trace sequences (see Section 3.2) is less than 0.5.

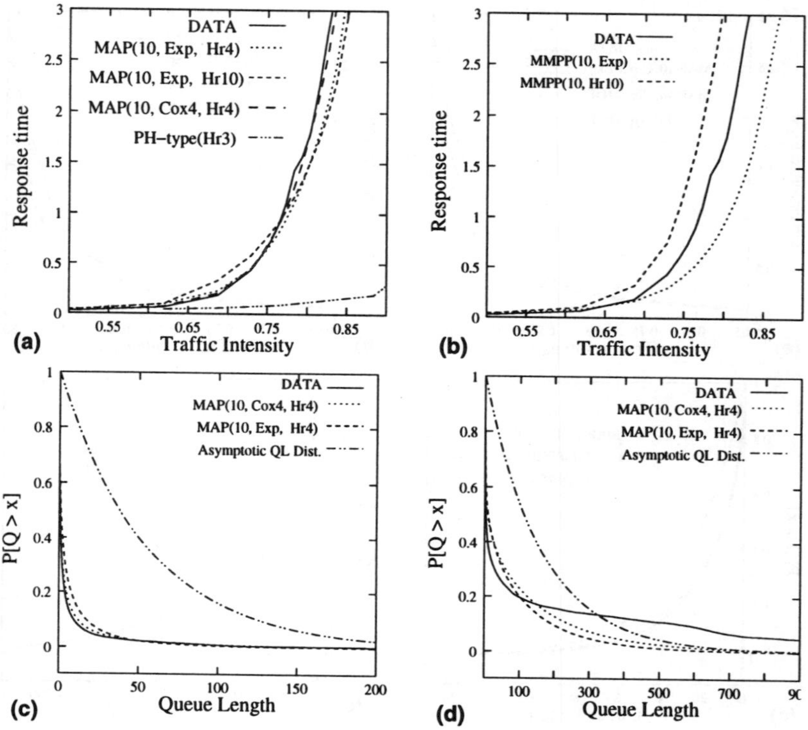

Figure 4. Response time as function of traffic intensity for (a) MAP models, (b) MMPP models; and Queue length tail distribution of fitted models for (c) moderate, (d) high system loads; all for Trace C.

Figure 4(b) is similar to Figure 4(a) but for a small subset of the fitted MMPP models. These results show that the MMPP models only roughly approximate the simulation curve. The best fit for Trace *C* is provided by the MAP(10,Cox4,Hr4) model with relative error less than 12%.

The queue length tail distributions for the two best fitting models, MAP(10,Cox4,Hr4) and MAP(10,Exp,Hr4), are plotted in Figure 4(c) for the moderate traffic intensity of 0.69, and in Figure 4(d) for the high traffic intensity of 0.83. These plots illustrate that the tail of the queue length distribution for Trace *C* obtained by simulation is rather heavy and that the MAP(10,Cox4,Hr4) model does a much better job of capturing the characteristics of this heavy tail up to a relatively large queue length value under both moderate and high traffic intensities. However, at high traffic intensities, the heavier tail of the simulation results eventually crosses from below that of the MAP(10,Cox4,Hr4) model, beyond which it decays much more slowly than all other tail distribution curves. This further explains why the expected response times under the MAP model underestimate those obtained from simulation at heavier loads. On the one hand, as previously noted, the range of traffic intensities that cover mean response time values from $\mathsf{E}[S]$ to $100 \times \mathsf{E}[S]$ represent by far the set of traffic intensities that might be of interest in practice. On the other hand, more accurate response times under the MAP model at heavier loads can be obtained by increasing the number of (control and/or phase-type) states in the underlying Markov chain.

We note that the maximum response time value plotted for the MAP(10,Cox4,Hr4) model represents $\rho = 0.83$, $\widehat{\rho} = 0.84$ and $\eta = 0.99$. This illustrates and quantifies the degree to which the latter two variables provide a better measure of effective system load than the standard traffic intensity ρ for Trace C, at least from a practical perspective.

5 Conclusions

In this paper we presented an initial study that considers a MAP/PH/1 queue whose interarrival time and service time processes are fitted to measurement data from actual commercial Web sites. Our fitting methodology is based on the use of standard HMM methods to identify and characterize the dependence structure of, and some of the variability in, these large data sets, together with additional statistical analysis of the data sets and the EM method for fitting phase-type distributions to construct different instances of MAPs.

The results of our many numerical experiments are quite promising, demonstrating that, contrary to common beliefs, Markovian models and matrix-analytic methods can be used to efficiently and accurately capture the complexities of commercial Web site data sets that exhibit a wide range of dependence structures and variabilities. Moreover, the size of the MAP models required to achieve such results over the spectrum of system loads of interest from a practical perspective is relatively small and manageable, at least for the Web data sets considered in our study.

Finally, the fitting methodology presented in this paper simply leverages standard HMM methods in a direct manner, essentially treating these procedures as a "black box". We are currently exploring the extension and tailoring of these methods from first principles for our specific purposes.

Acknowledgments

This work was performed during a summer internship by Alma Riska at the IBM T.J. Watson Research Center. The software tools used in our study were developed by Alma Riska under grants CCR-0098278 and EIA-9974992. We thank Jim Challenger for performing the detailed measurements described in Section 2.1. We further acknowledge the use of an implementation[22] of the EM algorithm developed in [5], and we thank the authors for making their code publically available via the World Wide Web. Lastly, we thank the referees for their helpful comments on an early draft of this paper.

References

1. Walter Willinger, Murad S. Taqqu, Robert Sherman, and Daniel V. Wilson. Self-similarity through high-variability: Statistical analysis of Ethernet LAN traffic at the source level. *IEEE/ACM Transactions on Networking*, 5(1):71–86, February 1997.
2. Mark S. Squillante, David D. Yao, and Li Zhang. Web traffic modeling and web server performance analysis. In *Proceedings of the IEEE Conference on Decision and Control*, December 1999.
3. Bo Friis Nielsen. Modelling long-range dependent and heavy-tailed phenomena by matrix analytic methods. In *Advances in Algorithmic Methods for Stochastic Models,* G. Latouche and P. Taylor (eds.), pages 265–278. Notable Publications, 2000.
4. Zhen Liu, Nicolas Niclausse, and Cesar Jalpa-Villanueva. Traffic model and performance evaluation of Web servers. *Performance Evaluation*, 46:77–100, October 2001.
5. Soren Asmussen, Olle Nerman, and Marita Olsson. Fitting phase-type distributions via the EM algorithm. *Scandinavian Journal of Statistics*, 23:419–441, 1996.
6. Soren Asmussen. Phase-type distributions and related point processes: Fitting and recent advances. In Srinivas R. Chakravarthy and Attahiru S. Alfa, editors, *Matrix-Analytic Methods in Stochastic Models*, pages 137–149. Marcel Dekker, 1997.
7. Andreas Lang and Jeffrey L. Arthur. Parameter approximation for phase-type distributions. In Srinivas R. Chakravarthy and Attahiru S. Alfa, editors, *Matrix-Analytic Methods in Stochastic Models*, pages 151–206. Marcel Dekker, 1997.
8. Andras Horvath and Miklos Telek. Approximating heavy tailed behaviour with phase type distributions. In *Advances in Algorithmic Methods for*

Stochastic Models, G. Latouche and P. Taylor (eds.), pages 191–214. Notable Publications, 2000.

9. Kathleen S. Meier-Hellstern. A fitting algorithm for Markov-modulated Poisson processes having two arrival rate. *European Journal of Operations Research*, 29:370–377, 1987.
10. Tobias Ryden. Parameter estimation for Markov modulated Poisson processes. *Communications in Statistics-Stochastic Models*, 10(4):795–829, 1994.
11. A. Horvath, G. Rozsa, and M. Telek. A MAP fitting method to approximate real traffic behaviour. In *Proceedings of the IFIP Workshop on Performance Modelling and Evaluation of ATM & IP Networks*, pages 32/1–12, Ilkley, UK, 2000.
12. Lothar Breuer. Parameter estimation for a class of BMAPs. In *Advances in Algorithmic Methods for Stochastic Models,* G. Latouche and P. Taylor (eds.), pages 87–97. Notable Publications, 2000.
13. Lothar Breuer. An EM algorithm for batch Markovian arrival processes and its comparison to a simpler estimation procedure. Preprint, 2001.
14. Tobias Ryden. Statistical estimation for Markov-modulated Poisson processes and Markovian arrival processes. In *Advances in Algorithmic Methods for Stochastic Models,* G. Latouche and P. Taylor (eds.), pages 329–350. Notable Publications, 2000.
15. Cathy H. Xia, Zhen Liu, Mark S. Squillante, Li Zhang, and Naceur Malouch. Traffic modeling and performance analysis of commercial web sites. Preprint, 2001.
16. Arun K. Iyengar, Mark S. Squillante, and Li Zhang. Analysis and characterization of large-scale web server access patterns and performance. *World Wide Web*, 2, June 1999.
17. John D. C. Little. A proof of the queuing formula $L = \lambda W$. *Operations Research*, 9:383–387, 1961.
18. Marcel F. Neuts. The caudal characteristic curve of queues. *Advances in Applied Probability,* 18:221–254, 1986.
19. E. Seneta. *Non-Negative Matrices and Markov Chains.* Springer Verlag, New York, second edition, 1981.
20. Nigel G. Bean, Jian-Min Li, and Peter G. Taylor. Caudal characteristics of qbds with decomposable phase spaces. In *Advances in Algorithmic Methods for Stochastic Models,* G. Latouche and P. Taylor (eds.), pages 37–55. Notable Publications, 2000.
21. Marcel F. Neuts. *Matrix-Geometric Solutions in Stochastic Models: An Algorithmic Approach.* The Johns Hopkins University Press, 1981.
22. Marita Olsson. The EMpht-programme. Technical report, Depart-

ment of Mathematics, Chalmers University of Technology, June 1998. http://www.maths.lth.se/matstat/staff/asmus/pspapers.html.

23. L. R. Rabiner. A tutorial on hidden Markov models and selected applications in speech recognition. *Proceedings of the IEEE*, 77(2):257–286, February 1989.
24. B. Sin and J. H. Kim. Nonstationary hidden Markov model. *Signal Processing*, 46:31–46, 1995.
25. S.V.Vaseghi. State duration modeling in hidden Markov models. *Signal Processing*, 41:31–41, 1995.
26. Y. K. Park, C. K. Un, and O. W. Kwon. Modeling acoustic transitions in speech by modified hidden Markov models with state duration and state duration-dependent observation probabilities. *IEEE Transactions on Speech and Audio Processing*, 4(5):389–392, September 1996.
27. M. J. Rusell and R. K. Moore. Explicit modeling of state occupancy in hidden Markov models for automatic speech recognition. In *Proceedings of ICASSP85*, pages 5–8, 1985.
28. S. E. Levinson. Continuously variable duration hidden Markov models for automatic speech recognition. *Computer Speech and Language*, 1(1):29–45, 1986.
29. C. Mitchell and L. Jamieson. Modeling duration in a hidden Markov model with the exponential family. In *Proceedings of ICASSP93*, pages 331–334, 1993.
30. J. D. Ferguson. Variable duration models for speech. In *Proceedings of the Symposium on the Application of Hidden Markov Models to Text and Speech*, pages 143–179, October 1980.
31. Leonard Kleinrock. *Queueing Systems Volume I: Theory.* John Wiley and Sons, 1975.

ANALYSIS OF PARALLEL-SERVER QUEUES UNDER SPACESHARING AND TIMESHARING DISCIPLINES

JAY SETHURAMAN

IEOR Department, Columbia University, New York, NY 10027, USA
E-mail: jay@ieor.columbia.edu

MARK S. SQUILLANTE

IBM Thomas J. Watson Research Center, Yorktown Heights, NY 10598, USA
E-mail: mss@watson.ibm.com

We study a class of parallel-server queues under scheduling disciplines that share the servers in both space and time. A matrix-analytic analysis of these queues is derived, making it possible to efficiently compute exact solutions for small to moderate instances of the stochastic model and providing very accurate approximations for larger model instances. We then exploit these results to analyze fundamental properties of parallel-server queues under spacesharing and timesharing.

1 Introduction

Stochastic models and related queueing-theoretic results have played an important role in the design of scheduling disciplines of both theoretical and practical interest. This has been especially the case in single-server queues (e.g., Fife,[1] Schrage and Miller,[2] Schrage,[3] Wolff,[4,5] Sevcik,[6] Kleinrock,[7] among many others). Some fundamental principles also have been derived for parallel-server queues, but these results have been restricted to certain parallel environments, such as customers with serial service demands in parallel or distributed queues (e.g., Brumelle,[8] Stoyan,[9] Sethuraman and Squillante[10]) or fork-join parallel applications in distributed queues (e.g., Makowski and Nelson,[11] Chang et al.,[12] Squillante and Tsoukatos[13]).

With the significant advances of computer and network technology in recent years and the considerable development of parallel scientific and commercial applications, new classes of scheduling disciplines have emerged in practice each differing in the way the parallel servers are shared among the parallel jobs submitted for execution. This includes two basic classes of scheduling policies: *spacesharing* disciplines that share the servers in space by partitioning them among a set of jobs; and *timesharing* disciplines that share the server partitions by rotating them among a set of jobs in time. Our present study is motivated by these parallel computing environments, which make it possible to solve large and complex problems in a wide variety of scientific and commercial applications that are otherwise prohibitively expensive.

In this paper we introduce and study a class of parallel-server queues under spacesharing and timesharing disciplines, which we call (K, D) *parallel-server queues* where K and D respectively denote the degree of spacesharing and timesharing. We derive an analysis of such parallel-server queues based on matrix-analytic methods, yielding exact solutions of matrix-geometric form. The complexity of larger instances of these queues makes an exact solution computationally prohibitive, and thus we also derive a matrix-analytic analysis of (K, D) parallel-server queues that yields approximate solutions with significant reductions in computation complexity. A large number of numerical experiments are used to demonstrate the accuracy of our approach. We then exploit our results to establish fundamental properties of (K, D) parallel-server queues that are of both theoretical and practical interest.

The remainder of the paper is organized as follows. Section 2 presents the class of (K, D) parallel-server queues. A matrix-analytic analysis of these queues is derived in Section 3. Some of the results of our numerical experiments are presented in Section 4.

2 Parallel-Server Queue

We consider a parallel-server queue consisting of P identical servers that are partitioned into K sets of size $M = \lfloor P/K \rfloor$ or $M = \lceil P/K \rceil$ (the so called *spacesharing* of the servers). Parallel jobs arrive to this queue from an exogenous source at rate λ. When served at a partition, a parallel job simultaneously receives all M servers comprising the partition where its service requirements are a function of M and have mean μ_M^{-1}. The M servers of a partition are shared among a set of parallel jobs, up to a maximum cardinality of D, in a round-robin manner by rotating the time allocated to each job according to a positive time quantum with mean δ^{-1} (the so called *timesharing* of the server partitions). Each arrival is allocated an available timesharing slot at one of the server partitions that has been assigned fewer than D jobs, if any; when more than one server partition is assigned fewer than D jobs, the arrival is given to the server partition with the fewest assigned jobs, with ties broken arbitrarily. Otherwise, the arrival is placed in an FCFS infinite-capacity queue which is used to hold all parallel jobs that are waiting to be allocated a server partition. Upon a departure, the available timesharing slot is allocated to the job at the head of the queue, unless it is empty which can only happen when the system contains no more than $V = KD$ jobs. For clarity of exposition, we shall henceforth assume that K evenly divides P so that $M = \lfloor P/K \rfloor = \lceil P/K \rceil$, although our analysis can be extended in a straightforward manner to handle values of K for which this assumption does not hold.[14]

The parallel jobs are assumed to arrive according to a MAP having descriptors $(\mathcal{D}_0, \mathcal{D}_1)$ of order m^A. We assume that the job service times are i.i.d. following an order m^B PH-type distribution with parameters $(\underline{\beta}, \mathcal{S}^B)$ and finite expectation $\mu_M^{-1} = -\underline{\beta}(\mathcal{S}^B)^{-1}\mathbf{e}$, where $\mathbf{e}$ is the column vector of appropriate dimension containing all ones. The lengths of the timesharing quantums at each partition and the overheads for context-switching from one job to another are respectively assumed to be i.i.d. according to PH-type distributions with parameters $(\underline{\eta}, \mathcal{S}^D)$ and $(\underline{\zeta}, \mathcal{S}^C)$ of orders m^D and m^C having finite expectations $\delta^{-1} = -\underline{\eta}(\mathcal{S}^D)^{-1}\mathbf{e}$ and $\gamma^{-1} = -\underline{\zeta}(\mathcal{S}^C)^{-1}\mathbf{e}$. From a practical viewpoint, it is important for our stochastic model to allow (essentially) general distributions for the timesharing quantums and the context-switch overheads at each partition due to the inherent variability of these operations in large and/or distributed parallel computers; this also makes our analysis more general from a theoretical perspective. All of the above stochastic sequences are assumed to be mutually independent. Without loss of generality, we define a *timeplexing cycle* for a given server partition to be the interval of time between successive quanta for one of the jobs assigned to the partition. Thus, the timeplexing cycle of a partition consists of a timesharing quantum for each of the d jobs assigned to the partition, $1 \leq d \leq D$, as well as d context-switch overheads for switching among this series of jobs.

The use of MAPs and PH-type distributions for all stochastic model parameters is of theoretical importance in that we exploit their properties to derive solutions of the general class of (K, D) parallel-server queues. It is also of practical importance in that, since the class of PH-type distributions is dense within the set of probability distributions on $[0, \infty)$, and since the class of MAPs provides a general framework for capturing the dependence structure and variability of a process, any stochastic process on this space for the parallel computing environments of interest can in principle be represented arbitrarily closely by a MAP or a PH-type distribution. Moreover, a considerable body of research has examined the fitting of PH-type distributions and MAPs to empirical data, and a number of algorithms have been developed for doing so; e.g., see [15,16,17,18,19] and the references cited therein. This includes recent work that has considered effectively approximating long-range dependent and heavy-tailed behaviors with instances of the classes of MAPs and PH-type distributions in order to analyze performance models. By appropriately setting the parameters of our stochastic model, various properties of a wide range of (K, D) parallel-server queues can be investigated.

Remarks. We consider a static partitioning of a set of homogeneous servers both in space (i.e., fixed K) and in time (i.e., fixed D), under a single-class workload where a parallel job can execute on any number of servers (refer to the speedup function defined and used in Section 4). Our goal is to gain insight into the fundamental properties of parallel-server queues under spacesharing and timesharing, which can be used in turn to design adaptive and dynamic control policies for spacesharing and timesharing in parallel computer systems. This is analogous to the progression of previous research in the area of pure spacesharing systems where stochastic models of static partitioning were used to facilitate the design of adaptive and dynamic spacesharing policies.[14] Moreover, a matrix-analytic analysis of such combined spacesharing and timesharing control policies, where different values of K and D are chosen and adjusted in an adaptive or dynamic manner, can be derived by exploiting the results in this paper together with those in [14].

Furthermore, our choice of the FCFS queueing discipline is motivated by the fact that this discipline tends to minimize the variance of waiting times for the parallel jobs.[20] The FCFS queueing discipline is also fairly standard in large-scale parallel computing environments.[21,22]

3 Matrix-Analysis Analysis

The (K, D) parallel-server queue of the previous section can be represented by a continuous-time stochastic process $\{X(t)\,;\, t \in \mathbb{R}_+\}$ on the state-space $\Omega = \bigcup_{i=0}^{\infty} \Omega^i$, where

$$X(t) = (\, I(t),\; J_A(t),\; \underline{J}_{B,1}(t),\; \ldots,\; \underline{J}_{B,K}(t)\,), \tag{1}$$

$$\underline{J}_{B,k}(t) = (\, Q_k(t),\; J_{k,1}(t),\; \ldots,\; J_{k,Q_k(t)}(t),\; L_k(t)\,), \quad k \in \{1,\ldots,K\}, \tag{2}$$

$I(t) \in \mathbb{Z}_+$ denotes the number of parallel jobs in the system at time t, $J_A(t)$ is based on the underlying Markov chain of the MAP with generator $\mathcal{D} = \mathcal{D}_0 + \mathcal{D}_1$, $Q_k(t) \in \{0,\ldots,D\}$ denotes the number of parallel jobs assigned to partition k at time t, $\sum_{k=1}^{K} Q_k(t) = \min\{I(t), V\}$, $J_{k,n}(t) \in \{1,\ldots,m^B\}$ denotes the phase of the service time process for the nth job assigned to partition k at time t, $n \in \{1,\ldots,Q_k(t)\}$, $L_k(t) \in \{1,\ldots,Q_k(t)\cdot(m^D + m^C)\}$ denotes the phase of the current timeplexing cycle for partition k at time t, and Ω^i denotes level i (i.e., the set of all states with $I(t) = i$), $t \in \mathbb{R}_+$, $k \in \{1,\ldots,K\}$, $i \in \mathbb{Z}_+$. Note that $I(t) - \sum_{k=1}^{K} Q_k(t)$ represents the number of parallel jobs waiting to receive a server partition. Furthermore, the value of $L_k(t)$ is used to keep track of which job assigned to partition k is being served at time t, such that the first job (in phase $J_{k,1}(t)$) is being served when $L_k(t) \in [1, m^D]$, no jobs are being served while $L_k(t) \in [m^D+1, m^D+m^C]$, the second job (in phase $J_{k,2}(t)$) is being served when $L_k(t) \in [m^D+m^C+1, 2m^D+m^C]$, and so on.

Let $x_{i,z} \in \Omega^i$, $z \in \{1, \ldots, Z_i\}$, $i \in \mathbb{Z}_+$, be a lexicographic ordering of the elements of level i, and define $Z \equiv \sum_{i=0}^{V-1} Z_i$, where Z_i denotes the cardinality of Ω^i. We then define

$$\boldsymbol{\pi} \equiv (\, \boldsymbol{\pi}_0,\ \boldsymbol{\pi}_1,\ \boldsymbol{\pi}_2,\ \ldots \,), \tag{3}$$

$$\boldsymbol{\pi}_i \equiv (\, \pi(x_{i,1}),\ \pi(x_{i,2}),\ \ldots,\ \pi(x_{i,Z_i}) \,), \quad i \in \mathbb{Z}_+, \tag{4}$$

$$\pi(x_{i,z}) \equiv \lim_{t\to\infty} \mathsf{P}\,[\, X(t) = x_{i,z} \,], \quad i \in \mathbb{Z}_+,\ x_{i,z} \in \Omega^i,\ z \in \{1, \ldots, Z_i\}. \tag{5}$$

Assuming the stochastic process $\{X(t)\,;\, t \in \mathbb{R}_+\}$ to be irreducible and positive recurrent, which shall be assumed henceforth, the stationary probability vector $\boldsymbol{\pi}$ is uniquely determined by solving $\boldsymbol{\pi}\mathbf{Q} = \mathbf{0}$ and $\boldsymbol{\pi}\mathbf{e} = 1$, where $\mathbf{Q}$ is the infinitesimal generator matrix for the process.

The generator matrix $\mathbf{Q}$, organized in the same order as the elements of the invariant vector $\boldsymbol{\pi}$, has a structure given by

$$\mathbf{Q} = \begin{bmatrix} B_{00} & B_{01} & \mathbf{0} & \mathbf{0} & \mathbf{0} & \ldots \\ B_{10} & B_{11} & A_0 & \mathbf{0} & \mathbf{0} & \ldots \\ \mathbf{0} & A_2 & A_1 & A_0 & \mathbf{0} & \ldots \\ \mathbf{0} & \mathbf{0} & A_2 & A_1 & A_0 & \ldots \\ \vdots & \vdots & \vdots & \vdots & \vdots & \ddots \end{bmatrix}, \tag{6}$$

where B_{00}, B_{01}, B_{10}, B_{11} and A_n, $n = 0, 1, 2$, are finite matrices of dimensions $Z \times Z$, $Z \times Z_V$, $Z_V \times Z$, $Z_V \times Z_V$ and $Z_V \times Z_V$, respectively. The matrices corresponding to the non-homogeneous boundary of Ω also have the structures

$$B_{00} = \begin{bmatrix} \Psi_0 & \Lambda_0 & \mathbf{0} & \mathbf{0} & \cdots & \mathbf{0} & \mathbf{0} \\ \Phi_1 & \Psi_1 & \Lambda_1 & \mathbf{0} & \cdots & & \\ \mathbf{0} & \Phi_2 & \Psi_2 & \Lambda_2 & \cdots & \vdots & \vdots \\ \vdots & \vdots & \vdots & \vdots & \ddots & & \\ \mathbf{0} & \mathbf{0} & \mathbf{0} & \mathbf{0} & \cdots & \Phi_{V-1} & \Psi_{V-1} \end{bmatrix}, \qquad B_{01} = \begin{bmatrix} \mathbf{0} \\ \mathbf{0} \\ \vdots \\ \mathbf{0} \\ \Lambda_{V-1} \end{bmatrix}, \tag{7}$$

$$B_{10} = \left[\, \mathbf{0}\ \mathbf{0} \cdots \mathbf{0}\ \Phi_V \,\right], \qquad B_{11} = \Psi_V, \tag{8}$$

where Φ_i, Ψ_i and Λ_i have dimensions $Z_i \times Z_{i-1}$, $Z_i \times Z_i$ and $Z_i \times Z_{i+1}$, respectively. The stationary probability vector $\boldsymbol{\pi}$ is then given by[23]

$$(\boldsymbol{\pi}_0,\ \boldsymbol{\pi}_1,\ \ldots,\ \boldsymbol{\pi}_V) \begin{bmatrix} B_{00} & B_{01} \\ B_{10} & B_{11} + RA_2 \end{bmatrix} = \mathbf{0}, \tag{9}$$

$$\boldsymbol{\pi}_{V+k} = \boldsymbol{\pi}_V R^k, \quad k \in \mathbb{Z}_+, \tag{10}$$

$$(\boldsymbol{\pi}_0,\ \boldsymbol{\pi}_1,\ \ldots,\ \boldsymbol{\pi}_{V-1})\,\mathbf{e} + \boldsymbol{\pi}_V (\mathbf{I} - R)^{-1} \mathbf{e} = 1, \tag{11}$$

where R is the minimal non-negative solution of the equation

$$R^2 A_2 + R A_1 + A_0 = \mathbf{0}, \tag{12}$$

and $\mathbf{I}$ is the identity matrix of order Z_V.

Instances of the stochastic process with even moderate values for the model parameters can cause the dimensions of the boundary submatrices in equation (6) to become quite large. Following the results derived in [24] for infinite QBDs, we next establish a theorem that significantly reduces the time and space complexities of computing equations (9) and (11). We also point out that the results in [24] for (both) infinite (and finite) QBDs were derived independently of the analogous results for finite QBDs obtained in [25].

Theorem 3.1 *Let $\mathbf{Q}$ be irreducible and in the form of (6) through (8). If the minimal non-negative solution R of equation (12) satisfies $sp(R) < 1$, and if there exists a positive probability vector $(\boldsymbol{\pi}_0, \ldots, \boldsymbol{\pi}_V)$ satisfying equation (9), then the components of this probability vector are given by*

$$\boldsymbol{\pi}_k = -\boldsymbol{\pi}_{k+1}\,\Phi_{k+1}\,\widetilde{R}_k^{-1}, \quad k \in \{0, \ldots, V-1\}, \tag{13}$$

$$\boldsymbol{\pi}_V = -\boldsymbol{\pi}_{V-1}\,\Lambda_{V-1}\,\widetilde{R}_V^{-1}, \tag{14}$$

where $\widetilde{R}_0 \equiv \Psi_0$, $\widetilde{R}_k \equiv \Psi_k - \Phi_k \widetilde{R}_{k-1}^{-1} \Lambda_{k-1}$, $k \in \{1, \ldots, V-1\}$, $\widetilde{R}_V \equiv \Psi_V + RA_2$. Furthermore, when $V > 1$, the vector $\boldsymbol{\pi}_{V-1}$ can be determined up to a multiplicative constant by solving

$$\boldsymbol{\pi}_{V-1}\left[\Psi_{V-1} - \Phi_{V-1}\,\widetilde{R}_{V-2}^{-1}\,\Lambda_{V-2} - \Lambda_{V-1}\widetilde{R}_V^{-1}\,\Phi_V\right] = \mathbf{0}, \tag{15}$$

$$\boldsymbol{\pi}_{V-1}\,\mathbf{e} = \theta, \quad \theta > 0. \tag{16}$$

Otherwise, when $V = 1$, the vector $\boldsymbol{\pi}_1$ can be determined up to a multiplicative constant by solving

$$\boldsymbol{\pi}_1\left[\widetilde{R}_1 - \Phi_1\,\widetilde{R}_0^{-1}\,\Lambda_0\right] = \mathbf{0}, \tag{17}$$

$$\boldsymbol{\pi}_1\,\mathbf{e} = \theta, \quad \theta > 0. \tag{18}$$

In either case, the vector $(\boldsymbol{\pi}_0, \boldsymbol{\pi}_1, \ldots, \boldsymbol{\pi}_V)$ then can be obtained from (13), (14) and the normalizing equation (11).

Proof: The stationary probability vector $\boldsymbol{\pi}$ is given by equations (9), (10) and (11). Substitution of (7) and (8) into equation (9) shows that the generator matrix for the boundary has the structure

$$\widetilde{\mathbf{Q}} = \begin{bmatrix} \Psi_0 & \Lambda_0 & \mathbf{0} & \mathbf{0} & \cdots & \mathbf{0} & \mathbf{0} & \mathbf{0} \\ \Phi_1 & \Psi_1 & \Lambda_1 & \mathbf{0} & \cdots & & & \mathbf{0} \\ \mathbf{0} & \Phi_2 & \Psi_2 & \Lambda_2 & \cdots & \vdots & \vdots & \vdots \\ \vdots & \vdots & \vdots & \vdots & \ddots & & & \mathbf{0} \\ \mathbf{0} & \mathbf{0} & \mathbf{0} & \mathbf{0} & \cdots & \Phi_{V-1} & \Psi_{V-1} & \Lambda_{V-1} \\ \mathbf{0} & \mathbf{0} & \mathbf{0} & \mathbf{0} & \cdots & \mathbf{0} & \Phi_V & \Psi_V + RA_2 \end{bmatrix}, \tag{19}$$

where we have made use of (10). The invariant vector $(\boldsymbol{\pi}_0, \ldots, \boldsymbol{\pi}_V)$ then satisfies, up to a multiplicative constant, the equations

$$\boldsymbol{\pi}_0 \Psi_0 + \boldsymbol{\pi}_1 \Phi_1 = \mathbf{0}, \tag{20}$$

$$\boldsymbol{\pi}_{k-1} \Lambda_{k-1} + \boldsymbol{\pi}_k \Psi_k + \boldsymbol{\pi}_{k+1} \Phi_{k+1} = \mathbf{0}, \quad k \in \{1, \ldots, V-1\}, \tag{21}$$

$$\boldsymbol{\pi}_{V-1} \Lambda_{V-1} + \boldsymbol{\pi}_V (\Psi_V + RA_2) = \mathbf{0}, \tag{22}$$

$$(\boldsymbol{\pi}_0, \cdots, \boldsymbol{\pi}_V)\,\mathbf{e} = \theta > 0. \tag{23}$$

Upon substituting $\widetilde{R}_0, \ldots, \widetilde{R}_V$ into (20) – (22), we obtain equations (13) and (14). The non-singularity of the matrices $\widetilde{R}_k$, $k \in \{1, \ldots, V\}$, follows from the properties of the irreducible generator matrix and the submatrices in (6) and (19). Substitution of (13) and (14) into (21) for $k = V - 1 > 0$ yields equation (15), which then can be used together with (16) to obtain the vector $\boldsymbol{\pi}_{V-1}$ up to a multiplicative constant when $V > 1$. Similarly, when $V = 1$, equation (17) is obtained by substituting (13) and $\widetilde{R}_V$ into (22). The remaining components of the vector $(\boldsymbol{\pi}_0, \ldots, \boldsymbol{\pi}_V)$ are determined up to the same multiplicative constant via recurrence using equations (13) and (14). The vector $\boldsymbol{\pi}$ is then uniquely determined by the normalizing equation (11). □

Using Theorem 3.1 for calculating the stationary probability vector $\boldsymbol{\pi}$ significantly reduces the computational complexity over that of numerically solving equations (9) and (11) directly. In particular, this approach makes it possible to obtain the boundary components of the invariant vector (up to a multiplicative constant) by solving $V + 1$ matrix equations of (time and space) complexity $\mathrm{O}(Z_i^2)$, $i \in \{0, \ldots, V\}$, as opposed to solving a single matrix equation of (time and space) complexity $\mathrm{O}((\sum_{i=0}^{V} Z_i)^2)$. This in turn makes it possible for us to compute solutions for fairly large instances of the (K, D) parallel-server queue that are otherwise prohibitively expensive. Moreover, the algorithm (based on Theorem 3.1) used to compute these solutions is numerically stable across a wide spectrum of model parameters. In fact, throughout all of the numerous experiments performed as part of our study, we encountered no numerical stability problems.

While the above exact solution of the class of (K, D) parallel-server queues can be computed in an efficient manner for small to moderate model parameter values, it can become prohibitively expensive to compute exact solutions for large parameter values due to an explosion in the size of the state space. We now consider a few approaches to address this problem, which also can be combined in various ways to obtain effective hybrid schemes.

3.1 Limited Processor-Sharing

One approach to addressing the state space explosion problem is based on approximating the timeplexing cycles for each server partition with a limited form of processor-sharing. Processor-sharing, which was first introduced by Kleinrock[7] in M/G/1 queues to approximate the timesharing of computer systems, is the limiting case of round-robin scheduling where the quantum lengths goes to 0. Thus, when the system contains n jobs, all n jobs are serviced simultaneously each receiving $1/n^{\text{th}}$ of the server capacity.

Here we consider a limited form of processor-sharing to approximate the timesharing of a server partition in the sense that at most D jobs share a partition. Note that, in the absence of context-switching overheads, this approach is exact in the limit as the quantum lengths tend to 0. More generally, our approach also involves extending the service time distribution for each job to include the context-switching overhead that it will be "charged" as a function of the quantum length. When the timesharing quantums and context-switching overheads are relatively small, this approximation should provide accurate results (which is indeed supported by our numerical experiments).

The first step of our approach based on limited processor-sharing consists of an offline analysis, which is used to construct a modified service time distribution of PH-type. This modification is intended to account for the total context-switch overhead "charged" to each job during its sojourn in the system, where a context-switch overhead is charged to a job each time it incurs a context switch as it is being served. With the exception of no timesharing because there is only one job at the partition of interest (which can be handled directly as described below), the total context-switch overhead charged to a job depends upon its service time and the quantum-length and context-switch-overhead distributions. The analysis is therefore quite simple, exploiting the properties of PH-type distributions and (in some cases) involving an application of Wald's equation.

Consider a generic job in the system, and let N be the random variable representing the number of context switches charged to this job before it departs the system. If we can determine the distribution of N, then the modified service time distribution $G(\cdot)$ can be obtained by

$$G(t) \;=\; \sum_{n} p_N(n)\; F^{(n)*} * S(t), \tag{24}$$

where $p_N(n)$ is the probability of the event $\{N = n\}$, F is the distribution of the context-switch overhead, $F^{(n)*}(t)$ denotes the n-fold convolution of F with itself, $*$ is the convolution operator, and S is the original service time distribution. In other words, for each $N = n$, the modified service time distribution is simply the convolution of the original service time with the n-fold convolution of the context-switch overhead with itself. This computation via equation (24) is easily accomplished as the result of (combinations of) known closure properties for the family of PH-type distributions,[26] yielding a distribution G of PH-type. We then can either use this PH-type distribution directly, or construct a more compact PH-type distribution by matching as many moments (and/or density function) of G as are of interest, using any of the best known methods for doing so; e.g., see [15,16,17] and the references cited therein. In either case, we use $(\underline{\beta}', \mathcal{S}'^B)$ to denote the order m'^B parameters of the resulting PH-type distribution having mean ${\mu'}_M^{-1} = -\underline{\beta}'(\mathcal{S}'^B)^{-1}\mathbf{e}$.

Alternatively, we can compute the first two moments of N from which we can obtain the first and second moments of the modified service time distribution $G(\cdot)$. Let $Y_1, Y_2, \ldots$ be a sequence of i.i.d. random variables such that $Y_n \stackrel{d}{=} Y$, $n = 1, 2, \ldots$, where Y is the generic random variable following the quantum-length distribution. Let the random variable N be a stopping time for this sequence such that the event $\{N = n\}$ occurs if

$$Y_1 + Y_2 + \ldots + Y_{n-1} < U,$$
$$Y_1 + Y_2 + \ldots + Y_n \geq U,$$

where U is the generic random variable following the original service time distribution. In other words, $N \equiv \sup\{n : Y_1 + Y_2 + \ldots + Y_n \geq U\}$. Define $S_N = Y_1 + Y_2 + \ldots + Y_N$. Then the first two moments of N and S_N can be determined by a simple application of Wald's equation, which we state in a form that is most useful to us here.

Theorem 3.2 (Wald) *Let N, Y_n, $n = 1, 2, \ldots$, and S_N be as defined above. Then*

$$\mathsf{E}[S_N] = \mathsf{E}[N]\,\mathsf{E}[Y], \tag{25}$$
$$\mathsf{Var}\,(S_N) = \mathsf{E}[N]\,\mathsf{Var}\,(Y) \;+\; (\mathsf{E}[Y])^2\,\mathsf{Var}\,(N)\,. \tag{26}$$

The values of $\mathsf{E}[N]$ and $\mathsf{Var}(N)$ can be computed from equations (25) and (26) together with the known first two moments of U and Y. This in turn can be used with the first two moments of the context-switch overhead to obtain the first two moments of the distribution G. We next construct a PH-type distribution with parameters $(\underline{\beta}', \mathcal{S}'^B)$ of order m'^B to match these first two moments of G using any of the best known methods. The mean of this PH-type distribution is given by $\mu'^{-1}_M = \mu_M^{-1} + \mathsf{E}[N]\gamma^{-1} = -\underline{\beta}'(\mathcal{S}'^B)^{-1}\mathbf{e}$.

Using either of the above approaches to construct the modified service-time distribution that implicitly includes the context-switching overhead charged to each job during its execution, the corresponding parallel-server queue can be represented by a continuous-time stochastic process $\{X_{\mathrm{PS}}(t)\,;\, t \in \mathbb{R}_+\}$ on the state-space $\Omega^i_{PS} = \bigcup_{i=0}^{\infty} \Omega^i_{PS}$, where

$$X_{\mathrm{PS}}(t) = (\, I(t),\ J_A(t),\ \underline{J}_{B,1}(t),\ \ldots,\ \underline{J}_{B,K}(t)\,), \tag{27}$$

$$\underline{J}_{B,k}(t) = (\, Q_k(t),\ J_{k,1}(t),\ \ldots,\ J_{k,m'^B}(t)\,), \quad k \in \{1,\ldots,K\}, \tag{28}$$

$I(t) \in \mathbb{Z}_+$ denotes the number of parallel jobs in the system at time t, $J_A(t)$ is based on the underlying Markov chain of the MAP with generator $\mathcal{D} = \mathcal{D}_0 + \mathcal{D}_1$, $Q_k(t) \in \{0,\ldots,D\}$ denotes the number of parallel jobs assigned to partition k at time t, $\sum_{k=1}^{K} Q_k(t) = \min\{I(t), V\}$, $J_{k,n}(t) \in \{0,\ldots,Q_k(t)\}$ denotes the number of jobs assigned to partition k at time t whose service time process is in phase n, $\sum_{n=1}^{m'^B} J_{k,n}(t) = Q_k(t)$, and Ω^i_{PS} denotes level i, $t \in \mathbb{R}_+$, $k \in \{1,\ldots,K\}$, $n \in \{1,\ldots,m'^B\}$, $i \in \mathbb{Z}_+$. The processor-sharing at each server partition k is taken into account by scaling the corresponding service rates by the number of jobs assigned to the partition $Q_k(t)$, $k \in \{1,\ldots,K\}$. Given this formulation, then all of the above matrix-analytic analysis for the stochastic process $\{X(t)\,;\, t \in \mathbb{R}_+\}$ applies directly to the process $\{X_{\mathrm{PS}}(t)\,;\, t \in \mathbb{R}_+\}$ with the appropriate (and straightforward) changes of variables (e.g., Ω^i is replaced by Ω^i_{PS}).

Remark. Note that a job incurs a context switch in the original stochastic system only when at least one other job has been assigned to the same partition. Thus, when an arrival finds an empty partition (i.e., $Q_k(t) = 0$), its service time follows the original service time distribution (without context-switching overhead), and upon a second arrival to the partition during its execution (i.e., $Q_k(t) = 2$), the initial job switches to the appropriate position in the constructed service time distribution (with context-switching overhead). Similarly, when a partition contains two jobs and one of them departs (i.e., $Q_k(t) : 2 \to 1$), the remaining job switches from the constructed service time distribution (with context-switching overhead) to the appropriate position in the original service time distribution (without context-switching overhead). The exact details of incorporating such effects in the formulation of $\{X_{\rm PS}(t)\,;\,t\in\mathbb{R}_+\}$ depend upon the PH-type distributions used for the service times, context-switch overheads and quantum lengths, and upon the approach used to construct the modified service-time distribution. One simple approach consists of constructing the modified service-time distribution so that each phase of the service time process captures in an appropriate manner the execution progress made up to the current time. Assuming the original service-time distribution also has this property, then transitions from $Q_k(t) = 2$ to $Q_k(t) = 1$, and vice versa, can be handled in a straightforward fashion by maintaining the execution progress made up to the time of this transition via the current phase. (For example, letting j_o and j_m be phases of the original and modified service-time distributions that represent 50% completion of the corresponding service times, then a transition from $Q_k(t) = 1$ to $Q_k(t) = 2$ would cause the current phase j_o of the service time process for the job in execution to go to j_m.) It is important to note that the context-switch overhead charged to a job should be small relative to its service requirements, i.e., $U \gg NW$ where W is the generic random variable following the context-switch overhead distribution. Moreover, the above effects are more pronounced at relatively light loads where a non-negligible amount of time is spent in transitions between $Q_k(t) = 1$ and $Q_k(t) = 2$.

3.2 Timesharing Decomposition

Another approach to address the problem of state space explosion in (K, D) parallel-server queues with large model parameter values is based on a key observation that timesharing simply modifies the departure process of the queue, or equivalently the effective service time process of the queue, whenever server partitions serve multiple parallel jobs, in a manner that can be approximated via an offline first-passage time analysis.

Consider level $V + 1$ of a single partition version of the original stochastic process $\{X(t)\,;\,t\in\mathbb{R}_+\}$ in which arrivals are ignored and all transitions to level V are redirected to a single absorbing state $\mathcal{A}$, i.e., the Markov process $\{\widehat{X}(t)\,;\,t\in\mathbb{R}_+\}$ consisting of $\widehat{X}(t) = (I(t) = V + 1, Q(t) = D, J_1(t), \ldots, J_{Q(t)}(t), L(t))$ and $\widehat{X}(t) = (\mathcal{A})$, $t\in\mathbb{R}_+$. This Markov process represents the details of the timeplexing cycle of a single partition in the homogeneous portion of the original process. Let $\widehat{\mathcal{Q}}_1$ denote the generator matrix for this Markov process, where

$$\widehat{\mathcal{Q}}_1 \;=\; \begin{bmatrix} \widehat{\mathcal{S}}^B & \widehat{\underline{S}}^B \\ \mathbf{0} & 0 \end{bmatrix}. \tag{29}$$

Note that there is a unique vector $(J_1(t), \ldots, J_D(t))$ for each non-absorbing state in this Markov process with $L(t)\,\mathbf{mod}\,(m^D + m^C) = 1$. An initial probability vector $(\underline{\alpha}_1, 0)$ is defined to have a value of $\prod_{n=1}^{D} \beta_{J_n(t)}$ for each of these states, and to have a value of 0 otherwise. Then the distribution of the time until absorption in this Markov process is, by construction, a PH-type distribution with parameters $(\underline{\alpha}_1, \widehat{\mathcal{S}}^B)$. Moreover, the nth moment of this distribution is given by $(-1)^n\, n!\, (\underline{\alpha}_1 (\widehat{\mathcal{S}}^B)^{-n} \mathbf{e})$, $n \geq 0$.

We next construct a PH-type distribution with parameters $(\underline{\beta}_D, \mathcal{S}^B_D)$ of order m^B_D to match as many of these moments (and/or density function) of the PH-type distribution with parameters $(\underline{\alpha}_1, \widehat{\mathcal{S}}^B)$ as are of interest, using any of the best known methods for doing so; e.g., see [15,16,17] and the references cited therein. While we can directly use the PH-type distribution with parameters $(\underline{\alpha}_1, \widehat{\mathcal{S}}^B)$, our goal is to significantly reduce the complexity of the process and thus we use the more compact PH-type distribution with parameters $(\underline{\beta}_D, \mathcal{S}^B_D)$ to approximate the PH-type distribution with parameters $(\underline{\alpha}_1, \widehat{\mathcal{S}}^B)$. Repeating the above analysis for $Q(t) = D-1, \ldots, 2$, yields the corresponding PH-type distributions with parameters $(\underline{\beta}_d, \mathcal{S}^B_d)$ of order m^B_d, $d \in \{2, \ldots, D-1\}$. For the case of $Q(t) = 1$, we set the corresponding order m^B_1 parameters $(\underline{\beta}_1, \mathcal{S}^B_1)$ to be identical to the order m^B parameters $(\underline{\beta}, \mathcal{S}^B)$ of the original service time distribution. We refer to this collection of PH-type distributions as the *compound service time distributions* for server partitions with d jobs, $d \in \{1, \ldots, D\}$.

The corresponding parallel-server queue then can be represented by a continuous-time stochastic process $\{X_{\mathrm{TD}}(t)\,;\, t \in \mathbb{R}_+\}$ on the state-space $\Omega^i_{TD} = \bigcup_{i=0}^{\infty} \Omega^i_{TD}$, where

$$X_{\mathrm{TD}}(t) = (\, I(t),\ J_A(t),\ \underline{J}_{B,1}(t),\ \ldots,\ \underline{J}_{B,D}(t)\,), \tag{30}$$

$$\underline{J}_{B,d}(t) = (\, P_d(t),\ J_{d,1}(t),\ \ldots,\ J_{d,m^B_d}(t)\,), \quad d \in \{1, \ldots, D\}, \tag{31}$$

$I(t) \in \mathbb{Z}_+$ denotes the number of parallel jobs in the system at time t, $J_A(t)$ is based on the underlying Markov chain of the MAP with generator $\mathcal{D} = \mathcal{D}_0 + \mathcal{D}_1$, $P_d(t) \in \{0, \ldots, K\}$ denotes the number of server partitions assigned d parallel jobs at time t, $\sum_{d=1}^{D} P_d(t) \leq K$, $\sum_{d=1}^{D} dP_d(t) = \min\{I(t), V\}$, $J_{d,n}(t) \in \{0, \ldots, P_d(t)\}$ denotes the number of partitions assigned d jobs at time t whose compound service time process is in phase n, $\sum_{n=1}^{m^B_d} J_{d,n}(t) = P_d(t)$, and Ω^i_{TD} denotes level i, $t \in \mathbb{R}_+$, $d \in \{1, \ldots, D\}$, $n \in \{1, \ldots, m^B_d\}$, $i \in \mathbb{Z}_+$.

We expect that this approach will provide accurate results in the limiting heavy-traffic regime in which the (K, D) parallel-server queue asymptotically behaves like the corresponding MAP/PH/K queue with PH-type service time distribution having parameters $(\underline{\beta}_D, \mathcal{S}^B_D)$, for sufficiently large quantum lengths (which is indeed supported by our numerical experiments). However, one of the potential problems with this approach, when not operating in the limiting heavy-traffic regime, results from the fact that each compound service time distribution primarily captures the departure time of the first parallel job to complete service under the timesharing of d i.i.d. jobs without capturing all of the progress made by the remaining $d-1$ jobs during this period of time, $2 \leq d \leq D$. To address this problem, we extend the above analysis to capture in more detail the effects of timesharing on all of these d jobs.

Consider again the above Markov process $\{\widehat{X}(t)\,;\, t \in \mathbb{R}_+\}$ consisting of $\widehat{X}(t) = (I(t) = V+1, Q(t) = D, J_1(t), \ldots, J_{Q(t)}(t), L(t))$ and $\widehat{X}(t) = (\mathcal{A})$, $t \in \mathbb{R}_+$, with generator $\widehat{\mathcal{Q}}_1$ given in equation (29). We recursively define for $\ell = 2, \ldots, \mathcal{L}$, the generator matrix

$$\widehat{\mathcal{Q}}_\ell \;=\; \widehat{\mathcal{S}}^B + \underline{\widehat{\mathcal{S}}}^B \cdot \underline{\alpha}_{\ell-1}, \tag{32}$$

where $\underline{\alpha}_1$ is as defined above and $\underline{\alpha}_\ell$ is the invariant probability vector of the generator $\widehat{\mathcal{Q}}_\ell$, i.e., the (unique) solution of $\underline{\alpha}_\ell \widehat{\mathcal{Q}}_\ell = \mathbf{0}$ and $\underline{\alpha}_\ell \mathbf{e} = 1$, $\ell \in \{2, \ldots, \mathcal{L}\}$. The matrix $\widehat{\mathcal{Q}}_\ell$ is the generator for the PH-type renewal process corresponding to the PH-type distribution having parameters $(\underline{\alpha}_{\ell-1}, \widehat{\mathcal{S}}^B)$, $\ell \in \{2, \ldots, \mathcal{L}\}$. Moreover, the distribution of the time until absorption in the Markov process with initial probability vector $(\underline{\alpha}_\ell, 0)$ and generator $\widehat{\mathcal{Q}}_1$ is a PH-type distribution with parameters $(\underline{\alpha}_\ell, \widehat{\mathcal{S}}^B)$ and n^{th} moment $(-1)^n\, n!\, (\underline{\alpha}_\ell (\widehat{\mathcal{S}}^B)^{-n} \mathbf{e})$, $\ell \in \{1, \ldots, \mathcal{L}\}$, $n \geq 0$.

We next construct a set of PH-type distributions having parameters $(\underline{\beta}_{D,\ell}, \mathcal{S}^B_{D,\ell})$ of order $m^B_{D,\ell}$ to match as many of the moments (and/or density function) of the set of PH-type distributions having parameters $(\underline{\alpha}_\ell, \widehat{\mathcal{S}}^B)$ as are of interest, $\ell \in \{1, \ldots, \mathcal{L}\}$. Once again, we use the more compact PH-type distribution with parameters $(\underline{\beta}_{D,\ell}, \mathcal{S}^B_{D,\ell})$ to approximate the PH-type distribution with parameters $(\underline{\alpha}_\ell, \widehat{\mathcal{S}}^B)$. Repeating the above analysis for $Q(t) = D-1, \ldots, 1$, yields the corresponding set of PH-type distributions with parameters $(\underline{\beta}_{d,\ell}, \mathcal{S}^B_{d,\ell})$ of order $m^B_{d,\ell}$, $d \in \{1, \ldots, D-1\}$, $\ell \in \{1, \ldots, \mathcal{L}\}$. This yields an even larger collection of compound service time distributions for server partitions with d jobs, $d \in \{1, \ldots, D\}$.

The corresponding parallel-server queue then can be represented by a continuous-time stochastic process $\{X_{\mathrm{TD}}(t)\,;\, t \in \mathbb{R}_+\}$ on the state-space $\Omega^i_{TD} = \bigcup_{i=0}^{\infty} \Omega^i_{TD}$, where

$$X_{\mathrm{TD}}(t) = (\, I(t),\; J_A(t),\; \underline{J}_{B,1}(t),\; \ldots,\; \underline{J}_{B,D}(t)\,), \tag{33}$$

$$\underline{J}_{B,d}(t) = (\, \underline{J}^1_{B,d}(t),\; \ldots,\; \underline{J}^{\mathcal{L}}_{B,d}(t)\,), \quad d \in \{1, \ldots, D\}, \tag{34}$$

$$\underline{J}^{\ell}_{B,d}(t) = (\, P^{\ell}_{d}(t),\ J^{\ell}_{d,1}(t),\ \ldots,\ J^{\ell}_{d,m^B_d}(t)\,), \quad \ell \in \{1,\ldots,\mathcal{L}\}, \tag{35}$$

$I(t) \in \mathbb{Z}_+$ denotes the number of parallel jobs in the system at time t, $J_A(t)$ is based on the underlying Markov chain of the MAP with generator $\mathcal{D} = \mathcal{D}_0 + \mathcal{D}_1$, $P^{\ell}_{d}(t) \in \{0,\ldots,K\}$ denotes the number of server partitions assigned d parallel jobs at time t whose compound service time process is of type ℓ, $\sum_{\ell=1}^{\mathcal{L}} \sum_{d=1}^{D} P^{\ell}_{d}(t) \le K$, $\sum_{\ell=1}^{\mathcal{L}} \sum_{d=1}^{D} dP^{\ell}_{d}(t) = \min\{I(t), V\}$, and $J^{\ell}_{d,n}(t) \in \{0,\ldots,P^{\ell}_{d}(t)\}$ denotes the number of partitions assigned d jobs at time t whose compound service time process of type ℓ is in phase n, $\sum_{n=1}^{m^B_d} J^{\ell}_{d,n}(t) = P^{\ell}_{d}(t)$, $t \in \mathbb{R}_+$, $d \in \{1,\ldots,D\}$, $\ell \in \{1,\ldots,\mathcal{L}\}$, $n \in \{1,\ldots,m^B_d\}$.

Note that, at any time t, each server partition assigned d jobs is operating according to one of the $\mathcal{L}$ compound service time distributions of PH-type with parameters $(\underline{\beta}_{d,\ell}, \mathcal{S}^B_{d,\ell})$ of order $m^B_{d,\ell}$, $d \in \{2,\ldots,D\}$, $\ell \in \{1,\ldots,\mathcal{L}\}$. Upon a departure when $I(t) > V = KD$, such a partition switches to operate according to the nth compound service time distribution of PH-type with parameters $(\underline{\beta}_{D,n}, \mathcal{S}^B_{D,n})$ of order $m^B_{D,n}$, where $n = (\ell+1) \,\mathbf{mod}\, \mathcal{L}$. Similarly, upon a departure when $I(t) \le V = KD$, a server partition assigned d jobs and operating according to the order $m^B_{d,\ell}$ PH-type compound service time distribution with parameters $(\underline{\beta}_{d,\ell}, \mathcal{S}^B_{d,\ell})$ then switches to operate according to the PH-type compound service time distribution with parameters $(\underline{\beta}_{d-1,n}, \mathcal{S}^B_{d-1,n})$ of order $m^B_{d-1,n}$, where $n = (\ell+1) \,\mathbf{mod}\, \mathcal{L}$, $d \in \{2,\ldots,D\}$, $\ell \in \{1,\ldots,\mathcal{L}\}$. Recall that we have the single PH-type distribution of order m^B with parameters $(\underline{\beta}, \mathcal{S}^B)$ of the original service time distribution when $d - 1 = 1$. Transitions to higher and lower levels within the boundary take into account the progress made by the corresponding set of jobs up to the time of this transition via the current phase of the service time process.

Then all of the matrix-analytic analysis for the stochastic process $\{X(t)\,;\, t \in \mathbb{R}_+\}$ derived earlier in this section applies directly to the process $\{X_{\text{TD}}(t)\,;\, t \in \mathbb{R}_+\}$ with the appropriate (and straightforward) changes of variables (e.g., Ω^i is replaced by Ω^i_{TD}). From a practical viewpoint, a proper choice for the value of $\mathcal{L}$ depends upon a number of different factors. Given the parallel computing environments of interest, however, we expect to be able to obtain accurate approximations with relatively small values of $\mathcal{L}$.

3.3 Performance Measures

Once the stationary probability vector $\boldsymbol{\pi}$ has been obtained, we can compute various performance measures of interest. Specifically, the first two moments of the number of parallel jobs in the system can be respectively expressed as

$$\begin{aligned}\mathsf{E}[\mathcal{N}] &= \sum_{k=1}^{V-1} k\boldsymbol{\pi}_k\mathbf{e} + V\boldsymbol{\pi}_V \sum_{k=0}^{\infty} R^k\mathbf{e} + \boldsymbol{\pi}_V \sum_{k=0}^{\infty} kR^k\mathbf{e}, \\ &= \sum_{k=1}^{V-1} k\boldsymbol{\pi}_k\mathbf{e} + V\boldsymbol{\pi}_V(\mathbf{I}-R)^{-1}\mathbf{e} + \boldsymbol{\pi}_V R(\mathbf{I}-R)^{-2}\mathbf{e}, \end{aligned} \tag{36}$$

and

$$\begin{aligned}\mathsf{E}[\mathcal{N}^2] &= \sum_{k=1}^{V-1} k^2\boldsymbol{\pi}_k\mathbf{e} + V^2\boldsymbol{\pi}_V \sum_{k=0}^{\infty} R^k\mathbf{e} + 2V\boldsymbol{\pi}_V \sum_{k=0}^{\infty} kR^k\mathbf{e} + \boldsymbol{\pi}_V \sum_{k=0}^{\infty} k^2R^k\mathbf{e}, \\ &= \sum_{k=1}^{V-1} k^2\boldsymbol{\pi}_k\mathbf{e} + V^2\boldsymbol{\pi}_V(\mathbf{I}-R)^{-1}\mathbf{e} + 2V\boldsymbol{\pi}_V R(\mathbf{I}-R)^{-2}\mathbf{e} \\ &\qquad + \boldsymbol{\pi}_V(R^2+R)(\mathbf{I}-R)^{-3}\mathbf{e}. \end{aligned} \tag{37}$$

Higher moments are obtained in a similarly straightforward fashion.

The expected sojourn time of a parallel job then can be calculated using Little's law[27] and (36), which yields

$$\mathsf{E}[\mathcal{T}] = \lambda^{-1}\left(\sum_{k=1}^{V-1} k\boldsymbol{\pi}_k\mathbf{e} + V\boldsymbol{\pi}_V(\mathbf{I}-R)^{-1}\mathbf{e} + \boldsymbol{\pi}_V R(\mathbf{I}-R)^{-2}\mathbf{e}\right). \tag{38}$$

The second moment of job sojourn time can be obtained from generalizations of Little's law[28,29,30] and equation (37), with higher moments calculated in a similar manner from these extensions of Little's law and the corresponding expression for $\mathsf{E}[\mathcal{N}^k]$. In particular, when the interarrival time process $\{\tau_n\,;\ n \in \mathbb{Z}^+\}$ is a renewal process, we have

$$\lambda \int_0^{\infty} \mathcal{M}(t)\,\mathsf{P}[\,\mathcal{T} > t\,]\,dt = \mathsf{E}[\mathcal{N}^2] - \mathsf{E}[\mathcal{N}] \tag{39}$$

where $\mathcal{M}(t)$ denotes the renewal function for the sequence $\{\tau_n\}$. As a specific example, for the case of exponential interarrival times, this yields

$$\mathsf{E}[\mathcal{T}^2] = 2\lambda^{-2}(\mathsf{E}[\mathcal{N}^2] - \mathsf{E}[\mathcal{N}]). \tag{40}$$

The expected slowdown of a parallel job is defined to be the ratio of the expected sojourn time to the expected service time, which can be directly calculated as $\mathsf{E}[\mathcal{T}]/\mu_M^{-1}$. Higher moments of slowdown can be obtained in the same manner from higher moments of the sojourn and service times.

Another performance measure of interest is the long-run proportion of time that the system spends context-switching. This can be expressed as

$$\begin{aligned}\widetilde{p}_C &= (\pi_0, \ldots, \pi_{V-1})\, v_b^C + \sum_{k=0}^{\infty} \pi_{V+k}\, v_r^C, \\ &= (\pi_0, \ldots, \pi_{V-1})\, v_b^C + \pi_V (\mathbf{I} - R)^{-1} v_r^C, \end{aligned} \tag{41}$$

where the nth position of the vector v_b^C (v_r^C) contains a 1 if the corresponding state of the boundary (level V) represents when the server partitions are performing context-switches, and otherwise contains a 0.

4 Numerical Examples

Our primary focus in this paper is on the matrix-analytic analysis of (K, D) parallel-server queues. In this section, however, we briefly exploit our exact and approximate solutions to establish fundamental properties for such parallel-server queues that are of both theoretical and practical interest. These numerical experiments also are used to validate and compare the accuracy of our approximate matrix-analytic solutions. Our results are then used in turn to obtain bounds and approximations on the optimal values of K^* and D^*.

While our present study is primarily of theoretical interest, we take advantage of measurement data appearing in the computer science literature to examine the wide range of model parameter settings that are also of practical interest. In particular, studies of parallel workloads at large-scale computing environments executing parallel scientific and engineering applications demonstrate various degrees of application parallelism and high variability in job interarrival and service times, with coefficients of variation spanning the ranges $1 \le cv^A < 4$ and $1 \le cv^B < 10$, respectively.[21,22] A very large number of numerical experiments were conducted based on the analysis of Section 3 under a general parallel application speedup function (defined below) and various combinations of the model parameters that cover the range of statistics reported in these measurement studies. A small representative sample of our results are presented in this section, consisting of the parameter settings: $P = 128$, $K \in \{1, 2, 4, \ldots, 64, 128\}$, $D \in \{1, 2, 4, 8\}$, $cv^A = 1$, $cv^B = 4$, $\mu_1^{-1} = 1000$, $\delta^{-1} \in \{0.1, 10, 20, 50, 100, 200, 500\}$, $\gamma^{-1} = 1$.

Another important aspect of these experiments concerns the parallel jobs comprising the queueing system workload, and in particular the efficiency with which this set of jobs utilizes the servers allocated to them. The speedup function $\mathbf{S}(M)$ used in our study for this purpose is defined as the ratio of the expected service time of a parallel job on a single server to the expected service time of a parallel job on M servers, $1 \le M \le P$. Specifically, we exploit a popular class of speedup functions that have the form (e.g., see [31])

$$\mathbf{S}(M) \;=\; \frac{\mathsf{E}[A + B]}{\mathsf{E}[A + B/M]}, \qquad M \in \{1, \ldots, P\}, \tag{42}$$

where $A + B$ represents the service time of a parallel job on a single server, A represents the *sequential* component of this service time, B represents the *parallel* component of this service time, and A and B are two random variables which are assumed to be independent. Let $\widehat{S}_M = A + B/M$ be a generic random variable that represents the service time of a parallel job on M servers and that follows the original PH-type service time distribution for the (K, D) parallel-server queue with $M = P/K$, having mean $\mu_M^{-1} = \mathsf{E}[\widehat{S}_M]$ and second moment $\mathsf{E}[\widehat{S}_M^2]$, $M \in \{1, \ldots, P\}$ (see Section 2). From equation (42), we have $\mu_M^{-1} = \mu_1^{-1}/\mathbf{S}(M)$. For practical reasons, we restrict our attention to sublinear speedup functions, and thus $A > 0$.

4.1 Accuracy of Approximations

We now consider the accuracy of the limited processor-sharing (LPS) and timesharing decomposition (TSD) approximate solutions of Section 3, which provide significant reductions in computation complexity (time and space) over exact solutions. A representative sample of our results are provided in Figure 1. The leftmost and rightmost graphs plot the relative errors of both the LPS and TSD approximations in comparison with simulation as a function of the mean quantum length δ^{-1}, for $\rho = 0.4$ and $\rho = 0.9$, respectively. In each case, we use the parameter settings $K = 8$ and $D = 4$.

Our results show that the LPS approximation provides very accurate results when the timesharing quantum lengths are small, as expected. Recall that, in the absence of context-switching overheads, the LPS approach is exact in the limit as the quantum lengths tend to 0. The results in Figure 1 further suggest that our approach for extending the service time distribution of each job to include context-switching overhead provides a very accurate approximation especially when the context-switching overheads are small relative to everything else. We further observe that the accuracy of the LPS approximation degrades as the quantum lengths increase, again as expected.

Our results also show that the initial TSD approximation of Section 3.2 provides very accurate results when the timesharing quantum lengths are large. In fact, this TSD approach becomes exact in the limit as the quantum lengths tend to infinity because the system reduces to FCFS without timesharing (and with a single context switch incurred at the completion of

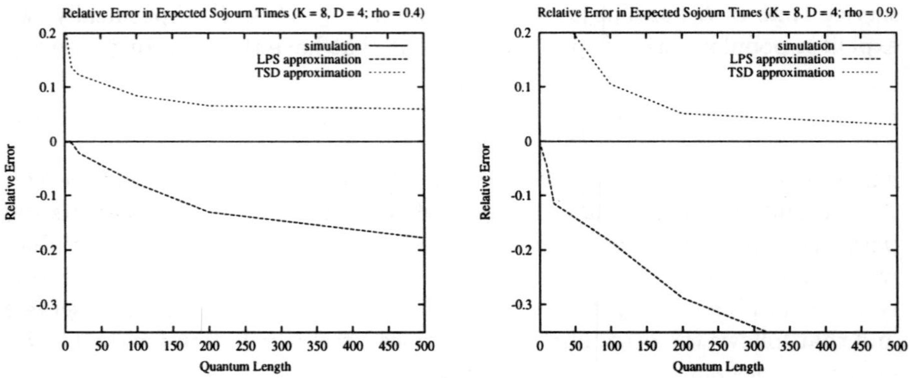

Figure 1. Relative errors in the expected sojourn times for both approximations.

service). We further observe that the accuracy of the initial TSD approximation improves as the traffic intensity increases, for sufficiently large quantum lengths. In the limiting heavy-traffic regime as the traffic intensity goes to its critical value, the relative errors for this TSD approximation tend to become negligible. On the other hand, when the timesharing quantum lengths are small, the initial TSD approximation provides much poorer results. This is due to the fact that each compound service time distribution primarily captures the departure time of the first parallel job to complete service under the timesharing of d i.i.d. jobs without capturing all of the progress made by the remaining $d-1$ jobs during this period of time, $2 \leq d \leq D$. The restarting of the service of these remaining $d-1$ jobs occurs more frequently under smaller quantum lengths, and this in turn exacerbates the performance impact of such errors especially at high traffic intensities.

It is important to point out, however, that the more general TSD approximation of Section 3.2 provides very accurate results for all traffic intensities when the timesharing quantums and context-switching overheads are not too small. These improved results are obtained by extending the initial TSD approach to capture in more detail the effects of timesharing on all of the jobs via sets of $\mathcal{L}$ compound service time distributions. Our results further suggest that very accurate approximations can be obtained with relatively small values of $\mathcal{L}$, at least for the parallel computing environments of interest.

4.2 Performance Results

We next exploit our matrix-analytic analysis to investigate fundamental properties of (K, D) parallel-server queues. A representative sample of the primary trends observed from our numerical experiments are provided in Figure 2. The upper figures plot the expected sojourn time as a function of the arrival rate λ, whereas the lower figures plot the variance of sojourn times as a function of λ. In the leftmost figures we set $D = 1$ and vary K, and in the rightmost figures we set $K = 4$ and vary D.

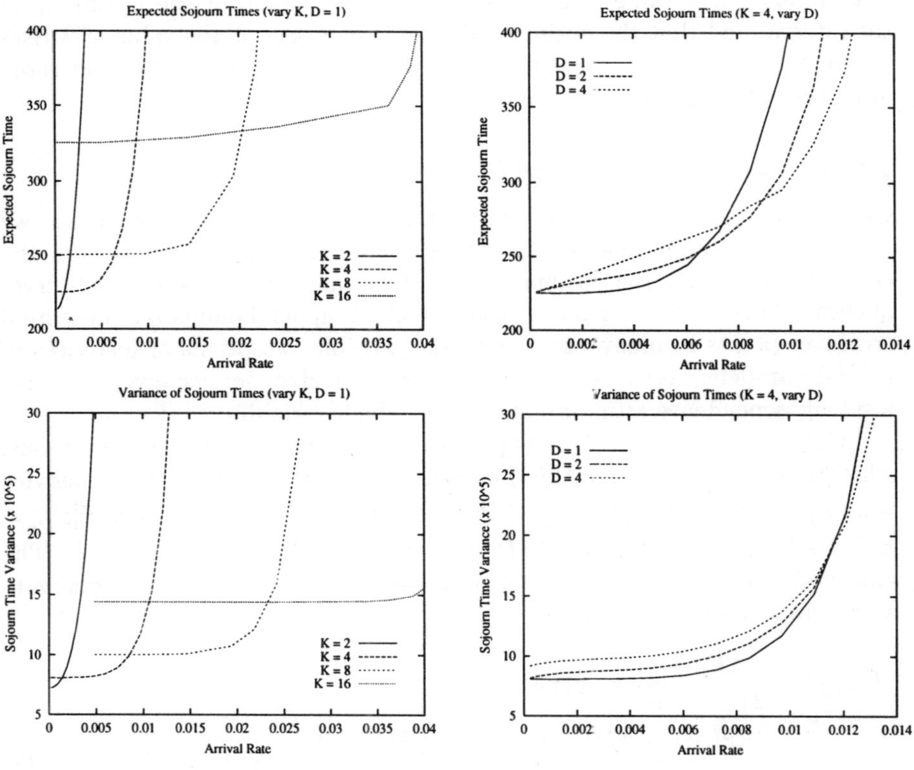

Figure 2. Primary trends for the first and second sojourn time moments.

We observe from both the expected sojourn time and sojourn time variance results that the optimal value of K^* tends to increase as the arrival rate λ rises and that the optimal value of D^* tends to increase as the arrival rate rises for a specific value of K. In fact, the results of many numerical solutions of (K, D) parallel-server queues under a wide variety of model parameter values based on our matrix-analytic analysis demonstrate that each value of K is optimal with respect to the expected sojourn time over an interval of λ values, where $K^* = 1$ in the interval $\lambda \in (0, \lambda_1)$, $K^* = 2$ in the interval $\lambda \in [\lambda_1, \lambda_2)$, $\ldots$, $K^* = P$ in the interval $\lambda \in [\lambda_{P/2}, \lambda_P)$, and λ_P denotes the critical point for the parallel-server queue. These results are obtained under the assumption that the speedup of parallel jobs is sublinear in the number of servers allocated, which is most appropriate in practice as noted above. Within each such interval of λ values for which a fixed value of K is optimal, we often find that a very small value of D (often $D = 1$) is optimal at the beginning of the interval and that the optimal value of D increases as the value of λ rises within the interval. The specific details of these trends depend considerably upon the other model parameters.

4.3 Bounds on Optimal K and D

Given some of the results and observations in the previous subsections, we now derive a simple analysis to obtain bounds and approximations on the crossover points for the optimal values of K^* and D^* in the parallel-server queues. These crossover point results will be upper bounds on the actual crossover points in many cases, with the possible exceptions of instances of (K, D) parallel-server queues where the arrival and/or service processes have a strong dependence structure and/or very high variability.

Consider the $(K = 1, D = 1)$ parallel-server queue under Poisson arrivals, which is equivalent to an M/G/1 queue with arrival rate λ, mean service time $\mu_P^{-1} = \mathsf{E}[\widehat{S}_P] = \mathsf{E}[A] + \mathsf{E}[B]/P$, second moment of service time $\mathsf{E}[\widehat{S}_P^2] = \mathsf{E}[A^2] + \mathsf{E}[B^2]/P^2 + 2\mathsf{E}[A]\mathsf{E}[B]/P$, and traffic intensity $\rho_P = \lambda \mathsf{E}[\widehat{S}_P]$. The expected sojourn time in this queue is given by the well-known Pollaczek-Khinchin formula,[32] and thus

$$\begin{aligned} \mathsf{E}[\widehat{\mathcal{T}}_{1,1}] &= \frac{\lambda\,\mathsf{E}[\widehat{S}_P^2]}{2\,(1 - \rho_P)} + \mathsf{E}[\widehat{S}_P], \\ &= \frac{\lambda\,(\mathsf{E}[A^2] + \mathsf{E}[B^2]/P^2 + 2\mathsf{E}[A]\mathsf{E}[B]/P)}{2\,(1 - \lambda(\mathsf{E}[A] + \mathsf{E}[B]/P))} + \mathsf{E}[A] + \mathsf{E}[B]/P. \quad (43) \end{aligned}$$

Consider for comparison the corresponding $(K = 2, D = 1)$ parallel-server queue in which each of the two server partitions is approximated by an M/G/1 queue with arrival rate $\lambda/2$, mean service time $\mu_{P/2}^{-1} = \mathsf{E}[\widehat{S}_{P/2}] = \mathsf{E}[A] + 2\mathsf{E}[B]/P$, second moment of service time $\mathsf{E}[\widehat{S}_{P/2}^2] = \mathsf{E}[A^2] + 4\mathsf{E}[B^2]/P^2 + 4\mathsf{E}[A]\mathsf{E}[B]/P$, and traffic intensity $\rho_{P/2} = (\lambda/2)\mathsf{E}[\widehat{S}_{P/2}]$. The corresponding expected sojourn time is then given by

$$\mathsf{E}[\widehat{\mathcal{T}}_{2,1}] = \frac{(\lambda/2)\,\mathsf{E}[\widehat{S}^2_{P/2}]}{2\,(1\;-\;\rho_{P/2})} \;+\; \mathsf{E}[\widehat{S}_{P/2}],$$
$$= \frac{(\lambda/2)(\mathsf{E}[A^2] + 4\mathsf{E}[B^2]/P^2 + 4\mathsf{E}[A]\mathsf{E}[B]/P)}{2(1-(\lambda/2)(\mathsf{E}[A] + 2\mathsf{E}[B]/P))} + \mathsf{E}[A] + 2\mathsf{E}[B]/P. \quad (44)$$

By continuing in this manner for increasing K, we obtain the closed-form expression for $\mathsf{E}[\widehat{\mathcal{T}}_{4,1}]$

$$\mathsf{E}[\widehat{\mathcal{T}}_{4,1}] = \frac{(\lambda/4)(\mathsf{E}[A^2] + 16\mathsf{E}[B^2]/P^2 + 8\mathsf{E}[A]\mathsf{E}[B]/P)}{2(1-(\lambda/4)(\mathsf{E}[A] + 4\mathsf{E}[B]/P))} + \mathsf{E}[A] + 4\mathsf{E}[B]/P, \quad (45)$$

and so on for the remaining expected sojourn time formulae. In each case, the expressions solely involve the variables λ, $\mathsf{E}[A]$, $\mathsf{E}[B]$, $\mathsf{E}[A^2]$ and $\mathsf{E}[B^2]$.

Hence, for any given arrival rate, we can compare the expected sojourn times among the corresponding approximate queueing systems using these formulae and choose the value of K that yields the smallest expected sojourn time. These results then can be used in turn to obtain bounds and approximations on the crossover points for the optimal values of K^*. In particular, for many instances of the (K, D) parallel-server queues, we can directly calculate bounds and approximate values for the optimal crossover points λ_{K^*} such that $K^* = 1$ in the interval $\lambda \in (0, \lambda_1)$, $K^* = 2$ in the interval $\lambda \in [\lambda_1, \lambda_2)$, ..., and $K^* = P$ in the interval $\lambda \in [\lambda_{P/2}, \lambda_P)$. Upon setting equation (43) equal to equation (44) and solving for λ, we obtain a bound and approximation for the value of λ_1 where the system with $K = 1$ provides the lowest expected sojourn time over the interval $\lambda \in (0, \lambda_1)$. Similarly, setting equation (44) equal to (45) and solving for λ, a bound and approximation for the value of λ_2 is calculated such that the system with $K = 2$ provides the lowest expected sojourn time over the interval $\lambda \in [\lambda_1, \lambda_2)$. We can calculate the remaining bounds on and approximate values of the optimal crossover points λ_{K^*} by continuing in this manner for increasing K^*.

As previously observed, within each of these λ intervals for which a fixed value of K is optimal, the numerical solutions of many (K, D) parallel-server queues based on our matrix-analytic analysis demonstrate that a very small value of D (often $D = 1$) is optimal at the beginning of the interval and that the optimal value of D increases as the value of λ rises within the interval. To obtain bounds and approximations on the crossover points for the optimal values of D^* within the λ interval for a given K^*, we apply the approximation methods of Section 3.1 or 3.2 to construct the set of PH-type distributions with parameters $(\underline{\beta}_d, \mathcal{S}^B_d)$ of order m^B_d having mean $\mathsf{E}[\widehat{S}_{P/K^*,d}] = -(\underline{\beta}_d(\mathcal{S}^B_d)^{-1}\mathbf{e})$

and second moment $\mathsf{E}[\widehat{S}^2_{P/K^*,d}] = 2(\underline{\beta}_d(\mathcal{S}^B_d)^{-2}\mathbf{e})$, $d \in \{2,\ldots,D\}$. Then, from the Pollaczek-Khinchin formula,[32] the expected sojourn time of the corresponding M/G/1 queue can be expressed as

$$\mathsf{E}[\widehat{\mathcal{T}}_{K^*,d}] = \frac{(\lambda/K^*)\,\mathsf{E}[\widehat{S}^2_{P/K^*,d}]}{2\,(1\;-\;(\lambda/K^*)\mathsf{E}[\widehat{S}_{P/K^*,d}])} \;+\; \mathsf{E}[\widehat{S}_{P/K^*,d}], \tag{46}$$

$d \in \{1,\ldots,D\}$. Hence, for any given arrival rate $\lambda \in [\lambda_{K^*-1}, \lambda_{K^*})$, we can use this expression to compare the expected sojourn times among the corresponding approximate queueing systems and choose the value of D that yields the smallest expected sojourn time. These results can be used in turn to obtain bounds and approximations on the crossover points for the optimal values of D^* within the λ interval for a given K^*.

Once again, the specific details of these trends depend upon the other model parameters. Hence, the above set of expected sojourn time formulae can be used for comparison to obtain bounds and approximations on the crossover points for the optimal values of K^* and D^*.

References

1. D. W. Fife. Scheduling with random arrivals and linear loss functions. *Management Science*, 11(3):429–437, 1965.
2. L. E. Schrage and L. W. Miller. The queue M/G/1 with the shortest remaining processing time discipline. *Operations Research*, 14:670–684, 1966.
3. L. E. Schrage. A proof of the optimality of the shortest remaining processing time discipline. *Operations Research*, 16:687–690, 1968.
4. Ronald W. Wolff. Time sharing with priorities. *SIAM Journal on Applied Mathematics*, 19(3):566–574, November 1970.
5. Ronald W. Wolff. Work-conserving priorities. *Journal of Applied Probability*, 7:327–337, 1970.
6. Kenneth C. Sevcik. Scheduling for minimum total loss using service time distributions. *Journal of the ACM*, 21(1):66–75, 1974.
7. Leonard Kleinrock. *Queueing Systems Volume II: Computer Applications.* John Wiley and Sons, 1976.
8. Shelby L. Brumelle. Some inequalities for parallel-server queues. *Operations Research*, 19:402–413, 1971.
9. Dietrich Stoyan. *Comparison Methods for Queues and Other Stochastic Models.* John Wiley and Sons, 1983. Edited with revisions by D. J. Daley.
10. Jay Sethuraman and Mark S. Squillante. Optimal stochastic scheduling in

multiclass parallel queues. In *Proceedings of the ACM Sigmetrics Conference on Measurement and Modeling of Computer Systems*, pages 93–102, June 1999.
11. Armand M. Makowski and Randolph D. Nelson. Optimal scheduling for a distributed parallel processing model. Technical Report RC 17449, IBM Research Division, February 1992.
12. Cheng-Shang Chang, Randolph D. Nelson, and David D. Yao. Optimal task scheduling on distributed parallel processors. *Performance Evaluation*, 20:207–221, May 1994.
13. Mark S. Squillante and Konstantinos P. Tsoukatos. Optimal scheduling of coarse-grained parallel applications. In *Proceedings of the Eighth SIAM Conference on Parallel Processing for Scientific Computing*, March 1997.
14. Mark S. Squillante. Matrix-analytic methods in stochastic parallel-server scheduling models. In *Advances in Matrix-Analytic Methods for Stochastic Models,* S. R. Chakravarthy and A. S. Alfa (eds.). Notable Publications, 1998.
15. Soren Asmussen. Phase-type distributions and related point processes: Fitting and recent advances. In Srinivas R. Chakravarthy and Attahiru S. Alfa, editors, *Matrix-Analytic Methods in Stochastic Models*, pages 137–149. Marcel Dekker, 1997.
16. Andras Horvath and Miklos Telek. Approximating heavy tailed behaviour with phase type distributions. In *Advances in Algorithmic Methods for Stochastic Models,* G. Latouche and P. Taylor (eds.), pages 191–214. Notable Publications, 2000.
17. Bo Friis Nielsen. Modelling long-range dependent and heavy-tailed phenomena by matrix analytic methods. In *Advances in Algorithmic Methods for Stochastic Models,* G. Latouche and P. Taylor (eds.), pages 265–278. Notable Publications, 2000.
18. Lothar Breuer. Parameter estimation for a class of BMAPs. In *Advances in Algorithmic Methods for Stochastic Models,* G. Latouche and P. Taylor (eds.), pages 87–97. Notable Publications, 2000.
19. Tobias Ryden. Statistical estimation for Markov-modulated Poisson processes and Markovian arrival processes. In *Advances in Algorithmic Methods for Stochastic Models,* G. Latouche and P. Taylor (eds.), pages 329–350. Notable Publications, 2000.
20. J. F. C. Kingman. The effect of queue discipline on waiting time variance. In *Proceedings of the Cambridge Philosophical Society*, volume 58, pages 163–164, 1962.
21. Dror G. Feitelson and Bill Nitzberg. Job characteristics of a production parallel scientific workload on the NASA Ames iPSC/860. In *Job Schedul-*

ing Strategies for Parallel Processing, D. G. Feitelson and L. Rudolph (eds.), pages 337–360. Springer-Verlag, 1995. Lecture Notes in Computer Science Vol. 949.

22. Steven G. Hotovy. Workload evolution on the Cornell Theory Center IBM SP2. In *Job Scheduling Strategies for Parallel Processing,* D. G. Feitelson and L. Rudolph (eds.), pages 27–40. Springer-Verlag, 1996. Lecture Notes in Computer Science Vol. 1162.
23. Marcel F. Neuts. *Matrix-Geometric Solutions in Stochastic Models: An Algorithmic Approach.* The Johns Hopkins University Press, 1981.
24. Mark S. Squillante. A matrix-analytic approach to a general class of $G/G/c$ queues. Technical report, IBM Research Division, May 1996.
25. D.P. Gaver, P.A. Jacobs, and G. Latouche. Finite birth-and-death models in randomly changing environments. *Advances in Applied Probability,* 16:715–731, 1984.
26. Guy Latouche and V. Ramaswami. *Introduction to Matrix Analytic Methods in Stochastic Modeling.* ASA-SIAM, Philadelphia, 1999.
27. John D. C. Little. A proof of the queuing formula $L = \lambda W$. *Operations Research*, 9:383–387, 1961.
28. Shelby L. Brumelle. A generalization of $L = \lambda W$ to moments of queue length and waiting times. *Operations Research*, 20:1127–1136, 1972.
29. Ward Whitt. A review of $L = \lambda W$ and extensions. *Queueing Systems,* 9:235–268, 1991. Also, 12:431–432, 1992.
30. Dimitris Bertsimas and Daisuke Nakazato. The distributional Little's law and its applications. *Operations Research*, 43(2):298–310, 1995.
31. Emilia Rosti, Giuseppe Serazzi, Evgenia Smirni, and Mark S. Squillante. Models of parallel applications with large computation and I/O requirements. *IEEE Transactions on Software Engineering*, March 2002.
32. Leonard Kleinrock. *Queueing Systems Volume I: Theory.* John Wiley and Sons, 1975.

ROBUSTNESS OF FS-ALOHA

B. VAN HOUDT AND C. BLONDIA

University of Antwerp, Department of Mathematics and Computer Science, Performance Analysis of Telecommunication Systems Research Group, Universiteitsplein, 1, B-2610 Antwerp - Belgium, Email: {vanhoudt,blondia}@uia.ua.ac.be

This paper evaluates the robustness of FS-ALOHA, a random access algorithm used to reserve uplink—that is, from the end user to the network—bandwidth in centralized wireless access networks. The performance of FS-ALOHA when subject to Poisson arrivals and operating on an error free channel was evaluated in Vazquez *et al* [1] by means of a Quasi-Birth-Death (QBD) Markov chain. In this paper we relax these assumptions and study discrete time batch Markovian arrivals on a channel with memoryless errors by means of a Markov chain of the GI/M/1 type. It is concluded that FS-ALOHA performs well under correlated and bursty arrivals and memoryless errors. However, error rates above $1/5T$, where T is a protocol parameter, can seriously increase the delays suffered on the contention channel and might even make the system unstable. Finally, it is concluded that implementing multiple instances of FS-ALOHA can significantly improve the delays and the robustness of the algorithm.

1 Introduction

There are, roughly speaking, two ways to transmit information on a communication channel that is shared among multiple users. Either, the protocol followed by the users avoids that two or more users transmit information at the same time, or it allows for simultaneous transmissions to occur. In the first case we refer to the channel as a contention free channel, in the latter case, the channel is referred to as a contention channel. Simultaneous transmissions are commonly known as collisions (between information) and any information that collides is considered lost, that is, the receiver is unable to retrieve the original information. Although collisions always result in the loss of information, there are many situations in which it is beneficial to use a protocol that allows for collisions to occur, e.g., when the number of users is large and each user uses the channel on a sporadic basis.

A protocol that operates on a contention channel is called a random access algorithm (RAA) or a random access protocol. The functionality of a RAA is often subdivided as follows:

- Controlling the transmission of new informantion. This task is referred to as the channel access protocol (CAP).

- Managing retransmission after a collision occured. A task that is referred to as the contention resolution algorithm (CRA).

Thus, a RAA is a combination of a CAP and a CRA. One way to classify RAAs is to subdivide them based on their CAP. In this case, there are two main categories: RAAs with free and RAAs with blocked access, meaning that either users that generate a new information packet transmit this information immediately, or they are blocked until a certain event occurs. A subclass of the RAAs with blocked access are the RAAs with grouped access. In this particular case, new arrivals are grouped based on their arrival time and packets belonging to a certain group are not allowed to make a first transmission attempt until all the packets belonging to the previous groups have been transmitted successfully. A packet is successfully transmitted if it did not collide with another packet.

FS-ALOHA is a RAA that can be regarded as a RAA with grouped access because the requests—we refer to information packets transmitted on the contention channel as requests—are grouped in Transmission Sets (TSs) so that just one TS attempts transmission at a time (i.e., a subset of all pending requests). Also, the requests belonging to a certain TS use a CRA, in the case of FS-ALOHA one uses slotted ALOHA[a], to gain access to the medium. Hence, FS-ALOHA combines the simplicity of slotted ALOHA with the efficiency obtained by grouping the requests that arrive at the mobile stations (MSs). Although FS-ALOHA was designed to reserve bandwidth in centralized wireless access networks, it can be used for the same purpose in hybrid fiber coaxial cable (HFC) networks as an alternative for the binary exponential backoff (BEB) algorithm.

A Quasi-Birth-Death (QBD) Markov chain that allowed the performance evaluation of FS-ALOHA on an error free channel subject to Poisson arrivals was developed in Vazquez *et al* [1]. This study indicated that FS-ALOHA is capable of guaranteeing low delay bounds and high throughput rates. Moreover, FS-ALOHA was shown to outperform ALOHA both in terms of delay and throughput. In this paper we address the robustness of FS-ALOHA and develop a Markov chain of the GI/M/1 type that allows us to evaluate FS-ALOHA on a channel with memoryless errors and D-BMAP arrivals. Thus, we can see how bursty and correlated arrivals, as well as errors, influence the performance of the algorithm. We also investigate whether some of the engineering rules, obtained from the study in Vazquez *et al* [1], still apply in such errorprone systems with bursty and correlated arrivals.

Although FS-ALOHA is not believed to be as powerful as a RAA with

[a]The slotted ALOHA algorithm is described in the next Section.

grouped access that uses a tree algorithm (see Bertsekas *et al* [2])—also known as a splitting algorithm or an algorithm of the CTM type—as its CRA, it presents an attractive tradeoff between simplicity, that is, the ease to implement the algorithm, and its performance. The fact that simplicity is indeed a major player in the standardization of any Medium Access Control (MAC) layer was demonstrated once more during the development of the DOCSIS standard for HFC networks. Finally, it should be noted that, from an information theoretical point of view, FS-ALOHA is a full-sensing algorithm[b]; hence, it belongs to the same class of algorithms as the RAA with grouped access that uses a tree algorithm as its CRA.

The remainder of this paper is structured as follows. FS-ALOHA is described in Section 2. An informal outline of the model is given in Section 3. Section 4 and 5 present the analytical models used to evaluate FS-ALOHA. Numerical results can be found in Section 6 and conclusions are drawn in Section 7.

2 FS-ALOHA Algorithm: a Review

In this section the operation of FS-ALOHA, and the environment in which it operates, are described in some detail, additional comments and discussions can be found in Vazquez *et al* [1]. Consider a cellular network with a centralized architecture, i.e., the area covered by the wireless access network is subdivided into a set of geographically distinct cells each with a diameter of approximately 100m. Each cell contains a base station (BS) serving a finite set of mobile stations (MSs). This BS is connected to a router, which supports mobility, realizing seamless access to the wired network. Two logically distinct communication channels (uplink and downlink) are used to support the information exchange between the BS and the MSs. Packets arriving at the BS are broadcasted downlink, while upstream packets must share the radio medium using a MAC protocol. The BS controls the access to the shared radio channel (uplink). A different frequency band is used for the uplink and downlink traffic (that is, the access technique is Frequency Division Duplex (FDD)).

Traffic on both the uplink and downlink channel is grouped into fixed length frames, with a length of L time slots, to reduce the battery consumption[c] (see Van Houdt *et al* [3]). The uplink and downlink frames are synchro-

[b]This means that a user requires feedback, that is, an indication that a packet collided or not, from the channel for each packet transmission attempt made on the channel and not merely for its own packet transmission attempts.

[c]The frame structure enables the BS to inform the MSs, at the start of the frame, about

nized in time, i.e., the header of a downlink frame is immediately followed by the start of an uplink frame (after a negligible round trip time that is captured within the guard times[d], see Figure 1). Each uplink frame consists of a fixed

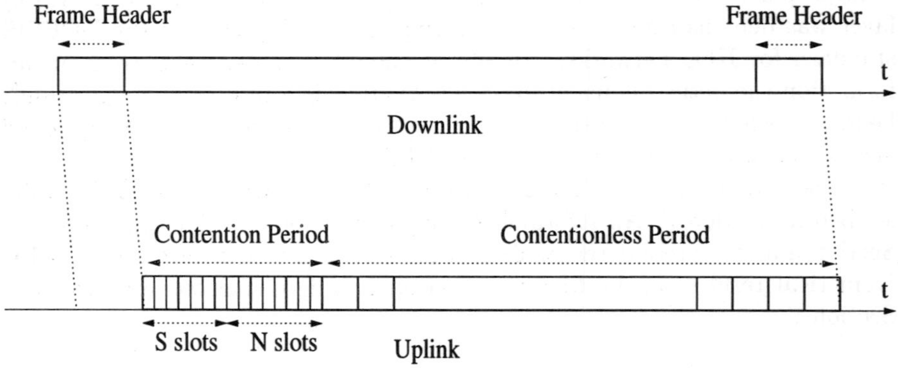

Figure 1. Frame Structure

length contentionless and a fixed length contention period, where the length of the contentionless period, in general, dominates that of the contention period. An MS is allowed to transmit in the contentionless period after receiving a permit from the BS. The BS distributes these permits among the MSs based on the requests it receives from the MSs and the existing QoS agreements between the end users and the network. These requests are used by MSs to declare their current bandwidth needs to the BS, e.g., by indicating how many packets they have ready for transmission. Requests are transmitted using the contention channel, unless the MS can piggyback it to a data packet for which a permit was already obtained, thereby reducing the load on the contention channel.

A request is generally much smaller than a data packet; therefore, slots

---

the destination addresses of the downlink packets within the frame. As a result, an MS can switch to the sleep mode for the remaining frame time, unless there is a packet destined for this MS.

[d]A guard time is a small time interval at the end of each time slot during which the MSs and the BS do not transmit information. Guard times are necessary to avoid that a collision can occur between a packet that is transmitted in time slot t and $t+1$. Indeed, any information transmitted, i.e., broadcasted, by an MS (or the BS) needs a small fraction of time to reach the other MSs, therefore, the guard time has to be larger than the maximum time required by an electromagnetic wave to travel from an arbitrary MS to any other MS. Given the small size of the cells (approximately 100m), we get a small guard time.

part of the contention period can be subdivided into k minislots (realistic values for k in a wireless medium are 1 to 3, in a wired medium higher values for k are possible). Each downlink frame starts with a frame header in which, among other things, the required feedback on the contention period of the previous uplink frame is given. This informs the MSs participating in the contention period whether there was a collision or whether the request was successfully received.

FS-ALOHA operates on the slots that are part of the fixed length contention period. Define T as the number of minislots part of the contention period of a frame. From hereon we refer to minislots as slots. In slotted ALOHA systems, an MS with a pending request will randomly choose one out of the T slots to send its request in the hope that no other MS with a pending request will choose the same slot. If an MS is unsuccessful, i.e., another MS also decided to transmit in this particular slot, it will retransmit the request in one of the T slots in the next frame. It is important to note that with slotted ALOHA, new requests join the competition immediately after being generated; hence, they are not blocked. FS-ALOHA on the contrary, divides the T slots of the contention period into two disjoint sets of S and N slots such that $T = S + N$. The operation of FS-ALOHA is as follows:

- Newly arrived requests are transmitted, for the first time, by randomly choosing one out of the S slots; this is the first set of S slots after the request was generated. If some of these transmissions were unsuccessful, because multiple MSs transmitted in the same slot, the unsuccessful requests are grouped into a Transmission Set (TS), which joins the back of the queue of TSs waiting to be served.

- The other N slots are used to serve the queue of backlogged TSs on a FIFO basis. A TS is served using slotted ALOHA, that is, all the requests part of the TS select one out of the N slots and transmit in this slot. The requests that were transmitted successfully leave the TS, the others retransmit in the N slots of the next frame using the same procedure. The service of a TS lasts until all the requests part of the TS have been successfully transmitted, in which case the service of the next TS, if there is another TS in the queue, starts service in the N slots of the next frame.

Hence, two parameters play an important role in FS-ALOHA:

- The number of $S \geq 1$ slots in a frame. These slots are used to transmit newly arrived requests; S determines the TS generation rate.

- The number of $N \geq 2$ slots in a frame. These slots are allocated to the service of the backlogged TSs in the distributed queue.

Notice, two requests that were generated in different frames can never be part of the same TS. Thus, it is said that the grouping of requests in Transmission Sets is based on a time period corresponding to the frame length. Therefore, FS-ALOHA can be regarded as a RAA with grouped access that uses Slotted ALOHA as its CRA, that is, the algorithm used to resolve the TSs is Slotted ALOHA. More details and extensions of FS-ALOHA can be found in Vazquez *et al* [1,4].

3 An Informal Outline of the Model

Before we proceed with a detailed description of the model, it might be useful to outline how to translate the operation of FS-ALOHA to a queueing system. We wish to evaluate the performance of FS-ALOHA under correlated and bursty arrivals, therefore, we assume that new requests generated by the MSs arrive according to a D-BMAP arrival process. The time unit of the D-BMAP is chosen to be one frame. Thus, provided that the D-BMAP is characterized by the matrices $D_0, D_1, \ldots$, there is a probability $(D_i)_{j_1,j_2}$ that i new request, each originating from a different MS, are generated in a frame, provided that the D-BMAP is in state j_1, resp. j_2, at the start, resp. end, of the frame. A first transmission attempt for each of these i new requests will take place in the S slots of this frame (that is, each of the i corresponding MSs selects one of the S slots and transmits its request in this particular slot). If all i are successful, meaning that each of the i requests was transmitted in a different slot, we state that there is no customer arrival (that is, no TS is being formed). Otherwise, the unsuccessful requests are grouped to form a TS and this transmission set is considered a customer of our queue. The number of requests part of a TS varies (each TS holds at least two requests); hence, we state that a customer is of type k if there are $k+1$ requests part of the TS. As a result, the input process of our queue can be regarded as a discrete time MMAP[K] arrival process that generates either zero or one customer during a time instance (the value of K is equal to maximum number of requests q_m that can be generated in a single frame minus one).

The service time of a customer of type k equals the number of frames required to successfully transmit each of the $k+1$ requests part of the TS. During each frame, each of the requests that remain in the TS will be transmitted in one of the N slots. Those requests that were successfully transmitted—recall that a request is successfully transmitted if it is the only request to select a particular slot—leave the TS. The unsuccessful requests remain in the TS. Thus, the progress of a service of a customer, i.e., TS, can be represented at the start of each frame by the number of requests that remain within the TS. Hence, the service time distribution of a type k customer can be represented as a discrete time phase type distribution, with matrix representation (m_k, T_k, α_k), where the phase represents the number of requests left in the TS. As a result, the queue holding the TSs is nothing but a MMAP[K]/PH[K]/1 queue. We could generalize the idea introduced in this paper to obtain a procedure that calculates the delay distribution of a type k customer in an arbitrary MMAP[K]/PH[K]/1 queue (see Van Houdt *et al* [5]) and apply this general approach to this particular queue. However, in this case we are not interested in the delay distribution of a type k customer, but in the delay of a request. Moreover, the service time distributions of all the customers are very similar. Indeed, we could define a matrix T such that the service time distribution of a type k customer is identical to the phase type distribution represented by (m, T, α_k). The integer m and the matrix T are equal to m_{max} and T_{max}, where $(m_{max}, T_{max}, \alpha_{max})$ was the representation of the service time distribution related to a TS with q_m requests. The entries of the vector α_k are identical to zero, except for the k-th entry which equals one.

4 Performance Evaluation of FS-ALOHA on an Error Free Channel

4.1 Analytical Model

In this section an exact analytical model is developed, allowing the computation of the delay density function associated to the request packets under the following conditions:

- We assume a D-BMAP request arrival process with a mean rate of λ arrivals per frame.
- The number of slots T for contention is fixed and within these T slots, $S > 0$ are used by the new arrivals and $N > 1$ are used for the service of the Transmission Sets in the queue.
- If there are no Transmission Sets in the queue nor in service, the total $T = S + N$ slots is used by new arrivals.
- The Bit Error Rate (BER) is assumed to be zero, this assumption is relaxed further on.

These assumptions are identical to Vazquez *et al* [1], except that we assume D-BMAP arrivals instead of Poisson arrivals. In the next section we will also relax the assumption on the BER. For Poisson arrivals one obtains a QBD Markov chain by observing the couple $(\hat{q}, \hat{Q})$ at the start of each frame, where $\hat{q}$ represents the number of requests in the TS that is currently in service (provided that a TS is in service) and $\hat{Q}$ is the number of TSs waiting in the distributed FIFO queue[e]. If we consider the same stochastic process for D-BMAP arrivals and add the current state of the D-BMAP, say $\hat{j}$, we no longer have a Markov chain. Therefore, a different approach is required; the basic idea is to remember the "age" of the TS currently in service instead of the number of TSs waiting in the TS queue[f]. The state of the system is modeled by the triple (q, j, Q), where

- $q \geq 2$ denotes the number of requests in the Transmission Set that is currently in service (if there is a Transmission Set in service).
- j denotes the state of the D-BMAP associated with the frame that follows the frame in which the Transmission Set currently in service was generated (if there is a Transmission Set in service, otherwise it is the state of the D-BMAP associated with the current frame).
- Q indicates how many frames ago the Transmission currently in service Set was created ($Q = 0$ if there is no Transmission Set in service).

For instance, $(q, j, Q) = (4, j, 3)$ indicates that 4 requests will attempt a transmission in the N slots of the current frame, say frame n. Each of these 4 stations has had at least 1, in frame $n - 3$, and at most 3 unsuccessful attempts in the previous 3 frames (depending on the service completion time of the previous TS) and the state j of the D-BMAP determines the number of requests that make use of the S slots in frame $n - 2$. If, for example, 2 of the 4 request are transmitted successfully (within the N slots of frame n), the new state, associated with frame $n + 1$, would be $(2, j, 4)$.

[e] Level 0 consists of one state that corresponds to the case where there are no TSs waiting in the queue and there is no TS in service, level $i > 0$ consists of multiple states that correspond to the case where there are $i - 1$ TSs waiting in the queue and a TS is in service (the j-th state of level i indicates that there are $j + 1$ requests left in the TS).

[f] This trick can also be used to obtain the waiting time distribution for each class of customers in a discrete time FCFS MMAP[K]/PH[K]/1 queue provided that the MMAP[K] arrival process has $D_J = 0$ for all strings J with a length $|J| > 1$ (i.e., the customers arrive one at a time). In this case one remembers: the age of the customer currently in service, its class type, the state of its service and the state of the MMAP[K] input process. Because the age can only increase by one at a time we obtain a GI/M/1 Type Markov chain by observing the system at each time slot. Moreover, in Van Houdt *et al* [5], we have generalized this technique to MMAP[K] arrival processes with batch arrivals.

Notice that this model can be used for Poisson arrivals as well. Moreover, although the model in Vazquez *et al* [1] uses a QBD Markov chain, the calculations required to obtain the delay distribution from the steady state probabilities are cumbersome. Whereas with this model, that uses a GI/M/1 type Markov chain, one obtains the delay distribution from the steady state probabilities by means of a simple formula (see Section 4.5).

4.2 Transition Matrix

The transitions in the system take place at the start of each frame. The maximum value of q, say q_m, corresponds to the highest possible i for which D_i contains entries that differ from zero, where D_i are the $l \times l$ matrices that characterize the input D-BMAP traffic. For D-BMAPs that do not posses such an index i or for D-BMAPs for which this index i is very large, we choose q_m such that the sum of the entries of the matrices $D_i, i > q_m$ is negligible. Therefore, the impact on the accuracy of the results is minimized. The range of j is equal to $\{j \mid 1 \leq j \leq l\}$. During a state transition, Q can never increase by more than one.

Therefore, the system can be described by a transition matrix P with the following structure:

$$P = \begin{bmatrix} B_1 & B_0 & 0 & 0 & 0 & \dots \\ B_2 & A_1 & A_0 & 0 & 0 & \dots \\ B_3 & A_2 & A_1 & A_0 & 0 & \dots \\ B_4 & A_3 & A_2 & A_1 & A_0 & \dots \\ \vdots & \vdots & \vdots & \ddots & \ddots & \ddots \end{bmatrix}, \tag{1}$$

where A_i are $l(q_m - 1) \times l(q_m - 1)$ matrices, $B_i, i > 1$, are $l(q_m - 1) \times l$ matrices, B_1 is an $l \times l$ matrix and B_0 is an $l \times l(q_m - 1)$ matrix.

The matrices B_0 and B_1 describe the system when the current frame is not serving a Transmission Set ($Q = 0$). This implies that the total of $T = S + N$ slots is used for new arrivals. B_0 describes the transitions when a Transmission Set is generated within these T slots, whereas B_1 describes the situation in which no Transmission Set is generated.

The matrices A_i and $B_i, i > 1$, hold the transition probabilities provided that a Transmission Set t is in service in the current frame. A_0 covers the case in which the service of the current Transmission Set t is not completed within the current frame. The transition probabilities held by the matrices $A_i, i > 0$, correspond to the following situation: the service of the current Transmission Set t is completed within the current frame, say frame n, and the first $i - 1$ frames following frame $n - Q$, i.e., the frame in which the

Transmission Set t was generated, do not generate a new Transmission Set, whereas frame $n - Q + i$ ($\leq n$) does generate a new Transmission Set. The matrices $B_i, i > 1$ on the other hand correspond to case where the service of the current Transmission set t is completed within the current frame, frame n, and the first $i - 1$ ($= Q$) frames following frame $n - Q$ do not generate a new Transmission Set (as a result the total of $T = S + N$ slots is used for new arrivals in frame $n + 1$).

4.3 Calculating the Transition Probabilities

In this subsection we indicate how to calculate the matrices A_i and B_i described above. Define $p_x(q, q')$, for $q \geq q'$, as the probability that in a set of q requests, $q - q'$ request are successful when a set of x slots is used to transmit the q request packets[g]. We are particularly interested in $p_S(q, q')$, $p_N(q, q')$ and $p_{S+N}(q, q')$. Von Mises [6] has shown, in 1939, that

$$p_x(q, q') = \sum_{v=q-q'}^{\min(q,x)} (-1)^{v+q-q'} C^v_{q-q'} C^x_v \frac{q!}{(q-v)!} \frac{(x-v)^{q-v}}{x^q}, \tag{2}$$

where C^r_s denotes the number of different ways to choose s from r different items. Equation 2 is numerically stable for the parameter ranges of interest ($x \leq 20$). It is also possible to calculate the $p_x(q, q')$ values recusively using the $p_{x-1}(q, q')$ values, thus, higher parameter values do not cause any problems.

Next, denote P_N as an $q_m - 1 \times q_m - 1$ matrix whose $(i, j)^{th}$ element equals $p_N(i + 1, j + 1)$. Let $P_{N,0}$ be a $q_m - 1 \times 1$ vector whose i^{th} component equals $p_N(i + 1, 0)$. In order to describe the matrices A_i and B_i we also define the matrices F_S, F_{S+N}, $E^k_S, 2 \leq k \leq q_m$, and $E^k_{S+N}, 2 \leq k \leq q_m$, as (these matrices are $l \times l$ matrices)

$$F_S = \sum_{i \geq 0} D_i \, p_S(i, 0) \tag{3}$$

$$F_{S+N} = \sum_{i \geq 0} D_i \, p_{S+N}(i, 0) \tag{4}$$

$$E^k_S = \sum_{i \geq k} D_i \, p_S(i, k), \tag{5}$$

$$E^k_{S+N} = \sum_{i \geq k} D_i \, p_{S+N}(i, k), \tag{6}$$

[g] This corresponds to the following combinatorial problem: provided that we, randomly, distribute q balls among a set of x urns, what is that probability that we have exactly $q - q'$ urns holding a single ball.

where the D-BMAP arrival process is characterized by the matrices D_i. Notice that $(E_S^k)_{j,j'}$ represents the probability that a new TS with k requests is generated in a frame where S slots are used for the new arrivals, thus, another TS is currently in service in the remaining N slots, and the D-BMAP governing the new arrivals makes a transition from state j to j'. F_S on the other hand holds the probabilities that no new TS is generated in a frame where S slots are used for new arrivals. Similar interpretations exist for the matrices F_{S+N} and E_{S+N}^k. The transition probability matrices A_i and B_i are then found as follows:

$$A_0 = P_N \otimes I_l, \tag{7}$$
$$A_i = P_{N,0} \otimes \left((F_S)^{i-1}\left[E_S^2\ E_S^3\ \ldots\ E_S^{q_m}\right]\right), \tag{8}$$
$$B_0 = \left[E_{S+N}^2\ E_{S+N}^3\ \ldots\ E_{S+N}^{q_m}\right], \tag{9}$$
$$B_1 = F_{S+N}, \tag{10}$$
$$B_i = P_{N,0} \otimes (F_S)^{i-1}, \tag{11}$$

where $\otimes$ denotes the Kronecker product between matrices and I_l the $l \times l$ unity matrix. Notice that the matrices A_i and B_i decrease to zero according to $(F_S)^i$. Looking at the probabilistic interpretation of F_S, it should be clear that, in general, the smaller the arrival rate λ the slower A_i and B_i decrease to zero. Therefore, the model is not suited for very small arrival rates λ (because this would imply that thousands of A_i and B_i matrices are needed to perform the calculations).

4.4 Calculating the Steady State Probabilities

Define $\pi_i^n(q,j), i > 0$, resp. $\pi_0^n(j)$, as the probability that the system is in state (q,j,i), resp. $(j,0)$, at time n, i.e., at the start of frame n. Let

$$\pi_0(j) = \lim_{n\to\infty} \pi_0^n(j), \tag{12}$$
$$\pi_i(q,j) = \lim_{n\to\infty} \pi_i^n(q,j). \tag{13}$$

Define the $1 \times l$ vector $\pi_0 = (\pi_0(1), \ldots, \pi_0(l))$ and the $1 \times l(q_m - 1)$ vectors $\pi_i = (\pi_i(2,1), \ldots, \pi_i(2,l),\ \pi_i(3,1), \ldots, \pi_i(3,l), \pi_i(4,1), \ldots, \pi_i(q_m,l))$, $i > 0$. From the transition matrix P (Equation 1) we see that the Markov chain is a generalized Markov chain of the $GI/M/1$ Type (see Neuts [7]). From such a positive recurrent Markov chain, we have $\pi_i = \pi_{i-1}R, i > 1$, where R is an $l(q_m - 1) \times l(q_m - 1)$ matrix that is the smallest nonnegative solution to the following equation:

$$R = \sum_{i\geq 0} R^i A_i. \tag{14}$$

This equation is solved by means of an iterative scheme (see Neuts [7]). In order to obtain π_0 and π_1 we solve the following equation

$$(\pi_0, \pi_1) = (\pi_0, \pi_1) \begin{bmatrix} B_1 & B_0 \\ \sum_{i\geq 2} R^{i-2} B_i & \sum_{i\geq 1} R^{i-1} A_i \end{bmatrix}. \qquad (15)$$

The vector (π_0, π_1) is normalized as $\pi_0 e_l + \pi_1 (I - R)^{-1} e_{l(q_m-1)} = 1$, where I is the unity matrix of size $l(q_m - 1)$ and e_i is an $i \times 1$ vector filled with ones. Theorem 1.5.1 in Neuts [7] states that the Markov chain with transition matrix P is positive recurrent if and only if the spectral radius $sp(R)$ of the matrix R, where R is the minimal nonnegative solution to Equation 14, is smaller than one and there exists a positive solution to Equation 15. It is not difficult to see that $A = \sum_{i\geq 0} A_i$ is irreducible[h], provided that the input D-BMAP is irreducible, therefore a simple condition exists to check whether $sp(R) < 1$ (see Neuts [7,8]). We could also check the positive recurrence by noticing that FS-ALOHA, when subject to D-BMAP arrivals, is equivalent to a discrete time MMAP[K]/PH[K]/1 queue with a generalized initial condition, where the MMAP[K] stands for a Markov chain with marked arrivals (see He [9]). The stability of such queues has been studied by He [10] in Theorem 7.1.

4.5 Calculating the Delay Density Function

Let D be the random variable that denotes the delay suffered by a request packet. We state that $D = 0$ if a request packet is successful during its first attempt. $D = i$ if a request packet is successful in frame $n + i$ provided that the first attempt took place in frame n. Using the steady state probabilities we easily find

$$P[D = i] = \sum_{q=2}^{q_m} \frac{(1 - 1/N)^{q-1} q}{\lambda} \sum_{j=1}^{l} \pi_i(q, j), \qquad (16)$$

for $i > 0$, with λ the arrival rate of the D-BMAP, i.e., the mean number of newly arriving request packets per frame. While $P[D = 0]$ is found as $1 - \sum_{i>0} P[D = i]$.

---

[h] After removing the possible (obvious) transient states of level $Q > 0$. Indeed, the states (q, j, Q), for $Q > 0$, are transient if the j-th entry of the vector $\theta \sum_{i\geq q} D_i$ equals zero, where θ is the stochastic stationary vector of $\sum_{i\geq 0} D_i$. It is not necessary to remove their corresponding rows and columns when calculating the steady state probabilities, because the algorithm outlined in Section 4.4 will automatically assign a probability zero to these states.

5 Performance Evaluation of FS-ALOHA on a Channel with Memoryless Errors

In this section we relax the assumption on the BER made in the previous section, and allow for memoryless errors to occur. From a practical point of view, Markovian errors would probably be more appropriate, but there seems to be no apparent way to incorporate such errors in the current model, even if we were to restrict ourselves to Poisson arrivals. Perhaps a short explanation is appropriate. If we assume Markovian errors, the number of requests in a TS depends, among other things, upon the error state related to the frame in which the TS is created. We define the error state as the state of the Markov chain governing the errors. This is similar to the model in the previous section where the number of requests in a TS depended, in a similar way, on the state of the arrival process. However, with Markovian errors the resolution of a TS with k requests is influenced by the error state, whereas this is not the case for the state associated to the arrival process. Thus, if we want to enrich the previous model with Markovian errors we need to keep track of the error state in the current frame, and of the error state related to the frame in which the TS currently in service was created; therefore, in order to obtain a Markov chain that observes the system at every frame time—a desirable property if we want to calculate the delay distribution with a simple formula from the steady state probabilities—we need to keep track of the entire history of the error state between these two time instances. This would clearly result in an explosion of the state space, unless the Markov chain has only one state, that is, if the errors are memoryless. Therefore, we restrict ourselves to memoryless errors and state that an error occurs in a slot with a probability $0 \leq e \leq 1$.

Errors occurring on the channel influence the transmissions as follows. If a slot holds a collision, that is, if two or more MSs transmit a request in the same slot, then the BS, correctly, interprets this slot as a collision, whether or not an error occurred in this slot. On the other hand, if a slot does not hold a collision and an error does occur in the slot, the BS will, incorrectly, interpret the slot as holding a collision. A slot that neither holds a collision or an error is correctly recognized by the BS. As a result, a single error in the slots dedicated to the new arrivals is sufficient to create a new TS; hence, TSs with zero or one request exist, as opposed to the model in the previous section. Also, the average number of frames required to resolve a TS with k requests increases due to the presence of errors. The service of a TS ends if the N slots, assigned to the service of TSs, do not hold an unsuccessful transmission nor an error.

It should be clear that the triple (q, j, Q) as defined in the previous section is still a Markov chain of the GI/M/1 type. However, the entries and the size, because TSs with zero or one request exist, of the matrices A_i and B_i have changed. These matrices will be denoted as $\tilde{A}_i$ and $\tilde{B}_i$ in order to avoid any confusion with the matrices of the previous section (this is also done for other matrices or vectors that appear in both sections).

First, define $p_x^E(q,q')$ as the probability that in a set of q requests, $q-q'$ are successful when a set of x slots is used to transmit the q request packets and this provided that at least one error occurs in these x slots. Because the errors are memoryless we have

$$p_x^E(q,q') = \sum_{k=1}^{x} C_k^x e^k (1-e)^{x-k} \sum_{v=\max(0,q'-k)}^{q'} p_x(q,v) \frac{C_{q'-v}^k C_{q-q'}^{x-k}}{C_{q-v}^x}, \tag{17}$$

where $p_x(q,q')$ was defined in Section 4.3 and e represents the probability that an arbitrary slot holds an error. Obviously, we are interested in $p_S^E(q,q')$, $p_N^E(q,q')$ and $p_{S+N}^E(q,q')$.

Next, denote P_N^E as a $q_m+1 \times q_m+1$ matrix whose $(i,j)^{th}$ element equals $p_N^E(i-1,j-1)$. $\tilde{P}_N$ is defined as a $q_m+1 \times q_m+1$ matrix whose first two columns are equal to zero and whose $(i,j)^{th}$ element, for $j>2$, equals $(1-e)^N p_N(i-1,j-1)$. The $q_m+1 \times 1$ vector $\tilde{P}_{N,0}$ has its ith entry equal to $(1-e)^N p_N(i-1,0)$. Finally, the $l \times l$ matrices $\tilde{F}_S$, $\tilde{F}_{S+N}$, $\tilde{E}_S^k$, for $0 \le k \le q_m$, and $\tilde{E}_{S+N}^k$, for $0 \le k \le q_m$, are defined as

$$\tilde{F}_S = \sum_{i \ge 0} D_i \; p_S(i,0) \; (1-e)^S \tag{18}$$

$$\tilde{F}_{S+N} = \sum_{i \ge 0} D_i \; p_{S+N}(i,0) \; (1-e)^{S+N}, \tag{19}$$

$$\tilde{E}_S^k = \sum_{i \ge k} D_i \; \left[1_{\{k>0\}} \; p_S(i,k) \; (1-e)^S + p_S^E(i,k)\right], \tag{20}$$

$$\tilde{E}_{S+N}^k = \sum_{i \ge k} D_i \; \left[1_{\{k>0\}} \; p_{S+N}(i,k) \; (1-e)^{S+N} + p_{S+N}^E(i,k)\right], \tag{21}$$

where $1_A = 1$ if A is true and 0 otherwise. Notice that $p_x(i,1) = 0$, therefore, it is sufficient to write $1_{\{k>0\}}$ instead of $1_{\{k>1\}}$. The matrices $\tilde{E}_x^k$ hold the probability that a new TS with $k \ge 0$ requests is generated in a frame where x slots are used for the new arrivals. $\tilde{F}_x$ on the other hand holds the probabilities that no new TS is generated. We are now in a position to specify the matrices

$\tilde{A}_i$ and $\tilde{B}_i$:

$$\tilde{A}_0 = (\tilde{P}_N + P_N^E) \otimes I_l, \tag{22}$$

$$\tilde{A}_i = \tilde{P}_{N,0} \otimes \left((\tilde{F}_S)^{i-1} \left[\tilde{E}_S^0 \ \tilde{E}_S^1 \ \ldots \ \tilde{E}_S^{q_m}\right]\right), \tag{23}$$

$$\tilde{B}_0 = \left[\tilde{E}_{S+N}^0 \ \tilde{E}_{S+N}^1 \ \ldots \ \tilde{E}_{S+N}^{q_m}\right], \tag{24}$$

$$\tilde{B}_1 = \tilde{F}_{S+N}, \tag{25}$$

$$\tilde{B}_i = \tilde{P}_{N,0} \otimes (\tilde{F}_S)^{i-1}, \tag{26}$$

where I_l is the $l \times l$ unity matrix. The steady state probabilities, denoted as $\tilde{\pi}_i$, are calculated in a similar manner as before. Finally, the delay distribution $P[\tilde{D} = i]$, for $i > 0$, is found as

$$P[\tilde{D} = i] = \sum_{q=0}^{q_m} \frac{(1-e)(1-1/N)^{q-1} q}{\lambda} \sum_{j=1}^{l} \tilde{\pi}_i(q, j). \tag{27}$$

$P[\tilde{D} = 0]$ is found as $1 - \sum_{i>0} P[\tilde{D} = i]$.

6 Numerical Results

In this section we explore the influence of correlation, burstiness, the number of $T = S + N$ slots and memoryless errors on the delay distribution of a request packet. A first, important, question that needs to be addressed is: What type of D-BMAP arrivals should be considered, that is, are of practical relevance ? Clearly, for any arrival rate λ and medium access protocol we can find a D-BMAP that causes delays as high as we like.

From Section 2 we know that if the traffic flow generated by an MS is very irregular, the MS is obliged to use the contention channel frequently. Therefore, depending on the characteristics of the traffic flow, we regard an MS as either being in a period where most the requests are piggybacked to the data packets transmitted in the contention free period, or in a period where the contention channel is used to transmit most of the requests. As a result, we will identify M different levels of activity, where a higher level indicates that more MSs are in a period where the contention channel is used frequently. We use M states to model these activity levels and state that the number of requests generated in a frame, by the arrival process, in state j is distributed binomially with parameters (jm, β), where m and β are parameters of the model. Hence, denoting $(D_i e)_j$, for $1 \leq j \leq M$, as the j-th component of D_i multiplied with e, an $M \times 1$ vector with all entries equal to one, results in

$$(D_i e)_j = C_i^{jm} \beta^n (1-\beta)^{jm-i}. \tag{28}$$

Transitions between these M states, occuring at the end of each frame, take place according to the following $M \times M$ transitions matrix P_M:

$$P_M = \begin{bmatrix} 1-\alpha^+ & \alpha^+ & 0 & \dots & 0 \\ \alpha^- & 1-\alpha^+-\alpha^- & \alpha^+ & \dots & 0 \\ 0 & \alpha^- & 1-\alpha^+-\alpha^- & \dots & 0 \\ \vdots & \vdots & \vdots & \ddots & \vdots \\ 0 & 0 & 0 & \dots & 1-\alpha^- \end{bmatrix}. \tag{29}$$

Therefore, $(D_i)_{j_1,j_2}$ equals $(D_ie)_{j_1}(P_M)_{j_1,j_2}$. Thus, the arrival process is characterized by the following five parameters: M, m, β, α^- and α^+. In this section, the parameters M and m are fixed at 6 and 5, whereas the parameter β is set such that de arrival rate λ is $0.2T$ requests per frame; hence, the throughput on the contention channel is 20% (provided that the Markov chain is positive recurrent). An average input rate of 20%, on a contention channel, is considered as realistic because higher values would imply that the number of contention slots T is underestimated by the network designer and the network would have great difficulties in guaranteeing any QoS, whatever protocol is used on the contention channel. Notice, with $M = 6$ and $m = 5$, the mean arrival rate related to state j is $5j/\beta$. Finally, it should be clear that this arrival process is an M-state D-BMAP.

6.1 Poisson Arrivals vs. D-BMAP Arrivals

In this section we compare the delay distribution of a request packet for Poisson and D-BMAP arrivals. For now, the bit error rate (BER) is equal to zero; hence, we use the model presented in Section 4. For the D-BMAP arrivals we fix $\alpha^+ = \alpha^- = 1/5$, therefore, the mean sojourn time in a state is 2.5 frames. The number of contention slots $T = S + N = 10$, whereas the number of S and N slots varies and is represented in the figures as (S, N). The results are presented in Figure 2.

A first, obvious, observation in Figure 2 is that the delays are larger for D-BMAP arrivals. This follows from the fact that for Poisson arrivals the mean arrival rate is always 2, whereas for the D-BMAP arrivals we have periods were the mean arrival rate is as low as $2/3.5 = 4/7$, being when the arrival process is in state 1, and periods were the mean arrival rate is as high as $24/7$, being when the arrival process is in state $M = 6$. A second observation is that the delay distribution decays exponentially[i], except for N small. To some

[i]This is not exactly true, what we mean here is that this seems to be the case if we consider the 1 to 10^{-10} region only.

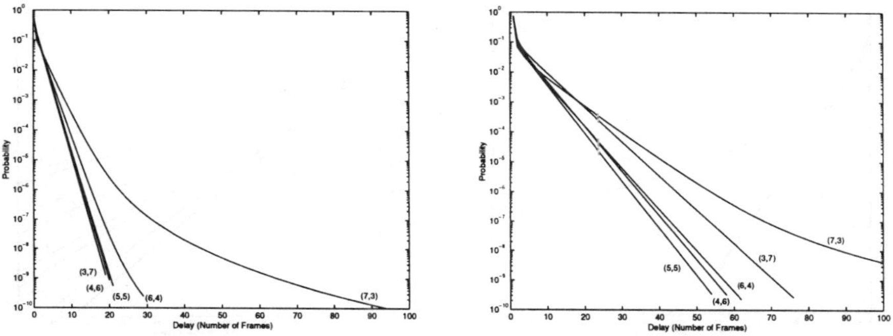

Figure 2. Delay distribution for $T = 10$, Left: Poisson arrivals ($\lambda = 2$), Right: D-BMAP arrivals ($M = 6$, $m = 5$, $\alpha^+ = \alpha^- = 1/5$ and β such that the arrival rate $\lambda = 2$).

extent, this can be explained by means of Equation 16, that is, if we forget about the q in Equation 16 and approximate $(1 - 1/N)^{q-1}$ by one, we get an exponential decay. Finally, in Vazquez *et al* [1], it was shown that, for Poisson arrivals, the best delays are obtained with $S \approx N$. Figure 2 seems to confirm the usefulness of this engineering rule, which is also based on the intuitive idea that $S \approx N$ provides the best balance between the TSs generation rate, related to S, and the TSs service times, related to N.

6.2 The Influence of the Number of Contention Slots (T)

Apart from checking whether the engineering rule concerning the number of S and N slots still applies, this section addresses the issue whether it is worth implementing parallel instances of FS-ALOHA in the contention period. With parallel instances we mean the following. Suppose that we have $T = T_1T_2$ contention slots, with $T_1 \geq 3$. Then, we could use T_2 instances of FS-ALOHA, that each use T_1 slots. New arrivals decide which instance they use based on their arrival time—that is, we partition the frame in T_2 subframes and any new arrival occurring in the i-th subframe, uses the i-th instance[j]. In this scenario we have T_2 distributed queues with TSs, instead of one. Clearly, implementing multiple instances increases the complexity of the algorithm, but perhaps the delay improvements outweigh the additional implementation effort.

Figure 3 presents the results for $T = S + N = 5$ and $T = 15$ contention

[j] Instead of using their arrival time, a request could also select the instance randomly. Given that the arrivals occur uniformly in a frame, these two scenarios are the same.

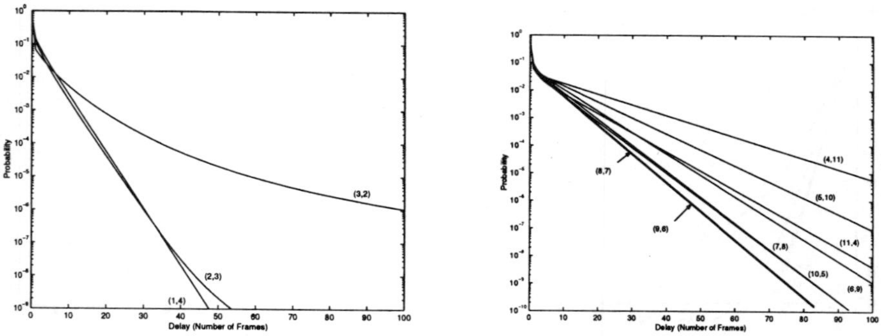

Figure 3. Delay distribution for D-BMAP arrivals ($M = 6$, $m = 5$, $\alpha^+ = \alpha^- = 1/5$), Left: $T = 5$ and β such that $\lambda = 1$, Right: $T = 15$ and β such that $\lambda = 3$.

slots[k]. The input process is the same as in the previous paragraph, except that β is chosen such that $\lambda = 0.2T$. For $T = 5$ the best results are found for N larger than S, whereas for $T = 15$ we get the best results for S slightly larger than T. In conclusion choosing $S \approx N$ seems like a useful rule of thumb. As far as the parallel instances are concerned, we can see by comparing the results for $T = 5$ and 15 that the delays can be reduced by a factor two using three instances with $T = 5$ instead of one with $T = 15$. Thus, if a network designer provisions a lot of contention slots, we suggest to implement more than one instance of FS-ALOHA.

6.3 Correlation and Burstiness

In this section we study the influence of the mean sojourn time on the delay distribution. We start with $\alpha^+ = \alpha^- = 1/2$ and decrease both gradually until 1/50, in which case the mean sojourn time in a state is 25 frames. The results are in presented in Figure 4, the other parameters are the same as in Section 6.1. From this figure we can conclude that the grouping strategy works well in limiting the delay increase due to the augmented correlation and burstiness.

---

[k]It should be noted that, provided that the arrivals occur uniformly in a frame, we can evaluate the performance of multiple instance by adapting the value of β appropriately. Indeed, it is easy to show that $\sum_{g \geq k} C_g^{mi} \beta^g (1-\beta)^{mi-g} \; T_2^{-k}(1 - T_2^{-1})^{g-k} = C_k^{mi}(\beta/T_2)^k(1-\beta/T_2)^{mi-k}$, where T_2 denotes the number of instances used.

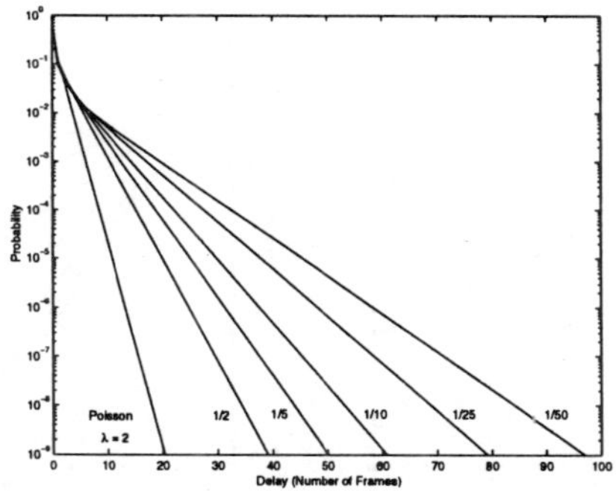

Figure 4. Delay distribution for D-BMAP arrivals ($M = 6$, $m = 5$, β such that $\lambda = 2$), $T = 10, S = N = 5$.

6.4 Errors on the Channel

In this section we investigate the influence of errors on the channel by means of the model presented in Section 5. We start by setting e, the probability that a slots holds an error, equal to $1/50, 1/100$ and $1/250$. It is hard to state whether such a value of e is an optimistic or pessimistic estimate as the probability of an error depends on the modulation scheme, the signal-to-noise ratio (SNR), the forward error control (FEC), length of a slot and much more (see Pahlavan *et al* [11]). For a wired channel it is safe to say that $e = 1/50$ is very pessimistic. We start by reproducing Figure 2 for $e = 0, 1/50, 1/100$ and $1/250$ and $S = N = 5$. Numerical experiments, omitted for brevity, show that errors have a similar impact on the delay for other choices of S and N, with $S + N = 10$ (actually, the impact of errors is slightly smaller for larger values of S).

The results are presented in Figure 5, where the curves for $e = 0$ where obtained with the model in Section 4. A first, obvious, observation is that the delay increases with increasing e. Moreover, the results show that the increase for Poisson arrivals is less compared to D-BMAP arrivals. Thus, models that study the impact of errors using Poisson arrivals are, from a practical point of view, somewhat optimistic. Therefore, we use D-BMAP arrivals for our remaining experiments. Finally, although the impact on the delay distribution is small for $e = 1/100$ or smaller, errors can seriously increase the delay for higher error rates (for $e = 1/20$ the delays are more than three times as high compared to $e = 0$). Therefore, if the modulation scheme, error codes, signal-to-noise ratio, ... cannot guarantee an error rate e less than $1/5T$, the

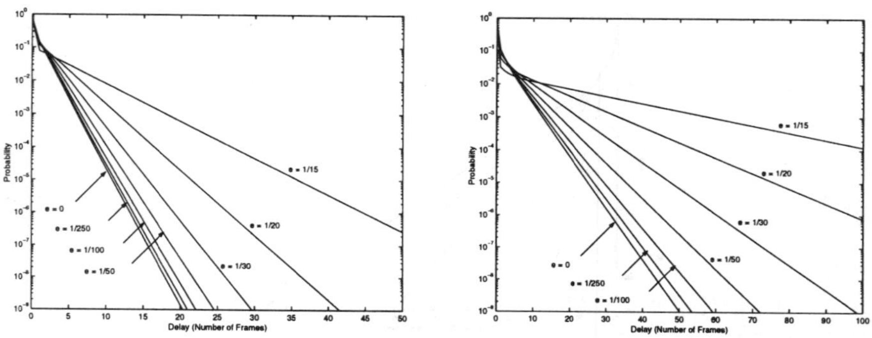

Figure 5. Delay distribution for $T = 10, S = N = 5$ and $e = 0, 1/250, 1/100, 1/50$, Left: Poisson arrivals ($\lambda = 2$), Right: D-BMAP arrivals ($M = 6$, $m = 5$, $\alpha^+ = \alpha^- = 1/5$ and β such that the arrival rate $\lambda = 2$).

performance of FS-ALOHA might degrade drastically.

This rule is confirmed by Figure 6, where we study FS-ALOHA for $T =$ 5 and 15. For $T = 15$ the Markov chain becomes transient for $e \geq 1/20$ (actually, the chain becomes unstable for e somewhere in between 1/20 and 1/21). For Poisson arrivals and $T = 15$ we get instability for $e \geq 1/19$, thus the instability is only slightly influenced by the arrival process and is mainly determined by the error rate.

These observations further indicate that the use of multiple instances of FS-ALOHA, each with a small value of T, is not only better in terms of the suffered delay, but also improves the sensitivity of the algorithm to errors. As a result, we strongly support the use of multiple instances for wireless networks, i.e., networks with high error rates.

7 Conclusions

In this paper we have evaluated the robustness of FS-ALOHA, a random access algorithm, by means of an GI/M/1 Type Markov chain. The robustness was investigated by relaxing prior assumptions made on the arrival process in Vazquez *et al* [1], that is, discrete time batch Markovian arrivals were considered as opposed to Poisson arrivals. Moreover, memoryless errors were also added to the channel. Using the analytical model, it is concluded that FS-ALOHA, in general, performs well under correlated and bursty arrivals and memoryless errors. However, error rates above $1/5T$, were T is a protocol parameter, can seriously increase the delays suffered on the contention chan-

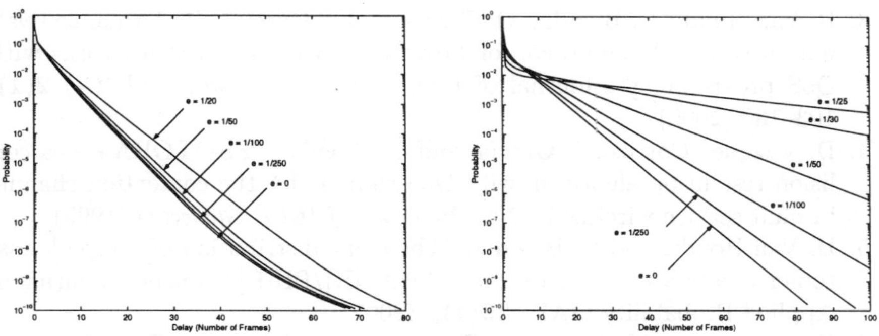

Figure 6. Delay distribution for D-BMAP arrivals ($M = 6$, $m = 5$, $\alpha^+ = \alpha^- = 1/5$ and β such that the arrival rate $\lambda = 0.2T$), $e = 0$ to $1/20$, Left: $T = 5, S = 2, N = 3$, Right: $T = 15, S = 8, N = 7$.

nel and might even make the system unstable at moderate arrival rates. It should be mentioned that FS-ALOHA++ introduced in Vazquez *et al* [4] might, to some extent, improve the stability of FS-ALOHA on a channel with errors, because FS-ALOHA++ services multiple TSs simultaneously, thereby reducing the penalty introduced by empty transmission sets. This and many other properties of FS-ALOHA++ are reported in Van Houdt *et al* [12], where we also use matrix analytic methods to obtain the performance measures of interest. Finally, it is concluded that implementing multiple instances of FS-ALOHA can significantly improve the delays and the robustness of the algorithm and is therefore advisible for wireless channels with high error rates.

Acknowledgements

B. Van Houdt is a postdoctoral Fellow of the FWO Flanders. We would like to thank the reviewers for their valuable comments and suggestions.

References

1. D. Vázquez Cortizo, J. García, C. Blondia, and B. Van Houdt, "FIFO by sets ALOHA (FS-ALOHA): a collision resolution algorithm for the contention channel in wireless ATM systems", Performance Evaluation **36-37**, 401-427 (1999)
2. D. Bertsekas, and R. Gallager, *Data Networks*, Prentice-Hall Int., Inc., 1992.

3. B. Van Houdt, C. Blondia, O. Casals, and J. García, "Performance Evaluation of a MAC protocol for broadband wireless ATM networks with QoS provisioning", Journal of Interconnection Networks (JOIN) **2(1)**, 103–130 (2000)
4. D. Vázquez Cortizo, J. García, and C. Blondia, "FS-ALOHA++, a collision resolution algorithm with QoS support for the contention channel in multiservice wireless LANs", In *Proc. of IEEE Globecom* (1999)
5. B. Van Houdt, and C. Blondia, "The delay distribution of a type k customer in a first come first served MMAP[K]/PH[K]/1 queue", Journal of Applied Probability (JAP) **39(1)**, (2002)
6. R. von Mises, *Mathematical Theory of Probability and Statistics*, Academic Press Inc., New York (1964)
7. M.F. Neuts, "Markov Chains with Applications in Queueing Theory, which have a matrix geometric invariant probability vector", Adv. Appl. Prob. (AAP) **10**, 185–212 (1978)
8. M.F. Neuts, *Matrix-Geometric Solutions in Stochastic Models, An Algorithmic Approach*, John Hopkins University Press (1981)
9. Q. He, "Classification of Markov Processes of Matrix M/G/1 type with a Tree Structure and its Applications to the MMAP[K]/G[K]/1 Queue", Stochastic Models **16(5)**, 407-434 (2000)
10. Q. He, "Classification of Markov Processes of M/G/1 type with a Tree Structure and its Applications to Queueing Models", O.R. Letters **26**, 67-80 (1999)
11. K. Pahlavan, and A.H. Levesque, *Wireless Information Networks*, John Wiley and Sons, Inc., New York"(1995)
12. B. Van Houdt, and C. Blondia, "Robustness properties of FS-ALOHA(++): a random access algorithm for dynamic bandwidth allocation", Journal on Special Topics in Mobile Networking and Applications (MONET) on Performance Evaluation of QoS Architectures in Mobile Networks (submitted) , (2002)

ACCURATE ESTIMATE OF SPECTRAL RADII OF RATE MATRICES OF GI/M/1 TYPE MARKOV CHAINS

QIANG YE

Department of Mathematics, University of Kentucky, Lexington, KY 40506, USA
E-mail: qye@ms.uky.edu

The spectral radius of the rate matrix of a GI/M/1 type Markov chain characterizes long term high level behavior of the system. In this paper, an analytic expression of the spectral radius is established for its estimation. An effective numerical procedure is proposed for its computation. Numerical examples are presented to illustrate the effectiveness of the estimation.

1 Introduction

We consider the GI/M/1 type Markov chain with an irreducible transition matrix

$$P = \begin{pmatrix} B_0 & A_0 & & & \\ B_1 & A_1 & A_0 & & \\ B_2 & A_2 & A_1 & A_0 & \\ \vdots & \vdots & \vdots & \vdots & \ddots \end{pmatrix} \tag{1}$$

where A_i, B_i, $(i = 0, 1, \cdots)$ are $m \times m$ nonnegative matrices with $m < \infty$ and $A \equiv \sum_{k=0}^{\infty} A_k$ is stochastic and irreducible. Let R be its rate matrix. Then R is the minimal nonnegative solution to the matrix equation (see Neuts [15])

$$R = \sum_{k=0}^{\infty} R^k A_k. \tag{2}$$

The rate matrix R and its spectral radius $\eta \equiv \rho(R)$ play an important role in characterizing and studying the Markov chain P [1,2,3,7,16].

It is well-known [15] that the Markov chain P is positive-recurrent if and only if

$$\eta = \rho(R) < 1 \tag{3}$$

and $\sum_{k=0}^{\infty} R^k B_k$ has a positive stationary probability vector. Another necessary and sufficient condition is obtained by replacing (3) with

$$\pi^T \beta > 1, \quad \text{with} \quad \beta \equiv \sum_{k=1}^{\infty} k A_k \mathbf{e} \tag{4}$$

where $\mathbf{e}$ is the column vector of all ones and π is the stationary probability vector of A (i.e. $\pi^T A = \pi^T$, $\pi^T \mathbf{e} = 1$). In this case, the stationary probability distribution of P is matrix-geometric, i.e., of the form $(x_0, x_0 R, x_0 R^2, \cdots)$. Then, $\eta = \lim_{k\to\infty} x_0^T R^{k+1}\mathbf{e}/x_0^T R^k \mathbf{e}$ is the ratio of the expected time spent at two consecutive high levels. Therefore, in addition to providing a test for positive recurrence of P, the value of η also characterizes how long it will take the queue to move back to lower levels when it is at a high level [13,16]. In particular, η is called caudal characteristic by Neuts [16] in studying its dependence on traffic intensity.

In practice, the caudal characteristic η is not easy to compute or even to estimate, requiring first solving for R and then computing its largest eigenvalue. On the other hand, given that (3) and (4) are equivalent and that $\pi^T\beta - 1$ is an indication of the overall trend of the queue towards lower level [13], it appears that there should exist some relationship between the values of η and $\pi^T\beta$. As $\pi^T\beta$ is much easier to compute, it is desirable to determine such a relation which can also be used to estimate η from $\pi^T\beta$. We also note that, in light of (3) and (4), each of $1-\eta$ and $\pi^T\beta - 1$ provides in a way a measure of nearness of the queue to being null-recurrent.

In this paper, we shall establish the following expression of η

$$\eta = 1 - 2\frac{\pi^T\beta - 1}{\pi^T A''(1)\mathbf{e} + 2\gamma^T(I - A - \mathbf{e}\pi^T)^{-1}\beta + 2(\pi^T\beta)^2} + \mathcal{O}((1-\eta)^2)$$

where $\gamma^T = \pi^T \sum_{k=1}^{\infty} kA_k$. This provides a connection between η and $\pi^T\beta$ as well as an easily computable estimate of η. With an error of order $(1-\eta)^2$, the estimate is particularly accurate when η is near 1. Furthermore, such a direct analytic expression of η will also have theoretical advantages over numerical computation of η when one studies the caudal characteristic with respect to certain parameters.

The paper is organized as follows. We shall present the detailed derivation of the above formula in Section 2 and discuss some numerical issues in Section 3. We present some examples to demonstrate the effectiveness of the estimate in Section 4.

2 Estimate of Spectral Radius

Throughout this paper we only consider the discrete case and we assume the Markov chain P is positive recurrent.

Let

$$A(t) = \sum_{k=0}^{\infty} t^k A_k \tag{1}$$

where $0 \le t \le 1$ and let $\chi(t)$ be the spectral radius of $A(t)$, i.e.

$$\chi(t) = \rho(A(t)). \tag{2}$$

It is shown in Neuts [15] that η is the unique root in $0 < t < 1$ to the equation

$$\chi(t) = t. \tag{3}$$

We shall use some properties of $\chi(t)$ to estimate η. Because $A(t)$ is irreducible, $\chi(t)$ is a simple eigenvalue of $A(t)$ for $0 < t \le 1$ and hence it is analytic on $0 < t < 1$ and continuous on $0 \le t \le 1$ (see [10] for example). It is established in [15] that $\chi'(1-) = \pi \sum_{k=1}^{\infty} kA_k e$. Here, we shall use $\chi''(1-)$, which is explicitly determined by Theorem 1 below.

Lemma 1 *Let π^T be the stationary probability vector of A. Then $I - A + \mathbf{e}\pi^T$ (and $I - A - \mathbf{e}\pi^T$ resp.) are invertible.*

Proof Since A is irreducible, 0 is a simple eigenvalue of $I - A$ with $\mathbf{e}$ as a right eigenvector and π as a left eigenvector. Let $I - A = PJP^{-1}$ be the Jordan canonical form of $I - A$ with $J = \text{diag}\{0, J_1\}$ and $P = [\mathbf{e}, P_1]$. Since the first row of P^{-1} is a left eigenvector and, by the normalization $P^{-1}P = I$, it must be π. Writing $P^{-1} = [\pi, P_2]^T$, we have $P\text{diag}\{1, 0, \cdots, 0\}P^{-1} = \mathbf{e}\pi^T$. Thus,

$$I - A \pm \mathbf{e}\pi^T = P\hat{J}P^{-1} \quad \text{with } \hat{J} = \text{diag}\{\pm 1, J_1\}.$$

So, the eigenvalues of $I - A \pm \mathbf{e}\pi^T$ are ± 1 and those of J_1 (i.e. the nonzero eigenvalues of $I - A$), which proves the lemma. □

The matrix $(I - A + \mathbf{e}\pi^T)^{-1}$ is called the fundamental matrix in the Markov chain theory (see [11] for example). Its computations are discussed in [8,9,14].

Theorem 1 *Let π^T be the stationary probability vector of A and*

$$\beta = A'(1)\mathbf{e} = \sum_{k=1}^{\infty} kA_k\mathbf{e}, \quad \gamma^T = \pi^T A'(1) = \pi^T \sum_{k=1}^{\infty} kA_k.$$

Then, we have

$$\chi''(1-) = \pi^T A''(1)\mathbf{e} + 2\gamma^T(I - A \pm \mathbf{e}\pi^T)^{-1}\beta \mp 2(\pi^T\beta)^2. \tag{4}$$

Proof For $0 < t \le 1$, let $v(t)$ be the right eigenvector and $u(t)$ be the left eigenvector of $A(t)$ corresponding to $\chi(t)$ with $u^T(t)v(t) = 1$, i.e.,

$$(A(t) - \chi(t)I)v(t) = 0, \quad u^T(t)(A(t) - \chi(t)I) = 0. \tag{5}$$

As $A(t)$ is irreducible and hence $\chi(t)$ is a simple eigenvalue for $0 < t < 1$, $\chi(t)$, $u(t)$ and $v(t)$ are analytic with respect to t (see [10] for example). Differentiating (5), we obtain

$$(A(t) - \chi(t)I)v'(t) = -(A'(t) - \chi'(t)I)v(t) \tag{6}$$

$$(u^T)'(t)(A(t) - \chi(t)I) = -u^T(t)(A'(t) - \chi'(t)I) \tag{7}$$

This leads to $\chi'(1-) = u(1)^T A'(1)v(1) = \gamma^T \mathbf{e} = \pi^T \beta$. Differentiating (6) again, we obtain

$$(A(t) - \chi(t)I)v''(t) + 2(A'(t) - \chi'(t)I)v'(t) + (A''(t) - \chi''(t)I)v(t) = 0. \tag{8}$$

To determine $\chi''(t)$, we multiply (8) from left by $u^T(t)$ and obtain

$$\begin{aligned} &u^T(t)(A''(t) - \chi''(t)I)v(t) \\ &= -u^T(t)(A(t) - \chi(t)I)v''(t) - 2u^T(t)(A'(t) - \chi'(t)I)v'(t) \\ &= 2(u^T)'(t)(A(t) - \chi(t)I)v'(t) \end{aligned}$$

where we have used (7). Then

$$\chi''(t) = u^T(t)A''(t)v(t) - 2(u^T)'(t)(A(t) - \chi(t)I)v'(t). \tag{9}$$

Now considering (6) at $t \to 1-$, we have

$$(A - I)v'(1-) = -(A'(1) - \chi'(1-)I)\mathbf{e} = -\beta + \chi'(1-)\mathbf{e}.$$

Then, it is easy to verify that

$$(I - A \pm \mathbf{e}\pi^T)v'(1-) = \beta + \alpha\mathbf{e},$$

where $\alpha = -\chi'(1-) \pm \pi^T v'(1-)$. By Lemma 1, $I - A \pm \mathbf{e}\pi^T$ is invertible and therefore

$$\begin{aligned} v'(1-) &= (I - A \pm \mathbf{e}\pi^T)^{-1}\beta + \alpha(I - A \pm \mathbf{e}\pi^T)^{-1}\mathbf{e} \\ &= (I - A \pm \mathbf{e}\pi^T)^{-1}\beta \pm \alpha\mathbf{e}, \end{aligned}$$

where we have used $(I - A \pm \mathbf{e}\pi^T)\mathbf{e} = \pm\mathbf{e}$. Similarly, using (7) at $t \to 1-$, we have

$$(u^T)'(1-)(I - A \pm \mathbf{e}\pi^T) = \gamma^T + \hat{\alpha}\pi^T,$$

where $\hat{\alpha} = -\chi'(1-) \pm (u^T)'(1-)\mathbf{e}$, and hence

$$(u^T)'(1-) = \gamma^T(I - A \pm \mathbf{e}\pi^T)^{-1} \pm \hat{\alpha}\pi^T.$$

Thus,

$$\begin{aligned}&(u^T)'(1-)(A-I)v'(1-)\\&=\gamma^T(I-A\pm\mathbf{e}\pi^T)^{-1}(A-I)(I-A\pm\mathbf{e}\pi^T)^{-1}\beta\\&=-\gamma^T(I-A\pm\mathbf{e}\pi^T)^{-1}\beta\pm\gamma^T(I-A\pm\mathbf{e}\pi^T)^{-1}\mathbf{e}\pi^T(I-A\pm\mathbf{e}\pi^T)^{-1}\beta\\&=-\gamma^T(I-A\pm\mathbf{e}\pi^T)^{-1}\beta\pm\gamma^T\mathbf{e}\pi^T\beta\\&=-\gamma^T(I-A\pm\mathbf{e}\pi^T)^{-1}\beta\pm(\pi^T\beta)^2\end{aligned}$$

where we have used again $(I-A\pm\mathbf{e}\pi^T)^{-1}\mathbf{e}=\pm\mathbf{e}$ and $\pi^T(I-A\pm\mathbf{e}\pi^T)^{-1}=\pm\pi^T$, as well as $\gamma^T\mathbf{e}=\pi^T\beta$. Finally, considering (9) at $t=1-$, we have

$$\begin{aligned}\chi''(1-)&=\pi^TA''(1)\mathbf{e}-2(u^T)'(1-)(A-I)v'(1-)\\&=\pi^TA''(1)\mathbf{e}+2\gamma^T(I-A\pm\mathbf{e}\pi^T)^{-1}\beta\mp2(\pi^T\beta)^2.\end{aligned}$$

□

From this theorem, we can derive the following expression of η.

Theorem 2 *Assume that $\chi^{(3)}(t)$ is bounded on $0<t<1$. The caudal characteristic η satisfies*

$$\eta-1=2\frac{1-\pi^T\beta}{\pi^TA''(1)\mathbf{e}+2\gamma^T(I-A\pm\mathbf{e}\pi^T)^{-1}\beta\mp2(\pi^T\beta)^2}+\mathcal{O}((1-\eta)^2).\quad(10)$$

Proof Consider the following Taylor expansion of $\chi(t)$ at a point t_0 with $\eta<t_0<1$

$$\chi(\eta)=\chi(t_0)+\chi'(t_0)(\eta-t_0)+\frac{1}{2}\chi''(t_0)(\eta-t_0)^2+\frac{1}{6}\chi^{(3)}(\psi)(\eta-t_0)^3$$

where $\eta\le\psi\le t_0$. Letting $t_0\to1-$ and using the assumption on $\chi^{(3)}(t)$, we obtain

$$\chi(\eta)=\chi(1)+\chi'(1-)(\eta-1)+\frac{1}{2}\chi''(1-)(\eta-1)^2+\mathcal{O}((\eta-1)^3).$$

Since $\eta=\chi(\eta)$, we have

$$1=\chi'(1-)+\frac{1}{2}\chi''(1-)(\eta-1)+\mathcal{O}((\eta-1)^2)$$

or

$$\eta-1=2\frac{1-\chi'(1-)}{\chi''(1-)}+\mathcal{O}((\eta-1)^2).$$

The proof is now completed by using Theorem 1 and $\chi'(1-)=\pi^T\beta$. □

Equation (10) provides a direct formulation of η in terms of π, β and γ, up to a term of order $(1-\eta)^2$. It also connects $\eta - 1$ to $1 - \pi^T\beta$, both of which describe in some sense how close the queue is to being null-recurrent. While the denominator of (10) contains an inverse, we shall show in the next section that it does not incur much extra computation than what is already needed to obtain π. Thus a numerical evaluation of (10) is no more difficult than $\pi^T\beta$ in any significant way.

We will use (10) to estimate η as

$$\eta \approx \zeta = 1 - 2\frac{\pi^T\beta - 1}{\pi^T A''(1)\mathbf{e} + 2\gamma^T(I - A \pm \mathbf{e}\pi^T)^{-1}\beta \mp 2(\pi^T\beta)^2}. \tag{11}$$

Clearly, $\zeta < 1$ also provides a condition for positive recurrence.

For a QBD model, $1 - \pi^T\beta = 1 - \pi^T A_1\mathbf{e} - 2\pi^T A_2\mathbf{e} = \pi^T A_0\mathbf{e} - \pi^T A_2\mathbf{e}$ and the above estimation simplifies to

$$\begin{aligned}\zeta &= 1 + 2\frac{\pi^T A_0\mathbf{e} - \pi^T A_2\mathbf{e}}{2\pi^T A_2\mathbf{e} + 2\gamma^T(I - A \pm \mathbf{e}\pi^T)^{-1}\beta \mp 2(\pi^T\beta)^2} \\ &= 1 + \frac{\pi^T A_0\mathbf{e}/\pi^T A_2\mathbf{e} - 1}{1 + \gamma^T(I - A \pm \mathbf{e}\pi^T)^{-1}\beta/\pi^T A_2\mathbf{e} \mp (\pi^T\beta)^2/\pi^T A_2\mathbf{e}}\end{aligned}$$

which also provides a connection to $\pi^T A_0\mathbf{e}/\pi^T A_2\mathbf{e}$.

3 Numerical Evaluation of Estimation Formula

In this section, we discuss issues related to efficient numerical evaluation of estimation (10). Clearly, one needs to compute the stationary probability vector π^T of A, which can be accomplished by applying the GTH-algorithm [6] to $I - A$. One also encounters the fundamental matrix $(I - A + \mathbf{e}\pi^T)^{-1}$, or rather solving a linear system with $I - A \pm \mathbf{e}\pi^T$ as the coefficient matrix. Heyman and O'Leary [8,9] have presented a stable method for this based on the factorization of $I - A$ obtained from the GTH-algorithm. Here, we present a similar but more direct approach also using the LU factorization of of $I - A$ as obtained by the GTH-algorithm. The following theorem shows how this can be done.

Theorem 3 *Let $LU = I - A$ be the LU-factorization of $I - A$ as obtained by the GTH-algorithm with the diagonals of U equal to* 1 *and let π^T be the stationary probability vector of A, i.e. $\pi^T L = 0$. Let $\hat{\pi}^T = \pi^T U^{-1}$, $\hat{\mathbf{e}} =$*

$[1,\cdots,1,0]^T$ *and*

$$\hat{L}_{\pm} = L \pm \begin{pmatrix} 0 \\ \vdots \\ 0 \\ \hat{\pi}^T \end{pmatrix}.$$

Then $\hat{L}_{\pm}$ *is an invertible lower triangular matrix with the* (m,m) *entry equal to* ± 1 *and*

$$\gamma^T(I - A \pm \mathbf{e}\pi^T)^{-1}\beta = x_1^T y_1 \mp \frac{(x_1^T y_2)(x_2^T y_1)}{1 \pm x_2^T y_2}. \tag{1}$$

where $x_1^T = \gamma^T U^{-1}$, $x_2^T = \hat{\pi}^T = \pi^T U^{-1}$, $y_1 = \hat{L}_{\pm}^{-1}\beta$ *and* $y_2 = \hat{L}_{\pm}^{-1}\hat{\mathbf{e}}$.

Proof We prove for $I - A - \mathbf{e}\pi^T$ only; the case for $I - A + \mathbf{e}\pi^T$ follows similarly. First, the GTH-algorithm produces a LU-factorization with

$$L = \begin{pmatrix} L_{11} & 0 \\ l_{21} & 0 \end{pmatrix}$$

where L_{11} is an $(m-1)\times(m-1)$ invertible matrix. Using $L(U\mathbf{e}) = (I-A)\mathbf{e} = 0$, we obtain $U\mathbf{e} = [0,\cdots,0,z]^T$ for some z. Since U is upper triangular with diagonals being 1, $z = 1$, i.e. $U\mathbf{e} = e_m$.

Now, we have

$$\begin{aligned} I - A - \mathbf{e}\pi^T &= (L - \mathbf{e}\pi^T U^{-1})U = (L - \mathbf{e}\hat{\pi}^T)U \\ &= (\hat{L}_- - \hat{\mathbf{e}}\hat{\pi}^T)U \end{aligned}$$

where $\hat{L}_- = L - [0,\cdots,0,\hat{\pi}]^T$ is still a lower triangular matrix as L is lower triangular. As the (m,m) entry of L is 0, the (m,m) entry of $\hat{L}_-$ is

$$-\hat{\pi}^T e_m = -\pi^T U^{-1} e_m = -\pi^T \mathbf{e} = -1$$

where we have used $U^{-1}e_m = \mathbf{e}$ from above. Therefore, $\hat{L}_-$ is invertible. Thus

$$\begin{aligned} \gamma^T(I - A - \mathbf{e}\pi^T)^{-1}\beta &= \gamma^T U^{-1}(\hat{L}_- - \hat{\mathbf{e}}\hat{\pi}^T)^{-1}\beta \\ &= \gamma^T U^{-1}\hat{L}_-^{-1}\beta + \frac{(\gamma^T U^{-1}\hat{L}_-^{-1}\hat{\mathbf{e}})(\hat{\pi}^T\hat{L}_-^{-1}\beta)}{1 - \hat{\pi}^T\hat{L}_-^{-1}\hat{\mathbf{e}}} \end{aligned}$$

where we have used the Sherman-Morrison formula [5]

$$c^T(\hat{L}_- - uv^T)^{-1}b = c^T\hat{L}_-^{-1}b + \frac{c^T\hat{L}_-^{-1}uv^T\hat{L}_-^{-1}b}{1 - v^T\hat{L}_-^{-1}u}.$$

This completes the proof. □

Once the GTH-algorithm has been invoked to compute π, computations of (1) involves solving a few lower or upper triangular systems, the cost of which is by an order smaller compared to the factorization part, i.e. the arithmetic operations used is of order $\mathcal{O}(m^2)$ versus $\mathcal{O}(m^3)$ of the GTH-algorithm.

Another implication of (1) is that the formulation based on $I - A - \mathbf{e}\pi^T$ might be numerically more stable than the one based on $I - A + \mathbf{e}\pi^T$. This is because with the $-$ sign, $\hat{L}_-$ has negative off-diagonal entries and positive diagonal entries except the last one. Thus, when computing $\hat{L}_-^{-1}b$ with $b \geq 0$ using the forward substitution, only addition operations are involved and therefore $\hat{L}_-^{-1}b$ are computed to higher accuracy. Of course subtractions will still be involved in computing $c^T\hat{L}_-^{-1}b$ later.

We summarize the process for computing π and (1) to be used in (10) in the following algorithm. We note that the algorithm includes the computation of the stationary probability distribution π, which is also needed in forming γ.

Algorithm 1:
Input: a stochastic matrix A, two positive vectors γ and β
Output: $\delta = \gamma^T(I - A - \mathbf{e}\pi^T)^{-1}\beta$ (and π as well)
Step 1: LU-factorization by GTH-algorithm
  For $k = 1, 2, \ldots, m-1$ do
    $A(k,k) = \sum_{j=k+1}^{n} A(k,j)$
    $A(k, k+1:m) = A(k, k+1:m)/A(k,k)$
    $A(k+1:m, k+1:m) = A(k+1:m, k+1:m)+$
      $+A(k+1:m, k)A(k, k+1:m)$
  EndDo
Step 2: Solve stationary vector $\pi^T L = 0$
  $\pi(m) = 1$
  For $k = m-1, m-2, \ldots, 1$ do
    $\pi(k) = (\sum_{j=k+1}^{n} A(j,k)\pi(j))/A(k,k)$
  EndDo
  $\pi = \pi / \sum_{j=1}^{n} \pi(j)$
Step 3: Solve $x_1^T = \gamma^T U^{-1}$, $x_2 = \pi^T U^{-1}$
  $x_1(1) = \gamma(1)$, $x_2(1) = \pi(1)$
  For $k = 2, 3, \ldots, m$ do
    $x_1(k) = \gamma(k) + \sum_{j=1}^{k-1} A(j,k)x_1(j)$
    $x_2(k) = \pi(k) + \sum_{j=1}^{k-1} A(j,k)x_2(j)$
  EndDo
Step 4: Solve $y_1 = \hat{L}^{-1}\beta$, $y_2 = \hat{L}^{-1}\hat{\mathbf{e}}$
  $y_1(1) = \beta(1)/A(1,1)$, $y_2(1) = 1/A(1,1)$

For $k = 2, 3, \ldots, m-1$ do
$y_1(k) = (\beta(k) + \sum_{j=1}^{k-1} A(k,j) y_1(j))/A(k,k)$
$y_2(k) = (1 + \sum_{j=1}^{k-1} A(k,j) y_2(j))/A(k,k)$
EndDo
$y_1(m) = -\beta(m) - \sum_{j=1}^{m-1}(A(k,j) + x_2(j)) y_1(j)$
$y_2(m) = -\sum_{j=1}^{m-1}(A(k,j) + x_2(j)) y_2(j)$
Step 5: Compute $\delta = \gamma^T(I - A - \mathbf{e}\pi^T)^{-1}\beta$
$\delta = x_1^T y_1 + (x_1^T y_2)(x_2^T y_1)/(1 - x_2^T y_2)$

4 Numerical Examples

In this section, we present several numerical examples of QBD models to verify our results and to show effectiveness of the spectral estimation. We shall compare the estimation ζ (11) with the value of η computed directly through explicitly computing R first by the Latouche-Ramaswami's logarithmic reduction algorithm [12] (using the refined implementation in [17]) and then the spectral radius by `eig` of MATLAB. All testing is carried out in MATLAB with the machine precision $\epsilon \approx 10^{-16}$.

Example 1: This example is a QBD process with two phases, which is used by Latouche and Ramaswami [12]. From the state $(i, 1)$, $i \geq 1$, the chain moves to $(i, 2)$ with probability p and to $(i-1, 1)$ with probability $1-p$; from the state $(i, 2)$, $i \geq 0$, the chain moves to $(i, 1)$ with probability $2p$ and to $(i+1, 2)$ with probability $1-2p$. The QBD is always positive recurrent with the traffic coefficient $\rho = \pi^T A_0 \mathbf{e}/\pi^T A_2 \mathbf{e} = (1-2p)/(2-2p)$ and $\eta = (1-2p)/(1-p)$. For this problem, we can explicitly compute the spectral radius of $A(t)$ to obtain

$$\chi(t) = \frac{1}{2}(1-p)t^2 + \frac{1}{2}(1-2p) + \frac{1}{2}\sqrt{\Delta}$$

where $\Delta = (1-p)^2 t^4 + 2(2p^2 + 3p - 1)t^2 + (1-2p)^2$. From this, we have

$$\chi''(1-) = 1 - p + \frac{5p^2 - 3p + 2}{3p} - \frac{2(3p+1)^2}{27p}.$$

We have used this to verify Theorem 1. Furthermore, we compare the spectral radius η with the estimate (11) for p in the range $[10^{-5}, 1/2]$ and we present the results by plotting them against p in Figure 1 (left). The solid line is for η and +-mark is for the estimate ζ. We have also plotted the traffic coefficient ρ in the dot line. The caudal characteristic curves as given by η and the estimate ζ are indistinguishable. To better see the estimate error, we plot in

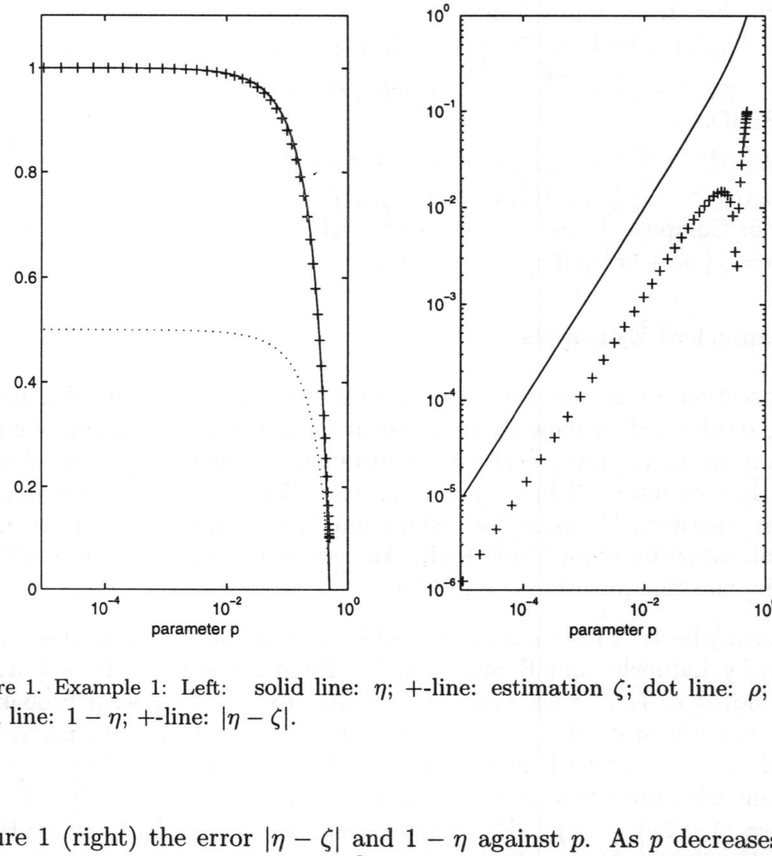

Figure 1. Example 1: Left: solid line: η; +-line: estimation ζ; dot line: ρ; Right: solid line: $1-\eta$; +-line: $|\eta-\zeta|$.

Figure 1 (right) the error $|\eta-\zeta|$ and $1-\eta$ against p. As p decreases to 0, we see initially $|\eta-\zeta| \sim \mathcal{O}((1-\eta)^2)$ but quickly when p is around 10^{-1}, this property is lost. This is due to the violation of the condition of Theorem 2 at $p=0$, i.e. $\chi(1)$ is an eigenvalue of multiplicity two and $\chi^{(3)}(t)$ does not remain bounded at $t=1-$ in this case.

Example 2: This is an example used by Latouche and Ramaswami [13] to study the caudal characteristic. It is an M/M/1 queue in a random environment with eight phases and the generator is

$$S = \begin{pmatrix} -1 & 1 & & & \\ & -1 & 1 & & \\ & & \cdot & \cdot & \\ & & & -1 & 1 \\ 1 & & & & -1 \end{pmatrix}.$$

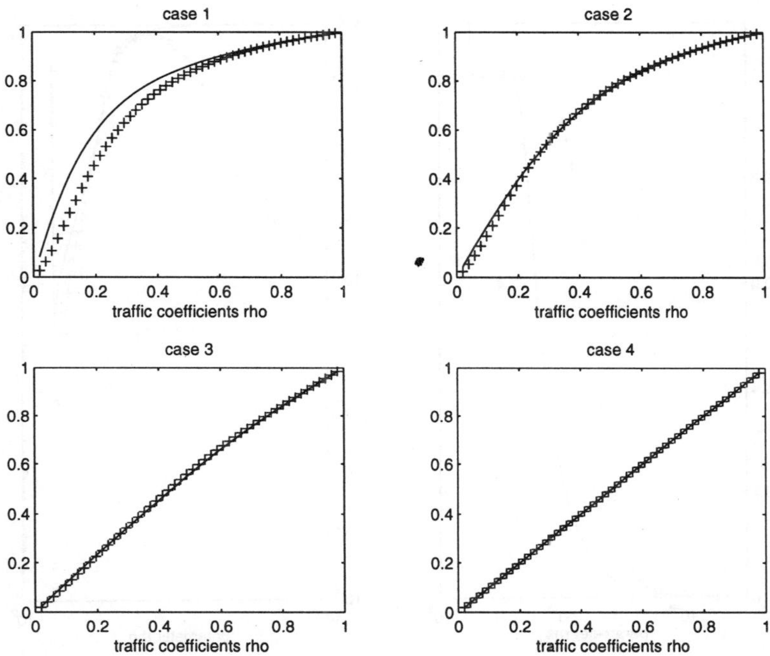

Figure 2. Example 2: solid line: η; +-line: estimation ζ.

As a continuous QBD model, it has $A_0' = diag(\mu)$, $A_2' = diag(\lambda)$ and $A_1' = S - A_0 - A_2$, where $\mu = [\mu_1, \cdots, \mu_8]$ is the vector of service rate and $\lambda = [\lambda_1, \cdots, \lambda_8]$ is the vector of arrival rate. Correspondingly, $A_0 = -{A_1'}^{-1}A_0'$, $A_1 = 0$ and $A_2 = -{A_1'}^{-1}A_2'$. Following [13] , we use in our test $\lambda = \rho[0.2; 0.2; 0.2; 0.2; 13; 1; 1; 0.2]$ and one of the following four cases for μ

1. $\mu = 2[1; 1; 1; 1; 1; 1; 1; 1]$;
2. $\mu = [1; 1; 1; 1; 5; 5; 1; 1]$;
3. $\mu = [0.4; 0.4; 0.4; 0.4; 10; 2; 2; 0.4]$;
4. $\mu = [0.2; 0.2; 0.2; 0.2; 13; 1; 1; 0.2]$.

For each of the four cases, we plot η in the solid lines and the estimate ζ in the +-mark against ρ in Figure 2. In all cases, the two curves match well at the end near 1 as expected. For cases 2-4, the two curves are indeed indistinguishable.

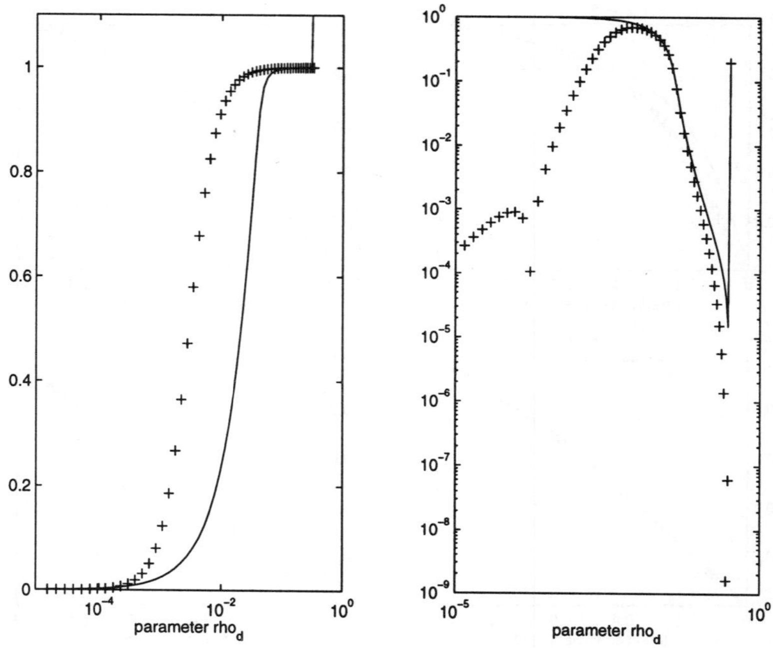

Figure 3. Example 3: Left: solid line: η; +-line: estimation ζ;; Right: solid line: $1-\eta$; + line: $|\eta-\zeta|$.

Example 3: This example is a continuous QBD model of a teletraffic system taken from Daigle and Lucantoni [4]. It is defined by 24×24 blocks A_0', A_1' and A_2' in which A_0', A_2' are diagonal such that $A_0' = 192\rho_d I$, $(A_2')_{jj} = 192(1 - j/24)$ for $0 \leq j \leq 23$, and A_1' is tridiagonal such that $(A_1')_{j,j+1} = ar(M-j)/M$ (for $0 \leq j \leq 22$) and $(A_1')_{j,j-1} = jr$ (for $1 \leq j \leq 23$). The corresponding discrete model equation is given by $A_0 = -{A_1'}^{-1}A_2'$, $A_1 = 0$ and $A_2 = -{A_1'}^{-1}A_0'$ (in our notation). We use in our test $r = 1/300$, $a = 18.244$ and $M = 256$ and consider ρ_d as a parameter ranging from 10^{-5} to 0.326 (the point at which the system lost the positive recurrent property as seen in Figure 3). Again, we plot η in the solid line and the estimate ζ in the +-mark against ρ_d in Figure 1 (left) and also plot in Figure 1 (right) the error $|\eta - \zeta|$ in the +-mark and $1 - \eta$ in solid line. We see that the error is small at the two ends but larger at the middle range.

Acknowledgments

I would like to thank Prof. Guy Latouche for bringing the references [8,9] to my attention. My research is supported in part by the National Science Foundation under Grant CCR-0098133.

References

1. J. Abate, L. Choudhury and W. Whitt, *Asymptotics for steady-state tail probablities in structured Markov queueing models,* Commun. Statist–Stochastic Models 1(1994):99-143.
2. A.S. Alfa, J. Xue and Q. Ye, *Perturbation theory for the asymptotic decay rates in the queues with Markovian arrival process and/or Markovian service process*, Queueing Systems: Theory and Applications 36(2000):287-301.
3. N.G. Bean, J.M. Li and P.G. Taylor, *Caudal characteristic of QBDs with decomposable phase spaces*, in Advances in Algorithmic Methods for Stochastic Models - Proceedings of MAM3, G. Latouche and P.G. Taylor (Editors), 2000, Notable Publications Inc. NJ. pp. 37-55.
4. J. Daigle and D. Lucantoni, *Queueing systems having phase-dependent arrival and service rates*, In Numerical Solution of Markov Chains, ed. W.J. Stewart, Marcel Dekker, New York, pp. 161-202.
5. J.W. Demmel, *Applied Numerical Linear Algebra,* SIAM, Philadelphia, 1997.
6. W.K. Grassmann, M.J. Taksar and D.P. Heyman, *Regenerative analysis and steady-state distributions for Markov chains,* Oper. Res. 33(1985):1107-1116.
7. A. Haegemans, G. Latouche and H. Leemans *How to interpret the condition number of the caudal characteristic of a QBD*, in Advances in Algorithmic Methods for Stochastic Models - Proceedings of MAM3, G. Latouche and P.G. Taylor (Editors), 2000, Notable Publications Inc. NJ. pp. 153-165.
8. D. P. Heyman and D. P. O'Leary, *What is fundamental for Markov chains: first passage times, fundamental matrices, and group generalized inverses*, in Computations With Markov Chains, W. J. Stewart, Eds., 1995, Kluwer Academic Publishers, Boston, MA, pp.151–161.
9. D. P. Heyman and D. P. O'Leary, *Overcoming instability in computing the fundamental matrix for a Markov chain*, SIAM J. Matrix Anal. Appl. 19(1998):534-540.
10. T. Kato, *A short introduction to perturbation theory for linear operators*

Springer-Verlag, New York, 1982.
11. J. G. Kemeny and J. L. Snell, *Finite markov chains*, Van Nostrand, Princeton, N. J., 1965
12. G. Latouche and V. Ramaswami, *A logarithmic reduction algorithm for quasi-birth-death processes,* J. Appl. Prob. 30(1993):650-674.
13. G. Latouche and V. Ramaswami, *Introduction to Matrix Analytic Methods in Stochastic Modeling,* SIAM, Philadelphia, 1999.
14. C.D. Meyer, *The role of the grouip generalized inverse in the theory of Markov chains,* SIAM Rev., 17(1995):43-464.
15. M.F. Neuts, *Matrix Geometric Solutions in Stochastic Models,* Johns Hopkins University Press, Baltimore, 1981.
16. M.F. Neuts, *The caudal characteristic curve of queues,* Adv. Appl. Prob. 18(1986):221-254.
17. Q. Ye, *On Latouche-Ramaswami's Logarithmic Reduction Algorithm for Quasi-birth-and-death Processes,* Commun. Statist–Stochastic Models, to appear.